Endocrinology of Social Relationships

ENDOCRINOLOGY OF SOCIAL RELATIONSHIPS

Edited by

PETER T. ELLISON

PETER B. GRAY

Harvard University Press
Cambridge, Massachusetts, and London, England

First Harvard University Press paperback edition, 2012

Library of Congress Cataloging-in-Publication Data

Endocrinology of social relationships / edited by Peter T. Ellison and Peter B. Gray.
    p. ; cm.
  Includes bibliographical references and index.
  ISBN 978-0-674-03117-3 (cloth : alk. paper)
  ISBN 978-0-674-06399-0 (pbk.)
1. Animal behavior—Endocrine aspects.   2. Human behavior—Endocrine aspects.
3. Psychoneuroendocrinology.   I. Ellison, Peter Thorpe.   II. Gray, Peter B., 1972-
[DNLM: 1. Behavioral Research—methods. 2. Neuroendocrinology—methods.
3. Behavior—physiology.   4. Hormones—physiology.   5. Social Behavior.
WL 105 E56 2009]
  QP356.45.E53   2009
    612.4—dc22   2008029736

Contents

Endocrinology of Social Relationships

Introduction

Peter B. Gray and Peter T. Ellison

Biology of Social Relationships

Many of us appreciate the central role that social relationships play in our lives. Through long- or short-term relationships, we find our mates. Through relationships with our parents and our offspring, we connect across generations. Through interactions with extended kin, we find essential help, especially in the reproductive realm. Through friendships, we forge ties that help us weather the day-to-day challenges we face. Through all these relationships, we broaden our knowledge and experience, buffer our vulnerabilities, and extend our capacities.

Of course, our relationships are not only sources of benefits. The contrast between male and female perspectives and expectations in social relationships provides a major source of angst and mystery. We find that our family and others close to us sometimes serve as our main sources of competition rather than cooperation. We bond with some in order to battle against others.

The impact of social relationships on our lives may be very familiar, but the scientific study of this domain of our lives is relatively new and rapidly growing. We are learning that there is a fundamental biology underlying the behavior that is expressed in social relationships and that this biology applies to us as much as it does to the birds outside our window or the monkeys we last saw at the zoo.

Many of the important advances in our understanding of the biology of social relationships can be traced to the 1950s. Classical work in ethology by Nobel laureates Konrad Lorenz and Niko Tinbergen, among others, drew attention to the innate mechanisms that served to establish relationships between potential mates and between parents and offspring. They also looked for phylogenetic relationships in patterns of behavior that would suggest evolutionary relationships. Robert Hinde's 1983 edited volume *Primate Social Relationships* gave us a working definition of social relationships as "repeated interactions that affect future ones." Hinde's own work developed many of the operational definitions and quantitative methods for studying social relationships in animals, launching an empirically robust research agenda that many others took up. He also conducted research on humans as well as nonhuman primates, establishing an important comparative perspective on human social relationships as well as embracing the changes in relationships that occur over the life span.

While empirically rich and experimentally insightful, the corpus of classical ethology lacked a solid connection to evolutionary theory. The field of "sociobiology," marked by E. O. Wilson's landmark 1975 volume, provided that connection and ushered in a period of remarkable research advances in the study of the biology of social behavior. At the core of many of these advances were new insights about the biology of social relationships. Kinship theory provided a new basis for understanding a wide range of behaviors from alarm calls to helpers at the nest. Parent-offspring conflict theory illuminated the ineluctable tension and divergence of interests between generations. The theory of reciprocal altruism provided a foundation for understanding the factors that shaped relationships among unrelated individuals.

Out of the confluence of ethology and sociobiology has emerged a new approach to the study of behavioral biology, including the biology of social relationships. This integrated approach embraces all four ways in which the question "why" can be answered in biology, originally proposed by Tinbergen over 40 years ago: in terms of phylogeny, development, mechanism, and function. The study of behavioral endocrinology, including the endocrinology of social relationships, might seem at first to stress mechanistic answers to the question "why," since there is a great deal of focus on the close connections between the endocrine and nervous systems and demonstrations of causal links between them. But all four levels of explanation are actually used at different times.

Phylogenetic explanations help us to understand the conservatism displayed in hormonal mechanisms. Prolactin is involved in mediating both behavioral and physiological aspects of parental behavior and offspring nurturance in a wide range of taxa including both birds and mammals. The role of gonadal steroids in synchronizing reproductive physiology and reproductive behavior, including courtship and mating behavior, is perhaps as ancient as metazoan life. In a familiar pattern, natural selection creates evolutionary novelty by tinkering with materials already at hand. When new hormones or receptors appear in an evolutionary lineage, they usually represent duplication and divergence from an existing hormone or receptor precursor, such as the evolution from prolactin of growth hormone and placental lactogen in mammals, or the elaboration of different classes of steroid receptors from an original estrogen receptor. One of the themes that will emerge in this volume is the conservation of hormonal mechanisms underlying social relationships across different taxa.

Developmental explanations are also fundamental to the study of the endocrinology of social relationships. The distinction between organizational and activational effects of hormones on behavior is perhaps the clearest example of this. The ontogeny of sexual differentiation in vertebrates begins during embryological development and usually exposes the developing nervous system to sexually dimorphic patterns of steroid hormones. There is convincing evidence across a broad range of taxa that this differential steroid exposure results in differential development of the nervous system itself, reflected in quantitative and qualitative variation in brain structures and their functional connections. As a result of this differential organization, the same pattern of steroid exposure in later life can have different behavioral consequences in individuals of different sex. There is also evidence that similar organizational differences can underlie quantitative variation in later hormonal responses even within the same sex. Other important examples of developmental effects on the endocrinology of social relationships include the elegant studies of Michael Meaney's group (Meaney et al., 1996; Meaney, 2001), showing the epigenetic effects of maternal care on the stress axis of young mice and rats.

But it is the domains of mechanistic (proximate) and functional (ultimate) explanations that are most on display in studies of the endocrinology of social relationships. Establishing a pattern of covariance between a particular hormonal profile and a particular behavior or set of behaviors is usually the first step toward establishing a causal connection. Further studies in the laboratory can sometimes elucidate specific pathways by which the causal relationships are mediated. Others can show, by surgical and pharmacological manipulation, how dose-response relationships are shaped. But understanding the functional role of hormone-behavior relationships usually requires attention to the natural ecology of the organism and the ways in which modification of the behavior in question affects its ultimate reproductive success. Because of the quality of the work that can be done at all of these explanatory levels, behavioral endocrinology provides one of the most integrative arenas for understanding behavior.

The subject of this book—the endocrinology of social relationships—places the integrative nature of behavioral endocrinology on display. While the contributors of different chapters place emphasis on different domains of the fourfold explanatory framework described above, all of them share an awareness of the importance of integrated explanations. The impact of social relationships on individual reproductive success is central to every chapter. And of course the role of hormones in mediating and supporting the formation, maintenance, and quality of social relationships is also implicit throughout. But phylogenetic and developmental perspectives are also broadly prevalent. Some chapters, particularly in the first section of the volume, are broadly comparative, explicitly invoking phylogenetic relationships in seeking patterns. Yet even the later chapters that focus on specific groups of animals implicitly build on phylogenetic causation by investigating in one species hormonal mechanisms that have been demonstrated in another. Similarly developmental perspectives are often in the foreground, particularly in chapters that explicitly adopt a life historical framework or in those that discuss sex differences. While it is difficult, in the space allowed, for any one chapter to shine the spotlight on all four levels of causation, the volume as a whole incorporates them all.

Aims of this Volume

We have two major goals in assembling this volume. The first is to help consolidate the rapidly advancing theoretical and empirical work on the endocrinology of social relationships. This body of research is inherently interdisciplinary. This diversity is demonstrated by the fact that the contributors to this volume come from a variety of disciplines: anthropology, psychology, psychiatry, and biology. We imagine a similarly diverse group of readers holding this book, including students and nonacademic readers of widely differing backgrounds and experiences. A consolidation of research on the endocrinology of social relationships helps pull a scattered literature together under one cover.

The topics included in the volume have been chosen to enable substantive discussions and to allow reasonably robust conclusions to be drawn. We did not want to edit a book on the black holes of science but rather one that speaks to what we know and that in turn suggests steps to take to fill in what we do not know. The best-studied social relationships in endocrinology are dyadic ones, mating relationships and parent-offspring relationships. These situations provide the clearest context for experimental research in the laboratory and are often the easiest to observe in the field as well. Competitive and coalitionary behavior among groups of individuals are also very important in the fabric of relationships among group-living species but are more difficult to capture under the lens of experiment or observation.

Nevertheless, choices had to be made. The taxonomic diversity on display in this volume is restricted. But this fact is not simply the result of a necessary trade-off between breadth and depth of coverage. It also reflects our second major goal, the desire to place significant emphasis on humans and the relevance of behavioral endocrinology to understanding our own social behavior. Thus the taxonomic range of the book is restricted to "higher" vertebrates, more to mammals than to birds, more to primates than to rodents. Some of the most important theoretical and empirical contributions have come from studies of birds and rodents, but some of the most exciting recent advances are being made in studies of wild primates and humans.

The particular focus on humans is warranted, we feel, for several reasons. Many who study nonhuman animals regularly extrapolate their findings to humans even when human data are lacking altogether or sparse. We believe that humans deserve attention as research subjects themselves. As the corpus of work on the endocrinology of human social relationships expands, it permits direct assessment rather than inferences from other species to ourselves. Being biological anthropologists, we, as editors, have a particular interest as well in human evolution and the factors that have shaped us as a species and that continue to organize our lives today. We are particularly excited by the growing opportunity to bring the rich, integrative perspective of behavioral endocrinology to bear on this, our own central professional concern.

That said, we still depend on a synthesis of nonhuman and human data because many of the causal relationships we would like to explore in humans simply are not logistically or ethically accessible. Researchers cannot perform the kinds of invasive studies in humans that would, for example, determine whether affiliative interaction with a friend or loved one results in increased oxytocin receptor staining in limbic system structures (by sacrifice of the study animal, followed by dissection and staining of brain tissue). Birds or apes in the wild are not appropriate for these types of studies, either. Even brain imaging studies have significant constraints that limit their usefulness.

Increasing attention to minimally invasive techniques of hormone measurement (for example, saliva or blood) has, however, led to enhanced success working with wild nonhuman animals and with humans alike. The study of individuals under controlled lab circumstances can be combined and complemented with more ecologically valid studies of members of the same species under seminatural and natural conditions to see how variables differing across such contexts matter. Many of these techniques and approaches are prominently on display in the chapters of this volume.

It is important, however, always to keep in mind the distinction between wild and captive regimes. This distinction can underlie differences in diet, activity levels, disease loads, and social contexts, among other things. We can gain better experimental control in lab settings when investigating the endocrinology of social relationships, but that may come at some expense of removing the animal from its typical environment. This principle applies, in general terms, to humans as well as to nonhuman primates and other animals. Human populations exhibit variation in these same factors—diet, activity patterns, social and economic contexts, and so forth—in ways that can shape the endocrine system and its links to social relationships. An important new frontier in human behavioral endocrinology, and one that is only beginning to be explored, involves the study of variation across societies, cultures, and ecological contexts.

Phylogenetic causation is a two-edged sword in the effort to synthesize animal and human studies. The "animal model" approach recognizes the usefulness of nonhuman animals in providing experimental insight into biological phenomema that may represent highly conserved mechanisms, especially the more closely related that organism (nematode versus fruit fly versus rat) is to ourselves. Phylogeny in this way becomes a broad indicator of similarity and the justification for working with other animals in ways that might be logistically and ethically impossible among humans. At the same time, we must remain cognizant of the unique evolutionary trajectory giving rise to each species and the differences in the selective forces acting on them. Natural selection may make use of the same "parts" or mechanisms in different lineages, but it can put them to use in novel ways to serve novel functions. It is particularly important to remember that hormones are molecules that carry information, not molecules that catalyze chemical reactions or otherwise "cause" biochemical events. How the central nervous system of an animal makes use of the information carried by its hormones, how it chooses to modify its behavior to complement its physiology, is not necessarily constrained phylogenetically. Testosterone, for example, does not necessarily cause an individual, male or female, to be bold or aggressive or sexually motivated. It often does, however, convey to the central nervous system information about gender, maturity, reproductive state, physical condition, and perhaps social status as well. This information can be integrated with other sensory, somatic, and cognitive information to produce behavioral patterns that we associate with testosterone.

Why does an endocrinology of social relationships matter? We began this introduction by invoking the personal importance of social relationships in each of our lives. In addition to personal curiosity and the intellectual satisfaction that the pursuit of knowledge brings, there are a number of practical reasons to care. Social relationships can benefit, or sometimes harm, our health and longevity, and it is a worthwhile endeavor to try to discern the physiological pathways by which these effects on health and longevity arise. We also live in a world of endocrine interventions, from estrogen mimics in the environment to which we may be unwittingly exposed to pills, gels, and patches containing hormones that we may consciously apply to ourselves. These exposures may potentially impact our social relationships and behavior and deserve our consideration. An understanding of how the endocrine system contributes to the formation and maintenance of social bonds can also inform the development and design of medical or social interventions such as hospital birthing practices or adoption policies. As with most areas of basic research, practical considerations such as these are rarely the driving force behind research, but they provide glimpses of the ways in which advancing basic research can have tangible impact in our lives.

Organization of the Volume

The volume is structured in three parts. Part I covers a set of key theoretical and empirical contexts. Phyllis Lee reviews the evolutionary and ecological bases of animal social behavior, highlighting the impact on sex differences. Kim Wallen and Janice Hassett describe the key features of the endocrine system and how these causal players help account for variation in social relationships between and within species. Peter Ellison links ecological and endocrine interactions in shaping social behavior tied closely to reproduction. John Wingfield describes the ways animals adapt to changing environments, hearkening primarily to avian data but speaking more generally to adaptive social responses to environmental changes. Jane Lancaster and Hillard Kaplan describe how some of these and related principles have played out to shape the "human adaptive complex" of resource extraction and social organization in ways that call on male-female cooperative relationships in which children are raised. Collectively, these five chapters in Part I set the stage for integrating ecology, endocrinology, and social relationships, particularly those built around reproduction.

Part II of the volume focuses on some of the major nonhuman taxa for which substantial theoretical and empirical advances on the endocrinology of social relationships have been made. Sue Carter, Ericka Boone, Angela Grippo, Michael Ruscio, and Karen Bales detail the mechanisms and development of rodent social relationships, animals that have been highly influential in shaping research on the endocrinology of mammalian social relationships. Toni Ziegler and Charles Snowdon describe the neuroendocrinology of South American monkey social relationships. Lynn Fairbanks overviews the neuroendocrinology of group-living monkey relationships primarily from Africa and Asia. And Melissa Emery Thompson reviews the endocrinology of ape social behavior, organizing her chapter with a focus on the sex differences in these. These chapters in Part II cover research viewed as "classic" within this young subfield, while including active contributors who remain on the cutting edge of research in their taxonomic specialty.

Part III presents the current state of theoretical and empirical studies on the endocrinology of social relationships in humans. Matthew McIntyre and Carole Hooven highlight the organizational and activational effects of hormones on human sex differences in social relationships. James Roney examines the endocrine mechanisms entailed in the initiation of human mating relationships. Peter Gray and Benjamin Campbell review the cross-cultural evidence linking differences in human male testosterone levels to involvement in pair-bonding and paternal care. Alison Fleming and Andrea Gonzalez cover the neurobiology underlying human maternal relationships. Roxanne Sanchez, Jeffrey Parkin, Jennie Chen, and Peter Gray review the role of oxytocin and vasopressin in human social behavior. Sari van Anders covers human diversity in sexual orientation and partnering and its underlying endocrine basis. Pablo Nepomnaschy and Mark Flinn review the role of children's stress responses in light of early socioecological context. Empirical topics chosen for this part represent those with substantive research, the majority of which has only been conducted within the past 10 years. Human data tend to involve more observational, cross-sectional, and correlational designs, compared with their nonhuman animal counterparts. Still, these recent human findings, complemented by general theoretical overviews and more experimental research on nonhuman animals, illustrate the ways in which endocrine mechanisms underlie human social relationships.

We should note that the individual contributors to this volume do not always agree on all particulars. These differences are concentrated on areas of theory and interpretation more than on empirical facts. The reader should not be alarmed at encountering differences of opinion. Most vigorously growing fields are fueled by areas of controversy and dispute, hallmarks of the developing frontier of a discipline as opposed to the broader consensus that may develop regarding basic principles and core concepts. Areas of disagreement and contention are, in fact, "areas to watch" for the emergence of new ideas and observations. They provide a sense of direction to the field as a whole and an intimation of things to come.

In the end, space limitations do play a role. This volume is not comprehensive. There are gaps in coverage that may seem glaring to those knowledgeable of the field. But hopefully they will find richness in other areas that they may not have expected. What we are sure of is the exciting future of this field: the study of the biology of social relationships, the hormonal mechanisms that help to shape that biology, and the relevance of that biology to understanding ourselves.

Acknowledgments

Thanks to all of the contributors to this volume for their efforts in making this book possible. They are researchers on the front lines, engaged in the lab work and fieldwork that generate the data, and the thinkers who synthesize the key concepts and findings. Thanks to the individuals from numerous populations around the world who have participated in our research. Thanks, too, to all those whose efforts helped shepherd an initial concept to print. Particular thanks are due to Tom Steiner, who assisted with the fine-grained details of content editing; to Rob Durette, who created a Web site accessible to contributors that proved useful in tracking the evolution of this volume; and to the University of Nevada at Las Vegas and Harvard University, for providing supportive environments throughout this process. Last, yet foremost, thanks to our families—especially Megan, Sophie, and Stella, Pippi, Sam, and Silas—and to our friends who make reflecting on social relationships not just an academic exercise but the most meaningful engagement of all.

Theoretical and Empirical Context

CHAPTER ONE

Evolution and Ecological Diversity in Animal Mating and Parenting Systems

Phyllis C. Lee

Both REPRODUCTION AND RELATIONSHIPS have obvious hormonal underpinnings as well as hormonal consequences, and these act to influence behavior. In this chapter, I review mate choice and mating systems from a behavioral perspective rather than from an endocrinological one. I concentrate on animals that reproduce several times over a life span and discuss how mating systems relate to the observed diversity of parental care and social systems. Although species differ considerably in their reproductive natural history, my focus is not on this detailed variation but on the evolved commonalties underlying the larger patterns of mating and parenting strategies for enhancing reproductive output.

Parental care is a mechanism for maximizing offspring growth and survival and is set against the competing parental interests of minimizing the time between reproductive events and ensuring an individual parent's survival from one event to the next (see Clutton-Brock, 1991; Charnov, 2002). This perspective is firmly rooted in Lack's (1968) studies of the reproductive ecology of birds, and trade-offs between offspring and parental mortality and reproductive effort have been extensively modeled mathematically (Trivers, 1974; Maynard Smith, 1977; Parker and MacNair, 1978; Lazarus and Inglis, 1986; Key and Aiello, 2000). In Charnov's (2002) dimensionless life history cube, the relative size of the offspring (which reflects offspring survival for parental resources invested), reproductive effort per unit adult mortality, and reproductive life span as a function of time to reach maturity clearly link reproductive strategies with mortality in a series of trade-offs. Recent work on care allocation suggests that mortality can be considered from two perspectives; the first is that of offspring mortality resulting from variation in intrinsic features of the individual allocating care or under the control of the caregiver in terms of rate and quality of allocation—care-dependent mortality. This mortality source contrasts with care-independent sources of mortality, such as exposure to novel diseases, predation, or catastrophic environmental events (see Pennington and Harpending, 1988; Lycett, Henzi, and Barrett, 1998). The relative importance of these distinctive sources of offspring mortality underlies parental investment (PI) decisions. When parental investment affects survival probabilities, differential levels of investment will have variable payoffs, and PI will be related to offspring needs for investment as well as a parent's capacity to invest. When, however, mortality is independent of care allocation, PI reflects only the parent's willingness to invest. There are, of course, areas of overlap between these sources of mortality. For example, passive immunity to some bacteria can be acquired by suckling, and thus responses to disease can be mediated by care allocation. Individuals who grow well due to care allocation may generally be less susceptible to disease or environmental insults.

A life history perspective can also be linked with an understanding of the social context dynamics of families and other social groupings (for example, Emlen, 1997). Key to descriptions of parental care is an understanding of constraints due to sex differences in the costs of reproductive events and how these vary over the life span of the reproductively active organism.

Mating systems are typically categorized by the partners that males and females copulate with during a reproductive event or, somewhat confusingly, over a reproductive life span. Most mating systems are descriptions of the male behavioral component—the distribution and dynamics of females in mating tend to be overlooked, at least in typological contexts, and referred to only with respect to extra-pair copulations (see Gowaty, 2004). Conventional definitions of mating systems are presented in Table 1.1, where the female viewpoint has been added to show the extent of potential variation when mating systems are examined from both male and female perspectives.

Costs of Reproduction

Sexual reproduction itself has significant costs. The first of these are the genetic costs associated with an individual reduction in replication rate (by comparison to parthenogenesis or budding) and in the generation of inefficiencies and errors during recombination. However, these are offset

Table 1.1 Conventional definitions of mating systems (after Emlen and Oring, 1977), with potential benefits to each sex noted as well as kin structure to groups.

Mating System Typology	Male Perspective	Female Perspective
MONOGAMY: An individual male and individual female mate with only one partner per breeding season.		
FACULTATIVE MONOGAMY: Males and females occur at low densities and are so spaced out that only a single member of the opposite sex is available for mating. Neither sex avoids copulating with others in succession, should any opportunity arise (promiscuity).	Paternity cannot be assured.	No resource provision; help with infant rearing or protection is available but not essential.
OBLIGATE MONOGAMY: A solitary female cannot rear young without aid from others, and males provide such parental care.	Reproductive success increases enough to compensate for mating opportunities lost by staying with one female.	Help with infant rearing or protection is essential. Avoid competition for resources with other females who could potentially share infant care and protection.
POLYGYNY: Individual males may mate with more than one female per breeding season. Males can economically monopolize access to several females.		
FEMALE DEFENSE POLYGYNY: A. Females live in groups defended permanently by one or more associated males. Males are ensured access to several sexually active females, paternity certainty attempted.		
	Unimale harem: One male mates with more than one female. Competition is high but temporally limited to the context of initial control of female units or occasional takeover attempts.	*Female kin bonded:* Ability to retain female kin in philopatric unit to maximize resource control in the context of female-female competition; potential to manipulate the success of the harem-holding male; protection from harassment or infanticidal males.

(continued)

Table 1.1 *(continued)*

Mating System Typology	Male Perspective	Female Perspective
	Multimale unit: Male matings occur with multiple females—can be promiscuous or monopolized. Competition is high and constant in the context of mating.	*Female non-kin bonded:* Minimal female-female competition when in non-kin units; protection from harassment or infanticidal males.

B. Females live in permanent groups that are only associated with males during breeding seasons. Male opportunities to mate are too limited over time within a single group to constrain roving.

Mating System Typology	Male Perspective	Female Perspective
	Roving males; contest and sperm competition paramount; no paternal investment; low probability of infanticide.	*Female kin-bonded:* Shared infant care and communal female resource defense possible. *Female non-kin bonded:* Similarity in physiological and social requirements for females and their infants—temporary units in a fission-fusion context are possible.

RESOURCE DEFENSE POLYGYNY: Females do not live in permanent groups but are spatially concentrated at food, water, nesting, or breeding sites or in relation to other resources that some males can control.

Mating System Typology	Male Perspective	Female Perspective
	The amount of resources controlled by the male should correlate with opportunities to copulate. Male-male competition is intense for best sites, with high turnover in territory possession. Males encourage visits or residence by maximal numbers of females.	Ability to assess male quality from their control of resources. Females make the best of a bad job when forced to share their male's resources with other females. If females form kin units, they can share infant care and defend against the settlement of additional females.

POLYANDRY: Individual females mate with more than one male per breeding season.

Males in kin group: Retention of resources that cannot be divided but can be shared with other males. Several males provide protection or territorial defense.

Infant or resource protection from groups of males without sharing resources with other females.

Males solitary: Inability to control females, resources, or females' access to other males.

Access to multiple males during each reproductive cycle to maximize selection for specific male attributes.

by the evolutionary potential of sex since recombination provides a mechanism for resisting environmental changes and especially those due to parasites and pathogens (see Paland and Lynch, 2006). Thus sex can "speed up" evolution, especially in dynamic ecologies or under pressure from pathogens.

Other costs of sex are those associated with mating activities. In all species that sexually reproduce, time and energy are invested in securing mates. The behavior and elaborate ornamentation associated with sexual reproduction could be dispensed with in a parthenogenetically reproducing individual. Reproductive activity can reduce longevity in fruit flies (Partridge, 1988), sheep (Festa-Bianchet, 1989), and humans (Kirkwood and Rose, 1991). The production of gametes is energetically costly (Maynard Smith, 1978); and the development of secondary sexual characteristics requires time, exposing individuals to risks of mortality as well as additional growth and physical development costs (Clutton-Brock, 1994; Setchell and Lee, 2004). The act of sexual reproduction tends to require some associated courtship behavior that can be extremely energetically expensive and risky (Halliday, 1994; see below). Finally, sexual reproduction also exposes individuals to transmissible pathogens and parasites during mating (Nunn and Altizer, 2004).

Since females are defined as the gender with large gametes and males as the gender with small gametes, there is a disproportion in the costs of sexual reproduction between the genders (Maynard Smith, 1978; Halliday, 1994). Having established an evolutionary advantage to sexual reproduction, this fundamental asymmetry (even when corrected for numbers at delivery; for example, Dewsbury, 1982) will dictate the metabolic but not necessarily the behavioral or life history costs to each gender. As first discussed by Bateman (1948) and modeled by Trivers (1972), a male's reproductive success depends on the number of receptive mates he obtains access to. Securing additional matings should have a reproductive payoff at any point in time—or at least within the breeding season, if one exists. A female's reproductive success, by contrast, does not depend on the number of mates but rather on her capacity to sustain the costs of the reproductive event. The sex that invests more in the young becomes a resource for which other members of the less parental sex compete, producing different degrees of variation in the reproductive success of males and females. These observations form the basis of much fundamental theory as to reproductive costs, mating strategies, and the ecology of parental care systems.

Reproductive costs are also dependent on postmating investment, which varies among species and depends on fertilization mode. For species with external fertilization, such as insects, amphibians, reptiles, and fish, either sex can differentially bear the costs of reproduction and therefore be the object of competition. This leads to a diversity of mating strategies and parenting systems that far exceed those seen in mammals and birds (see Gowaty, 2004). For species with internal fertilization and external embryonic development, such as birds, biparental care is relatively common in the context of energy constraints; the costs of feeding hatchlings in order to sustain their growth and survival may require the combined efforts of two parents (Drent and Daan, 1981; Daan and Tinbergen, 1997) and, in many cases, additional helpers at the nest (Emlen, 1997). Where one parent alone can meet the growth needs of the offspring, male care can result. Implications of this diversity of investment strategies for mating systems are seen in the extremes of mate choice mechanisms used by birds (see Wingfield, this volume). For species with internal fertilization and lactation (mammals), parental nurture is overwhelmingly female dominated, and there tends to be significant variation in the reproductive output between males and females.

Competition, Sexual Selection, and Mate Choice

Sexual selection was defined by Darwin (1871) as the differential success of individuals in the context of reproduction, and he questioned how morphological features, for example, bright coloration or large antlers, evolved when they may be detrimental to the individuals who carry these characteristics. Fisher (1930) suggested that the opposite sex was initially attracted to features that have survival value and that these characteristics become exaggerated via rapid selection. Females who mate with attractive males will have attractive sons (and choosy daughters), provided that attractiveness is inherited, and this will provide a selective context for runaway sexual selection (Andersson, 1994). Selection for such potentially deleterious but reproductively advantageous traits could conceivably be applied to either sex, as males may also develop preferences for extreme female traits, such as sexual swellings in anthropoid primates. In addition, for either sex, signaling fitness could be advantageous. For example, well-maintained, brightly colored plumage might indicate flying efficiency, or it might indicate that the individual can afford the time and energy to maintain plumage or that it is of sufficient quality to carry a parasite load without a fitness reduction (Hamilton and Zuk, 1982; Møller, 1990). A review of the mechanisms, processes, and evolutionary outcomes of sexual selection are beyond the scope of this chapter (but see Kappeler and van Schaik, 2004), and it is mentioned here with respect to implications for reproductive and parenting strategies.

Male-male reproductive competition takes a large number of forms. Behavioral and hormonal suppression of reproductive function are possible (Keverne, 1979; Kraus, Heistermann, and Kappeler, 1999). Males compete via sperm (see Birkhead and Parker, 2005), and competition structures the size and shape of sperm (Gomendio, Harcourt, and Roldan, 1998; Anderson and Dixson, 2002). Competition also underlies much morphology adapted for postcopulatory sperm removal, as well as penile morphology and testes size (Dixson, 1987). Mating plugs, sperm repellents, and prolonged copulations are alternative forms of postcopulatory competition.

One of the more conspicuous forms of male-male competition is that of direct aggressive interactions that limit the access of other males to a female or a group of females. These interactions are associated with traits that improve success in fights, such as large size, strength, weaponry, agility, or elaborate threat signals. Much male-male competition reflects age, size, and condition-dependent strategies (for example, Coltman et al., 2002). For instance, in African elephants, older, larger males in good condition maintain a state of reproductive activity (musth) longer, are actively chosen by females, and have a higher probability of paternity (Moss, 1983; Poole, 1989; Poole, Lee, and Moss, in press); longevity and enhanced growth thus have reproductive payoffs. The ability to remain reproductively active for a large proportion of the breeding season, such as at lek sites (Kruijt and de Vos, 1988), is a further example of endurance rivalry. Males may compete to disperse quickly, and thus find sexually receptive females, or to signal their fitness to females during courtship, as well as via offers of food, territories, nest sites, or other resources needed by their mates for breeding (Halliday, 1994). In terms of life history outcomes, when direct male-male competition is more intense than female-female competition, males may suffer higher mortality as a consequence. However, as Clutton-Brock (2004) has noted, most of the additional mortality costs of sexual dimorphism are related to differentials in the costs of early growth between sons and daughters, with implications for parental energy costs and rearing strategies. Parents may thus be "forced" to allocate greater care to sons in sexually dimorphic mammals rather than allocating care on the basis of reproductive success returns, since such species exhibit extremes of care-dependent mortality.

There are also many alternative social strategies, and males use a variety of social tactics to gain access to females. Some males "sneak" copulations; other males act as satellites to territorial or harem-holding males. Alternative tactics can reflect bimaturism (for example, arrested development in orangutans: Atmoko and van Hooff, 2004) or genetically distinct growth morphs (salmon: Aubin-Horth et al., 2005). Males can form "special relationships" with individual females (for example, Smuts, 1985), enhancing their access to future mating opportunities as well as gaining the ability to protect their own offspring from possible infanticidal attacks via paternal behavior (Buchan et al., 2003). They can also acquire access to breeding females by joining an existing group as a subordinate follower who is tolerated by the harem holder. Among gelada baboons (Dunbar, 1984) the follower forms relationships with females who are often peripheral or subordinate. These relationships develop into full breeding relationships, and the follower can lead his females out of the old unit and form a new group. In gorillas, the harem holder typically passes on his harem intact to a follower who may be his son (Watts, 2000). In most cases, the follower could establish his own group rather than wait, but being a follower pays when unit sizes are very small and there are no surplus females in the population to provide the nucleus of a new group.

Mate guarding, such as forming consortships or friendships or controlling a harem, is an attempt to regulate paternity, although it by no means ensures paternity, since both sexes can indulge in extra-pair copulations. Females appear to seek out extra-pair copulations in order to obtain a variety of possible genetic benefits (Jennions and Petrie, 2000).

Males can cooperate to enhance their reproductive access (Noë, 1992). By forming male-male coalitions, males assist each other in gaining a higher rank than expected for their age or individual fighting ability, and these coalitions can also slow the rate of rank decline once males are past their prime. When social status or dominance is associated with success in "priority of access" competition (for example, Altmann et al., 1996), behavioral mechanisms to maximize status can be key determinants of reproductive success. Alternative modes of attaining rank exist for socially political species. For example, among baboons and macaques in multimale, female kin groups, males can (1) opt for high rank, move repeatedly, and compete individually for access to a number of females or (2) migrate only once and then concentrate on building long-term pair-bonds with particular females. Computer simulations of life history characteristics suggest those males who cannot achieve high rank are likely to do relatively badly in terms of lifetime reproductive output (van Noordwijk and van Schaik, 2004). Males potentially can pursue these alternative strategies at different stages in life and use different conditional rules. The high-gain / high-risk dominance-based strategy is associated with young and physically powerful individuals who optimize the time when they challenge others, while the low-gain / low-risk strategy is seen when males are older or when a demographic context of many males increases the risks of contest. Male lifetime reproductive success can be determined by the rank he attains early in a reproductive career and the length of his reproductive life span.

Finally, postcopulatory competition can take a variety of forms, from sperm competition, as discussed above, to the removal of competitors' offspring. Induced abortions and a return to sexual activity as a result of reproductive competition have been suggested for species as diverse as rodents, baboons, and equids (see Rubenstein and Hack, 2004). By far the most common form of postcopulatory male-male competition is infanticide, causing the premature death of one male's offspring and the reinsemination of the mother by the infanticidal male. Recent phylogenetic reviews of the prevalence of infanticide as a mode of reproductive competition have demonstrated its presence in 21 species of rats and mice, 66 species of carnivores, 11 species of ungulates, 21 species of pinnipeds, and 58 species of primates (van Schaik and Janson, 2000). Male infanticide is associated with life history traits such as low rates of care-independent infant mortality and with demographic and social opportunities such as high male replacement rate (harem takeovers) and reduced or energetically limited female body size and therefore low power relative to that of the male. Consequences of infanticide are an extreme skew in reproductive success among males, rapid rates of female return to sexual activity, and possibly female philopatry.

All these different tactics are part of male mating strategies; males use whichever yields the higher reproductive success. Different tactics do not necessarily produce equal fitness gains, but they may reflect individual males "making the best of a bad job." For example, in gray seals, dominant males compete intensively to hold small beach territories and attempt to monopolize mating with females as they haul out to pup. Paternity determinations on pups (Amos et al., 1995) found that only 1% of males fathered more pups than expected by simple proportions of males to females. Many of the pups clearly were fathered by nonterritorial males. Why, then, do males risk fights and expend energy in holding territories? The distribution, density, and age structure of males determine payoffs at any point in time, and thus extreme reproductive skew in favor of harem males remains a possibility worth the risks (see Le Boeuf and Reiter, 1988). The key issues here are that environments and individuals both vary in quality and over time and thus that the variance among males determines their access to the variable resources.

What are males choosing when they choose mates? Female fecundibility, high reproductive value (often associated with younger ages in many mammals), large body size or stature, high status, reproductive experience or parity and thus older females, and health and vigor are all possible attributes that make individual females attractive. "Attractiveness" thus appears to relate to offspring production ability and investment quality and can be an honest signal of genetic and developmental experiences. One example of a potentially honest signal is that of the perineal sex skin swellings seen in the majority of female catarrhine lineages (Dixson, 1998). These do not necessarily signal ovulation (and indeed may act as a mechanism to confuse paternity expectations: Heistermann et al., 2001) but rather seem to be an indicator of reproductive competence, as well as general size and health. Since female body size is correlated with swelling size, but swelling size is also independently associated with interbirth intervals in species such as baboons, swellings can potentially provide males with some relative assessments of female fertility histories (Domb and Pagel, 2001; Zinner et al., 2004).

Females also choose specific males for a variety of reasons, many of which are similar to those males find attractive. Again, health, high immune competence, and capacity to sustain parasite loads are reflected in a large body size achieved via rapid early growth, in bright coloration, in vigorous displays, in social dominance or resource-holding potential, and in the ability to provide resources; all are attractive traits to females. Furthermore, as noted above, preferences for "paternal" or noninfanticidal males, choices for friends, allies, and social partners, and a preference for novelty as a mechanism of avoiding mating with kin are seen among highly social species. Using cues such as MHC (major histocompatibility complex) gene products, females may be able to select males with lower degrees of relatedness (Potts, Manning, and Wakeland, 1994).

In contrast to contexts where males exhibit preferences for younger females with high reproductive value, longevity can be an important signal to females of male quality, leading to a preference for mating with older males in some species (for example, elephants, Moss, 1983; Poole, Lee, and Moss, in press; black grouse: de Vos, 1983). Other traits such as holding territories or lekking in locations with low predator risk (for example, Uganda cob: Deutsch, 1994) may influence female choice during an estrus event. Females also may selectively choose a mate that contributes to their offspring's fitness. Even in fruit flies, males who are better competitors are chosen (Partridge, 1988), as this is a heritable trait. Among zebra finches (Burley, 1981) and peacocks (Petrie, Halliday, and Sanders, 1991), males who were arbitrarily chosen helped more, and their chicks had higher growth rates and survival.

Mating Systems, Parenting, and Social Evolution

While it clearly pays both sexes to choose genetically and behaviorally attractive mates, the interests of males and females diverge in relation to their reproductive strategies, as outlined above. Female reproductive strategies are strongly influenced by ecological factors affecting fertility—energy for their reproduction is limiting; for males, access to receptive females is more often the major constraint. Among social species, both males and females experience constraints as well as advantages due to the specific social context. While female-female competition can result in the hormonal suppression of fertility (Abbott, 1993), the aggressive manipulation of group composition via eviction (for example, mole rats: Faulkes and Bennett, 2001), female-female cooperation via helpers and allomothers (elephants: Lee, 1987; lions: Packer, Lewis, and Pusey, 1992), and kin group contribution to infant protection (meerkats: Clutton-Brock et al., 2001) can enhance infant survival and thus female reproductive success. These latter contexts arise when females have few opportunities to disperse, limited success if they do disperse or form a new group, and high rates of care-independent infant mortality due to predation. As outlined above, the nature, timing, and cause of infant mortality are determinants of the structure of cooperation and competition among females. When mortality is care dependent and due to a failure intrinsic to the mother (low physical condition or low status), then changing levels of care allocation in response to individual condition are predicted (Lee, Majluf, and Gordon, 1991; Fairbanks and McGuire, 1995; Lycett, Henzi, and Barrett, 1998). By contrast, when mortality is care independent and extrinsic to the mother, then mothers may not evolve investment strategies. Rather, they may facultatively respond to local environments simply with variable maternal styles. It is worth noting that at least some of the confusion in the parental effort literature may result from the muddling of these two modes of mortality into a single perspective on parental investment, thus obscuring the relationship between PI and mating systems.

The form of the mating system (see Table 1.1) has long been held to be determined by the resource base (Jarman, 1974; Wrangham, 1980), which facilitates either females grouping to defend food or females dispersing to acquire food individually. Local ecology and the consequent female grouping mode therefore define what mating tactics will be productive for males. If females are dispersed, they cannot be controlled without expenditure of time and energy by a male (for example wildebeest, where males have small territories that females rove through). If one or more males can control resources (resource defense polygyny), then they may be able to control a cluster of females using that resource. Such systems are relatively rare among mammals and may be exemplified by the male seals controlling a patch of beach next to female pupping areas. Alternatively, males can attempt to control female clusters (female defense polygyny), but only if the resource base permits aggregation among females. In such a case, either the males need to aggressively herd or control females, or they need to offer some advantage in the form of protection from predation (plains zebra: Rubenstein and Hack, 2004) or infanticide (gorillas: Watts, 2000). Whether one male controls the females alone or several males cooperate depends on (1) the size of the female unit (Dunbar, 1984; Andelman, 1986), (2) the relative risk of predation (cercopithecines: Hill and Lee, 1998), and (3) potential for loss of control of paternity due to seasonal reproduction, synchrony, or influxes of neighboring males.

Polygyny is almost always advantageous to a male, as it increases his access to females, whereas females "prefer" monogamy (Davies, 1989). Under the predictions of the polygyny threshold model (Orians, 1969), the male who is the first to obtain a mate should also have priority of access to a second or third mate. Females that arrive later and settle in successively less valuable breeding situations should raise successively fewer young. If two females arrive at the same time, and one chooses to be a second female in a polygynous territory and the other a monogamous female, then they should have equal reproductive success up to some maximum controllable number of females. However, consider the first female; she always suffers a cost due to sharing resources in a polygynous relationship. Therefore, when resources limit female reproductive success, females should try to keep other females out. Only when females obtain an independent advantage from co-residing with other females, such as shared infant care (Borgerhoff-Mulder and Milton, 1985) or protection from infanticide, will polygyny be advantageous to the second "wife."

Avoidance of inbreeding has long been a structuring principle underlying social group composition and stability. As the mechanism for minimizing breeding contacts between parent and offspring, natal dispersal occurs in most birds and mammals (Greenwood, 1980; Chepko-Sade and Halpin, 1987; Pusey and Packer, 1987). There are a few exceptions (killer whales: Baird, 2000; bottlenose dolphins: Connor et al., 2000) to the "rule" that prior to breeding one or both sexes disperse from the area or group where their parents reside. Among primates (baboons: Alberts and Altmann, 1995; chimpanzees: Morin et al., 1994; Gagneux et al., 2001), social carnivores (lions: Spong et al., 2002; hyenas: Van Horn et al., 2004;

coatis: Gompper, Gittleman, and Wayne, 1998), seals (Amos et al., 1995), cetaceans (Amos, Scholotterer, and Tautz, 1993), and birds (for example, Petrie, Krupa, and Burke, 1999; Coltman, 2005; Double et al., 2005), movement of individuals between groups or geographic areas increases genetic heterogeneity, although neighboring individuals tend to be related. Thus kin clustering remains an important feature of sociality, as suggested by Emlen (1997). Knowledge of how nonrandom genetic patterns or clusters of kin over space arise through dispersal behavior is leading to new models of social evolution.

Several predictions arise from natal dispersal. The first is that the sex that disperses should reach reproductive maturity before the death (or dispersal) of the opposite-sexed parent. The tenure of breeding males in female philopatric groups is shorter than the time to reproductive maturity of their daughters (Clutton-Brock, 1989). When males are the philopatric sex and female dispersal is the norm, females may also occasionally mate with males outside their residence groups. Reproduction in the natal group

A. Routes from solitary foraging to stable social units

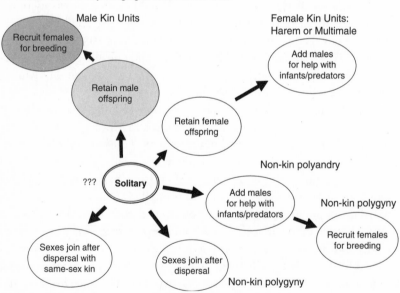

Figure 1.1. A (above) and B (opposite page)   Routes between social states with a solitary forager as the starting point. The different possible pathways are illustrated. Some of the intermediate states are likely to be unstable (with rapid movement between adjacent states) or transitory (stable for short periods), while some could become stable states in rare circumstances. No assumptions are made about which routes are most likely.

by the typically dispersing sex is rare before dispersal. Behaviorally, there is inhibition of parent-offspring mating, the avoidance of natal males by natal females during adolescence, incitation of copulations with novel partners, and immigration encouraged by the resident sex. There may also be hormonal suppression of reproduction within natal groups.

The outcome of dispersal mechanisms can be seen in the social structure of groups. When same-sexed individuals co-reside in a temporally and spatially distinct unit, then the social system that emerges is that of a single-sex kin unit. Onto these kin units formed by the initial dispersal or retention of one sex are mapped opposite-sexed individuals, who themselves may be kin, depending on dispersal strategy and life history traits such as large litter size or rapid rates of reproduction. Ultimately, a variety of social systems, each with specific mating and parenting opportunities, can arise from different selective routes (Figures 1.1A and 1.1B). Such models of social system evolution, based on the emergence of units from a solitary foraging individual (Müller and Thalmann, 2000), on principles of kinship and parental investment (Lee, 1994), and on intergroup reciprocity and exchange (Foley and Lee, 1989; Rodseth et al., 1991), demonstrate the power of relating the reproductive interests and strategies of each sex to the opportunities for aggregation, dispersal, and reproductive cooperation (for example, Clutton-Brock, 2002).

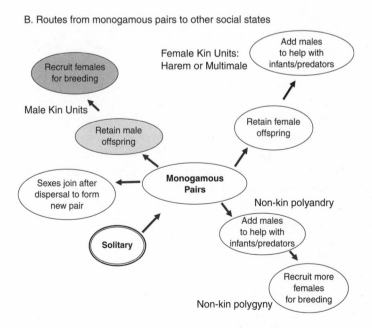

B. Routes from monogamous pairs to other social states

Are primates special? They have intrigued modelers of social systems and social evolution since Darwin, and indeed, they represent an order with extremely high diversity in social structure and mating systems, as well as with particularly slow life histories. Primates are, however, relatively infrequently directly compared with other groups of mammals or birds, and thus we still can ask whether primates are special in several aspects of sociality. The first of these is the relatively constant co-association of males with females among the haplorhine primates, compared to ungulates, carnivores, or even rodents (Lee, 1994; van Schaik, 2000). Strepsirrhines, by contrast, appear to follow a more general mammalian trend of loose female matrilines or dynamic associations among males and females. Co-association between the sexes in haplorhines has been linked specifically to avoidance of infanticide (van Schaik, 2000) as well as to the economic (energetic and time) capacity for males to control groupings of females (see extensive models and discussions in Kappeler, 2000). A final pattern that emerges is that of infant care (Lee, 1989). If, as discussed above, females can share with others the energy costs of infant care and minimize mortality risks from extrinsic and care-dependent sources, then grouping of some sort is expected (see Jennions and Macdonald, 1994). In birds and some mammals (such as those with specifically high risks of infant mortality due to infanticide), male partnerships with females are a mechanism to provide this care.

Despite significant differences in life history traits, primates do not differ in these sharing of care contexts. Thus pair living, as in owl monkeys, klipspringers, or jackals, or extended families with membership that varies over time due to reproductive dispersal, such as in callitrichids, meerkats, or ground squirrels, provide opportunities for shared infant care. Matrilines in the absence of males, such as those of elephants, coatis, and perhaps mouse lemurs, are common mammalian forms for sharing infant care, as expected, given high degrees of relatedness and the potential for reciprocity. Matrilines with the addition of permanent males—for example, in buffalos, lions, or many cercopithecid primates—provide two sources of help: that from female kin and that from males defending against infanticide or predation. By contrast, in groups with non-kin females, such as plains zebras, gorillas, or howler monkeys, or fission-fusion groups, such as those of orangutans, chimpanzees, or bottlenosed dolphins, males are at least available to help, while non-kin females can still reciprocate via coordination in forming foraging or play groups and thus sharing some elements of infant care or facilitating opportunities for learning.

Primates differ from other mammals not in the proportion of groups where infant care is shared (Figure 1.2) but in the proportion of groups with stable and very long-lasting male-female associations. As noted throughout this chapter, the reproductive ecology of females, their access to energy, the evolved life history traits of infant growth, when and how risks of infant mortality can be minimized, and whether risks derive from predation, infanticide, or maternal care all determine the shape of the mating and parenting strategies within the social context.

We can ask, again, whether in the mammalian or in the anthropoid primate contexts, are humans and their care systems special? As detailed in Chapter 5 and throughout this volume, the human system of mating

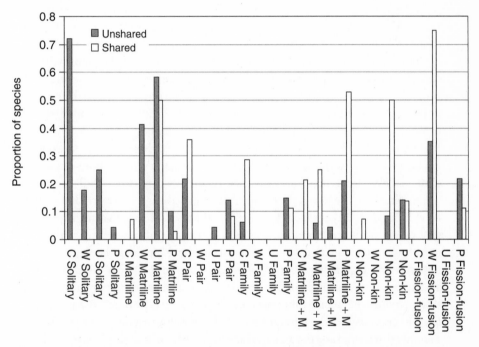

Figure 1.2   Proportion of species where some form of shared infant care is observed compared to those with no shared care among carnivores (*C*; n = 78); cetaceans (*W*; n = 21); ungulates (*U*; n = 28), including elephants; and primates (*P*; n = 156). Data on group types and life histories from Lee (1994) and Lee and Kappeler (2003); data on presence/absence of shared infant care compiled by McAuliffe (2004). Group types coded as solitary, matrilines, pairs, extended family with resident male, matrilines with one or more males, non-kin females with one or more males, and fission-fusion units with variable associations between known individuals of both sexes. No phylogenetic corrections are applied, as the pattern at evolutionary branch tips is of interest.

and parenting relies on behavioral components rare in other mammals: for example, division of labor, provisioning, and intergenerational transfer of food and foraging knowledge. Some behavioral care components are generalized across anthropoids—for example, lactation durations responsive to local environmental quality and variability as well as infant needs for growth (Lee, 1999). Others are found in nonprimate species—for example, provisioning and the division of foraging tasks seen in many carnivores—suggesting these elements of human behavior are associated with foraging on large packets that can be shared and transported. Others, however, such as information exchange over time and space, are unique to modern human patterns of brain development and growth and impact on all elements of the human life history (Kaplan et al., 2000).

Three concepts emerge from heuristic models of social evolution derived from comparisons among mammals. First, sociality is defined by consistent maintained relationships between individuals. Thus cognitive evolution can also be predicted as an outcome of maintained groups, with communication, memory, exchange, and reciprocity operating to structure relationships among individuals within stable groups. While the cognitive capacities of some species have received considerable attention (for example, primates: Whiten and Byrne, 1997; Dunbar, 1998; dolphins: Herman, 2002; elephants: McComb et al., 2000), these are less seldom explicitly linked to the nature of evolved parenting systems (but see Gittleman, 1994; Keverne, Martel, and Nevison, 1996; Barton, 1999).

Second, there are limits to the number of possible social states that can be exhibited by any species (Lee, 1994). Why social space varies among species is poorly understood, although avoidance of inbreeding via the need for prereproductive dispersal of one (or both) sex may be an important constraint. Furthermore, while there may be limits on the number and types of social outcomes, there are numerous alternative routes to these "adaptive" social outcomes. Phylogenetic constraints on sociality need to be considered in greater detail (DiFiore and Rendall, 1994). We thus need to separate the constraints on forms of sociality (outcomes) from constraints on the routes (mechanisms such as mating systems) to the same social state before we can better understand the causation and function of sociality.

Finally, as has been shown repeatedly for the past 25 years, the fundamental principle underlying social evolution is that of female reproductive energetics. Predation and other sources of environmental risk of mortality are important influences shaping social systems, while the causes and timing of infant mortality are also significant factors for the form of sociality expressed. Social options may vary depending on whether the costs of infants can be shared among individuals—can care-dependent mortality be manipulated via helpers? Whether nepotistic or despotic competitive systems (van Schaik, 1989; Isbell and Young, 2002) emerge is most likely to be a secondary consequence of the nature of the resource base rather than parenting contexts. Fundamentally, the resource base determines the mode of female sociality via reproductive energetic strategies, while males map onto the female reproductive distributions, and thus mating systems can be viewed as secondary outcomes rather than determinants of social systems.

Future models and adaptive explanations of social evolution need to move beyond male-focused competitive mating systems. These will require a far greater understanding of female reproductive ecology, female mate choice, and parental care allocation systems (see also Emlen, 1997; Gowaty, 2004). We need to be able to challenge our preconceptions derived from three decades of observations biased toward competition and understand more fully parenting in a social and ecological context. The chapters in this volume add to this perspective.

Acknowledgments

Many thanks to Peter Gray and Peter Ellison, for the invitation to write this chapter, and to Jane Lancaster, for her insightful comments on the draft.

CHAPTER TWO

Neuroendocrine Mechanisms Underlying Social Relationships

Kim Wallen and Janice Hassett

THE HORMONAL SYSTEM (Figure 2.1), together with its interactions with the genome and nervous system, provides a remarkable and energetically economical method of integrating diverse sources of input into relatively simple messages transmitted throughout the body via the circulatory system. While traditionally these hormonal signals were viewed in the context of the endocrine system, those ductless glands that secrete their products directly into the blood, it is now apparent that the nervous and endocrine systems are fully integrated and that both participate in a highly reciprocal manner to transduce information about the individual's environment, physical and social, and changes in internal state resulting from environmental input. We describe here some mechanisms by which hormones affect, and in turn are affected by, social relationships. A complete description of these mechanisms is not yet possible; we still have only a partial understanding of how this integration is achieved, and we do not even yet know the complete scope of social information that affects the neuroendocrine system or is in turn affected by it. However, a number of principles of hormone action have emerged that form the basis for developing a more complete understanding. These principles will be described; then the physiological mechanisms that regulate neuroendocrine responsiveness to social context will be presented; and finally, these will be applied specifically to the steroid system and to the oxytocin and vasopressin neuropeptide system. The intent of this chapter is to provide a functional framework for understanding hormone action and how it modulates social behavior and in turn is modulated by social behavior, social context, and social history. Further discussion of endocrine signaling, particularly with reference to steroid hormones, is also included in Chapter 3.

Frank Beach, some 30 years ago, in describing the new field of behavioral endocrinology (Beach, 1975) articulated a number of relationships between hormones and behavior that are as relevant today as when first proposed. Investigating hormone-behavior interactions typically starts by assessing covariation between changes in the hormones of interest and changes in behavior. This correlational approach has been highly successful and remains the initial approach for investigation of any potential relationship between hormones and behavior. For example, the role of estradiol and progesterone on female rodent sexual receptivity was first inferred by describing the covariation between ovarian anatomy and the

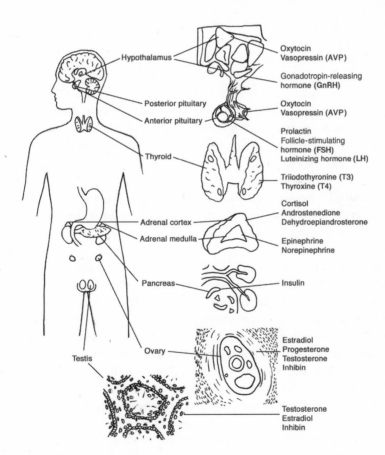

Figure 2.1   The major components of the human endocrine system that are discussed in this volume. Most of these have homologues in other vertebrates as well.

onset of female sexual receptivity (Young, Myers, and Dempsey, 1933; Young, Dempsey, and Myers, 1935). Though steroid hormones could not be measured at that time, knowing which class of steroids the ovary produced at each stage of the ovarian cycle allowed development of specific hypotheses concerning the steroid hormones involved and their sequence and timing in inducing female sexual receptivity. This ultimately resulted in discovery of effective hormonal replacement therapies that fully restored female sexual behavior following ovariectomy (Dempsey, Hertz, and Young, 1936).

Today studies of covariation are the staple of endocrinological studies of social behavior in natural, field populations of animals (Aujard et al., 1998; Mooring et al., 2004) and humans (Bruce, Davis, and Gunnar, 2002; Gunnar and Donzella, 2002) and continue to provide meaningful leads to more mechanistic studies not possible under field conditions. One problem with correlational studies is that the direction of the relationship between hormones and behavior is often unknown. As Beach described, hormones can both cause behavioral change or be changed themselves as a result of behavior. Thus hormone-behavior relations are inherently interactional. This interrelationship between hormones and behavior and behavior and hormones means that correlations between hormones and behavior do not often indicate the causal direction. This is particularly likely to be the case in social behavior, where hormones bias social interactions but are similarly modulated by social behavior, such as winning or losing, social status, and stress or other social challenges. Thus without explicit experiments in which hormones are manipulated and behavioral change monitored, it is difficult to go from covariation to causation.

Identifying the causal effects of hormones is complicated by the fact that hormones operate through two markedly different sets of processes: organization and activation. Organizational effects are those actions of hormones that permanently or semipermanently alter an individual's potential to show sexually dimorphic behavior by actions during periods of developmental sensitivity. By contrast, activational effects are transitory and reversible and typically occur postpubertally and in adulthood (Phoenix et al., 1959; Young, Goy, and Phoenix, 1964).

Because the effects of organizational actions of hormones are often only evident months or years after the hormone has acted to organize a particular pattern of behavior, it is often difficult to distinguish between a current activational effect of a hormone, which might be reflected in current circulating levels of hormones, and an organizational effect of the hormone that would likely bear little or no relationship to current hormone levels. In addition, organizational effects of hormones, particularly on sexually differentiated behavior, act through two distinct processes themselves.

Mammalian sexual differentiation is biased in a female direction (Jost, 1970) such that morphogenic processes are geared toward sexually differentiating female end points more easily than they do male end points. Some have thus referred to the female path of sexual differentiation as the "default" path, meaning it is the pathway most easily followed and that male characteristics are imposed on an essentially female template. While female traits require specific morphogenic processes, they are the end points that are created unless additional alterations produce males. This concept is valuable, as it implies that blocking a process necessary to produce a male trait results in a female phenotypic trait instead. By contrast, when a female process is absent or blocked, a male characteristic does not arise. Thus, male differentiation requires two specific processes— suppression of female characteristics and the addition of male characteristics that allow the male phenotype to alter an essentially female-biased differentiation. In birds and other vertebrate species where the female is the heterogametic sex, the bias is unclear, and the processes by which hormones organize behavior are even less completely understood than are those in mammals. Thus what is described here applies to mammals and may or may not apply to other vertebrates.

Historically, there has been no consistency in the terms used to describe the processes underlying the organization of sexual dimorphic behavior, with *feminization, masculinization, defeminization,* and *demasculinization* all being used, resulting in significant confusion. However, the terminology need not be complicated, as it is clear that there are two processes necessary to create a male: defeminization and masculinization. Defeminization suppresses female-like characteristics that would otherwise arise, whereas masculinization imposes male-like characteristics on the developing organism. Except in very limited domains, there is no evidence for feminization, that a female phenotype is imposed on an essentially neutral anlage, or for demasculinization, in which an existing masculine characteristic is removed or eliminated. Demasculinization has been demonstrated in birds, where some male characteristics are the default condition, but not in mammals.

Thus masculinization refers to the production of male-typical characteristics. Anatomically, these would be the presence of penis, scrotum, testes, and internal Wolffian duct derivatives. Behaviorally, masculinization refers to the behavioral characteristics normally associated with a male physical phenotype: it is primarily seen in the differentiation of male copulatory behaviors facilitating gamete transfer but also applies to social behavior such as patterns of play, aggression, and even affiliation. Additionally, masculinization may describe the differentiation of cognitive systems that underlie social behavioral differences between males and females.

Defeminization, by contrast, refers to the suppression of already existing female-typical characteristics or prevention of their development. Anatomically this results in the suppression of the development of uterus, fallopian tubes, portions of the vagina, and vaginal opening. Behaviorally, defeminization suppresses sexual receptivity, termed *receptive defeminization*. In monkeys, defeminization appears to suppress interest in sexual initiation, termed *proceptive defeminization*. However, because investigations of sex differences in social and cognitive domains are still not well developed, we do not yet know which patterns of social cognition might be the default and which might be suppressed in males.

These processes of masculinization and defeminization do not act on the same developmental substrate in all species, as species vary widely in the extent of their development at birth. Species that are born with their eyes closed and lacking the capacity to function relatively independently at birth are termed *altricial;* more completely physically developed offspring are termed *precocial* (Wallen and Baum, 2002).

Many of the most often studied species are those where sexual differentiation is only partially completed at birth. In these species a significant portion of the differentiating process occurs when the offspring is no longer attached to the maternal circulation. Although the gonad and duct systems have differentiated in utero, the external genitalia have only begun to differentiate, and distinguishing males from females typically requires measuring the distance from the penile/clitoral glans to the anus, the anogenital distance, which is longer in males than in females.

Whether a species has altricial young follows no clear phylogeny. Species as diverse as rats and ferrets produce altricial young. While altricial offspring is a characteristic of many laboratory rodents, not all rodents are altricial, the guinea pig being that principal exception. It appears in some orders, such as the carnivores, that altricial offspring are the rule, with carnivores from ferrets to lions producing altricial young. Similarly, in bird species this dichotomy is widely evident.

Altricial and precocial species probably reflect two different reproductive strategies that are expressed in different patterns of neural development (Gaillard et al., 1997). In general, precocial species are characterized by a greater brain-to-body-weight ratio than are altricial species (Pagel and Harvey, 1989). In addition, a substantially greater portion of brain development occurs in utero in precocial than in altricial species. This difference in development may reflect a life history difference in that altricial species produce young rapidly, with short maturation times and rapid brain growth (Lewin, 1988). In contrast, precocial species develop more slowly, even though both species may reach maturity at comparable times (Tessitore and Brunjes, 1988). For our purposes, the altricial-precocial distinction is potentially important for understanding the effects of hormones on social behavior, as the specific hormonal processes that organize sexually dimorphic patterns of behavior differ markedly between altricial and precocial species (Wallen and Baum, 2002).

These basic processes likely apply to the two major groups of behaviorally active hormones in social behavior: steroid hormones and peptide hormones. In some domains, much more is known about steroid hormone action; however, with regard to peptide hormones, specifically oxytocin and vasopressin, social behavioral effects have been the primary behavioral end points studied. Steroid and peptide hormones exert their influences on social behavior both independently and in concert. They also use similar mechanisms of hormone-specific receptors and regional differences in receptor expression. In the sections that follow, we first describe the behaviorally relevant mechanisms that regulate steroids and their relation to social behavior and then offer a similar description for peptide hormones. Later chapters in this volume address the specific relationships between these hormonal control mechanisms and specific social behaviors. Here we present the mechanisms that one must consider to understand hormone action.

Steroids

Steroid hormones are relatively small molecules, ranging between 18 and 21 carbons; all are derived from cholesterol, a 27-carbon compound. They are generally highly lipophilic, passing easily through cell walls and thus moving easily through tissues in the body independently of their transmission through the circulatory system. This capacity to pass through cells results in the externalization of steroids in saliva, on the body surface, in sweat, and in feces. Were one able to image all of the steroids and steroid metabolites as light, our bodies would likely appear to glow in the dark. Radiating steroids makes them a potential medium of communication between individuals, and many of the putative mammalian pheromones are steroids or steroid derivatives.

The biological activity of steroids is determined by two aspects of their chemical makeup: (1) the number of carbons present, which determines the steroid family to which a specific steroid belongs, and then specific bonds; and (2) the side groups occurring at particularly biologically active carbons within the structure. Steroids are somewhat unique as hormones in that they can be readily metabolized to more than one active form, unlike most peptides, which are likely to become biologically inactive if their structure is altered. This system of active transformations of steroids can make it difficult to identify with certainty which specific steroid produces a particular biological or behavioral change. It is possible, for example, to administer a steroid that is then transformed into a steroid completely different from the family of steroids that constitutes the actual biologically active form. Such changes can occur throughout the body, or they could occur within a discrete, localized population of cells where it is not possible to detect the change in steroid form. In addition, as in the case of peptide hormones, these transformations can render steroids biologically inactive, increase their excretion rate by making them water soluble, or increase the ease with which they pass through cell walls.

Enzymatic Conversion of Steroids

There are four families of steroids based on the number of carbons they contain: progestins, with 21 carbons; corticoids, with 21 carbons; androgens, with 19 carbons; and estrogens, with 18 carbons. All steroids share 18 carbons in common, and all steroids start as a 21-carbon progestin—pregnenolone—and then are modified to form all the other steroids. No matter what steroid is measured, at one point in its life, it was pregnenolone. Transformations of steroids are promoted by specific enzymes that make specific changes in the steroid molecule biologically meaningful. Enzymes can be thought of as organic catalysts in that they make a specific reaction proceed at a rate that produces meaningful product but are not themselves consumed in the reaction, being recycled to promote the transformation of another molecule of steroid. It is beyond the scope of this chapter to detail all of the specific reactions catalyzed by specific enzymes; however, some major transformations will be described where appropriate.

Steroids are produced de novo from cholesterol in two primary sites in the body, the gonads and the adrenal cortex. In addition, the placenta in pregnant females can convert cholesterol to steroids. Recent evidence, first in birds (Holloway and Clayton, 2001) and now in rats (Micevych et al., 2007), suggests that local regions in the brain possess the enzyme necessary to transform cholesterol into pregnenolone, raising the possibility that the brain produces steroids internally for its own use. How such intracerebral steroids mesh with the signaling of gonadal and adrenocortical steroids is not yet known. It is likely that steroid signaling is substantially more complex and possibly more precise than is currently suspected.

Steroid transformations always proceed from more carbons to less carbons, and there are no processes by which steroids gain carbons. Thus progestins are precursors for androgens, which are precursors for estrogens. The corticoids, glucocorticoids, and mineralocorticoids are all derived from progestins and do not themselves serve as precursors for other steroids. Corticoid production is controlled by a single enzyme, 21-hydroxylase, which occurs in only the adrenal cortex, making this the only place in the body where corticoids can be created. The lack of 21-hydroxlyase has profound effects on the individual and can result in death as a result of adrenal insufficiency. Reduced effectiveness of this enzyme is also possible, not producing a frank disease state but reducing the capacity of the individual to mobilize a corticoid response to environmental challenge.

The progestins, androgens, and estrogens are considered the classic sex steroids, and forms of each are produced by both sexes. Thus the primary sex difference in steroids is not which classes of steroids are produced but, rather, the relative amounts produced within each sex. The sex differences in production are typically on an order of magnitude or more. The following estimates deserve some degree of caution, as they are based on findings from industrialized societies and thus may not reflect the entire and more common range of variation of most human populations. Peak testosterone levels in women are on the order of 0.5 ng/ml, whereas typical male levels would be 10 ng/ml or more. Similarly, in humans, male estradiol levels would be in the 20 pg/ml to 40 pg/ml range, whereas female midcycle levels average 300 pg/ml to 600 pg/ml. The magnitude of the sex differences often results in little attention being paid to the hormone levels in the sex that produces less of a given hormone. However, it is increasingly evident that even low levels of hormones can have profound effects. This is particularly true of estrogens because of the crucial nature of the conversion of androgens to estrogens in both males and females.

Production of the high levels of estrogens seen in the female's ovarian cycle first requires the production of significant levels of androgens, primarily androstenedione and testosterone in ovarian thecal cells (Hedge, Colby, and Goodman, 1987). Only a small proportion of this ovarian androgen is actually secreted into the bloodstream, with the majority being taken up by granulosa cells, which contain the enzymes necessary to cleave the 19th carbon from the androgenic precursor and alter the 18-carbon product to create estradiol or estrone. Although this is done with an enzyme complex, the enzymes necessary are routinely referred to as *aromatase* and are present in a variety of tissues (ovarian, testicular, and fat) and neural cells. Essential for the production of ovarian estrogens, aromatase is also critically involved in the production of intracellular estrogens within neural target tissues. By expressing aromatase within specific nuclei within the brain, it is possible for tissues seeing the same testosterone message circulating in the blood to produce markedly different intracellular consequences as the testosterone is converted to estrogens within the cells. The aromatization of androgens to estrogens is of critical importance to male sexual differentiation in a number of species, but it is not crucial to some species where androgens act directly to masculinize the developing fetus (Wallen and Baum, 2002). In those species where it is important, notably rats, mice, hamsters, and ferrets, microamounts of estrogens within neural cells, not detectable in the circulating blood, have dramatic effects on the developing nervous system. Similarly, in these same species, intracellular aromatization of androgens is also necessary for the activation of masculine sexual behavior (Wallen and Baum, 2002); in others, such as the guinea pig and the monkey, it is not. In addition to demonstrating regional and species differences in how hormones act, these findings demonstrate how difficult it can be to infer causal actions of steroids from blood levels or from excreted levels of hormones, as the active agents may only occur intracellularly and what is measured in the blood may be relatively high levels of precursor hormones whose actual circulating levels do not correlate with the active intracellular levels that are created through enzymatic action.

Feedback Control of Steroids

Steroid production, in addition to being affected by the presence and amounts of specific enzymes, is also the result of specific feedback mechanisms that can either maintain a homeostatic level of specific steroids through negative feedback or marshal dramatically increased levels of steroid through positive feedback. In vertebrate species, the hypothalamus secretes releasing hormones that act on the anterior pituitary, which in turn secretes trophic hormones that act on the target endocrine gland that secretes the steroid products. The hypothalamic-pituitary-gonadal (HPG) axis and the hypothalamic-pituitary-adrenal (HPA) axis use different releasing and trophic hormones to regulate the production and release of steroids from the gonad and the adrenal cortex, respectively, but use a similar negative feedback mechanism in which one or more products of the target of the endocrine gland feed back to suppress secretion of the trophic hormone from the anterior pituitary. Control of ovarian steroid secretion also includes a positive feedback system in which ovarian estradiol and progesterone stimulate, rather than suppress, the secretion of trophic hormones. Only the ovary has such a positive feedback loop.

HPG AXIS

In the HPG axis a decapeptide, gonadotropin-releasing hormone (GnRH; also called luteinizing hormone-releasing hormone, LHRH, or luteinizing hormone-releasing factor, LRF), originating in cell bodies of the anterior hypothalamus that terminate in the medial basal hypothalamus, is released directly into the portal circulation going to the anterior pituitary (Guillemin, 1980). GnRH stimulates the anterior pituitary to release glycoprotein gonadotropins; follicle-stimulating hormone (FSH), which stimulates gamete growth in the gonad; and luteinizing hormone (LH), which stimulates the production of steroids in the gonads of both males and females.

These trophic hormones are maintained at relatively constant levels in males by the negative feedback of testosterone on GnRH secretion or directly on the anterior pituitary. Both estradiol and testosterone produce the same negative feedback on LH in males, and evidence supports local aromatization of testosterone to estrogen as important in primate male negative feedback (Hayes et al., 2000; Rochira et al., 2006). FSH secretion in primates is under primary control of the testicular protein inhibin, which can suppress FSH secretion independent of testosterone levels, allowing independent regulation of gamete and steroid production within the testes (Tilbrook and Clarke, 2001).

The negative feedback regulation of gonadotropin secretion in females is similar to that of males, except that circulating estrogens and progesterone maintain negative feedback, whereas androgens play no role (Couzinet and Schaison, 1993). The ovarian steroids act at both the level of the hypothalamus and directly on the anterior pituitary. As the ovarian follicle grows under constant stimulation from FSH, its production of estradiol increases with the size of the follicle (Apter et al., 1987). This increased estradiol secretion eventually exceeds a threshold level, and the feedback control switches from negative feedback to positive feedback in which estradiol now increases LH and FSH secretion (Hoff, Quigley, and Yen, 1983). Following the switch from negative feedback to positive feedback, estrogens continue to increase gonadotropin secretion, with the final LH surge initiated by elevated preovulatory progesterone (Hoff, Quigley, and

Yen, 1983; Batista et al., 1992). Ovulation follows the pronounced go-nadotropin surge within 24 to 36 hours, which in some species is followed by a luteal phase characterized by elevated progesterone secreted by the corpus luteum (Hedge, Colby, and Goodman, 1987).

Ultimately, the corpus luteum atrophies, and the ovarian cycle ends. In menstrual primates, this results in sloughing of the uterine endometrium and menstruation, but in most species, there is no specific marker of the end of the ovarian cycle; if the female does not become pregnant, in gestating species the cycle restarts. In egg-laying species, gamete shedding may end the annual ovarian cycles. In nesting birds, incubating an egg clutch suppresses further ovarian cycles, though ovulation may continue if eggs are removed from the nest (Millam, Zhang, and el Halawani, 1996).

Behavioral inputs affect many aspects of ovarian cycles with some species, like the musk shrew relying completely on the male's behavior to induce ovarian cycles (Rissman, 1990). In this species, females do not even go through puberty unless they are exposed to a male. Remarkably, males can induce the start of ovarian cycles with as little as 20 minutes of exposure to the females. In other species, male sexual behavior is necessary to induce corpus luteum formation (rats and other altricial mammals) or ovulation (ferrets, cats, and camels); in primates and precocial species, all components of ovarian cycles occur spontaneously without access to a male (Crews, 1987). In males, behavior affects testicular function. In rodent species, mating can result in elevated testosterone (T) in males, while social defeat can result in suppressed testicular function (Kamel et al., 1975; Kudryavtseva, Amstislavskaya, and Kucheryavy, 2004). Similarly, in monkeys, social defeat decreases T secretion for weeks after defeat, but exposure to a sexually active female almost immediately increases T secretion (Rose, Bernstein, and Gordon, 1975; Vandenbergh and Post, 1976). In human males, winning or losing an athletic competition can also alter testicular function (Booth et al., 1989; Edwards, Wetzel, and Wyner, 2006), suggesting that social modulation of gonadal function may be an evolved mechanism for relating reproductive output to social conditions and social status.

HPA AXIS

The adrenal has two anatomical divisions, the medulla, which is a mixture of neural and endocrine tissue and which secretes catecholamines, and the cortex, which secretes a wide range of steroids. The medulla is primarily activated through neural input and represents the high-speed-response aspect of the body's stress system. Epinephrine and norepinephrine secreted by the adrenal medulla chromaffin cells can cause the adrenal cortex to release corticoids acutely (Delbende et al., 1992). However, corticoid release from HPA is primarily regulated through negative feedback control, where glucocorticoids, but not mineralocorticoids, from the adrenal cortex regulate the secretion of the peptide corticotropin-releasing hormone (CRH) from the ventromedial hypothalamus. CRH causes the release of the polypeptide adrenocorticotropic hormone (ACTH) from the anterior pituitary. ACTH in turn stimulates the adrenal cortex to produce a wide range of steroids.

While the corticoids are unique to the adrenal cortex, it also produces and secretes the classical sex steroids, the amount and distribution of which varies across species. Thus the adrenal cortex can produce biologically significant amounts of progesterone, androgens, and estrogens in addition to glucocorticoids and mineralocorticoids (Hedge, Colby, and Goodman, 1987). This steroid negative feedback pathway is independent of the HPG negative feedback pathway in that different releasing hormones, trophic hormones, and steroids are involved in the HPA and HPG negative feedback systems. Thus it is possible for adrenocortical function and gonadal function to be regulated separately.

Separate regulation does not, however, mean that steroids secreted by the adrenal cortex cannot affect gonadal function. For example, the significant amounts of progesterone secreted from the rat adrenal cortex can serve as a substrate for testicular production of androgens, increasing testicular output above what would result from LH stimulation of the testes (Feek et al., 1985; Feek, Tuzi, and Edwards, 1989). In this manner, increased levels of stress hormones can also be associated with increased androgen levels from the testes. In another example of interplay between the adrenal cortex and gonads, suppression of adrenal cortical steroid secretion using the synthetic glucocorticoid dexamethasone in female rhesus monkeys resulted in significantly higher levels of estradiol, suggesting that the adrenal steroids exerted low-level negative feedback on the HPG axis (Lovejoy and Wallen, 1990).

Conclusion

Modulation of steroid secretion is possible at the level of the brain, anterior pituitary, and gonad. Mechanisms exist that can regulate the output of these systems in concert or independently. This makes complex responses to unique environmental challenges possible while also allowing consistent responses to predictable changes in the environment. Steroid action in the brain likely reflects steroid concentrations within specific target cells. These concentrations in turn reflect the pattern of enzymes within these target cells, making possible marked regional differences in intracellular steroid types and levels in response to a similar hormonal message in the circulation. This localized capacity for alterations in the type and amount of steroids within tissues means that our primary measurement for studies of hormones and behavior—circulating levels of steroids—is often likely not to fully reflect the hormones of most importance in modulating behavior. Thus when circulating levels of steroids correlate poorly with behavioral changes, it is not possible currently to know whether or not this truly reflects the fact that steroids do not modulate the behavior under study in a predictable manner or whether the relevant levels are intracellular and not yet readily available for inspection. The evidence of consistent covariation between steroid levels and behavioral change seen in a variety of contexts suggests that for at least some hormone-behavior relationships, blood, urine, or fecal levels can serve as a useful proxy for levels within the relevant tissues. However, when such relationships are not seen, it is not yet possible to know exactly how to interpret that absence.

Peptides

While steroid hormone structures consist of rings of interconnected carbon atoms, peptide hormones are composed of strings of amino acids. Also in contrast to steroid hormones, which are under the control of the hypothalamus and anterior pituitary and specific releasing and trophic hormones, two peptide hormones important to mammalian social behavior, oxytocin (OT) and vasopressin (also known as arginine vasopressin, or AVP), remain under the control of the hypothalamus but are released from the neural processes of hypothalamic neuroendocrine cells terminating in the posterior pituitary. Though steroid hormones easily cross the blood-brain barrier, OT and AVP do not, meaning that central and peripheral release of the peptides can occur independently. In a final contrast to steroid hormones, OT and AVP do not diffuse out of cells but are released through action potentials, in a fashion similar to that for classic neurotransmitters. Thus the cells they are produced in are often referred to as *neurosecretory,* and the peptides themselves are referred to as *neuropeptides.* OT and AVP are not unique as neuropeptides—some estimate that up to 200 different peptide molecules also function like neurotransmitters (Holmgren and Jensen, 2001).

Peptide Synthesis

OT and AVP are closely related nonapeptides, with different amino acids in only two of nine positions. They are encoded by the genome, and their respective genes are situated closely together, with a short intergenic region between them (Murphy and Wells, 2003). Both are synthesized in cells of the paraventricular nucleus (PVN) and the supraoptic nucleus (SON) of the hypothalamus. AVP is also produced within the suprachiasmatic nucleus of the hypothalamus, and outside the hypothalamus, it is produced in the medial amygdala and the bed nucleus of the stria terminalis. Production in these extrahypothalamic areas is dependent on androgens and may account for some sex differences in the brain (De Vries and Miller, 1998). OT and AVP are synthesized, like other peptides, in a series of cleavage events: first, DNA in the cell nucleus encodes messenger RNA, or mRNA, which allows polypeptides to be produced on ribosomes in the rough endoplasmic reticulum (Holmgren and Jensen, 2001; Landgraf and Neumann, 2004). This polypeptide, however, is not OT or AVP but a precursor, referred to generally as a prepropeptide. When part of this prepropeptide, the signal sequence, is cleaved by a signal peptidase enzyme (von Eggelkraut-Gottanka and Beck-Sickinger, 2004), a propeptide is formed. The propeptide includes a neurophysin, which is thought to be a carrier molecule for the active peptide (Richter, 1983; Murphy and Wells, 2003). OT is associated with neurophysin I and AVP with neurophysin II (Ivell, Schmale, and Richter, 1983; Murphy and Wells, 2003). Cleavage of the propeptide by prohormone convertase enzymes (von Eggelkraut-Gottanka and Beck-Sickinger, 2004) results in formation of the active peptide. These additional components of the prepropeptide and propeptide are necessary for passage into and through the endoplasmic reticulum (Landgraf and Neumann, 2004). In the case of OT and AVP, it is the propeptide that is packaged into granules by the Golgi complex for axonal transport (Brownstein, Russell, and Gainer, 1980; de Bree and Burbach, 1998; Murphy and Wells, 2003). The final cleavage step is completed prior to release of neuropeptide from the nerve terminal or dendrites (Holmgren and Jensen, 2001).

While OT has just one known receptor, AVP has three known specific receptors (Carter, 2003), though most central effects are thought to occur through interactions with one of these, the V1a receptor (Young, 1999). The peptides are unique to mammals, though most vertebrates possess structurally similar peptides (Gimpl and Fahrenholz, 2001), and their function is thought to be conserved functionally across taxa as well (Lim and Young, 2006). For a detailed review of gene structure and regulation of OT and AVP and other mechanistic details, see Zingg (1996) and Gimpl and Fahrenholz (2001).

Mechanisms of Action

Classically, OT was thought to be involved primarily in behaviors related to parturition and suckling, as OT facilitated the smooth muscle movements involved in both. More recently, researchers hypothesized that OT might also be involved in smooth muscle movements of sexual activity. OT has many other peripheral functions that will not be reviewed here, but for a review of these see (Gimpl and Fahrenholz (2001). Neural and behavioral research has revealed a tremendous breadth in the central effects of OT. Broadly, central OT's main effect on social behavior is thought to be the facilitation of the formation of *social bonds.* A large variety of behaviors fall under this umbrella term. Social bonds can be maternal, infant, peer, or pair, and affected behaviors include sexual behavior, parenting behavior, feeding, grooming, aggression, trust, responses to stress and anxiety-inducing conditions, and social recognition.

Like OT, AVP also has a broad range of classically known functions. Peripheral effects are on blood pressure, cardiovascular function, and kidney function—generally, it is central to the sympathetic nervous system (Carter, 1998). Centrally, it has been known for its role in attention, learning, and memory (Carter, 1998). Like OT, it is involved in facilitating the formation of social bonds (Carter, 1998). Certainly both OT and AVP have central effects that are not related to the formation of social bonds, such as in general learning and memory for cognitive tasks (Bruins, Hijman, and Van Ree, 1992) and other functions (Insel, 1992), but these are not the focus here.

A number of factors complicate investigations into the functions of OT and AVP. First, as peptide hormones, they do not cross the blood-brain barrier; thus there is potential for a separation of central and peripheral release (Meisenberg and Simmons, 1983b). Further evidence for the independence of central and peripheral release comes from the finding that different populations of neurons terminate in different regions. Magnocellular neurons terminate in the posterior pituitary, releasing OT or AVP into the peripheral bloodstream (Gimpl and Fahrenholz, 2001; Lim and Young, 2006). Parvocellular neurons terminate in other regions of the central nervous system (Gimpl and Fahrenholz, 2001; Lim and Young, 2006). Thus, central and peripheral release of either peptide may be coordinated or may occur independently, and there is evidence for both independent and coordinated release. As an example of independent effects, researchers examining the effects of OT on social recognition have found different dose-dependent relationships for OT administered in the periphery and for OT administered in central regions (Ferguson, Young, and Insel, 2002). Conversely, for suckling rat mothers, parallel increases of OT in the SON, in limbic regions, and in the bloodstream have been reported (Landgraf, 1995). Thus, specific function may determine the relationship between peripheral and central peptide levels, and how the peptide is best measured depends on the research question of interest. We cannot presume that changes in peptides in the brain will be reflected systemically.

Because of these separable effects, methodologies used in OT and AVP research deserve closer examination. Most research, whether in humans or in nonhuman animals, administers the peptides exogenously and examines effects, somehow blocks the effects of the peptide (using a receptor blocker or a gene knockout model), or measures endogenous levels of the hormones and relates this to an experimental or natural condition. Each of these different methodologies presents its own set of complications. When hormone is administered centrally, for example, it is unlikely that we will be able to replicate the natural patterns of release in the brain across space and time (Landgraf, 1995). Knockout animals generally lack the "knocked-out" gene for their entire lifetime; thus developmental effects of peptides or receptors can confound findings (Carter, 2003). One common way to measure endogenous brain peptide levels is through cerebral spinal fluid (CSF), but some researchers note that hormone levels in the CSF are probably not of biological significance (Landgraf, 1995). While some may see measures taken in specific brain regions as a good solution to this, it is further argued that such measures also are not optimal, since the amounts of peptide in a region are affected by factors such as synthesis and transport and may not be reflective of local release of hormone (Landgraf, 1995). Some have suggested that microdialysis procedures, used either centrally or peripherally, are a good alternative solution and overcome the problems with earlier methods (Neumann et al., 1993; Landgraf, 1995; Landgraf et al., 1998).

Interest in studying central AVP and OT function directly in humans has increased dramatically in recent years, but such research presents with tremendous methodological limitations (McCarthy and Altemus, 1997). Again, because peptide hormones do not cross the blood-brain barrier, measuring peptide hormones in the blood or urine is not a good proxy for central release. However, we also cannot obtain CSF measures without very invasive procedures, and better methodologies, such as microdialysis, can only be used with nonhuman animals. Likewise, administering the peptides systemically will not tell us much about central effects. Most recently, the preferred method of administration of OT or

AVP to humans has been via use of intranasal sprays. In individuals receiving diagnostic lumbar punctures, intranasal administration of a synthetic (desglycinamide-AVP, or DGAVP) AVP resulted in detection in plasma and in the CSF (Born, Pietrowsky, and Fehm, 1998). This effect has also been found using AVP and healthy human participants (Born et al., 2002). Recent studies have also implicated intranasally administered OT in central nervous system effects (Heinrichs et al., 2003; Kirsch et al., 2005; Kosfeld et al., 2005). Thus the data suggest that this noninvasive method of administration allows the peptide access to the brain, though not independently of access to the peripheral system (Born, Pietrowsky, and Fehm, 1998).

Species Specificity

While the general structure and function of OT and AVP may be conserved across taxa, there are effects that are not species general; so generalizations across species can be problematic. Attention to the behavioral ecology of species, such as those that are naturally polygynous or monogamous, has revealed some patterns in these data, as has attention to the distribution of OT and AVP receptors in the brain. Still, assumptions about effects across species—or assumptions about humans from research in rodents or other animals—must be treated with reasonable caution. Species differences in behavioral effects of OT and AVP and receptor-binding distributions have presented a considerable obstacle in our understanding of what OT and AVP affect across species. Gimpl and Fahrenholz (2001) reviewed OT's effects on a variety of behaviors in rats, mice, prairie voles, sheep, and humans, demonstrating that there were many cases in which OT would result in an increase or decrease in a behavior in one species and have an opposite or null effect in another species. For example, their review revealed that OT increased female sexual behavior in the rat, had no effect in the mouse, both increased and decreased such behavior in the prairie vole, decreased behavior in sheep, and increased female sexual behavior in humans. Such inconsistencies across species were found for 7 of the 14 behaviors reviewed (Gimpl and Fahrenholz, 2001).

Receptor distributions can vary considerably even in very closely related species. Montane voles and prairie voles, both of the genus *Microtus,* show nearly complementary patterns of OT and AVP receptor distributions (Insel, 1992). These different patterns may reflect the fact that the two vole species have considerably different patterns of affiliation. Prairie voles are monogamous with high maternal and paternal care, spending large amounts of time in close contact with one another and showing stress responses to social isolation. The montane vole is polygamous with minimal parental care, spends little time in contact with others, and shows little response to social isolation. Interestingly, the same differences in OT and AVP receptor distribution have been found in pine voles, whose behavioral patterns are much like those of prairie voles, and meadow voles, who behave similarly to montane voles (Insel, 1992). These findings strongly suggest that it is differences in receptor distribution that contribute to these differences in social behavior (Carter, 1998; Lim and Young, 2006). All of the studied vole species have the same distribution of OT projections in the brain, which has allowed manipulations of receptor distribution, revealing that social behavior is affected by these distributions.

Additional cross-species comparisons have yielded a striking pattern of receptor distributions that seems to align well with behavioral patterns of different species. For example, the common marmoset, a monogamous, biparental New World monkey species, has OT and AVP receptor distributions more similar to the prairie vole and the pine vole than to the rhesus monkey. In turn, the OT and AVP receptor distributions of the rhesus monkey are more like those of the montane or the meadow vole (Young, 1999).

Can differences in behavior explain all of the differences in OT and AVP receptor distributions seen across species? Some puzzling species differences in receptor distribution remain (Tribollet et al., 1992), and additional comparative research is needed to determine whether there are other behavioral patterns that might help to explain such differences. Tribollet and colleagues (1992) noted, for example, that in parts of the ventromedial nucleus of the hypothalamus, there were many OT receptors in the rat and the guinea pig but none in the marmoset or in the rabbit. While the marmoset showed AVP receptors in these brain areas, the rabbit showed neither receptor type. If we hypothesized behavioral differences between these different species, we would then have to hypothesize another class of behavioral differences to explain receptor distribution differences in the dentate gyrus and cornu ammon is (CA) fields of the hippocampus, because there the rabbit shows OT binding, whereas the guinea pig and the marmoset do not (Tribollet et al., 1992). Since OT's central effects are diverse, this is a possibility, but further research is necessary.

Should we expect OT and AVP to have the same kinds of effects on social bonds in humans as they do in rodents, sheep, and some primates? Humans in general are less tied to hormonal effects than are many species—we can, for example, mate at any time, without regard to season or hormonal state (Wallen, 1990, 2001). Curley and Keverne (2005) suggest that, at least for apes and Old World nonhuman primates, hormones remain important but that there has been a considerable degree of release from hormonal determination for social behaviors and the formation of social bonds in general. So while rats and sheep rely on OT to elicit maternal behaviors, maternal behaviors are observed in Old World monkeys, apes, and humans even before puberty. Does this mean that neither OT nor AVP regulates social bonds in humans in any way? No. Recent research in humans has been able to measure endogenous hormone levels and deliver hormone exogenously, revealing that basal or reactive OT and AVP levels can vary based on early social experiences and that administration of the peptides can affect a variety of social behaviors and cognition.

Peptide Interdependencies

Peptide actions, like steroid actions, are affected by behavior, developmental timing, and other hormone actions. Similar to the organizational effects discussed for gonadal steroid hormones, there may be critical periods for organizational effects of OT and AVP for normal social functioning (Boer et al., 1994; Cushing and Kramer, 2005). One example is that how OT receptors are distributed and their numbers undergo great changes during development—in rats, the first major change is during the third week of postnatal life. A second period of major change occurs at puberty (Tribollet et al., 1992). The earlier period is marked primarily by a regression of receptors in brain regions where they had been seen at high density. At puberty, receptors are found in new brain regions (Tribollet et al., 1992). We do not have direct evidence that such changes occur in humans, as the only study of receptor distributions in humans was done in adults who died at 40 years of age or older (Loup et al., 1991). However, one study in children suggests the possibility of contributions of both peptides to a critical period for normal social behavior. This study examined children who spent time in orphanages as infants, thus experiencing early neglect, and who had been adopted into families. These children were compared to children who had been raised in typical family environments for OT and AVP levels before and after engagement in physical contact with their mothers or with an unfamiliar female (Fries et al., 2005). All measures were obtained from urine samples. Baseline AVP levels differed between groups, with lower levels in the children who had experienced early neglect. AVP levels did not differ between groups after physical interaction with either the mother or the unfamiliar female. While baseline OT levels did not differ between groups, OT levels after physical contact with the mother differed, with typically raised children showing higher OT levels than those children who had experienced early neglect (Fries et al., 2005). While larger studies are needed to replicate this study and examine effects of time in adopted families and time in the neglectful environment to better understand individual differences, this study suggests a reciprocal relationship between OT, AVP, and responses to the social environment. Such effects have been more directly demonstrated in rats, with evidence that the type of mothering that a female rat pup receives affects OT receptor expression in specific brain regions, which in turn affects the mothering behavior of that female as an adult (Pedersen and Boccia, 2002).

The effects of OT and AVP must be examined in the context of interactions with other steroids and neurotransmitters, relationships that are not yet completely understood. One of the most studied relationships has been between OT and estrogen. Early studies revealed that estrogen modulated OT receptor distributions in some brain regions. However, later studies have revealed that OT may have modulatory influences over estrogens as well. Estrogen is thought to be more involved with OT function than androgens, primarily because an estrogen-response element has been found on OT genes in rats (Bale and Dorsa, 1997). One of the best-known influences of estrogen on OT is in the ventromedial nucleus of the hypothalamus in rats, where estrogen upregulates the number of OT receptors and also facilitates their transport to regions around the nucleus (Coirini, Johnson, and McEwen, 1989; Coirini et al., 1991; Johnson, 1992). Early studies showed upregulation of OT receptors by estrogen in other brain regions, but only using levels of estrogen that were at supraphysiological levels (de Kloet et al., 1986). Interestingly, this effect has been reported to occur in both male and female rats (Tribollet et al., 1990; Coirini et al., 1992; Johnson, 1992). Effects of estrogen on OT, however, do seem to vary across species: there is no effect on oxytocin receptor (OTR) binding in prairie voles, and in mice, estrogen decreases the density of OTR in the ventromedial nucleus (VMN) (Young, 1999). Demonstrated influences of OT on estrogen have mainly been developmental, with neonatal manipulations of OT affecting estrogen recepton alpha (ER-alpha) expression in prairie voles and pubertal timing in rats (Withuhn, Kramer, and Cushing, 2003; Yamamoto, Carter, and Cushing, 2006).

OT-estrogen interactions are not, however, the only known interactions. OT has also been shown to play a role in regulating the gonadotropin LH in female rats and mice (Robinson and Evans, 1990). It may also regulate other hormones of the HPA axis, including CRH and ACTH (Gimpl and Fahrenholz, 2001). Glucocorticoids are thought to modulate OT and AVP receptors (Carter, 1998; Gimpl and Fahrenholz, 2001). While classical neurotransmitters such as norepinephrine, dopamine, serotonin, acetylcholine, glutamate, and gamma-aminobutyric acid, present in nerve fibers of nuclei in the hypothalamus, are known to affect firing rates of OT and AVP neurons, subsequently affecting release, these influences are only known to occur for the peripheral system (Ludwig, 1995). In addition, several peptides have been found in magnocellular OT and AVP neurons in rats—such co-localization suggests the modulatory potential of other peptides, but again, only for the peripheral system. However, in sheep, OT increases in the olfactory bulb are thought to modulate release of norepinephrine, acetylcholine, glutamate, and gamma-aminobutyric acid (GABA) (Lim and Young, 2006), so central interactions in other species are probable. Last, because OT and AVP are quite similar, differing only in two amino acids, they have the potential for either agonistic or antagonistic effects on the other's receptors (Carter, 1998). Thus, while we assume that effects of OT are the result of interactions at the OT receptor and that effects of AVP are the result of interactions at AVP receptors, this may not always be the case. Such crossover in receptor binding may also complicate the findings from gene knockout studies.

As a result of these known steroid-peptide interactions, where there are sex steroid influences, we should also expect sex differences in OT and AVP effects within a species. Sex differences in OT and AVP receptor distributions and function are a final factor complicating the study of these peptides. Many have suggested that the experimental evidence points to the differential importance of OT versus AVP in males and females, with OT being more central to social bonding functions in females, and AVP more central to similar functions in males (Winslow et al., 1993; Gimpl and Fahrenholz, 2001; Goodson and Bass, 2001; Cushing and Kramer, 2005). Throughout this section, the complications offered by receptor distribution variation have been a recurring theme, showing dependencies on development, species, and sex. Thus, this is a critical area for continuing research, as is careful attention to methodological details to ensure that measures of peptides accurately reflect central, and not merely systemic, levels.

Conclusion

Both steroid and peptide systems have a wide range of mechanisms that allow them to affect social behavior and in turn be affected by social behavior. The tools that investigators have to unravel these hormone-behavior interactions are increasingly sophisticated. These have resulted in identifying a number of potential mechanisms that can affect behavior. Yet, surprisingly, we have still not yet unlocked the secret of exactly how circulating levels of hormones translate into an unambiguous message at the target tissue. The exact relationship between the intracellular message seen in the brain and the magnitude of hormonal changes in the body remains elusive. Hormonal effects on behavior, on one hand, can be remarkably precise, the induction of lordosis in female rodents being a principal example of this precision. On the other hand, hormonal effects can be strikingly easily suppressed or modified by social context and experience (Wallen, 1990, 2001).

The mechanisms described in this chapter provide some of the tools necessary to understand the wide range of effects that hormones have on social behavior. Still, they reflect a field that, for all its apparent sophistication, is still in its infancy in understanding how the highly efficient and conserved system of hormonal communication has evolved to coordinate multiple organ systems, integrate the external environment with internal states, and produce coordinated and complicated behavioral responses to changing social contexts.

CHAPTER THREE

Social Relationships and Reproductive Ecology

Peter T. Ellison

Hormones and Information Flow

The vertebrate endocrine system is one of three closely related and over-lapping systems of cellular communication by which the integration of biological functions is maintained (the other two being the nervous system and the immune system). The key features of the endocrine system that distinguish it from its sister regulatory systems are the diffuse nature of its signals and its capacity to regulate both gene expression and cellular metabolism. The first key feature—diffuse signaling—distinguishes endocrine signals from the point-to-point signaling of the central nervous system (CNS). The second key feature of the endocrine system—its ability to regulate gene expression in addition to regulating ongoing enzymatic activity within target cells—gives the endocrine system a key role in developmental regulation and the unfolding of an organism's life history. Many of the key differences between closely related species—humans and chimpanzees, for instance—appear to be the consequences of differences in gene expression (in timing, sequence, intensity, and kind) rather than in genetic complement per se (Patterson et al., 2006). Hence an important role for the endocrine system in regulating species differences and even speciation itself seems likely.

The signals propagated by the vertebrate endocrine system can be classified in several ways. One of the traditional schemes emphasizes the distance or separation between source and target cells, distinguishing classical endocrine (specific gland to distant target through the circulatory system), paracrine (between cells in close proximity without the necessity of humoral transport), autocrine (influences on the source cell itself that are mediated by hormone receptors), and even, for some researchers, exocrine or pheremonal signals (signals propagated by one organism affecting a different conspecific organism through specific receptors whose expression is controlled by the receiving organism). A different classification scheme distinguishes the chemical and biological properties of the signal molecules themselves. Important distinctions can be drawn between steroid, protein, and peptide hormones, for example, as noted in Chapter 2. A third way to classify the various hormones of the endocrine system is by their receptors. All hormone receptors are themselves proteins and as such are derived from evolvable, coded genomic elements. Synapomorphies in the array of hormone receptors allows them to be clustered in phylogenetically related "families" and "superfamilies" that share important structural features by common descent, even as it is possible to group protein hormones themselves. The nuclear receptor superfamily, for example, can be traced back to the origin of multicellular, metazoan life (Owen and Zelent, 2000; Schwabe and Teichmann, 2004). This origin makes sense, since it is also the origin of the need for differentially regulating gene expression in distinct cells of the same organism: each cell's fate was, at this branch point in the history of life, no longer its own.

A final classificatory scheme, and one that pertains very much to the rest of this chapter and the rest of this book, organizes the endocrine system in terms of the major routes of information flow between major parts of the organism. This scheme can be rendered in greater or lesser detail. For our purposes, we will adopt a macroview, distinguishing only the central nervous system and the soma as nonoverlapping domains within the organism. We can then recognize four main pathways for information to follow: (i) CNS to soma; (ii) soma to CNS; (iii) soma to soma; and (iv) CNS to CNS (Figure 3.1).

Protein and peptide hormones dominate the channels that carry information along pathway (i) (CNS-soma), with the hypothalamic-pituitary axis providing the most important link. The endocrine hypothalamus is often referred to as a "transducer" of neural signals into endocrine signals, releasing small peptide hormones either into the hypophyseal portal system, to reach the anterior pituitary, or directly into the systemic circulation from the posterior pituitary. The pituitary is often referred to as the "master gland" of the body, since its tropic hormones target a variety of other endocrine glands. Researchers who concentrate on the flow of information along this pathway, from CNS to soma, naturally conceive of a causal hierarchy that features the brain as prime mover and the body

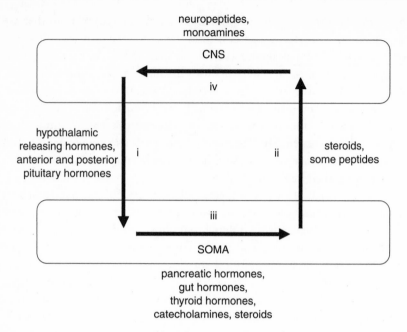

Figure 3.1   Four channels of information flow between the central nervous system (CNS) and the soma: *i,* CNS to soma; *ii,* soma to CNS; *iii,* soma to soma; and *iv,* CNS to CNS.

as falling under its control. In the hypothalamic-pituitary-ovarian (HPO) axis, for example, the hypothalamus is referred to as "controlling" the pituitary, which in turn "controls" the ovary.

Steroids dominate the channels that carry information flow along pathway (ii) (soma-CNS). This is due in part to their chemical properties as small, lipid-soluble molecules that allow them to readily cross the blood-brain barrier. It is thus no surprise that steroids feature prominently in the field of behavioral endocrinology since they provide the clearest examples of hormones that are measurable in peripheral tissues and fluids with demonstrable effects on behavior. Those who attend to information flow along this pathway often conceive of a different hierarchy of causal control from the traditional brain-to-body hierarchy. Rather than being the passive servant of higher CNS control, the ovary (for example) can be seen as the controlling organ of the HPO axis, modulating hypothalamic releasing hormone secretion and pituitary response via changing levels of estrogen and progestagen secretion. There is also an increasing list of peptide hormones that originate in the periphery, such as leptin and ghrelin, with important targets in those parts of the hypothalamus, such as the arcuate nucleus, that are most exposed to the peripheral circulation by way of the cerebral ventricles and cerebrospinal fluid or via the median eminence.

A number of protein and peptide hormones primarily carry information along pathway (iii) (soma-soma). Among these are the various digestive tract hormones that help to regulate energy and substrate assimilation and allocation, such as glucagon, cholecystokinin, and insulin. Steroid hormones are also important users of the soma-soma pathway, helping to integrate activity throughout the various parts of the reproductive system, for example.

Pathway (iv), CNS-CNS, might well be excluded from the catalog of endocrine system pathways and simply recognized as synonymous with the CNS itself. The reason to include it in this classification is because some of the same molecules that carry information along one of the other pathways also appear along this pathway. For example, monoamines such as dopamine and norepinepherine function within the CNS as neurotransmitters but are also released into the peripheral circulation from the adrenal medulla, where they function as hormones. Particularly important to this volume are the nonapeptides oxytocin and vasopressin (and their cognates in other organisms). Both of these molecules are released by magnocellular neurons of the paraventricular and supraoptic nuclei of the hypothalamus into the peripheral circulation via the posterior pituitary and function as hormones passing along the CNS-soma pathway. But they are also released from parvocellular neurons of the same nuclei into other areas of the brain, functioning as neuromodulators flowing along the CNS-CNS pathway. As noted in Chapter 2, there is considerable isolation of these two pathways by the blood-brain barrier, so that the information carried by a peptide messenger in one is not necessarily correlated with the information carried in the other pathway by the same messenger. Interactions between steroids and centrally active nonapeptides may provide a more robust linkage of central nonapetide signaling with peripheral endocrine signals. Similar cautions must be kept in mind regarding other protein and peptide messenger molecules that may be identifiable within the CNS. Various members of the gut hormone group, for example, are detectable within the CNS, but probably as a result of local production rather than transport across the blood-brain barrier. Levels of these molecules measured in the periphery cannot be assumed to reflect central levels.

Behavioral endocrinology as a field is concerned with the way hormones influence, and are influenced in turn by, behavior. Thus it is concerned with the flow of information between the CNS and the soma that helps to integrate behavior—the primary domain of the CNS—and physiology—the primary domain of the endocrine system. Any introductory course will emphasize that information flows in both directions between these two domains and that causal arrows also always point, at least potentially, in both directions. Reproductive ecology is a relatively new field that is concerned with the way in which the reproductive physiology of an organism adjusts in relation to aspects of ecology, environment, and life history. The central goal of reproductive ecology is to understand the regulation of reproductive effort (Ellison, 2001, 2003) or the allocation of time and energy to reproduction at the expense of other biological functions. How fast to mature? When to begin reproduction? How often to reproduce and how many offspring to reproduce at one time? How to modulate reproduction with age? These are all life history questions relating to the allocation of reproductive effort. The central insight of reproductive ecology is recognizing that the optimal answers to these questions depend on many features of the organism and its interaction with its environment that are highly variable and yet clearly patterned. Uncovering and understanding those patterns is the work of reproductive ecology.

Behavioral endocrinology and reproductive ecology thus have an inextricable relationship with one another. Information about environmental circumstances "enters" an organism both through its CNS and through its soma. That information must be integrated in one or the other or both domains and then used to regulate reproductive effort. Regulation of reproductive effort can occur through the regulation of physiology and through the regulation of behavior. In either case, information again must flow within and between the two organismal compartments for optimal outcomes to be achieved. Both fields are intimately concerned with the ways in which the states and functions of brain and body are integrated. The behavioral endocrinologist and the reproductive ecologist often encounter each other studying the same phenomena from different ends of the same conduits, the information channels connecting behavior and physiology.

Watching the Steroid Channel

An important principle of endocrinology is that hormones only carry information. They do not by themselves cause changes in cells, tissues, organs, or whole organisms. They do not serve as enzymes, chemical catalysts, poisons, or disruptors of homeostatic balance. They simply carry information about the state of the organism. It is up to the receiving cell both to attend to that information and to "decide" how to react to it. Many of the most powerful effects studied in behavioral endocrinology are mediated by steroid hormones carrying information about reproductive state from the gonads to the central nervous system. Major differences in behavior can be observed, depending on the information flowing in this channel. "Organizational effects" that condition subsequent responses to gonadal steroid information occur early in development in most mammals and birds when gonadal steroid levels carry basic information about gonadal sex to the brain. Later, the presence of adult gonadal steroid levels informs the brain about the organism's reproductive state. Appropriate responses can involve territorial aggression and courtship behaviors in males, estrus behavior or maternal behavior in females. The appropriateness of these behaviors depends on the reproductive state of the organism, and the flow of information in the form of steroid hormones helps to synchronize reproductive behavior with reproductive state. The manipulation of reproductive state, either by gross interventions such as castration or by subtle interventions such as the quantitative manipulation of hormone levels or the use of antagonists and agonists to disrupt information flow, has clear effects on the reproductive behavior of affected individuals.

The most familiar of these steroid-mediated relationships between reproductive state and reproductive behavior are what we might refer to as "categorical" effects. The information that is flowing might be translated as "I am male" or "I am reproductively mature" or "I am about to ovulate" or "I have nursing offspring." Each of these categorical states is often associated with a more or less distinct behavioral program that can sometimes appear and disappear with switchlike clarity. But in addition there are subtler variations in reproductive state that are associated with what we might call "quantitative" effects. A reproductively mature, cycling female mammal may, for example, vary in her fecundity or probability of conception. This quantitative variation in state within the categorical condition of "mature, cycling female" may be associated with quantitative variation in her reproductive behavior revealed, for example, by measures of proceptivity or receptivity. Potential male partners may be sensitive to this variation both in its behavioral manifestations and in its physiological correlates. In some cases, it may be to the female's advantage to accentuate the signals associated with such quantitative state variation. There is good reason to believe, for example, that variation in the size of sexual swellings in female chimpanzees and savannah baboons carries such information (Domb and Pagel, 2001; Emery and Whitten, 2003).

The information coded by gonadal steroids is of central importance in reproductive ecology as well. It is important in both its categorical and its quantitative content. But while the field of behavioral endocrinology is primarily interested in the steroid information flowing along pathways (ii) and (iv), from soma to CNS and from CNS to CNS, the field of reproductive ecology is also concerned with the information flowing along pathways (i) and (iii), from CNS to soma and from soma to soma. The central issue in reproductive ecology is the management of reproductive effort, and that effort can be allocated both behaviorally and somatically. Behavioral reproductive effort includes things like courtship and mating behavior, territorial defense and intrasex aggression, egg brooding, and offspring feeding. Somatic investment includes things like egg production, gestation, and lactation—all forms of female somatic effort. But males also exhibit important forms of somatic reproductive effort including the production of weapons and ornaments, the mobilization of fat reserves during the breeding season, and the accumulation and maintenance of muscle mass. From an evolutionary perspective, all organisms are devoted to capturing energy from the environment and turning that energy into more organisms like themselves. Reproductive ecologists concentrate on the ways in which the allocation of energy to reproduction is mediated and organized, regardless of whether the mediation is behavioral or somatic.

While behavioral endocrinology has traditionally focused on nonhuman animals, perhaps because of an initial dependence on rather invasive experimental manipulations, and has only recently and tentatively expanded to include humans, reproductive ecology has developed in the other direction. Investigation of human reproduction in relation to ecology may have been foreshadowed in some clinical (example, Bullen et al., 1985) and demographic studies (Howell, 1979), but its major advances were made when techniques for noninvasive monitoring of reproductive hormones were developed and adapted to field conditions (Ellison, 1988; K. Campbell, 1994; Worthman and Stallings, 1994; Whitten, Brockman, and Stavisky, 1998; O'Connor, Brindle, et al., 2003). These techniques made it possible to gather new information on variability in human reproductive function that was temporally more detailed and more extensive than had been possible previously, allowing an assessment of the influence of ecological factors on that functioning. Humans are more tractable subjects in this sort of research than wild animals, since they can agree to participate in frequent and longitudinal biological sampling. Only recently have comparable methods been successfully extended to some free-ranging animal populations, obviating the need for invasive and disrupting methods of capture and sampling (Knott, 1997, 1998; Whitten, Brockman, and Stavisky, 1998; Emery Thompson, 2005).

Not surprisingly, given the theoretical concern with energy allocation in reproductive effort, many of the first insights in reproductive ecology concerned the impact of energetic factors on reproductive function. Female ovarian function was shown to vary in response to energy balance and energy expenditure, not just in a qualitative way (for example, amenorrhea induced by high levels of energy expenditure: Bullen et al., 1985) but also in a quantitative way. Ovarian steroid levels have been observed to rise under conditions of positive energy balance and to fall under conditions of negative energy balance in a wide range of populations (Ellison et al., 1993), and such variation in ovarian steroid production has been linked to variation in female fecundity (Lipson and Ellison, 1996; Venners et al., 2006). Correlations have also been established between energetic conditions in the womb and during infancy and childhood on the levels of adult ovarian function under common energetic conditions (Jasienska et al., 2006) and the responsiveness of adult ovarian function to changes in energetic conditions (Jasienska, Thune, and Ellison, 2006). The postpartum resumption of ovarian function in lactating women has also been shown to depend not only on the frequency and intensity of nursing but also on maternal energy balance (Ellison and Valeggia, 2003; Valeggia and Ellison, 2004). All of these observations can be subsumed nicely into a theory of human reproductive ecology that predicts a continuous relationship between metabolic energy availability and female fecundity as a primary modulator of female reproductive effort (Ellison, 2001, 2003).

More recently a similar logic has been extended to variation in male testosterone levels (Bribiescas, 2001a, 2006; Ellison, 2001, 2003). In this case, however, testosterone levels are not expected to correlate with male fecundity but rather with male reproductive effort. In humans, and indeed in many other mammals, the production and maintenance of muscle mass is an energetically expensive undertaking representing a substantial investment when integrated over a male's lifetime. Bribiescas (2001a, 2006) argues convincingly that this investment in muscle should be viewed as somatic reproductive effort, promoting an individual male's success in the arena of male-male competition for mating opportunities as well as signaling to potential mates a capacity for physical productivity in the procurement of subsistence and protection of mate and offspring. Testosterone serves as a primary modulator of muscle anabolism and maintenance. Gonadal maturation in males produces the dramatic dichotomy in muscle mass that distinguishes male from female strength and body composition. (Although there are clear differences in patterns of fat dep-osition between the sexes, the absolute amounts of fat carried by the av-erage male and female, corrected for height, are not very different. The lean body mass of comparably sized males and females, however, can vary dramatically. The fact that females have a higher percent body fat than males is thus more a consequence of differences in muscle mass that in fat per se.) The decline in testosterone levels with age is closely associ-ated with the decline in lean body mass (Vermeulen, Goemaere, and Kaufman, 1999; Campbell, O'Rourke, and Lipson, 2003; Lukas, Camp-bell, and Ellison, 2004). Between populations, restricted energy availabil-ity is associated with lower testosterone levels, smaller adult body size, and lower overall metabolic rates, suggesting that when energy availabil-ity is constrained, male somatic reproductive effort is downregulated, much as occurs in females. The primary difference between the sexes is in the level of sensitivity to change in energetic conditions and the time course of the response.

In addition to modulating somatic reproductive effort, there is ample evidence that testosterone helps to modulate behavioral reproductive ef-fort as well. Testosterone supports male libido, which increases dramati-cally at puberty and decreases more gradually with advancing age. Although more controversial, correlations between testosterone and in-dices of male aggressiveness, boldness, and self-image are frequently re-ported (Mazur and Booth, 1998; Dabbs and Dabbs, 2000; King et al., 2005; Archer, 2006; van Bokhoven et al., 2006), and the response of testosterone to victory or defeat in many forms of male-male competition closely resembles that reported to follow dominance interactions in cap-tive and wild primates (Rose, Holaday, and Bernstein, 1971; Eaton and Resko, 1974; Mazur and Booth, 1998; Muller and Wrangham, 2001). We will return to this subject in the next section when we consider the en-docrinology of social relationships from the perspective of reproductive ecology.

Increasing attention is being given to the mechanisms by which energy availability modulates reproductive effort and gonadal steroid produc-tion in particular. Those who approach this question from the perspec-tive of conventional endocrinology or clinical research tend to adopt a familiar "top-down" causal model with the CNS playing the central reg-ulatory role, our pathway (i). Information on energy availability is as-sumed to make its way to the CNS and there to result in changed patterns of releasing hormone secretion by the hypothalamus, which in turn alters pituitary gonadotropin release, ultimately affecting gonadal steroid production. Pathway (i) is a familiar construct and readily invoked. How the information of energy availability makes its way through pathway (ii) from soma to CNS has been more difficult to uncover. There has recently been great enthusiasm for leptin as a potential vehicle for this information along pathway (ii), but there is still considerable confusion about what information leptin actually encodes and what, if any, central response it generates. At a gross level, leptin levels reflect adipose tissue mass (Hauner, 2005; Trayhurn, Bing, and Wood, 2006). But leptin levels do not simply encode fat mass, in any case. Steroid milieu strongly alters leptin-to-fat-mass ratios, as does energy balance. Particularly problematic is the accumulating evidence that non-Western populations show very different ranges and patterns of leptin secretion from Western urban populations (Bribiescas, 2001b, 2005; Bribiescas and Hickey, 2006).

Other candidates have been proposed as carriers of information about energy availability along pathway (ii), but in a theoretical sense, it is not clear why this pathway, or pathway (i), need be involved at all. Information flows from the soma to the CNS when it is needed to inform behavior. Information flows from the CNS to the soma when cognitive or sensory information has implications for physiology (photoperiod information helping to regulate seasonal breeding physiology, for example). But when information on somatic state is needed to properly regulate somatic function, it does not need to travel up through pathway (ii), only to then travel down through pathway (i). It can simply flow through pathway (iii) from one part of the soma to another. Insulin regulation of blood glucose levels is a classic example of information flowing through pathway (iii). Elevated glucose levels are detected by the pancreas, which secretes insulin as a response. Elevated insulin levels are detected by liver, muscle, and fat cells (among others), which take up glucose from the bloodstream in response, storing it as fat or glycogen. There is no need for the CNS to be involved in this loop, and it is not.

Similar pathways have now been elucidated for information about energy availability to be conveyed directly to the gonads through pathway (iii). The ovary expresses receptors for insulin, growth hormone, and insulinlike growth factor I, all of which convey information about energy availability (Poretsky and Kalin, 1987; Serio and Forti, 1989; Franks, 1998; Franks et al., 1999; Hull and Harvey, 2000). The ovary responds to this information by upregulating steroid production when energy is abundant and downregulating it when energy is restricted (Ellison, 2001). Insulin appears to achieve this effect by synergizing with gonadotropins in stimulating ovarian steroid secretion. Either insulin or follicle-stimulating hormone (FSH), for example, is capable of stimulating ovarian steroid production acting alone, but their combined effect is much greater than the sum of their individual effects (Nahum, Thong, and Hillier, 1995; Willis and Franks, 1995; Willis et al., 1996).

The role of pathway (iii) in the case of postpartum resumption of ovarian function provides a good example. One of the primary signals that maintains milk production in a lactating woman comes from the nursing activity of her infant. Afferent nerves carry signals from the nipple to the mother's CNS, where, by mechanisms that are still not fully elucidated, they influence the release of gonadotropin-releasing hormone (GnRH) from the hypothalamus (Ellison, 1995; McNeilly, 2001). The pulsatile pattern of GnRH production must fall in the appropriate range to stimulate pituitary release of gonadotropins (FSH and luteinizing hormone [LH]). Thus the "hypothalamic pulse generator" represents an important permissive regulator of the HPO axis, an "on/off" switch, if you will. Frequent nursing by an infant keeps the switch in the "off" position. At the same time, prolactin release from the anterior pituitary is stimulated, providing a metabolic signal that supports continued milk production in the breast. Short (1987) has elegantly expounded the adaptive significance of linking continued milk production, on the one hand, and suppression of the HPO axis, on the other, to the presence of a nursing stimulus. The information that an infant exists who is still energetically dependent on maternal milk results in a pattern of maternal reproductive effort that allocates energy to milk production for the infant and away from a subsequent pregnancy. If the infant should die or be weaned, the signal is removed, and maternal reproductive effort can shift away from milk production toward a new pregnancy. The coupling of milk production and ovarian suppression thus functions as a "natural contraceptive" or "natural birth-spacing mechanism," switching maternal reproductive effort from investment in current to future offspring.

As elegant as this picture is, it is incomplete. It does not satisfactorily explain the tremendous heterogeneity that exists among lactating women in the timing of the resumption of ovarian function postpartum. Within any given population, there is a very poor to nonexistent correlation between the intensity of nursing and the resumption of ovarian activity (Huffman et al., 1987; Lewis et al., 1991; Tay, Glasier, and McNeilly, 1996). Nor is there a consistent pattern between populations. Some populations with among the highest recorded frequencies of nursing (example, the !Kung San of Botswana: Konner and Worthman, 1980; and the Gainj of highland Papua New Guinea: Wood et al., 1985) have among the longest reported average durations of lactational amenorrhea for noncontracepting populations, while others with comparable nursing patterns (example, the Toba of northern Argentina: Valeggia and Ellison, 2004; the Amele of lowland Papua New Guinea: Worthman et al., 1993; and La Leche League members in Boston: Elias et al., 1986) have among the shortest. Furthermore, in almost every noncontracepting population the majority of lactating mothers resume ovarian cycling while still nursing their infants, so the separation of these forms of reproductive effort is not by any means absolute.

From the perspective of reproductive ecology, there is an important element missing from the nursing-intensity mechanism: energy availability. It can easily be inserted, at least conceptually, by assuming that nursing intensity and the rate of milk withdrawal from the breast provide information on the energetic demand of the infant. When this demand is high, so the logic would run, potential additional investment in a new pregnancy should be avoided. When the energetic demand of the infant is sufficiently low, potential additional investment can be allowed. The crux of the issue, however, is in the term *sufficiently low*. To determine this it is not sufficient to know the absolute level of energy demanded by the nursing infant. We must also know the amount of metabolic energy available to the mother. Nothing in the nursing-intensity mechanism provides that information. Yet it clearly seems to be important. For one thing, it appears to account for the paradoxical between-population variance in the duration of lactational amenorrhea among the high-intensity-nursing populations referred to above. The !Kung and the Gainj are subject to chronic energy restriction evidenced by very late ages at maturation and very low body mass indices as adults, whereas the Toba, Amele, and La Leche League mothers all live in conditions of chronically high energy availability. The same given level of energy demand from a nursing infant may be "sufficiently low" to allow the resumption of ovarian function in women with abundant metabolic energy availability but still be above the threshold for women with low metabolic energy availability.

Where would we look for a signal of maternal energy availability? Although it is conceivable that such a signal could flow up to the CNS via pathway (ii), perhaps be converted within the CNS through pathway (iv), and then be sent down to the ovary via pathway (i), it is much more likely to flow directly to the ovary through pathway (iii). Claudia Valeggia and I have investigated one potential messenger along this pathway, insulin, as a likely modulator of the postpartum resumption of ovarian function. We studied 70 Toba mothers living in the village of Namqom outside Formosa in northern Argentina. As mentioned above, Toba women are quite well nourished, with high energy intakes and low expenditures (Ellison and Valeggia, 2003; Valeggia and Ellison, 2004).

Their babies are quite large at birth (Faulkner, Valeggia, and Ellison, 2000), and they nurse them very frequently, usually until they become pregnant again. Nevertheless, the average duration of lactational amenorrhea is 10.2 ± 4.3 months, quite short for such an intensive pattern of breastfeeding. We collected morning urine samples every week from these mothers (as well as other biological and behavioral information) in order to measure both metabolites of ovarian steroids (to monitor ovarian function) and C-peptide of insulin, a fragment of the proinsulin molecule that is removed whenever a molecule of insulin is produced. C-peptide is secreted in the urine in amounts that reflect both its production and insulin's (which occur in a 1:1 ratio). Since baseline insulin levels rise and fall in relation to metabolic energy availability, we can interpret the rise and fall of C-peptide in urine in the same way. Previous work had established the role of insulin in synergizing with gonadotropins to stimulate ovarian steroid production. Thus we hypothesized that as individual Toba women experienced an increase in metabolic energy availability, their insulin levels would rise, and that when these levels were

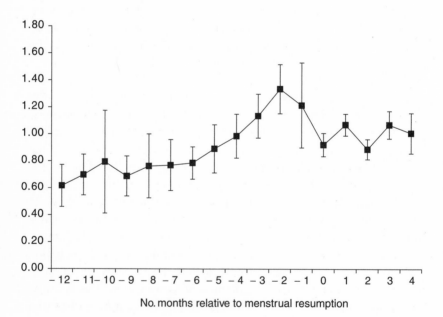

No. months relative to menstrual resumption

Figure 3.2   Mean ($\pm$ standard error) values for C-peptide, indexed to creatinine and standardized for each individual on their postresumption average, for 70 nursing Toba women, aligned on the month of menstrual resumption. Adapted from Ellison and Valeggia, 2003.

high enough, they would trigger the resumption of ovarian function. Energy demand by the nursing infant would certainly be one factor in determining metabolic energy availability, but not the only one. Insulin dynamics, reflected in C-peptide dynamics, should provide a better signal, we reasoned, one that integrates other aspects of maternal energy intake and expenditure.

Our hypothesis was beautifully confirmed by the data (Ellison and Valeggia, 2003). Early in the lactational period, C-peptides are low (60%) relative to each woman's normal level. But as time passes, the C-peptide levels rise at a rate and tempo that vary between women. This rise occurs sooner after birth in some women and later in others. But when we align the data not on the date of the infant's birth but on the date the mother's menses resume, this variance largely disappears (Figure 3.2). C-peptide levels rise steeply for all women over the six months or so prior to menstrual resumption, peaking at a level more than 30% higher than normal one to two months before menstrual resumption. This peak, we imagine, represents the level that restarts ovarian steroid production. It is also interesting to note that estradiol increases the sensitivity of adipose tissue to insulin, so the increase in estradiol that is triggered by the insulin peak also makes the adipose tissues more responsive to insulin, allowing the unusually high insulin levels to once again fall to their normal level for each woman after her ovarian function has resumed. Our proposed mechanism is also supported by evidence of a close association between both increasing body weight and increasing estrogen levels and the rise in C-peptide over the six months prior to menstrual resumption. On the other hand, no close relationship exists between any measure of nursing intensity and any of these variables.

Social Relationships and Reproductive Effort

Reproductive ecology has developed a very powerful perspective on the regulation of reproductive effort. Many of the insights first developed to understand human reproductive ecology are now being extended to non-human primates (Knott, 1998, 2001; Muller and Wrangham, 2001; Emery Thompson, 2005; Sherry and Ellison, 2007). Mechanisms governing the allocation of energy play a central role in regulating reproductive effort, with information carried by hormones both between the soma and the CNS and within the soma itself. But energy availability is not the only factor that determines the optimal allocation of reproductive effort. The nature and quality of social relationships are also extremely important, especially for social species like humans and many of our relatives. Other

chapters in this volume will explore the endocrinology of social relationships in detail. In this section, I will simply indicate how the endocrinology of social relationships may be incorporated into the larger
theoretical domain of reproductive ecology. In doing so I will focus on
four types of social relationships: dominance relationships, pair-bonds,
parent-offspring relationships, and social support.

Dominance relationships play an important role in the modulation of
reproductive effort in many species, including many group-living primates. These relationships have important significance for the optimal allocation of reproductive effort since they often determine access to
resources and access to mating opportunities. The determinants of dominance are not simple. Age and physical prowess are often important but
rarely sufficient to determine dominance status. Experience also emerges
as a key variable whenever longitudinal studies of dominance relationships are conducted. An individual animal's dominance rank may change
and may even do so rapidly under some circumstances, but it does not
fluctuate arbitrarily. Animals also act in ways that imply an understanding of their dominance status, avoiding confrontations with more dominant individuals, asserting themselves in relation to less dominant
individuals. Modulating behavior in this way minimizes the costs associated with unsuccessful dominance challenges while exploiting the advantages of current rank.

Much of the information needed to modulate behavior in relation to
dominance relationships no doubt remains within the CNS in the form
of memories and learned responses. But there are hormonal correlates of
dominance relationships as well. These are most clearly documented for
males of many group-living species. In captive rhesus monkeys the testosterone levels of lower-ranking males are lower than those of top-ranking
males (Rose, Holaday, and Bernstein, 1971). When new males are caged
together, they engage in a series of aggressive encounters that result in
the establishment of dominance relationships between them. When they
do so, the testosterone level of the more dominant animal rises, whereas
that of the less dominant falls (Rose, Holaday, and Bernstein, 1971;
Williams and Bernstein, 1983). Similar testosterone dynamics have now
been documented among free-living chimpanzees in Uganda (Muller and
Wrangham, 2001). Turnover in rank relationships as a result of the formation of newly successful coalitions are reflected in the rise and fall of
the testosterone levels of the males in the group.

Social dominance and subordination would also appear to most observers to be a feature of most human societies. However, social rank is
clearly not simple or unidimensional among humans. The same individual

may have very different "ranks" in different contexts, such as the workplace, the residential community, school, athletic teams, and various other peer groups. Nor are there simple consensus methods for determining dominance relationships analogous to the displacement matrix widely used in primatology. Nevertheless, there are suggestions that hormonal correlates of dominance relationships exist in humans, too (Mazur and Booth, 1998). The outcome of competitive encounters between human males is often associated with changes in testosterone that are reminiscent of those observed in nonhuman primates, an observation that has been made in activities as diverse as wrestling matches (Elias, 1981) and chess matches (Mazur, Booth, and Dabbs, 1992) and even among sports fans observing the victory or defeat of their favorite teams (Dabbs and Dabbs, 2000).

The significance of testosterone changes in males in response to dominance interactions is more difficult to determine. It is possible, though, that increases in testosterone that occur after successful encounters help to elicit increased reproductive effort, both behavioral and somatic. Nisbett (1996) has documented an association between testosterone response to an affront in human men and various behavioral measures of the degree of indignation experienced and the inclination to respond aggressively. In addition to supporting a tendency to aggressive response to social challenge, increased testosterone levels may affect a male's assessment of his own probability of success in new encounters (Wrangham, 1999). It may also support elevated libido and so support the pursuit of mating opportunities. Finally, even transient rises in testosterone have been shown to affect glucose uptake by muscle cells, while chronic elevation of testosterone is a potent anabolic stimulus (Wolfe et al., 2000).

Mating relationships are clearly integral to reproductive effort. Indeed, reproductive effort is often divided into mating effort and parenting effort, the former constituting allocations of time and energy to the procurement of mates and mating opportunities. The endocrinology of mating relationships has long been a focus of behavioral endocrinology. Elegant methods for assessing such things as proceptivity, receptivity, and attractiveness have been developed and applied to studies of a variety of species, including nonhuman primates (Dixson, 1998). Hormonal signals play important roles in coordinating mating behavior with reproductive state so that reproductive effort can be most efficiently expended. This category of social relationship seems therefore to require little in the way of additional explication in order to fit within the framework of reproductive ecology. However, within this category, pair-bonds deserve some special attention, in part because they are characteristic of our own species.

The term *pair-bond* is used to refer to an exclusive (or nearly so) mating relationship between one male and one female, *monogamy* being a close synonym. Pair-bonds need not be permanent but ordinarily should last long enough for at least one cycle of breeding to be recognized as such. In terms of reproductive ecology, pair-bonds represent a close linkage between the reproductive effort of the male and female so that the success of each depends on the effort of the other as much as on its own effort. This leads to a close alignment of evolutionary interests between the pair, at least for the duration of the bond. Such an alignment of interest with an unrelated individual is an unusual occurrence in evolutionary terms, so it is not surprising to find many special mechanisms that serve to support pair-bonds where they occur.

One such mechanism appears to be mediated by arginine vasopressin (AVP) and by oxytocin (OT), closely related peptide hormones released by the posterior pituitary. OT carries information to the soma along pathway (i), triggering milk letdown, for example, in response to nursing stimuli in lactating females. It can also convey information through the soma-soma pathway (iii), as in the stimulation of uterine contractions during labor or orgasm. OT also appears to convey information back to the CNS via pathway (ii), helping to support maternal behavior in laboratory rodents, for example, though it is not clear how the information crosses the blood-brain barrier (see Chapter 2). In some species, AVP appears to support pair-bond formation, particularly in males, perhaps in this case carrying information through pathway (iv). Different, closely related species of voles *(Microtus)* either do or do not form pair-bonds. Among those that do, males express AVP receptors in the brain, and activation of those receptors appears to facilitate pair-bond formation, whereas blocking the receptors inhibits it. This action appears to be an example of information flowing along the CNS-CNS pathway (iv) and so is perhaps better classified as neuromodulatory action than endocrine. Many more details of these mechanisms are provided in Chapters 6 and 14 in this volume.

In many birds, formation of pair-bonds requires a shift in behavioral program for males from a suite of behaviors appropriate to territorial defense and the attraction of a mate to a suite of behaviors appropriate to egg brooding and hatchling feeding. The former suite of behaviors seems to be supported by relatively high testosterone levels, higher than the minimum necessary to support sperm production. These behaviors include vigorous display (including singing in many species), territorial vigilance, and aggression toward intruding males. These behaviors undermine the parental behavior needed to brood eggs and feed nestlings, however, distracting the male's attention and effort away from the nest and its contents. Not surprisingly, males of pair-bonding bird species have been shown to undergo an endocrinological shift to lower testosterone levels in parallel with the behavioral shift from territorial defense and mate attraction to parental behavior. Manipulations that evoke territorial responses in nesting males, such as playing the song of an invading male, both undermine parental behavior and lead to an increase in testosterone. At the same time, hormonal manipulations to artificially raise or lower testosterone can elicit either the territorial or the parental suite of behaviors. Behavior and hormones shift in parallel regardless of the cause of the shift. These mechanisms are described in more detail in Chapters 4 and 12 in this volume.

This pattern of shift in male behavioral program associated with bond formation and parenting and the hormonal shift that accompanies it is not confined to birds. Similar patterns have been reported for many pair-bonding mammals, including nonhuman primates. Callitrichid primates (marmosets and tamarins) are prominent examples of species where pair-bonds are formed and heavily male parental investment is observed. Shifts in male testosterone and prolactin levels are also reported in these species in association with the shift to the parental role (Gubernick and Nelson, 1989; Ziegler, Wegner, and Snowdon, 1996; Mota and Sousa, 2000; Ziegler, 2000; da Silva Mota, Franci, and de Sousa, 2006). Recently evidence has even begun to accumulate suggesting that lower testosterone levels may be typical of human males in stable mating relationships and perhaps even lower levels in men who are fathers of infant children (Gray et al., 2002; Gray, Campbell, et al., 2004; Gray, Chapman, et al., 2004; Gray, Yang, and Pope, 2006). Once again, these hormonal/behavioral shifts make sense in the context of reproductive ecology as aspects of the regulation of reproductive effort. These studies are also presented in greater detail in Chapter 12 in this volume.

From the female perspective the presence of a stable male partner is a special case of social support, support of a conspecific that can have a positive impact on the female's own reproductive success. Social support in this sense synergizes with a female's own reproductive effort, making it more effective. Social support is not restricted to male breeding partners by any means. Support can come from adult relatives, older offspring, and even unrelated individuals. Where social support has a strong positive effect on a female's reproductive success, we would expect to find her own reproductive effort linked to the presence and quality of that support. Examples of this linkage include macaques and baboons who live in groups composed of matrilineal relatives. Female fecundity in these species is strongly related to the size and rank of a female's matrilineage within the group. In many of these species a negative version of social support, social harassment, is also observed with negative impact on a female's fecundity. Older offspring can often be important sources of social support to reproductive females. Such "helpers at the nest" have been reported for a wide range of species ranging from birds to rodents to primates (Ziegler, 2000).

The importance of social support to human reproduction is an open and somewhat controversial issue. The impact of social support at key moments, such as the effect of a companion during labor and delivery, has been well documented (Sosa et al., 1980; Klaus and Kennell, 1997). But a more general effect of the presence and quality of social support on female fecundity has not been established. Wasser and Barash (1983) have made a strong theoretical argument for the importance of social support as a predictor of the probability of success in any reproductive attempt, at least in formative human environments. But many of the examples used to support this thesis can also be viewed as examples of the effect of energy availability on female fecundity. Similarly, Belsky, Steinberg, and Draper (1991) have argued that certain social stresses in childhood, particularly father absence, are predictive of poor reproductive probabilities in adulthood. They cite evidence for a shift in developmental program under such circumstances toward an accelerated maturation that represents a different allocation of reproductive effort appropriate to the expected conditions in adulthood.

Closely related to the hypothesis that social support has an important impact on the fecundity of human females is the hypothesis that psychosocial stress in general has a suppressive effect on ovarian function (Sapolsky, 1990; Wasser and Place, 2001). While the impact of traumatic stress on the HPO axis has been clearly documented in several instances, the effect of moderate psychosocial stress, such as might accompany concern for the presence and predictability of social support, on the HPO axis has not been shown. For example, a group of colleagues and I (Ellison et al., 2007) carefully looked for an effect of moderate anxiety, either acute or chronic, on ovarian function using a very sensitive and powerful experimental design but without finding any evidence for an effect. It is possible that social support is too amorphous a variable in human societies for a clear effect on reproductive physiology to be demonstrated. Or it may be that the effect of social support is subsumed in other effects, such as the effect of energy availability, and so is not readily observed when those other effects are controlled.

Summary

What does seem clear, I hope, is that the endocrinology of social relationships is intimately connected to reproductive ecology. The reason that social relationships trigger endocrine signals is to help regulate reproductive effort. Testosterone does not rise in the victor of a competitive encounter to make him feel good. He would feel good even if he had his testes removed. Indeed, there is no reason to believe that the "thrill of victory and the agony of defeat" are any less acute in women that in men. The functional reason for the rise in testosterone (which occurs in men but not in women) is to help modulate reproductive effort in an appropriate direction. Similarly, all the other hormonal correlates of social relationships that are documented in this volume can be understood functionally as mechanisms regulating reproductive. Where an explanation is not provided in these terms, the student should challenge himself or herself to formulate one. Researchers in turn should think of the endocrinology of social relationships as part of a larger domain of life history biology known as reproductive ecology.

CHAPTER FOUR

Hormone-Behavior Interrelationships in a Changing Environment

John C. Wingfield

A LL ORGANISMS live in changing environments in which morphology, physiology, and behavior must be regulated to cope with local conditions. In recent decades global climate change has had a dramatic impact on the patterns of migrations, breeding seasons, and other aspects of the life cycles of organisms (Berthold et al., 1992; Thomas and Lennon, 1999; Both and Visser, 2001; Pulido et al., 2001; Walther et al., 2002). These effects are ongoing, and the outcomes for biodiversity, at the population level, and for the cellular and molecular mechanisms by which animals cope with a changing environment remain unknown. There is an urgent need to understand the mechanisms by which these organisms cope with a changing physical environment and the roles social interactions have in modulating those responses. Behavioral responses to global change and the hormone-behavior interactions that underlie them are particularly unknown.

Changes in the physical environment can be grouped into two major types. The first group includes predictable changes such as night/day rhythms, seasons, and tidal rhythms that determine the timing of stages in the predictable life cycle such as migration, reproduction, and molting. These are regulated by predictive cues such as annual photoperiod, rainfall, temperature, and food availability (Figure. 4.1; see Wingfield, Jacobs, Soma, et al., 1999; Wingfield, Jacobs, Tramontin, et al., 1999; Wingfield, 2006). Climate change is resulting in temporal shifts in many of these cues but not photoperiod. Thus the possibility exists for those species that rely on photoperiod changes to time events in the life cycle to become asynchronous with environmental conditions conducive for migrating, breeding, molting, and so on. Others that are more flexible would be able to adjust to these changes.

The second type of environmental change involves unpredictable events such as storms, changes in predator numbers, and associated consequences concerning access by individuals to, for example, food and shelter. How organisms deal with such perturbations of their environment that cannot be predicted, at least in the long term, is an important question in environmental biology (Wingfield et al., 1998). Global climate change is resulting in increasing intensity and numbers of severe storms, and human disturbance is invading virtually all habitats on earth. The combination of the two (climate change and human disturbance;

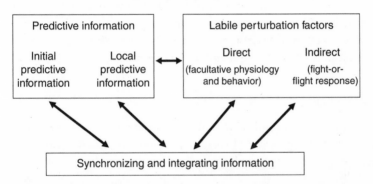

Classification of environmental factors
regulating life cycles

Figure 4.1   There are three major types of environmental factors that can regulate changes in morphology, physiology, and behavior of organisms. *Predictive information* allows individuals to anticipate changes in environment (such as seasons) so that an appropriate life history stage (such as breeding, molting, or migrating) can be developed and expressed at the appropriate time. There are two subtypes of predictive information—*initial predictive* (long-term) and *local predictive* (short-term) cues. *Labile perturbation factors* are the result of unpredictable events in the environment that trigger the emergency life history stage that allows an individual to cope with the potential stress of the perturbation. There are also two subtypes of this class of environmental information. The first are *direct* perturbation factors (such as storms) that trigger facultative physiology and behavior. The second are *indirect* perturbation factors (such as attacks by predators) that trigger rapid coping mechanisms such as the "fight-or-flight" response. The third major type of environmental factor—*synchronizing and integrating information*—includes all the social interactions among individuals that coordinate and bring together many aspects of the predictable life cycle *and* responses to perturbation factors. Modified after Wingfield, 2006.

see Travis, 2003) is referred to here as *global change*. How the mechanisms underlying responses to both predictable and unpredictable types of environmental change may contribute to phenotypic flexibility (that is, the ability to adjust to global change) or inflexibility (that is, the inability to adjust to global change) is a key question for the future (Wingfield, 2005).

Throughout an individual's life cycle, adjustments in response to environmental change are also influenced by social interactions—even in species that may spend much of their lives in isolation (Figure 4.1; see Wingfield, 2006). Although environmental signals may result in the activation of appropriate behaviors (example, during the reproductive season), or their deactivation, usually through hormone signals, it is also clear that social interactions can influence responsiveness to other environmental cues and hormone secretions (example, Balthazart, 1983; Harding, 1983; Wingfield and Silverin, 2002; Wingfield, 2006). Although social systems and environmental change are complex, there are clear interactions between environmental signals and expression of behavior through neuroendocrine and endocrine secretions.

Environmental Control and Sociality in Life Cycles

At this point it is useful to consider the spectrum of environmental signals, from the physical and social environments, that organisms use to adjust morphology, physiology, and behavior to cope with local conditions throughout the life cycle. Environmental signals affect the progression of development stages through ontogeny as well as through adult life history stages. Those signals affecting ontogeny can be from the physical environment, from social interactions (example, Francis et al., 1999; Meaney, 2001; Weaver, Cervoni, et al., 2004), and from maternal and, to a lesser extent, paternal and sibling effects (vom Saal, 1989; Schwabl, Mock, and Gieg, 1997; Palanza et al., 1999; Strasser and Schwabl, 2004; see Figure 4.2, modified after Wingfield, 2006). It is important to understand that ontogeny of morphology, physiology, and behavior results in a phenotype that is more or less fixed for the remainder of an organism's adult life (that is, ontogenetic life history stages are not reversible, Figure 4.2). In contrast, adult life history stages also show development of morphological, physiological, and behavioral traits characteristic of that stage. But these traits are reversible insofar as they can be terminated and regressed as the next life history stage develops (Figure 4.2). However, all these traits can be developed again when the life history stage is expressed in the next year of the life cycle. Adult life history

Ontogenetic stages (not repeatable)

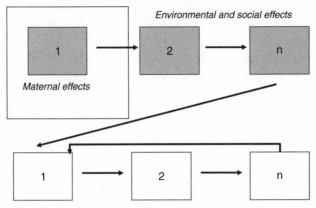

Environmental and social effects

Adult life history stages (repeatable)

Figure 4.2   The life cycles of all vertebrates consist of ontogenetic stages that occur once, resulting in the development of the adult individual. This sequence is then followed by the adult life history stages, each of which also involves a development phase (see Figure 4.3) and a termination phase. A fundamental distinction here is that ontogenetic stages are fixed and not repeatable, whereas the adult life history stages are expressed in a one-way sequence but are repeatable (usually on an annual basis). This strongly suggests that the mechanisms by which environmental factors regulate ontogenetic and adult life history stages are also fundamentally different. Environmental and social factors affect both but in different ways. Note also that ontogenetic stages can be strongly influenced by maternal—and to a lesser extent, paternal—effects. Modified after Wingfield, 2006.

stages are thus reversible and can be repeated (Wingfield, 2006). Given that ontogenetic stages are not reversible and repeatable, whereas adult life history stages are, suggests that the control mechanisms underlying development of these stages are fundamentally different.

What environmental signals do organisms use to time events in the life cycle, and how are those signals transduced into neuroendocrine secretions that regulate responses (Follett, 1984; Wingfield, Jacobs, Tramontin, et al., 1999; Wingfield and Silverin, 2002)? One environmental signal, the annual change in day length (photoperiod), has been investigated in very great detail in a wide spectrum of organisms from plants to vertebrates (example, Follett, 1984; Gwinner, 1986, 1996; Nicholls, Goldsmith, and Dawson, 1988; Dawson et al., 2001). Other environmental signals are essential to orchestrate responses to the predictable environment and the unpredictable. However, how they affect neuroendocrine and endocrine mechanisms and how they are integrated with social signals are much less well understood (Wingfield, Jacobs, Soma, et al., 1999; Wingfield, Jacobs, Tramontin, et al., 1999). The literature on a vast array of known environmental signals that can regulate almost all aspects of life cycles is extremely complex and bewildering. A classification of these environmental cues emanating from both the physical and social environments would be useful to begin teasing apart mechanisms (Wingfield, Jacobs, Soma, et al., 1999; Wingfield, Jacobs, Tramontin, et al., 1999). Basically, there are three major groups of environmental signals (Figure 4.1): *predictive information* (from the physical environment), *labile perturbation factors* (from the unpredictable environment), and *synchronizing and integrating information* (from the social environment). It is also important to acknowledge that the internal environment of an individual (example, body condition, age, parasite load, and personality) has a major influence on how an individual may respond to the three types of environmental information. Although the spectrum of environmental signals and the interrelationships with internal environment are very complex, this classification "framework" may have heuristic value in directing future investigations.

The predictive information group of environmental signals has two major subtypes (Figure 4.1): *initial predictive information* provides accurate, long-term predictive cues allowing preparation for onset of life history stages such as breeding, migrating, and hibernating several weeks or even months in advance. An example is the annual change in photoperiod. Initial predictive cues are integrated with *local predictive information* that provides rapid adjustment in relation to local phenology. For example, spring may be early and warm in one year but late and cool in another, thus influencing many life history traits such as reproductive development, migration, or emergence from hibernation. Examples of these local predictive cues are temperature, availability of food, and rainfall, and they can act in two ways (see Figure 4.3): to speed up development or termination of a life history trait (accelerators) or to slow it down (inhibitors).

A second group of environmental signals known as *labile perturbation factors* (Figure 4.1), also known as *modifying information*, includes all the unpredictable events in the environment that have the potential to be disruptive to normal progression of the life cycle (Wingfield, Jacobs, Soma, et al., 1999; Wingfield, Jacobs, Tramontin, et al., 1999). This group also has two major subtypes: *direct perturbation factors* include

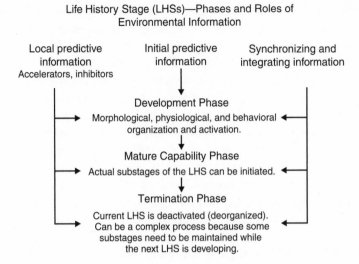

Figure 4.3   Each adult life history stage has three phases: *development phase,* in which morphological, physiological, and behavioral traits characteristic of that stage are organized and activated; *mature capability phase,* in which characteristic substages can be initiated (for example, onset of estrous cycling, parental care); and the *termination phase,* when morphological, physiological, and behavioral traits characteristic of that stage are deorganized and deactivated. *Initial predictive information* triggers development of life history stages, maintains them through mature capability, and then terminates that stage at the appropriate time. *Local predictive information* can speed this process up *(accelerators)* or slow them down *(inhibitors),* according to local conditions (for example, early versus late springs, autumn). Social interactions that make up the *synchronizing and integrating* type of information can influence all of these phases and how individuals respond to initial predictive and local predictive information. Note that labile perturbation factors redirect the individual away from the normal life history stages and trigger the emergency life history stage. This allows a number of coping mechanisms to be expressed so that the individual survives the perturbation in the best condition possible. After the perturbation passes, the individual will return to the normal sequence of life history stages. From Wingfield, Jacobs, Tramontin, et al., 1999; Wingfield, 2006.

chronic severe weather events or an increase in predator numbers that forces an individual to abandon its normal life history stage temporarily because of reduced resources such as food or access to those resources. This temporary abandonment of the normal life history stage involves components of the emergency life history stage that facilitate survival until the perturbation passes (Wingfield, Jacobs, Soma, et al., 1999;

Wingfield, Jacobs, Tramontin, et al., 1999; Wingfield, 2003). The second subtype, *indirect perturbation factors,* includes acute disruptive events, such as sudden storms, or brief predation attempts (such as loss of a mate or nest and offspring) that result in a rapid, emergency, "fight or flight" response (Figure 4.1), but the perturbation passes within a few minutes. The individual then readjusts its life history stage accordingly (example, find a new mate, breed again). Note that labile perturbation factors result in rapid changes in behavior and physiology *during* and *after* the environmental event, whereas responses to predictive information occur *before* the change in life history stage (that is, in anticipation of a future event; Wingfield, 2006).

The third group of environmental signals, *synchronizing and integrating information,* includes behavioral interactions among groups, inter- and intrasexual, and between adults and young. They coordinate changes in behavior with life history stages (example, Wilczynski, Allison, and Marler, 1993; Rhen and Crews, 2002) and precisely synchronize the behavior of a pair or group through the life cycle (Figure 4.3; see Balthazart, 1983; Harding, 1983; Wingfield, Jacobs, Soma, et al., 1999; Wingfield, Jacobs, Tramontin, et al., 1999; Wingfield, 2006). This process includes social plasticity (Bass and Grober, 2001), social recognition, and the ability to remember conspecifics (Ferguson, Young, and Insel, 2002). Synchronizing and integrating information impinges on all aspects of an animal's life (Figure 4.3) and has profound implications for physiology and endocrine control mechanisms.

It is possible to organize synchronizing and integrating information into subtypes that are involved both in the predictable life cycle and in response to unpredictable labile perturbation factors. Social interactions have profound effects on:

1. Maternal and paternal effects. Social interactions during ontogeny may have dramatic influences on the phenotype that reaches maturity (Figure 4.2). For example, maternal androgens can influence the plumage coloration of male house sparrows *(Passer domesticus)* once they become mature (Evans, Goldsmith, and Norris, 2000). Another example is the effect of stress during development, which can markedly influence the adrenocortical responses to stress in adults (Meaney, 2001; Weaver, Diorio, et al., 2004).
2. Development of adult life history stages, particularly regulation of physiology and social status (Figure 4.3). Examples are the well-known effects of male singing and courtship on ovarian development in female birds and the effects of courtship behavior of

female birds on testicular development and testosterone secretion (Balthazart, 1983; Wingfield, Whaling, and Marler, 1994). Social cues can also have an important role in determining how an individual responds to perturbation factors including stress. Dominants and/or subordinates may be susceptible to high adrenocortical responses to stress, depending on the energetic costs of maintaining rank (Goymann and Wingfield, 2004; Creel, 2005).

3. Regulation of onset of the life history stage (that is, mature capability; Figure 4.3; see Wingfield, Whaling, and Marler, 1994; Wingfield, 2006). There are many well known examples in which the sexual behavior of one sex can stimulate the onset of breeding in the other sex (example, Hinde, 1965; Lehrman, 1965; Wallen and Schneider, 1999). A similar action in response to labile perturbation factors would be the effects of coping style, such as reactive or proactive behavioral phenotypes in response to a predator (example, Koolhaas et al., 1999). The presence of predators may eventually result in the suppression of a life history stage such as breeding (see also Boonstra, 2005).

4. Regulation of support functions (such as territorial behavior). Agonistic behavior related to acquisition and maintenance of a territory in both the breeding and nonbreeding seasons can have dramatic effects on life cycles in general. For example, maintaining a territory is usually essential for a male to attract a female and breed successfully (Harding, 1983; Wingfield, Whaling, and Marler, 1994; Wingfield, 2006). Territoriality may also determine whether an individual stays to endure a perturbation factor or whether it leaves to seek a sanctuary elsewhere (Wingfield et al., 1998; Wingfield and Ramenofsky, 1999).

5. Integration of parental care to extend beyond the breeding life history stage to include molt, migration, and wintering strategies. The presence of offspring may have a profound influence on when an individual may molt as well as migrate (Wingfield and Silverin, 2002). Parental status can also influence adrenocortical responses to perturbations of the environment. There is growing evidence that the presence of young may result in marked suppression of adrenocortical responses to acute stress (example, O'Reilly and Wingfield, 1995; Wingfield and Sapolsky, 2003).

All of the social systems within a life history stage, as well as mating systems, and so on, can be included here (see Wingfield, 2006, for an in-depth review). Behavioral interactions are also essential in unpredictable social hierarchical situations such as change in dominance status; the effects of predators on daily and seasonal routines (Boonstra et al., 1998; Krebs et al., 2001); and perturbations of the environment that change, for example, foraging behavior and social interactions. The latter now includes human disturbance (Walker, Boersma, and Wingfield, 2005).

Using this framework of environmental signals in ontogenetic and adult life cycle contexts, we will go on to focus on social interactions, how they influence integration and synchronization of life cycles, and what common themes may be present.

Hormone-Behavior Interrelationships and Predictable Life Cycles

Vertebrates have varying numbers of life history stages throughout their predictable life cycles including breeding, migrations, molts, and often a nonbreeding stage such as hibernation (example, Wingfield, 2006). Behavioral patterns also change from stage to stage, some being unique to a particular life history stage and others expressed in several or all life history stages. Examples of behavioral patterns mostly unique to a life history stage are parental behavior when breeding and long-distance locomotor movement when migrating. Behavioral patterns expressed in several or all life history stages include aggression in social hierarchies, over resources such as food and territories, and against predators. Those behavioral patterns expressed uniquely within a life history stage may be regulated by a suite of hormones typical of that stage (example, reproductive hormones within the breeding stage). However, those that are expressed in several life history stages pose interesting questions about control mechanisms because the hormones involved may not always be secreted throughout the life cycle. For example, aggression has often been associated with the sex steroid hormone testosterone in most vertebrates (Balthazart, 1983; Harding, 1983; 2005; Adkins-Regan, 2005; Nelson, Wingfield et al., 2005). This makes sense when aggression is expressed in reproductive contexts, but is aggression expressed in nonbreeding life history stages also controlled by testosterone? This poses potential problems because secretion of testosterone and all its effects on reproductive morphology, physiology, and behavior (example, sexual) would be inappropriate in nonbreeding stages (Wingfield, Lynn, and Soma, 2001; Wingfield et al., 2005; Soma, 2006). There are many examples of how hormonal regulation of behavior in one context would be untenable for expression of that behavioral pattern in another life history stage, but how this is overcome has received very little attention. A well-known example is that of control of aggression throughout the life cycle.

Correlates of Aggression and Testosterone in Social
Interactions: The Challenge Hypothesis

Over the past 25 years, many studies have attempted to correlate circulating levels of testosterone, and its metabolites, with social interactions of all kinds. These have included a growing number of investigations in the field or in seminatural conditions in captivity. There are now extensive data for all of the major vertebrate groups, and selected examples are summarized below.

There is extensive evidence that testosterone activates aggression associated with male-male competition over territories and mates (Balthazart, 1983; Harding, 1983). However, correlations of plasma levels of testosterone with expression of territorial aggression when breeding are highly variable (Wingfield and Ramenofsky, 1985; Wingfield et al., 1990; Wingfield, Lynn, and Soma, 2001). It was suggested that baseline levels of testosterone during the breeding season are sufficient for development and maintenance of morphological, physiological, and behavioral components of the male reproductive system, but they do not necessarily correlate with actual expression of territorial aggression. Field investigations of circulating testosterone levels in various vertebrate species (now numbering over 160; Hirschenhauser et al., 2005) reveal transient surges to much higher concentrations that are correlated with periods of heightened male-male competition. Plasma levels of testosterone are particularly high when establishing a territory, being challenged by another male, or mate guarding. On the other hand, experimental manipulations revealed that high circulating levels of testosterone for long periods may incur a "cost" such as reduced male paternal care and lower reproductive success (Silverin, 1980; Hegner and Wingfield, 1987). This led to the "challenge hypothesis," which states: high plasma levels of testosterone occur during periods of social instability in the breeding season (resulting from male-male competition for territories and mates) but are at a lower breeding baseline in stable social conditions thus allowing paternal care to be expressed (Wingfield et al., 1990; Wingfield, Lynn, and Soma, 2001).

As one might expect, the results of field investigations showed consistent relationships of patterns of testosterone levels with mating systems and breeding strategies. Examples include the two male phenotypes of tree lizards, *Urosaurus ornatus,* associated with alternative male reproductive tactics—one territorial, the other nonterritorial (Thompson and Moore, 1991; Hews et al., 1997). Nonterritorial males display elevated plasma corticosterone and depressed plasma testosterone levels, whereas territorial males show no change in plasma hormone levels the day after an encounter with a conspecfic (Knapp and Moore, 1996). An important physiological difference between the morphs is the plasma steroid binding globulin capacity that appears to transport both androgens and corticosterone. Binding capacity of this protein in territorial males in significantly greater than in nonterritorial males (Jennings et al., 2000). Higher plasma levels of unbound corticosterone in the nonterritorial males, especially when plasma corticosterone levels are elevated, could trigger a decrease in plasma testosterone. Differences in free versus bound corticosterone may explain morph differences in testosterone responses to similar patterns of corticosterone levels.

Hormones play an important role in advertisement calling in anuran amphibians. Such calling is important for males of most anuran species and serves as an advertisement to females and a warning involved in territory defense from other males. Time spent calling is related to mating success, but calling behavior is also energetically costly for anurans (Bucher, Ryan, and Bartholomew, 1982). In several anuran species, plasma levels of both testosterone and corticosterone are elevated during bouts of calling, and corticosterone is positively correlated with both calling rate and relative energy expended in the call (Emerson and Hess, 2001). Emerson (2001) proposed an extension of the challenge hypothesis, termed the Energetics-Hormone Vocalization (EHV) model, to explain these relationships of calling behavior and testosterone and corticosterone levels in male anurans. It is proposed that calling behavior results in an increase in plasma testosterone accompanied by an increase in plasma corticosterone levels due to the energetic demands of the vocalization. Levels of both hormones increase until plasma corticosterone triggers a short-term stress response, following which plasma testosterone levels decline (Emerson, 2001). The EHV model may be extended to superficially paradoxical differences in the relationship between testosterone, corticosterone, and aggression in other taxa of reptiles and amphibians and possibly vertebrates in general (Romero, 2002; Moore and Jessop, 2003). When mating or courtship is energetically costly, a positive relationship between testosterone and corticosterone levels may occur. This needs to be tested in other species.

The concept that high levels of testosterone, above the breeding baseline, might have deleterious effects has received considerable attention in the past 20 years. There is now evidence that sustained high plasma concentrations of testosterone not only may result in elevated energetic costs and reduced paternal care but also may incur greater injury rates, increased mortality, reduced effectiveness of the immune system, lower mating success, and so on. Many elegant studies have been summarized by Ketterson and colleagues (Ketterson, Nolan, Wolf, et al., 1992; Ketterson, Nolan, Cawthorn, et al., 1996). Furthermore, there is growing evidence that patterns of testosterone in tropical species that may have long breeding seasons are very different from northern species (Goymann et al., 2004). Tropical species with long breeding seasons tend to have extremely low levels of testosterone that generally do not change markedly with social challenges. However, as latitude and altitude increase, patterns of testosterone secretion and blood levels increase in amplitude and may be strongly influenced by social interactions. However, this too appears to be related to the degree of male parental care. In mid-latitude songbird species, if male parental care was essential (that is, the female could not compensate, and without paternal care, reproductive success was zero), then these males tend to have low plasma levels of testosterone during the parental phase, do not socially modulate testosterone levels in blood, and may even become insensitive to the behavioral effects of testosterone treatment on aggression (Hunt et al., 1997, 1999; Lynn et al., 2002; Lynn and Wingfield, 2003; Lynn, and Walker, and Wingfield, 2005). Behavioral insensitivity to testosterone appears to involve a decrease in aromatase activity in the telencephalon and diencephalons of male Lapland longspurs, *Calcarius lapponicus* (Soma, Sullivan, and Wingfield, 1999).

A major confound when comparing patterns of testosterone to aggression in natural settings is phylogeny. It is possible that most of the differences may be due to phylogenetic reasons rather than being a result of ecological factors per se. A meta-analysis of all avian investigations in relation to the challenge hypothesis (Hirschenhauser, Winkler, and Oliveira, 2003) tested predictions of the "trade-off" of male-male interactions resulting in an increase of testosterone secretion with male parental care that requires a decrease in testosterone (Wingfield et al., 1990). The hypothesis has been tested widely in fish (example, Sikkel, 1993; Oliveira, 1998; Oliveira et al., 2001) as well as in tetrapods including primates and humans (Archer, 2005). Focusing first on birds, Hirschenhauser, Winkler, and Oliveira (2003) revealed that after adjustment for phylogeny the overall prediction of an effect of male paternal care disappeared, but the effects of mating system and male-male interactions and possibly male participation in incubation persisted. Furthermore, Goymann et al. (2004), in a phylogenetically controlled analysis of patterns of testosterone in birds in relation to latitude and altitude, found that variable patterns in the tropics were related to environmental factors such as short breeding seasons rather than to phylogeny per se. As the numbers of species and populations studied under natural conditions increase, analyses of this sort will be critical to tease apart phylogeny and ecological constraints, leading to insight into how hormone-behavior interrelationships evolved. Testosterone patterns may vary according to mating success and testis size as well (Garamszegi et al., 2005). Hirschenhauser and Oliveira (2006) extended the meta-analyses to all vertebrates. Tremendous variation in presence and types of parental care obscured any relationship of testosterone pattern with paternal behavior, but a general relationship with mating systems and male-male competition persisted.

It is clear that given the large database of vertebrate species, including humans, studied in natural or seminatural settings, many future analyses taking phylogeny into account may show other relationships with ecological factors. These in turn generate hypotheses and predictions that can then be tested to determine mechanisms, especially at the cell and molecular levels. Recently, two reviews have proposed some important hypotheses that will likely drive many field and laboratory experiments. Reed et al. (2006) evaluated effects of elevated testosterone over nine breeding seasons in dark-eyed juncos, *Junco hyemalis*. Experimentally increased testosterone in males resulted in decreased survival but more extra-pair offspring. This effect on fitness was unexpected and led them to look at indirect effects. Nest success was similar for testosterone-treated and control males, but testosterone-treated males produced smaller offspring that had lower postfledging survival. From previous experiments, it is known that females prefer to mate with older males and gained greater reproductive success as a result. Testosterone-treated males were able to attract older females, but reproductive success was less. These combined effects may thus constrain evolution of a high testosterone male phenotype in juncos (Reed et al., 2006).

In an excellent review of how testosterone may mediate life history trade-offs because of its pleiotropic actions in male vertebrates (Hau, 2007), it is suggested that if actions of testosterone are conserved across vertebrates, the evolutionary constraint hypothesis applies (see also Ketterson, Nolan, Wolf, et al., 1992, Ketterson, Nolan, Cawthorn, et al., 1996; Reed et al., 2006). Thus, if actions of testosterone are similar across species and with gender, then patterns of testosterone secretion and levels in blood should vary accordingly. That is, temporal patterns of testosterone concentrations should match behavioral physiological changes. However, we know from many studies in the tropics and in various vertebrates that the actions of testosterone may be more varied, leading Hau (2007) to propose the evolutionary potential hypothesis in which mechanisms underlying testosterone action (converting enzymes such as aromatase in target cells, receptor types, and so on) and male traits affected by them evolve independently. She reviews several examples of both and suggests the evolutionary constraints hypothesis may be one end of a spectrum of testosterone-behavior interrelationships, and the evolutionary flexibility hypothesis at the other end, where many new mechanisms of testosterone-behavior interrelationships wait to be discovered. Note also that these hypotheses could be extended to actions of testosterone in female vertebrates and even to other hormone-behavior interrelationships.

These two hypotheses may also address another interesting problem of behavioral actions of testosterone, notably that of regulation of territorial aggression in the nonbreeding season. For example, does apparently identical territorial aggression expressed in different life history stages such as breeding and nonbreeding seasons have the same hormone control mechanisms? Song sparrows, *Melospiza melodia morphna*, of western Washington State and southwestern British Columbia are territorial year-round with some breeding pairs, many staying on their territory for more than one year (example, Arcese, 1989; Nordby, Campbell, and Beecher, 1999). Territorial aggression in response to simulated challenges using a live male decoy and playback of tape-recorded songs was maintained throughout the breeding season but declined markedly by molt (Wingfield and Hahn, 1993). Males remained on their territories during molt but did not respond to experimental challenge. However, after the molt period, male song sparrows began to sing and defend territories in late September and October (Arcese, 1989). Although song stereotypy in autumn was less than in spring (Smith et al., 1997), all other measures of aggression during the challenge appeared identical (Wingfield and Hahn, 1993). Other avian species with autumn territories in nonreproductive contexts tended to have baseline testosterone and/or luteinizing hormone (LH) levels compared with elevated concentrations of testosterone in breeding season (example, Burger and Millar, 1980; Logan and Wingfield, 1990; Schwabl and Kriner, 1991; Gwinner, Rödl, and Schwabl, 1994; Canoine and Gwinner, 2002). Tropical birds that are territorial throughout their reproductive life of several years showed similar patterns (Wikelski, Hau, and Wingfield, 1999).

In song sparrows, in further experimental manipulations in which males were removed from their territories in autumn, replacement males and their neighbors had undetectable plasma levels of testosterone, unlike when the experiment was conducted in spring (Wingfield, 1985, 1994a, 1994b). Furthermore, behavioral challenges in autumn had no effect on LH and testosterone levels, unlike in spring (Wingfield and Hahn,

1993; Wingfield, 1994a, 1994b; Soma and Wingfield, 2001). Even castrated male song sparrows were able to defend territories and respond aggressively to challenges in autumn equally as well as sham-operated males, suggesting that gonadal hormones in plasma are not required for autumnal territorial aggression (Wingfield, 1994a, 1994b). Experiments in the field in free-living song sparrows showed that testosterone appears to increase persistence of aggression following an intrusion, maximizing reproductive success. This would not be adaptive in autumn, when territories are less fixed and reproductive success is not at stake (Wingfield, 1994a, 1994b). It is important to bear in mind that testosterone also has marked effects on sexual behavior, the morphology of reproductive accessory organs, and other traits. Therefore, secretion of this hormone outside of the breeding season would likely be inappropriate (Wingfield, Jacobs, and Hillgarth, 1997).

In the breeding season of songbirds, territorial aggression is mediated by aromatization of testosterone to estradiol in target neurons of the brain (example, Schlinger, 1994; Balthazart et al., 1999). However, in the nonbreeding season, free-living European robins, *Erithacus rubecula,* treated with an antiandrogen (flutamide) in autumn and winter showed no effect on territorial aggression, suggesting that androgen receptors may not be involved at this time (Schwabl and Kriner, 1991). In European stonechats, *Saxicola torquata,* although territorial aggression was lowered by antiandrogen and an aromatase inhibitor (1,4,6-and rostatrien-3,17-dione [ATD]) in spring, the treatment had no effect in winter. This suggests that territorial aggression may be regulated differently (Canoine and Gwinner, 2002). In contrast, field experiments with male song sparrows using a combination of flutamide and ATD reduced territorial behavior in autumn and winter (Soma, Sullivan, and Wingfield, 1999; Soma, Tramontin, and Wingfield, 2000). Furthermore, fadrozole (a potent aromatase inhibitor) treatment alone resulted in a marked reduction of territorial aggression in the nonbreeding season (Soma, Tramontin, and Wingfield, 2000; Soma et al., 2000). These data suggest that estrogens are an important component of the regulation of territorial behavior outside the breeding season. If fadrozole-treated male song sparrows were then given estradiol implants, territorial aggression in response to simulated territorial intrusion (STI) was restored (Soma, Tramontin, and Wingfield, 2000). Note, however, that these results are not universal for all species, possibly suggesting that mechanisms underlying control of autumnal territorial aggression evolved independently multiple times (Canoine and Gwinner, 2002; Moore, Walker, and Wingfield, 2004) and consistent with the evolutionary flexibility hypothesis (Hau, 2007).

Several hypotheses have been put forward to explain how the potential "costs" of testosterone in the nonbreeding season could be avoided (Wingfield, Lynn, and Soma, 2001; Wingfield and Soma, 2002). The first hypothesis (the null hypothesis) is possible in species that have a circulating binding globulin that would buffer the animal from high blood levels of testosterone. Birds generally lack high-affinity and highly specific sex steroid–binding proteins in blood, suggesting that the null hypothesis is not true for songbirds (Wingfield, Lynn, and Soma, 2001). Increased sensitivity of the brain to extremely low concentrations of sex steroids during the nonbreeding season is also unlikely in song sparrows because many other testosterone-dependent traits do not develop. Moreover, castration in autumn had no effect on territorial aggression (Wingfield, 1994b). It is possible that low levels of androgens are secreted by the adrenal, but these might be ineffective. A decrease in the number of androgen receptors (ARs) is inferred from the low efficacy of testosterone to activate postbreeding singing and a reduction of hypothalamic aromatase activity (example, Hutchison, Steimer, and Jaggard, 1986; Nowicki and Ball, 1989; Schlinger and Callard, 1990; Silverin and Deviche, 1991; Gahr and Metzdorf, 1997; Ball, 1999; Soma, Sullivan, and Wingfield, 1999). On the other hand, aromatase activity in the ventromedial telencephalon (vmTEL) of song sparrows correlated with the expression of aggression, showing similar levels in the breeding and nonbreeding seasons and significantly lower levels during the molt when aggressive responses to challenges were lowest (Soma, Sullivan, and Wingfield, 1999). Expressions of estrogen receptor (ER) and aromatase-mRNA (messenger RNA) in the telencephalon of canaries, *Serinus canaries*, were higher in November than in April, whereas AR-mRNA expression did not change (Fusani et al., 2000). In spotted antbirds *(Hylophylax n. naevioides)*, testosterone regulates territorial aggression in breeding and nonbreeding seasons even when testosterone levels were generally low (Wikelski, Hau, and Wingfield, 1999; Hau et al., 2000). In nonbreeding season, when testosterone levels are extremely low, males showed elevated mRNA expression for ER-alpha receptor in the preoptic area and AR in the nucleus taeniae (Canoine et al., 2007). This is evidence for increased brain sensitivity to hormones when the circulating levels are at their lowest. Investigations of expression of ERα and ERβ are needed, however, before this hypothesis can be tested fully, but again, it appears that multiple mechanisms may have evolved to regulate sensitivity to extremely low levels of testosterone in the nonbreeding season.

High local concentrations of steroids in the brain that were independent of changes in circulating levels are thought to be a result of the brain being able to synthesize steroids de novo from cholesterol (example, Robel and Baulieu, 1995; Tsutsui and Yamakazi, 1995; Baulieu, 1998; Ukena et al., 1999). Steroidogenic enzymes (protein and mRNA activity) are expressed in many brain areas (Vanson, Arnold, and Schlinger, 1996; Schlinger et al., 1999; Soma, Sullivan, and Wingfield, 1999; Ukena et al., 1999). Thus it is possible that high levels of sex steroids can be generated in the brain independently of the gonads and adrenals. The circulating precursor hypothesis for regulation of autumnal aggression depends on steroidogenic enzymes in the brain converting circulating precursors to biologically active sex steroids (Wingfield, Lynn, and Soma, 2001). These precursors, such as dehydroepiandrosterone (DHEA), may be produced by the adrenal and then converted to an active hormone in the brain (Labrie et al., 1995; Vanson, Arnold, and Schlinger, 1996; Ukena et al., 1999). Plasma levels of DHEA in free-living male song sparrows were elevated during breeding, low when molting, and then increased again in autumn when singing and territorial aggression also increased (Soma and Wingfield, 2001). Furthermore, DHEA treatment of male song sparrows in autumn increased singing and growth of HVc, a song-control nucleus in the telencephalon (Soma et al., 2002), although the implants also increased plasma testosterone levels slightly compared with controls. It is also entirely possible that the brain of song sparrows may be able to synthesize biologically active sex steroids de novo from cholesterol, thus negating any need for circulating steroids (neural steroid hypothesis). If either is true, then an important question becomes: what regulates enzyme activity leading to sex steroid production in the brain of birds during autumnal aggression?

Facultative Responses of Social Systems to Unpredictable Environmental Elements

Environmental perturbation factors can disrupt the life cycle and are unpredictable in timing, duration, and intensity (Wingfield, 2003). Individuals must survive such perturbations in the best condition possible and activate facultative physiological and behavioral responses collectively called the "emergency life history stage." This assemblage of responses to a capricious environment has four major components (Sapolsky, Romero, and Munck, 2000; Wingfield and Romero, 2001; Romero, 2002; McEwen and Wingfield, 2003; Wingfield, 2003):

1. The fight-or-flight response involving very rapid behavioral responses to immediate threats such as predator attack or dominant conspecifics.

2. Proactive and reactive coping styles in which personality determines to varying extents how an individual copes with social stress. The responses can be genetic or a result of experience (or both) and allow individuals to deal with psychosocial stress among conspecifics. Coping styles have been classified in various ways, but a review by Koolhaas et al. (1999) suggests a generally applicable grouping for vertebrates. The proactive coping style is an active response to a social challenge involving aggression, whereas the reactive coping style is characterized by behavioral immobility and low aggression. Both strategies are successful, and many intermediate forms exist as well (Korte et al., 2005).

3. Sickness behavior as a suite of responses to wounding and infection that are regulated by cytokines of the immune system and by prostaglandins. The behavioral patterns of mammals and humans to infectious diseases include lethargy (as well as soporific behavior), depression, anorexia, increased threshold for thirst, reduction in grooming, and altered physical appearance. This suite of responses may not be a maladaptive response or an effect of debilitation but a behavioral strategy that, in conjunction with fever, combats viral and bacterial infections (Hart, 1988; Kent et al., 1992; Dantzer, 2000). Evidence is growing that similar mechanisms are present in nonmammalian vertebrates as well (example, Johnson et al., "Central," 1993 Johnson et al., "Sickness," 1993; Owen-Ashley and Wingfield, 2006; Owen-Ashley et al., 2006).

4. Facultative behavioral and physiological strategies that involve animals responding to perturbations in the field to redirect them away from the normal life history stages (example, breeding, migrating) into survival mode. Strategies include moving away (temporarily or permanently) from the source of the perturbation, finding a shelter to endure the perturbation, or a combination of both. In general, these components of the emergency life history stage allow the individual to respond to a perturbation and avoid chronic stress (Wingfield and Ramenofsky, 1999; Wingfield and Romero, 2001; Wingfield, 2003).

Social interactions during unpredictable perturbations of the environment can be chaotic, such as when attacked by a predator or in response to major changes in dominance hierarchies. They can also be highly directed and coordinated among individuals, such as in mass movements away from an area affected by drought or disease. Additionally, aggressive behavior is frequently expressed even when in an emergency life history stage

(Figure 4.1). Antipredator aggression (or against a dominant conspecific) is well known under such circumstances in many vertebrates. Also, irritable aggression may also be expressed, especially in relation to food shortages, competition for shelter, and so on. (see Wingfield et al., 2005, for more details). The hormonal basis of behavioral responses to perturbations of the environment needs much more experimental analysis, a topic very timely in an era of global change.

Allostasis, Allostatic Load, and Hormone-Behavior Interactions

The term *allostasis* refers to the maintainance of stability through change and has been introduced to explain how individuals adjust to both predictable and unpredictable events on a continuum (McEwen, 2000; McEwen and Wingfield, 2003). The allostasis concept basically addresses changes in homeostatic set points as the life cycle progresses and as unpredictable events dictate. The concept has some useful terms that are relevant to expression of behavior, particularly aggression, in response to environmental perturbations. *Allostatic load* is the cumulative cost to the body of allostasis, with *allostatic overload* (accompanied by elevated plasma levels of glucocorticoids among other mediators of allostasis) being a state in which daily food intake and/or body reserves cannot fuel the cumulative cost. It is at this point that glucocorticoid levels surge, and various components of the emergency life history stage may be triggered (McEwen and Wingfield, 2003). Concerning social interactions, aggression during competition for food and shelter may be frequent and intense, but the mechanisms remain poorly known. A role for glucocorticoids is possible (example, Ramos-Fernández et al., 2000; Kitaysky, Wingfield, and Piatt, 2001; Kitaysky et al., 2003). The normal life cycle and appropriate life history stages will be resumed when the perturbation passes.

Whereas cooperation, social support, and cohesion provide many advantages for living in social groups, social conflict and competition introduce disadvantages. Such social conflict may elevate allostatic load, which results in greater sensitivity to overload resulting from other perturbations, and this in turn is followed by increased levels of glucocorticoids. Because individuals in groups have different degrees of allostatic load as a result of social status, this may influence glucocorticoid levels of dominants and subordinates differently. Goymann and Wingfield (2004) analyzed data from the literature on free-ranging animals using phylogenetic independent contrasts of allostatic load and relative levels of glucocorticoids. The relative allostatic load of social status appeared

to predict whether dominant or subordinate members of a social unit secreted higher or lower levels of glucocorticoids (Goymann and Wingfield, 2004). It was proposed that if allostatic load of dominance rank
(the sum of acquisition and maintenance) is greater than that of being
subordinate, then dominants are significantly more likely to have elevated levels of glucocorticoids. On the other hand, if allostatic load of
social status is higher in subordinates, then they are significantly more
likely to express higher levels of glucocorticoids. Thus the costs of social
status do matter, especially in relation to how sensitive an individual may
be to other perturbations and allostatic overload (Goymann and Wingfield, 2004).

Conclusion

All organisms live in changeable environments—there is no habitat on
earth that is absolutely constant. Animals must cope with two types of
environmental change—predictable changes associated with seasons,
day/night, low tide/high tide, and so on, and unpredictable changes associated with environmental perturbations such as severe weather, food
shortages, changes in predators, and many relatively new perturbations
associated with human disturbance and climate change (global change).
How social interactions and hormone-behavior interactions allow animals to cope with environmental changes is only just beginning to be investigated in depth.

   In relation to predictable changes in the life cycle, organisms use environmental cues to trigger morphological, physiological, and behavioral
changes in anticipation of a future event. One intriguing recent development in relation to how hormone-behavior interactions change in relation to seasons concerns the regulation of aggression (particularly
territorial aggression) by testosterone and its aromatization to estradiol.
Excellent reviews of field and laboratory studies (Reed et al., 2006; Hau,
2007) posit two testable hypotheses. The evolutionary constraints hypothesis asserts that there are only certain ways in which testosterone
can act through enzymatic conversion and intracellular receptors. These
constraints suggest that temporal patterns of testosterone secretion and
plasma levels should vary according to patterns of behavior and mating
system. In contrast, the evolutionary flexibility hypothesis states that
there may be multiple ways in which testosterone may act, particularly in
the nonbreeding season, and patterns of secretion of testosterone and
plasma levels are not so important. Several examples have been tested in
relation to "avoiding the costs of testosterone," although much more

testing is needed. It should also be pointed out that these two hypotheses are relevant to other hormone-behavior interrelationships. They are also not necessarily mutually exclusive, and many intermediate examples may exist, reflecting the biodiversity of mechanisms underlying responses to environment.

Hormone-behavior interrelationships in response to perturbations of the environment are very well investigated in some contexts such as the fight-or-flight response, sickness behavior, and proactive/reactive coping styles, but much more work needs to be done in populations of animals in their natural environment. Facultative behavioral and physiological responses that allow individuals to cope with perturbations and stress are much less well studied. The concepts of allostasis and allostatic load may provide a starting point that allows us to model the demands of normal life history stages (such as reproduction and migration) with additional perturbations—stress. This area needs more investigation, especially in relation to global change—a potent mix of global climate change and human disturbance (Travis, 2003).

Acknowledgments

The author has been supported by a series of grants from the National Science Foundation Divisions of Integrative Biology and Neuroscience and the Office of Polar Programs (currently IBN-0317141) and by grant number RO1 MH65974-01 from the National Institute of Mental Health. I also acknowledge valuable support from a Russell F. Stark University Professorship (University of Washington) and a Simon F. Guggenheim Fellowship.

CHAPTER FIVE

The Endocrinology of the Human Adaptive Complex

Jane B. Lancaster and Hillard S. Kaplan

Introduction

The *human adaptive complex* is a coadapted complex of traits including (1) the life history of development, aging, and longevity; (2) diet and dietary physiology; (3) the energetics of reproduction; (4) social relationships among men and women; (5) intergenerational resource transfers; and (6) cooperation among related and unrelated individuals (Kaplan, 1997; Kaplan et al., 2000; Kaplan et al., 2001; Kaplan and Robson, 2002; Kaplan et al., 2003; Robson and Kaplan, 2003; Kaplan and Gurven, 2005; Gurven and Kaplan, 2006; Gurven and Walker, 2006; Kaplan, Gurven, and Lancaster, 2007). It describes a very specialized niche characterized by (1) the highest-quality, most nutrient-dense, largest-package-size food resources from both plants and animals; (2) learning-intensive, sometimes technology-intensive, and often cooperative food acquisition techniques; (3) a large brain to learn and store a great deal of context-dependent environmental information and to develop creative food acquisition techniques; (4) a long period of juvenile dependence to support brain development and learning; (5) low juvenile and even lower adult mortality rates, generating a long, productive life span and population age structure with a high ratio of adult producers to juvenile dependents; (6) a three-generational system of downward resource flows from grandparents to parents to children; (7) biparental investment, with men specializing in energetic support and women combining energetic support with direct care of children; (8) marriage and long-term reproductive unions; and (9) cooperative arrangements among kin and unrelated individuals to reduce variance in food availability through sharing and to more effectively acquire resources in group pursuits.

The publications cited above show that the majority of the foods consumed by contemporary hunter-gatherers worldwide are calorically dense hunted and extracted (taken from an embedded or protected matrix—underground, in shells, and so on) resources, accounting for 60% and 35% of calories, respectively. Extractive foraging and hunting proficiency generally does not peak until the midthirties, because they are learning and technique intensive. Hunting, in particular, demands great skills and knowledge that takes years to learn, with the amount of meat acquired per unit time more than doubling from age 20 to age 40 even though strength peaks in the early twenties. This learning-intensive foraging niche generates large calorie deficits until age 20 and great calorie surpluses later in life. This life history profile of hunter-gatherer productivity is only economically viable with a long expected adult life span.

This chapter is divided into two parts. The first section reviews components of the human adaptive complex and the evidence on which our understanding is based. The second section discusses directions for new research, with a focus on the uniquely human relationships between men and women and on their implications for the life history of the endocrine system.

Part I

Life Histories of Wild Chimpanzees and Human Foragers

To appreciate the implications of the human adaptive complex for the life histories of foragers and horticulturalists, it is useful to compare humans with the chimpanzee, another large-bodied, long-lived mammal and our closest living relative in phylogenetic terms. Table 5.1 presents major differences in five critical parameters of life history: survivorship to age of first reproduction, life expectancy at the beginning of the reproductive period, absolute and relative length of the postreproductive period, spacing between births of surviving offspring, and growth during the juvenile period (Lancaster et al., 2000). Human and chimpanzee life history parameters based on data from extant groups of hunter-gatherers and wild chimpanzees indicate that forager children experience higher survival to age 15 (60% versus 35%) and higher growth rates during the first 5 years of life (2.6 kg/yr versus 1.6 kg/yr). Chimpanzees, however, grow faster in both absolute and proportional weight gain between the ages of 5 and 10. The early high weight gain in humans may be the result

Table 5.1   Life history characteristics and diet of human foragers and chimpanzees.

Life History Characteristics	Human Foragers	Chimpanzees
Maximum life span	~120	~60
Probability of survival to age 15	0.6	0.35
Expected age of death at 15 (years)	54.1	29.7
Mean age first reproduction (years)	19.7	14.3
Mean age last reproduction (years)	39	27.7*
Interbirth interval (months)**	41.3	66.7
Mean weight at age 5 (kg)	15.7	10
Mean weight at age 10 (kg)	24.9	22.5
Composition of Diet (%)		
Collected	9	94
Extracted	31	4
Hunted	60	2

Contributions by Sex (%)	Men	Women	
Adult calories	68	32	Sexes independent
Adult protein	88	12	
Caloric support for offspring	97	3	
Protein support for offspring	100	0	

*Age of last reproduction for chimpanzee females was estimated as two years prior to the mean adult life expectancy.

**Mean interbirth interval following a surviving infant.

Sources: Kaplan et al., 2000; Lancaster et al., 2000.

of the earlier weaning age (2.5 years versus 5 years), followed by provisioning of highly processed and nutritious foods. Fast growth and weight gain during infancy and the early juvenile period may also represent an adaptation to support the energetic demands of brain growth development, since a significant portion of this weight gain is in the form of fat.

The chimpanzee juvenile period is shorter than that for humans, with age at first birth of chimpanzee females about 5 years earlier than among forager women. This is followed by a dramatically shorter adult life span for chimpanzees. At age 15, chimpanzee life expectancy is an additional 15 years, whereas foragers can expect to live an additional 38 years, having survived to age 15. Importantly, women spend more than a third of adult life in a postreproductive phase, whereas few chimpanzee females spend any time as postreproductive. The differences in overall survival probabilities and life span of the two species are striking: less than 10% of chimpanzees ever born survive to age 40, and virtually none survive past 50, whereas 45% of foragers do, and more than 15% of foragers born survive to age 70!

Finally, despite the fact that human juvenile and adolescent periods take longer, and that human infants are larger than chimpanzees at birth, forager women are characterized by higher fertility. The mean interbirth interval between offspring when the first survives to the birth of the second is 1.6 times longer among wild chimpanzees than among modern forager populations.

To summarize, human foragers show a juvenile period 1.4 times longer and a mean adult life span 2.5 times longer than chimpanzees. They experience higher survival at all ages postweaning but slower growth rates during midchildhood. Despite a long juvenile period, slower growth, and a long postreproductive life span, forager women achieve higher fertility than do chimpanzees.

Consumption and Productivity through the Life Course

Table 5.1 also demonstrates the overlap in component categories of the diets of foraging societies and chimpanzee communities as well as the wide differences in relative proportions (Lancaster et al., 2000). For example, hunted meat makes up about 2% of chimpanzee but 60% of forager diets. Chimpanzees rely on collected foods for 94% of their nutrition, especially ripe fruits. Such foods are nutritious and are neither hard to acquire nor learning intensive, at least relative to human resource pursuits. Humans depend on extracted or hunted foods for 91% of their diet. The data suggest that humans specialize in rare but nutrient-dense resources (meat, roots, nuts), whereas chimpanzees specialize in ripe fruit and fibrous plant parts. These fundamental differences in diet are reflected in gut morphology and food passage times in which chimpanzees experience rapid passage of bulky, fibrous meals processed in the large intestine, whereas humans process nutritionally dense, lower-volume meals amenable to slow digestion in the small intestine (Milton and Demment, 1988).

Figure 5.1 presents the survivorship and net food production through the life course of humans and chimpanzees (Kaplan and Lancaster, 2003). Humans consume more than they produce for the first third of their life course. In contrast, chimpanzees are self-supporting by the age of 5. Thus, human juveniles, unlike chimpanzee juveniles, have an evolutionary history of dependency on adults to provide their daily energy needs. Even more striking is the steady increase in productivity over consumption among humans into their thirties and early forties. Forager males begin to produce more than they consume in their late teens, but their peak productivity builds slowly from their early twenties until their mid to late thirties and then is sustained for 20 or more years at a level of approximately 6,500 kcal per day. In contrast, forager women vary greatly from group to group in energy production, depending on the demands of intensive child care (Hurtado and Hill, 1990). In some groups, they consume more than they produce until menopause, when they are freed from child care demands; in other groups, such as the Ache, they remain nutritionally dependent on men throughout their lives. The provisioning of reproductive women and children has a powerful effect on the production of children by humans by reducing the energy cost and health risk of lactation to the mother and by lifting the burden of self-feeding from the juvenile, thus permitting a shortened interbirth interval without an increase in juvenile mortality (Hawkes et al., 1998).

The human adaptation is both broad and flexible, in one sense, and very narrow and specialized, in another. It is broad in the sense that, as foragers, humans have existed successfully in virtually all of the earth's

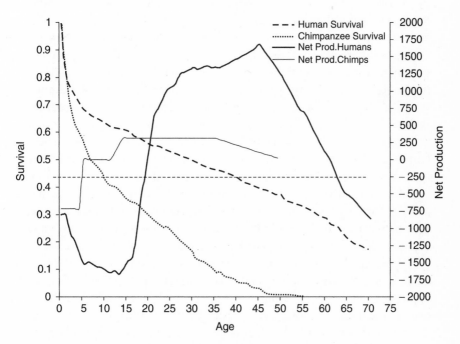

Figure 5.1 Survival and net food production: human foragers and chimpanzees. From Kaplan and Lancaster, 2003, 181.

major habitats. It is narrow and specialized in that it is based on a diet composed of nutrient-dense, difficult-to-acquire foods and a life history with a long, slow development, a heavy commitment to learning and intelligence, and an age profile of production shifted toward older ages. In order to achieve this diet, humans are very unproductive as children, have very costly brains, are extremely productive as adults, and engage in extensive food sharing both within and among age and sex classes.

The Sexual Division of Labor and the Acquisition of Skilled Performances

A feeding niche specializing in large, valuable food packages, and particularly hunting, promotes cooperation between men and women and high levels of male parental investment, because it favors sex-specific specialization in embodied capital investments and generates a complementarity between male and female inputs. The economic and reproductive cooperation between men and women facilitates provisioning of juveniles that both underwrites their embodied capital investments and acts to lower mortality during the juvenile and early adult periods. Cooperation between males and females also allows women to allocate more time to child care and improves their nutritional status, increasing both survival and reproductive rates. The nutritional dependence of multiple young of different ages favors sequential mating with the same individual, since it reduces conflicts between men and women over the allocation of food. Finally, large packages also appear to promote interfamilial food sharing. Food sharing assists recovery in times of illness and reduces the risk of food shortfalls due to both the vagaries of foraging luck and the variance in family size due to stochastic mortality and fertility. These buffers against mortality also favor a longer juvenile period and higher investment in other mechanisms to increase life span.

Unlike most other mammals, men in foraging societies provide the majority of the energy necessary to support reproduction. After subtracting their own consumption from total production, women supply an average of 3% of the calories to offspring, with men providing the remaining 97%, among the 10 foraging societies for which quantitative data on adult food production are available (Kaplan et al., 2000). Hunting, as opposed to gathering of animal protein in small packets, is largely incompatible with the evolved commitment among primate females to intensive mothering, carrying of infants, and lactation on demand in the service of high infant survival rates. First, hunting is often risky, involving rapid travel and encounters with dangerous prey. Second, it is often most efficiently practiced over relatively long periods of time, rather than in short stretches, due to search and travel costs. Third, it is extremely skill intensive, with improvements in return rate occurring over two decades of daily hunting. The first two qualities make hunting a high-cost activity for pregnant and lactating females. The third quality, in interaction with the first and second, generates life course effects such that gathering is a better option for females *even when they are not lactating,* and hunting is a better option for males (Gurven and Kaplan, 2006). Since women spend about 75% of their time during their reproductive years either nursing or more than three months' pregnant, it would not pay them to hunt because they never get enough practice to develop the skills to make it worthwhile, even when they are not nursing or pregnant or are postreproductive.

In our view the sex-based specialization in investment in skills over the life course (the sexual division of labor), together with the long period of parental investment leading to multiple child dependents, is directly responsible for high male parental investment, the universality of the marriage institution, and the extensive economic and reproductive cooperation among husbands and wives. This original specialization in different skills and in the procurement of different resources generated a complementarity between human men and women that is rare among mammals. The specialization generated two forms of complementarity. Hunted foods complemented gathered foods because protein, fat, and carbohydrates complement one another with respect to their nutritional functions (Hill, 1988). A second form of complementarity came from the fact that males tend to concentrate on large, shareable packages of food, such as meat, which shifts the optimal mix of activities for women, increasing their time spent in child care and lowering their efficiency demands in food production.

The end result of the human division of labor can be reduced to a few basic principles that reflect its core functions: men are allocated tasks that require strength, endurance, concentration, and risk taking. This places men in the role of both protectors and hunters, focusing their productive efforts on big game at the top of the food pyramid. Women are collectors, concentrating lower on the food pyramid to gather medium-quality food products such as the reproductive and storage organs of plants. This work is much lower in risk and thus compatible with lactation and the care of small children. The sexual division of labor goes far beyond hunting and gathering. It is typical of traditional societies as well. Brown (J. Brown, 1970) found that, in surveying the Human Relations Area Files, tasks assigned to women have very specific properties in line with the classic division of labor. In surveying the practice of crafts and other specializations in traditional societies, Brown found that only a limited number are not practiced by women: metallurgy; long-distance travel and trade; handling large, dangerous domestic animals; and warfare. Women's tasks such as gardening, market trade, and the crafts of pottery and cloth production have a set of practical characteristics: (1) they do not require rapt concentration, (2) they can be easily interrupted, (3) they do not require long-distance travel from home, and (4) they are not dangerous. In other words, they are all forms of productive labor that are compatible with the care of small children.

Human females evidence physiological and behavioral adaptations that are consistent with an evolutionary history involving extensive male parental investment. They both decrease metabolic rates and store fat during pregnancy, suggesting that they can lower work effort and are being provisioned (Sellen, 2006). During lactation, women in foraging societies decrease work effort and focus on high-quality care such as lactation and carrying (Hurtado et al., 1985; Lancaster et al., 2000; Gurven and Kaplan, 2006). In contrast, nonhuman primate females do not store appreciable fat, and they increase work effort during pregnancy and lactation; as a result, they have increased risk of mortality (Altmann, 1980; Cheney et al., 2004). Human specialization could not have evolved if women had not depended on men throughout human history for most of their food provision.

Part II

Sexual Division of Labor and Complementarity: Directions for Future Research

Given that the division of labor and long-term mateships have been such a critical feature of human reproductive strategies, it is likely that endocrine regulation of behavior, psychology, and physiology over the life course is designed to support the complementary roles of men and women. Such regulation should be sensitive to the costs and benefits of defection and desertion as numbers of dependent children change over time. It should also be responsive to the energetic demands of children and the complementary roles that men and women play in meeting those demands. The remainder of this chapter will discuss four areas of new research entailed by those considerations: (1) children's contribution to marital stability, (2) women's sexual preferences during conceptive ovulations in natural fertility populations, (3) male physical and psychological characteristics in relation to the division of labor, and (4) the process of acquisition of sexually dimorphic embodied capital.

PARENTAL INVESTMENT AND MATESHIP STABILITY

Marriage is probably the most complex cooperative relationship in which humans engage. It involves the production and processing of resources for familial consumption, distribution of those resources, the provision of child care, production and maintenance of belongings and residential amenities, and sexual rights and responsibilities. The ability to coordinate on the allocation and execution of those responsibilities (that is, the ability to "get along") is fundamental to successful marriage and appears to play a role in mate choice (Buss, 2003). In traditional societies, as well as contemporary, it is not uncommon to hear remarks about success and failure in coordinating and getting along as comments about why marriages succeed and fail.

A problem that people face in mate choice is that long-term, multiple dependency of offspring of differing ages—unlike reproduction in litters or in succession, with feeding independence established at weaning—makes mate switching much more costly. Once one has reproduced with a given partner, a change in partners can entail reduced investment in those previous children. Moreover, most mate choice occurs before economic abilities are proven. For example, at a marriage age of around 20, Ache and Tsimane men are only 25% and 50%, respectively, as proficient at hunting as they will be at their peak in their mid- to late thirties (Kaplan et al., 2001; Gurven, Kaplan, and Gutierrez, 2006). This issue remains true for most marriages even today. Excluding marriages based on inherited wealth, most others are formed and reproduction is begun before the full productive and reproductive potentials of the partners are clear.

From the perspective of both men and women, there are great gains from choosing a good partner, and there are great risks of economic and sexual defection. For the most part, it is a long-term choice with direct consequences for fitness. It is further complicated by the fact that partners contribute to fitness not only through behavior but also through genetic inputs, which can lead both to further complementarities or to conflicts of interest. Marriages redirect social interaction and cooperation not just within the pair-bond but across members of respective extended families.

Social capital, which can be defined as stored information by others about a person's characteristics, social status, and social support, is likely to play an important role in mate choice. Capital affecting perceptions about fairness, industriousness, loyalty, promiscuity, and economic abilities is likely to influence mate choice decisions by both men and women. Some of the same factors affecting the choice of production and sharing partners may also affect the choice of marriage partners (see Kaplan, Gurven, and Lancaster, 2007, for a discussion).

Extensive cooperation among human men and women would only make sense if the reproductive performance of spouses were linked. Women approach zero fecundability some 5 years before menopause (Holman and Wood, 2001). Forager men do have the option to continue reproducing with younger women but generally do not do so. For example, 83% of all last births to Ache women also represent a last birth for the fathers of these children (Gurven, Kaplan, and Gutierrez, 2006). An obvious reason for this is that the last child born to a couple still represents an additional 18 years of parental provisioning and investment. In traditional sedentary or herding societies, some men with resources do have the option of polygyny and can add new, younger wives as they age and so extend their own reproductive years. Nevertheless, the fact is that most marriages in time past and in the world today are either monogamous or serially monogamous; and even when polygyny is a cultural ideal, few men can attain it. Murdock's cross-cultural sample of 849 societies lists polygyny for 83.5% of traditional societies, but in 55.7% of these, polygyny is occasional (<20%), and only 27.8% are labeled as polygyny being common (>20%) (Lancaster and Kaplan, 1992).

We also know that the presence of children stabilizes marriage in societies ranging from traditional (Hill and Hurtado, 1996; Winking, 2006) to modern industrialized societies (Morgan, 1996; Anderson, Kaplan, and Lancaster, 2007). Betzig (1989) in a study of the causes of conjugal dissolution taken from authoritative accounts on 160 societies in the Standard Cross-Cultural Sample found that the two leading causes of divorce were infidelity and infertility, closely ranked at 88 and 75, respectively, with cruelty or maltreatment ranking third at 54. The first two speak directly to the issue of children as pivotal in marriage: the one in terms of the genetic relationship of children to their parents and the other to their very existence. Betzig proposes that the practice of trial mateships so common in both tribal societies and under modern conditions most often solidifies into full or common-law marriage with the production of children. It is interesting to note that the two largest ever studies of sexual behavior done in the United States and England in response to the HIV epidemic gave no information about the presence of children on marital fidelity in spite of the fact that their charge was to model sexual networks and the spread of disease (Johnson et al., 1994; Laumann et al., 1994). The authors do note that in a lifetime most sexual partnerships are accumulated before and after marriage and that the sexual revolution accounts for a historic increase in partners under those statuses. However, they made no attempt to distinguish between the sexual behaviors of marital partners *with* as opposed to *without* children.

The fact that humans are unique in raising multiple, nutritionally dependent offspring of differing ages also reduces the payoffs to defection and increases the benefits for men and women to link their economic and reproductive lives over the long run. Men and women who divorce and remarry during the time they are raising offspring will face conflicts of interest with new spouses over the division of resources (Anderson, Kaplan, and Lancaster, 1999). If they marry someone with children from a previous marriage, they may disagree with their new spouses over the allocation of food and care to their joint children, relative to children from previous marriages. These potential conflicts increase the benefits of spouses staying together and having all or most of their children together.

We know very little about the psychological and physiological processes underlying the formation of long-term pair-bonds and the role that children play in cementing those bonds. Prior to the birth of children, both men and women should be sensitive to cues about their partner's quality and ability to successfully accomplish their sex-specific role and their willingness to cooperate over the long run. During this period of evaluation, we might expect people to be more willing to consider alternative partners.

Winking and colleagues (Winking et al., 2007) found such an effect among Tsimane forager-horticulturists. Tsimane men were much more likely to have extramarital affairs in the first year or two of their marriages. Moreover, the likelihood of affairs diminishes radically with each additional child born to the couple. In spite of the fact that the reproductive value of men's wives decreases as they age, men are less likely to divert resources from their families and engage in affairs as both members of the couple age and have children. These results, especially if they are supported by further cross-cultural evidence in traditional societies, provide strong evidence against the mating effort or show-off hypothesis of men's work proposed by Hawkes and colleagues (Blurton Jones et al., 2000; Hawkes and Bliege Bird, 2002) and in favor of the view that the reproductive lives of men and women are linked through their complementary roles.

The endocrine supports of this system have yet to be investigated. Men appear to be responsive to the costs and benefits associated with both mating and parental effort, as they change over the life course. While there are very general changes in endocrine activity as men age, the particular life circumstances of men, such as entry into and out of bonded mateships and the birth of children, may also be reflected in and supported by endocrine responses (see Gray and Campbell, this volume). The relationship between the cognitive processing of costs and benefits, the motivation to engage in sexual activity in and out of mateships, and the feelings of parental concern have yet to be investigated in detail. We hypothesize that men's endocrine activity will play an important role in mediating those relationships.

A similar set of life history concerns should also govern women's behavior. For women, the presence of dependent children should be of particular importance, perhaps more so than whether or not they are mated. The mating strategies of women with and without dependent children should be very different, especially with respect to the multiple dependency of young of differing ages and condition. It raises the question of what the psychology of mate preference is for women who are mothers of young children in which the costs and benefits of seeking "good" genes in preference to male paternal investment are very different from those of childless women, regardless of whether or not they are mated. For that matter, throughout human evolutionary history, most sexually mature women who were available in the mating market already had children from previous relationships. Only young girls during the subfecund postadolescence period would be likely to have mate choice unfettered by the needs of dependents.

Over evolutionary and historical time, women more than men were likely to be single parents due to spousal death, divorce, or desertion (Vinovskis, 1990; Shackelford, Weekes-Shackelford, and Schmitt, 2005; Weekes-Shackelford, Easton, and Stone, 2007). Among the Ache, for example, death in childbirth was a rare event; the maternal death rate was 1 per 150 births, and the probability of death was higher for males from age 10 throughout adulthood (Hill and Hurtado, 1996). Such women should evidence a strong psychological preference for stable men with good parenting and provisioning skills, which would include the physical prowess to perform the demanding and high-risk tasks traditionally assumed by men, such as hunting large game, working at hard manual labor, risk taking, and protecting women and children from outside threat. These preferences cloud the stated division between being a good father and having good genes, since in times past being a good father and husband required good genes in terms of health and prowess.

Again, we suspect that there will be endocrine supports of this sensitivity to age and number of dependent children in the physiology of women. We also hypothesize that the quantity and quality of male support for children will also affect women's commitment to marriage and interest in alternative male suitors and that sensitivity to both dependency needs and male inputs will be mediated, at least partly, through endocrine controls. In the next section, we explore women's sexual preferences during conceptive ovulations in natural fertility populations in more detail.

WOMEN'S MATE CHOICE PREFERENCES DURING CONCEPTIVE OVULATIONS IN NATURAL FERTILITY POPULATIONS

In the past 15 years, evolutionary psychologists have produced a convincing body of articulated research on the evolution of human female mate choice (Gangestad, 2006; and Roney, this volume). The major findings of these studies are that women's sexual preferences vary according to where they are in their menstrual cycle. These preferences were found to fall in two phases, the nonovulatory parts of the cycle and a seven-day window leading up to and including ovulation. Because the research was on menstrual cycle effects, women who were on the pill, pregnant, or lactating were excluded from the sample. During the ovulatory window, women show strong preference for men with higher levels of testosterone, more symmetrical bodies and faces (Thornhill and Gangestad, 1999), more masculine faces (Penton-Voak et al., 1999; Johnston et al., 2001), a complementary major histocompatibility complex (Wedekind et al., 1995; Garver-Apgar et al., 2006), and personal histories of health (Gangestad and Thornhill, 2004). However, high levels of testosterone are associated not only with genetic quality but also with promiscuity. Men preferred during the nonovulatory window have lower levels of testosterone and are not as symmetrical or masculine but are more likely to indicate the ability to support a family. These results suggest that women have two kinds of sexualities linked to their menstrual cycle. During the ovulatory window, women prefer men with indicators of good genes, but during the nonovulatory parts of the cycle, they prefer men who are more likely to give paternal investment as measured by lower testosterone levels and access to resources. However, research on women in natural fertility populations representative of times past and most of the world today would greatly improve our understanding of the evolutionary forces and the endocrine architecture governing women's mating decisions during fertile menstrual cycles.

As established earlier, a major effect of the division of labor and the collaboration of human parents in the provisioning and care of their offspring is a parental condition unique to women: the dependency of multiple young of different ages (Lancaster, 1991, 1997; Draper, 1992). Figure 5.1 reveals the full impact of having multiple nutritionally dependent young with differing needs over many years. During infancy a mother provides both milk and care, but her older children between the ages of 5 and 15 are in a nutritional deficit of 1,500 calories on average. Unlike other primate offspring, human children need not self-feed at weaning. Figure 5.2 shows the major reduction in birth spacing for humans compared to the great apes. The mean interbirth interval between offspring when the first one survives to the birth of the second is 1.62 times longer among wild chimpanzees than among modern forager populations, thus reducing the interbirth interval to about 3 to 3.5 years for foragers and horticulturalists (Galdikas and Wood, 1990; Lancaster et al., 2000) and probably for natural fertility populations in general (Sellen, 2006). Consequently, because weaning long precedes nutritional independence for humans, the line between maternal investment by lactation for a toddler and the gestation of a subsequent child is less finely drawn.

Besides the dependency of multiple young of differing ages, women in natural fertility populations also have a characteristic hormonal profile during their reproductive span. Beginning with menarche at about age 16 and continuing to menopause around age 50, women spend most of this time either pregnant or lactating and not in menstrual cycling. In fact,

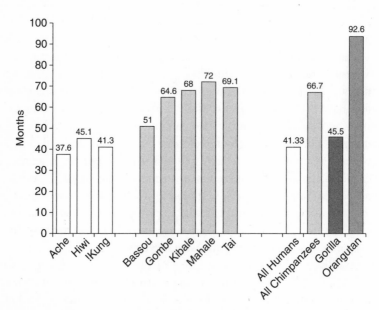

Figure 5.2　Birth spacing: human foragers and chimpanzees. After Lancaster et al., 2000.

menstrual cycling is a rare event; and even when these cycles occur, the majority are nonovulatory because most are during the first few years after menarche during the adolescent subfecundity period when ovulation is highly irregular or after age 45 when there is little likelihood of ovulation (Strassmann, 1997). Compared to the 350 to 450 cycles that women living in modern, low-fertility, contracepting societies can expect, women in natural fertility populations have 100 to 130 menstrual cycles in a lifetime, and perhaps only half of these will be ovulatory (Short, 1987; Eaton et al., 1994; Strassmann, 1997). In fact, such women will only ovulate repeatedly when they are sterile. Reproductive biologist R. V. Short was first to identify this pattern of excessive cycling today and to link repeated exposure to unopposed estrogens to the high rate of reproductive cancers in modern women. This insight has since been supported by research among a number of non-Western populations (Eaton et al., 1994; Strassmann, 1999).

Short also was very interested in the role of lactation as a natural contraceptive responsible for optimal birth spacing during most of human history and in the world today (Short, 1987, 1994). In natural fertility populations, (1) ovulatory menstrual cycles are very rare events, (2) the natural condition of women during their reproductive years is pregnancy or lactation, (3) lactation suppresses ovulation and controls birth spacing at the same time, and (4) most conceptions in natural fertility conditions occur during lactation (Short, 1987, 1994; Gray et al., 1990; Lewis et al., 1991; Kennedy and Visness, 1992). Well-nourished Australian and Scottish women who breast-fed exclusively for 6 months and on demand in the following months experienced on average 10 months of lactational amenorrhea, followed by 11 months of anovulatory menstrual cycles (Lewis et al., 1991). Similar results have been published in a meta-analysis of eight countries by Kennedy (Kennedy and Visness, 1992). Lactation, then, is a good contraceptive during the first year but not later, and most often the resumption of ovulation is forewarned by a menstrual cycle. However, the duration of lactation under natural fertility is much longer than 1 year, and the resumption of fecundity is not predicated on weaning (Lunn et al., 1984; Worthman et al., 1993; Tracer, 1996; Vitzthum et al., 2000; Ellison and Valeggia, 2003; Valeggia and Ellison, 2004). Sellen (Sellen and Smay, 2001) reviewed a sample of 113 ethnographic and demographic reports on the duration of breastfeeding published between 1873 and 1998. He found that duration of breastfeeding was 29.0 ± 10 months, with a median of 29.5 and a mode of 30.0. This, in combination with typical interbirth intervals of 3 years, supports Short's contention that most children are conceived while their mothers are breast-feeding a previous infant. The exception would be a firstborn child or a replacement baby whose older sibling died during its first 2 years of life.

These studies indicate that weaning in natural fertility populations is initiated by the recognition of a new conception and that conception is not dependent on weaning. In fact, researchers often report that infants are breast-fed until the second trimester of their successors' gestation—time enough for the implantation and viability of the next pregnancy to be assured (Tracer, 1996; Valeggia and Ellison, 2004; Nepomnaschy, 2007). Such a pattern of continued lactation during the first trimester of gestation of a subsequent child is highly adaptive for humans. The first trimester of gestation is the lowest in terms of energy demands on the mother, but it is also the highest in terms of pregnancy loss. Hormone assays indicate that mean early pregnancy loss nears 75% for women of reproductive age, and most of these have chromosomal or other defects (Holman and Wood, 2001). Selection would not favor sacrificing the interests of the suckling in the interests of a not-yet-conceived sibling until a successful pregnancy is firmly established.

The implication of this research on lactation and the resumption of ovulation in natural fertility societies is that the majority of children are conceived by mothers who are lactating and ovulating at the same time. In other words, conception occurs under a much richer hormonal environment than is usually considered in which levels of estrogen and progesterone are high enough for ovulation and implantation and levels of prolactin and oxytocin are high enough to sustain lactation. The main effect of prolactin during the postpartum period is to maintain milk production. Its levels surge following each sucking episode and remain chronically elevated as long as suckling continues. It inhibits pusatile secretion of gonadotropin-releasing hormone (GnRH), which, in turn, suppresses ovarian steroid production. According to Short, prolactin orders up the next meal, and oxytocin serves the current meal (Short, 1994). Oxytocin's classic role is to stimulate smooth muscle contraction in such vital reproductive processes as child birth, uterine contraction after birth, milk letdown, and orgasm. Another of its functions is to suppress cortisol and so reduce stress and foster a general sense of calm (Uvnäs-Moburg, 1998, 2003). It is the contentment hormone of nursing mothers. Oxytocin in concert with dopamine supports emotional attachment and bonding and conditions a preference for social partners whether a baby, mate, or pet (Uvnäs-Moburg, 2003; and Sanchez et al., this volume).

This research on lactation and conception in natural fertility populations brings to the front an obvious question. Since most pregnancies have been and are to mothers, and most often mothers of infants, research on female sexual preference in relation to the menstrual cycle needs to be done on the excluded categories: women on the pill, pregnant, and lactating. There is no published research on mating preferences of women who have been nursing for over a year, and yet these would be the typical condition of most conceptions in natural fertility populations. A research agenda that is more inclusive is in order. Given the importance of maintaining continued male parental investment for existing offspring, we hypothesize that variation in mate choice between the ovulatory and nonovulatory portions of the menstrual cycle identified in Western populations among nulliparous women and parous women who are not lactating would be greatly attenuated in natural fertility populations.

THE IMPACT OF THE SEXUAL DIVISION OF LABOR ON MALE SIZE AND MUSCULARITY

The tasks in the sexual division of labor that fell to human males have one thing in common: they cannot be done carrying an infant or being accompanied by a toddler. They demand the physical prowess of hunting large game, handling dangerous domestic animals, working at hard manual labor, risk taking in general, and protecting women and children from outside threat. In other words, men must be risk takers both as producers and as protectors in order to play their complementary role in the division of labor. Dabbs and Dabbs (2000) link these roles to the hormone testosterone, which facilitates male performance as "heroes, lovers, and rogues." What they do not do is to link these behaviors to being effective fathers. They and others see these male behaviors as the products of sexual selection (antithetical to fatherhood when defined only in terms of caregiving) and not male behaviors that need to complement what women can do in the raising of families.

Sexual dimorphism in muscularity and body size is an ancient feature in the anthropoid primates and is usually understood as sexually selected adaptations for intrasexual competition over access to females. However, according to Plavcan and van Schaik (1997), although monogamous anthropoids show low degrees of weight dimorphism, as would be predicted by the action of sexual selection, polygynous anthropoids show high variation in weight dimorphism that is not associated with measures of mating system or operational sex ratio. We know that human body size has become progressively smaller in the past 50,000 years—probably linked with the rise of technology that has accompanied human evolution and progressively distanced the human body from the harshest demands of its environment (Ruff, 2000, 2006). Tall human populations have a higher degree of stature dimorphism than do short ones, so the more recent prehistoric reduction in stature and size has probably also been associated with a reduction in sexual dimorphism in the tallest populations. Attempts to link population differences in the degree of sexual dimorphism in body size to mating systems in modern humans have thus far failed. Wolfe and Gray (1982) compared size dimorphism to the number of polygynous marriages and also the level of male parental investment in traditional societies, but no cross-population correlations could be found.

There is reason to believe that sexual dimorphism in humans may be due to factors other than sexual selection. One feature of human sexual dimorphism that we can link directly to the division of labor is stored body fat in the hips, thighs, and buttocks of women that is accessed during lactation (Rebuffé-Scrive et al., 1985; Lancaster, 1986; P. Brown, 1991; Sohlstrom and Forsum, 1995). Sex differences in body fat are more pronounced in humans than in stature or body size. Humans show only mild sexual dimorphism in variables like stature: males are only 5–9% taller than females and 11–20% larger in body mass (P. Brown, 1991; Cachel, 2006). Although males and females may carry proportionally similar pounds of fat, the sites of deposition are very different. Males deposit most of their fat around the midline. In contrast, women have fatter skinfold thicknesses at four sites on the trunk and five sites on the arms and legs, with the thighs being 45% fatter than males (P. Brown, 1991). The sites are readily accessed for energy via the hormones of lactation (Rebuffé-Scrive et al., 1985). Unlike the vast majority of nonhuman primate females, who must support lactation through their feeding effort, human females frequently reduce work effort during pregnancy and lactation (Altmann, 1980; Hurtado et al., 1992; Lancaster et al., 2000). They can rely on men for energetic support, as well as other kinswomen who are not lactating. These extra sources of energetic support, coupled with the critical energetic demands of human postnatal brain growth, may be why human women store body fat, presumably at the expense of physiological efficiency for foraging (for example, in terms of thermal regulation and the costs of carrying nonproductive mass).

Similarly, the greater muscularity and body mass of men need not be a selected adaptation for male-male violence. Comparing the rates of violence in chimpanzees and humans gives support to the idea that male-male physical competition over females within the social group is vastly less important in humans. Wrangham and his associates compared the rates of lethal violence between chimpanzees and human subsistence societies and found them similar (Wrangham, Wilson, and Muller, 2006). In sharp contrast, chimpanzees had rates of within-group nonlethal physical aggression between two and three orders of magnitude higher than humans. Although preliminary data, these results indicate a major reduction in male-male violence within human groups and supports Boehm's hypothesis on the evolution of human egalitarianism (Boehm, 1999).

It may be that selection on the hormonal regulation of human male growth, in terms of both stature and mass, is more the result of productive efficiency in acquiring nutrients to support reproduction and dependent children than of direct male competition. The effects of longitude and rainfall on male body size and form, well known to physical anthropologists (Bergman's rule that body size increases with latitude and Allen's rule that limbs are shorter in relation to trunk length in colder climates), provide some partial support for this view. Among the Ache, hunting efficiency peaks at the average male body height, with shorter and taller men having lower hourly return rates (Hill et al., n.d.). The full impact of the workload required to support multiple young of differing ages is modeled in Figure 5.3, illustrating the cumulative numbers of dependent infants, children, juveniles, and adolescents for Ache and Ju/'hoansi mothers (Gurven and Walker, 2006). The load peaks between the ages of 35 and 40 years but remains substantial well into the mother's fifties, especially in the higher-fertility Ache. Leonard (2003) compares the total energy expenditures and physical activity levels of adult men and women in 11 non-Western groups. For both men and women, rural subsistence groups have the highest expenditure levels. Men of rural subsistence groups have average energy expenditure levels of 12.5 MJ/day, and women, 10.0. When physical activity is compared, men have values of 2.02 PALs (a measure of physical activity levels), compared to 1.83 for women. Furthermore, the declines in physical activity and energy expenditure associated with modernization are more pronounced in men than women. Testosterone is necessary to sustain the physical capacity in musculature, lung, and heart function for such high levels of male labor into the late fifties. Ellison and his associates have found that between the ages of 45 and 60 there is little difference between human populations in male testosterone (I) levels. Figure 5.4 illustrates that the proverbial drop in T levels with aging is found only in groups that start at a very high level, such as males in the United States, and that subsistence populations show little age-related decline compared to U.S. and European men (Ellison et al., 2002; Campbell, Gray, and and Ellison, 2007).

In summary, sexual dimorphism in men and women appears to have direct connections to the sexual division of labor. Human males show

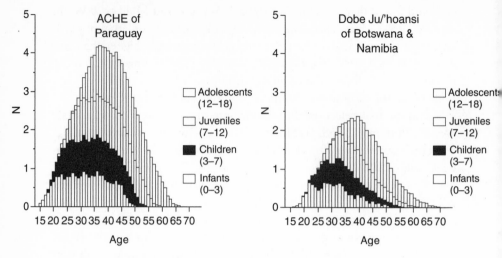

Figure 5.3 Number of dependents at different maternal ages for two forager groups: Ache and Ju'hoansi *(inset)*. After Gurven and Walker, 2006.

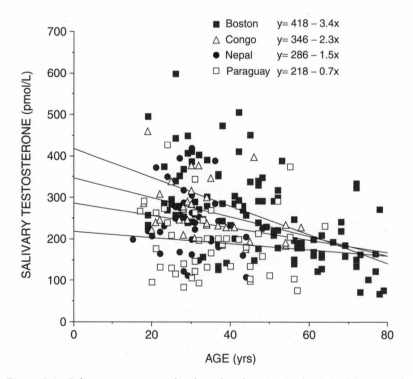

Figure 5.4 Salivary testosterone levels and male aging in the United States and three subsistence societies. Courtesy of P. Ellison.

adaptations for sustained and heavy work effort that facilitates women's care of children and children's learning, whereas women's sexually dimorphic characteristics are based on a lowered work effort during the reproductive years and specialized fat storage depots that support the growth of the human brain.

THE ONTOGENY OF COMPLEMENTARITY
IN THE SEXUAL DIVISION OF LABOR

The division of labor allows the two sexes to divide up the productive work so that each can master a single set of complementary skills. Even during infancy, humans display sex-differentiated behavior, and by middle childhood these divergences have become clearly pronounced (Geary, 1998). Aversion to physical proximity with the opposite sex rapidly increases during middle childhood, going from a tolerance of one foot at age 5 to seven feet by age 13. Puberty has the opposite effect. This social segregation is most pronounced in situations where children form their own social groups and are not monitored by adults (Geary, 1998). Preferences for specific activities also exist so that girls may prefer pretend play and boys rough-and-tumble play, which leads to spontaneous segregation by sex (Bjorklund and Pellegrini, 2002). Just as important is the strong preference during middle childhood to associate with same-sex peers, and time spent in social play peaks at this age (Pellegrini and Smith, 1998). The pattern described here suggests that juveniles are preparing for adult behaviors that are not directly related to finding a mate or reproduction. The physiological and behavioral preparations to find and attract mates do not begin in earnest until puberty (Lancaster, 1986). In fact, they would be major distractions to the acquisition and practice of adult foraging skills. B. Campbell (2006) has developed a model in which the physiological event of adrenarche plays an important role in launching a number of years of sex-specific learning. Adrenarche marks the start of middle childhood and is characterized by slowed growth and adrenal production of the steroid dehydroepiandrosterone sulfate (DHEAS) in both sexes. DHEAS levels climb steeply during middle childhood, from about $2\,\mu mol/L$ to nearly $8\,\mu mol/L$ by age 14. DHEAS is converted to both testosterone and estradiol by peripheral tissues and is an important source of these hormones before gonadal production at puberty begins. What is particularly interesting is that the level of DHEAS in the two sexes is identical, and it is not until after puberty that males begin to maintain higher levels than females for the rest of their adult lives.

Middle childhood, initiated by adrenarche, is uniquely related to human life history adaptations by serving as a time for juveniles to engage in activities related to sex-specific skills necessary for entry into the adult human ecological and social niche formed by the division of labor. The brain has nearly attained its adult size, suggesting that during middle childhood time and energy are invested in brain maturation or programming rather than in growth. Bock and Johnson (2004) in a study of village children's activities in the Okavango Delta in Botswana found that children play at productive tasks such as grinding maize or catching fish; but as they gain competency in a task, they spend less time playing at it. For example, girls will pretend to grind maize into meal, but as they become more proficient, they spend less time pretending and more time doing. The evolution of the life history phase of middle childhood establishes the time in life when the sex-specific physical, cognitive, and social skills necessary to the adult productive world are established in a developmental context free of the demands of self-feeding and the distractions of reproductive hormones and mate seeking characteristic of adolescence.

Conclusion

Each facet of the human adaptive complex and its component parts is directly related to, and dependent on, the sexual division of labor between men and women and intergenerational transfers across the life cycle. Unlike other species, the very entrance of *Homo sapiens* into its adaptive niche is based on the specialization of the two sexes in procuring food from two different levels of the food chain, a separate allocation of risk taking and care of young, the feeding of juveniles, and food sharing that widens the diet of both sexes and reduces variability in the food supply. Consonant with the investment in the acquisition of specialized and complementary skill sets by the two sexes is a long life course characterized by a long period of dependence in the first part of life and an even longer adulthood characterized by high production and transfers to children and grandchildren.

It should not be surprising, then, that we should find sex-specific adaptations in physiology and biology that reflect this mutual interdependence and stabilize such critical relationships. In this chapter, we have focused on four aspects of these adaptations: the stabilizing effect of offspring on (1) parental relationships and on (2) mate choice preferences; (3) complementarity in sexual dimorphism; and (4) the evolution of middle childhood, a uniquely human stage of development in which sex-specific skills are acquired free of the burden of growth and self-feeding and the distractions of pubertal hormones and adolescent mate seeking.

The relationships among skill investments, the long human life span, and intergenerational transfers have been the focus of several other papers by the authors and their collaborators (Kaplan, 1997; Kaplan and Lancaster, 2000; Kaplan et al., 2000. Lancaster et al., 2000; Kaplan and Robson, 2002; Robson and Kaplan, 2003; Gurven and Kaplan, 2006). We expect a similar sensitivity to life stage in the hormonal regulation of physiology and behavior and that the optimal mix of investments in survival, maintenance, and reproductive effort will reflect selection to reach the grandparental phase of the human life cycle. This could be the subject of another chapter.

Human evolutionary ecology identifies differences between human groups as largely environmentally induced variation in the expression of basically similar genotypes. This view sees facultative responses to environmental differences as the essential human adaptation to socioecological variation and change. Much of the discussion of this chapter has focused on the original human adaptive complex. However, we know there have been major socioecological changes since humans claimed some 50,000 years ago the top of the food chain as their own by means of a restructuring of the human life course and the relationship between the sexes (Lancaster and Kaplan, 1992). Paramount among these are (1) the development of extrasomatic wealth and the monopolization thereof that led first to variance in male quality and then to social stratification, (2) the growth of technology that reduces the value of sexual dimorphism in the division of labor by making male and female labor and skill sets substitutable, and (3) a reduction in fertility and family size so that active reproduction and child care compose increasingly smaller portions of the total human life span. Certainly these trends are not universal, but they are increasingly felt in world societies.

In spite of these major shifts in human experience, critical features of human reproductive partnerships such as marriage, sexual commitment, and the dedication of adults to rear offspring continue. In the Western world, monogamy has lost many of its variants such as concubinage and minor secondary families (Betzig, 1986, 1992, 1993, 1995; Betzig and Weber, 1993). Parental investment strategies have also shifted toward greater focus on the endowment of embodied capital through education and training rather than inheritance (Lancaster, 1997; Kaplan and Lancaster, 2000; Lancaster et al., 2000). The embodiment of capital appears to be much more labor-intensive than is immediately obvious. Although Western societies have focused on universal education provided at all levels, school attendance does not ensure education. Parental involvement is now recognized as a key feature in child educational development, and parental time and commitment are not readily shared between families (Kaplan, Lancaster, and Anderson, 1998; Anderson, Kaplan, and Lancaster, 1999, 2001, 2007; Anderson, 2000). While the socioecological context of the original evolution of the human adaptive complex has radically altered, the life history and reproductive patterns of the division of labor (marriage partnerships and parental investment) are still in place, along with the endocrinology and biology that support them.

Acknowledgments

The authors wish to thank the many people who have generously given their time to read and critique this chapter, to supply needed references, to permit reproduction of figures from their publications, and to offer personal communications regarding their unpublished research. We apologize to those whose names have been omitted accidentally. The following is heartwarming documentation of the cooperative and collegial nature of the modern scientific endeavor: Jeanne Altmann, Rosalind Arden, Ben Campbell, Colleen Costin, Peter Ellison, David Geary, Linda Gottfredson, Peter Gray, Michael Gurven, Phyllis Lee, William Leonard, Tanya Mueller, Pablo Nepomnaschy, Beverly Strassman, David Tracer, Robert Walker, Claudia Valeggia, and Virginia Vitzthum.

Kaplan acknowledges support from the following research grants: the National Science Foundation (BCS-0422690) and the National Institutes of Health / National Institute on Aging (#1R01AG024119-01).

Social Relationships among Nonhuman Animals

CHAPTER SIX

The Endocrinology of Social Relationships in Rodents

C. Sue Carter, Ericka Boone, Angela J. Grippo,
Michael Ruscio, and Karen L. Bales

Introduction

Social interactions can be either negative, including agonistic or aggressive behavior, or positive, including social proximity and engagement, the formation of social bonds, and the care of young. Positive relationships are of particular importance to both individual survival and reproductive success or fitness. In addition, social behaviors can be either nonselective (or indiscriminate) or selective. Selective behaviors necessary for social bonds are directed toward specific individuals and therefore require individual recognition and social memory (Insel and Fernald, 2004). The attachment between a mother and her infant is the best-documented form of social bonding in mammals (Carter and Keverne, 2002). However, adults also can form social bonds that may, in some cases, last for a lifetime. The birth process and the postpartum period have served as prototypes for understanding the endocrinology of social bonding, and there is increasing evidence for parallels between the mechanisms for social bonding between infants and adults and between adults and other adults.

The endocrine changes responsible for sociality are most readily appreciated in the context of their functions, which include reproduction and coping with the stress of life. Sociality and especially social bonds can provide a sense of safety, can reduce anxiety, and may both directly and indirectly influence physical and mental health. The proximate causes of social behaviors also are influenced by sex differences in reproductive demands and may differ in males and females (Palanza, 2001;

Morgan, Schulkin, and Pfaff, 2004; Carter, 2007). Mammalian reproductive and coping strategies involve either active forms of response, including approach or avoidance, or passive responses, including freezing or immobilization. Although there are many exceptions, males tend to show more active behaviors, while the response of females tends to be more passive (Koolhaas et al., 2001; Palanza, 2001). These patterns are consistent with sex differences in reproductive functions, including the necessity for female mammals to assume immobile postures, at least briefly, during sexual behavior, birth, and nursing. Such patterns also are consistent with the actions of neuropeptides, including oxytocin (OT), vasopressin, and corticotropin-releasing hormone (CRF).

Social Systems as Context for Social Behavior

In biology it is common to categorize vertebrate species based on mating systems. In the absence of definitive evidence, this is usually based on observations of living arrangements, such as nest occupancy, and the number of sexual partners presumed to be available. The most common mating system in mammals is polygamy (many mates) or, more specifically, polygyny (many female mates). The less common alternative is monogamy (one mate) (Kleiman, 1977), while polyandry (many male mates) is rarest. These categories are far from precise, tend to emphasize male behavioral patterns, and are only infrequently based on data across the life span of an individual.

More useful to understanding social behavior, especially at the level of its behavioral endocrinology, have been species classifications based on social systems. *Social monogamy,* or *cooperative breeding,* is now commonly used to refer to a social system or social organization formed around a male and female pair that, in some cases, also involves cohabitation with extended families. In these cases, pair-bonds and other forms of stable relationships may endure beyond the periods of sexual interaction, sometimes lasting for a lifetime. In the absence of other partners, or when mate guarding is successful, social monogamy can promote sexual exclusivity; however, based on DNA fingerprint studies, *social monogamy* and *sexual monogamy* are not synonymous. Determinations of paternity have revealed that sexual exclusivity is not necessarily a reliable trait of species that live in pairs; thus sexual monogamy and social monogamy are not synonymous, and it is social systems that seem most predictive of neuroendocrine mechanisms for social behavior.

Male parental behavior is typically not seen in polygynous species. In contrast, in socially monogamous or cooperatively breeding species, both parents and the extended family may exhibit parental behavior. The proximate causes of parental responses by reproductively naive animals toward infants or younger offspring, also known as *alloparental behavior,* have been most extensively studied in socially monogamous species.

Among the diverse mammalian taxa that share the traits of social monogamy are canids, several New World primates, including tamarins and marmosets, titi monkeys, and even a few small rodents, including prairie voles. Prairie voles can easily be reared and studied under laboratory conditions and thus have provided a valuable model for the analysis of the endocrinology of social behaviors, including pair-bonding, male parental behavior, and alloparenting (Carter, DeVries, and Getz, 1995; Carter and Roberts, 1997). For these reasons, many of the examples presented here come from our work with prairie voles.

Endocrinology and Social Behavior

Mammalian neuroendocrine systems utilize hormones and related neurochemicals that are necessary to meet the novel demands of mammalian reproduction, as well as stress management. Among the hormonal elements implicated directly or indirectly in social behaviors are gonadal steroids, including estrogen, progesterone, and testosterone. Also implicated in social behavior are hormones of the hypothalamic-pituitary-adrenal (HPA) axis, including CRF produced primarily in the nervous system, and steroids of adrenal origin, including corticosterone or cortisol. The extensive functions of steroids in reproduction or reproductive behaviors are reviewed elsewhere (Pfaff, Frohlich, and Morgan, 2002; Becker et al., 2005) and will be highlighted here only as they relate to research on sociality.

Of particular importance to social behaviors are centrally active neuropeptides including oxytocin, vasopressin, and CRF. The genes responsible for these ancient hormones are believed to have evolved initially to regulate cellular processes such as water balance and homeostasis (Chang and Hsu, 2004), with secondary roles in reproduction. However, in modern mammals neuropeptides (often originating primarily in the brain) regulate birth and lactation, as well as related social and sexual behaviors, defensive behaviors, and stress-coping responses.

Most behavioral studies of the actions of hormones are based on motor responses, such as approach to or contact with a partner or infant, and sometimes on specific immobility postures such as lordosis or kyphosis. Hormones also can mediate sensory perceptions as well as emotional and motivational states and reactivity. The latter may be indexed by changes in other hormonal systems, including changes in hormones of the HPA axis (most often using blood levels of corticosterone and cortisol) or autonomic responses.

Oxytocin and Vasopressin

At the center of mammalian birth, lactation, and several features of sociality is the brain peptide known as oxytocin (from the Greek for "swift birth"). The oxytocin molecule consists of nine amino acids, configured as a ring and tail. Oxytocin is a small neuropeptide, primarily of brain origin. Two hypothalamic nuclei, the paraventricular nucleus (PVN) and supraoptic nucleus (SON), are the main sources of oxytocin. Oxytocin from the PVN and SON is transported to the posterior pituitary (outside of the blood-brain barrier), where it is released into the general circulation. Significant amounts of oxytocin from the PVN are released into, and act directly on, the brain. In addition, smaller amounts of oxytocin are synthesized in other tissues, including the reproductive organs and immune system. Oxytocin induces uterine contractions and also the muscle contractions necessary for milk ejection.

Oxytocin has a broad range of behavioral effects at least in part through actions on brain receptors. For example, oxytocin can facilitate social approach and contact (Witt, Winslow, and Insel, 1992) and social recognition (Winslow and Insel, 2004) and is required for social bond formation (Williams et al., 1994; Cho et al., 1999). Oxytocin also can be a powerful anxiety-reducing (anxiolytic) agent (Uvnäs-Moberg, 1998). Oxytocin is capable of downregulating defensive reactions toward either unfamiliar adults or infants (Carter, 1998). The ability to remain physically immobile without fear or anxiety is critical to both female sexual behavior and maternal behavior (Porges, 2003b). Oxytocin also regulates the autonomic nervous system and specifically may upregulate the activity of the parasympathetic system, with potentially beneficial consequences for social behaviors and emotional regulation (Porges, 2003a). There is even recent evidence from human studies that oxytocin can increase the capacity of individuals to show "trust" (Kosfeld et al., 2005). Thus, systems reliant on oxytocin may be part of a broad neural system capable of mediating the protective effects of positive social interactions and social support. Oxytocin also has the capacity to synergize with dopamine, through which it may play a role in reinforcing social bonds (Aragona et al., 2003; Aragona et al., 2006) and other forms of social interactions (Fleming, 2005).

A second neuropeptide of central importance to social behavior is vasopressin (De Vries and Simerly, 2002). Vasopressin is similar in structure to oxytocin, consisting of nine amino acids, of which seven are shared with oxytocin and two differ. The structural similarities and differences between vasopressin and oxytocin allow each hormone to interact with the other receptors. Both molecules can influence behavior, although in some cases the functional effects are in opposite directions, possibly in part because these neuropeptides may be natural antagonists for each other. Vasopressin also is primarily of hypothalamic origin, although usually not made in the same cells that synthesize oxytocin. As with oxytocin, one of the basic actions of vasopressin may be to allow individuals to interact without fear. However, in contrast to the immobilizing actions of oxytocin, centrally active vasopressin is generally associated with mobility and the activation of the sympathoadrenal systems that support active motor behavior. Moderate levels of vasopressin may reduce anxiety, permitting animals to engage in social behavior with unfamiliar conspecifics. However, it is also possible—especially at higher doses—that vasopressin might increase anxiety and defensive aggression.

Receptors for both oxytocin and vasopressin are localized in areas of the nervous system, and especially in brain stem regions, that play a role in reproductive, social, and adaptive behaviors (Witt, 1997; Young, 1999; Gimpl and Fahrenholz, 2001; Insel and Young, 2001) and in the regulation of the HPA and autonomic nervous system (Sawchenko and Swanson, 1985; Neumann et al., 2001; Porges, 2001). The oxytocin receptor (OTR) belongs to the G protein-coupled receptor (GPCR) superfamily (Kimura et al., 2003; Zingg and Laporte, 2003). There is only one known form of the OTR. OTRs are found throughout brain regions that have been implicated in social behavior and reactions to stressors and also in many other tissues including the uterus and breast. Oxytocin binds preferentially to the OTR. The expression of the OTR is increased by other reproductive hormones including estrogen. Expression of, or binding to, the OTR also can be influenced by many factors, including various steroid hormones, which could in turn have direct consequences for social behavior.

Vasopressin has at least three receptor subtypes (De Vries and Simerly, 2002; Ring, 2005) and can bind to the OTR. Among the vasopressin receptors, the V1a receptor (V1aR), also a GPCR, is found in greatest abundance in the brain and other neural tissue. The V1aR is associated with cardiovascular functions as well as behavior. Vasopressin, acting at the V1aR, plays a central role in modulating social bonding (Winslow et al., 1993) and facilitating positive social behaviors (Cho et al., 1999). Vasopressin also has a role in modulating nonsocial behaviors such as emotionality. Stress-induced increases in vasopressin can act to amplify the actions of CRF (DeBold et al., 1984). Vasopressin also is capable of increasing certain forms of repetitive and defensive behaviors (Winslow et al., 1993; Ferris, 2000; Albers, Karom, and Smith, 2002; Bielsky et al., 2005; Ebner, Wotjak, et al., 2005).

Variations in the V1a receptor may have profound behavioral and emotional consequences. V1aR knockout (V1aRKO) male (but not female) mice show impaired social recognition. Reexpressing V1aR via site-specific injections of V1aR viral vectors resulted in a marked increase in V1aR-related functions. Viral vector–mediated increases in V1aR have even been shown to increase social behavior among other socially promiscuous mammals, including increasing the capacity to be social (that is, affiliative behaviors) and to form pair-bonds (Lim and Young, 2004; Lim et al., 2004). Polymorphisms within the 5' promoter region of the V1aR, in the form of tandem repeats, are associated in prairie voles with increased levels of V1a receptor binding in neural areas critical to male pair-bonding behavior (Hammock and Young, 2005). However, a comparative study of a number of mammalian species (Fink et al., 2006) found that these tandem repeats were not unique to monogamous species. This suggests that different mechanisms may be operating in different taxonomic groups to produce the convergent occurrence of the traits associated with social monogamy.

Sex Differences in the Effects of Hormones and Stressful Experiences on Social Behavior

Sex differences in the behavioral effects of hormones such as vasopressin and oxytocin may be due to sex differences in the synthesis and availability of these peptides. There is some evidence from both animal and human studies that females are less dependent than males on vasopressin (Table 6.1); in addition, vasopressin is androgen dependent in the central nervous system (CNS) (Winslow et al., 1993; Bielsky, Hu, and Young, 2005). In general, vasopressin seems to be of particular importance to social behavior and stress management in males (Carter, 2007). Sex differences in the production of, and effects of, vasopressin are most apparent in certain brain regions including the extended amygdala and lateral septum (De Vries and Simerly, 2002). Projections from sexually dimorphic areas to less dimorphic regions might influence many functions including social behaviors and reactions to stressors, helping to generate sex differences in both physiology and behavior.

In humans, elevations in central vasopressin, induced by intranasal infusions, exert behavioral consequences possibly by increasing behavioral

Table 6.1 Summary of sex differences and treatment effects on social behaviors in the monogamous prairie vole.

Social Behavior/Treatment	Male	Female	Reference
Alloparental behavior/baseline	High	Low-moderate	Bales, Pfeifer, Carter, 2004; Lonstein and De Vries, 1999b; Roberts et al., 1998
Alloparental behavior/stress	Increased	No change	Bales, Lewis-Reese, et al., 2007
Alloparental behavior/developmental exposure to OTA	Decreased	No change	Bales et al., 2004
Parental behavior/baseline	Higher licking and grooming/exploratory behavior	More time arched over pups	Lonstein and De Vries 1999a
Pair-bonding behavior/baseline	Longer to form pair-bond	Shorter to form pair-bond	DeVries and Carter, 1999; Williams, Catania, and Carter, 1992
Pair-bonding behavior/stress	Facilitation of bond formation	Delay of bond formation	DeVries et al., 1995, 1996
Pair-bonding behavior/adrenalectomy	Delay of bond formation	Facilitation of bond formation	DeVries et al., 1996
Pair-bonding behavior/CRF	Facilitation of bond formation	Not tested	DeVries et al., 2002
Pair-bonding behavior/developmental exposure to OT	Facilitation of bond formation	Facilitation of bond formation but at higher dosage than required in males	Bales and Carter, n.d.
Same-sex aggression/baseline	Higher	Lower	Bales and Carter, 2003a
Same-sex aggression/developmental exposure to OT	No change	Higher (after 4-hr exposure to a male)	Bales and Carter, 2003a
Same-sex aggression/developmental exposure to VP	Higher	No change (although reduced by VP antagonist)	Stribley and Carter, 1999

CRF = corticotropin-releasing factor; OT = oxytocin; OTA = oxytocin antagonist; VP = vasopressin.

or emotional reactivity to normally irrelevant stimuli or nonthreatening social stimuli. Consistent with the animal literature, the behavioral effects of exogenous vasopressin were sexually dimorphic in human subjects. Men given additional vasopressin showed increases in corregator muscle activity, a component of frowning, and also rated neutral facial expression as more "unfriendly." In contrast, females given vasopressin smiled more and reported more positive, affiliative responses to unfamiliar neutral faces (Thompson et al., 2004).

In contrast, estrogen can promote the synthesis of oxytocin and the expression of OTRs (Richard and Zingg, 1990; Zingg et al., 1995; Zingg and Laporte, 2003). Since estrogen is generally higher in females, this might be expected to have sexually dimorphic consequences for physiological and behavioral functions. Several of the functions normally served by vasopressin in males may rely on oxytocin in females, which might explain sex differences not only in social behaviors but also in stress management (Carter, 2007). A lack of dependence on vasopressin also could serve to protect females from overreacting to stimuli, such as those encountered during social interactions. Conversely, social behaviors that are more strongly dependent on androgens and thus on vasopressin, including defensive or territorial behaviors, may be of particular importance in males.

Stress and hormones of the HPA axis react to social experiences and also have effects on social behavior and anxiety. Adrenal hormones or stressful experiences increase oxytocin receptor binding in the amygdala (Liberzon and Young, 1997); increased oxytocinergic activity in this region might in turn reduce fear responses, possibly permitting social behavior during periods of stress. In prairie voles, basal or nonstressed levels of corticosterone tend to be high, and the HPA axis in prairie voles also is very reactive to stress—especially when compared to domestic rats and nonmonogamous voles (Carter et al., 1995).

Stressful experiences, mediated through hormones of the HPA axis, modulate social bonding (DeVries et al., 1995, 1996; DeVries et al., 2002). The effects of stress on pair-bond formation also differ in males and females. In females, exposure to a stressor inhibited female-male pair-bonding (DeVries et al., 1996). However, in males exposed to a brief stressor or treated with stress hormones, pair-bond formation was facilitated. Centrally administered CRF, which stimulates the HPA axis, also increased pair-bonding in male prairie voles (DeVries et al., 2002). However, the relationship between these hormones and behavior is complex and may differ according to the amount or duration of exposure to hormones of the HPA axis. Moderate amounts of activation and "stress" hormones might drive a tendency toward increased affiliation, while very high levels of these same factors might inhibit sociality or increase defensive aggression. Thus, once again, biological and behavioral contexts are critical to understanding the consequences of different hormonal experiences.

Parental Behavior

In the immediate postpartum period, females of most mammalian species show high levels of maternal behavior. This literature is reviewed in more depth elsewhere (Carter and Keverne, 2002; Numan and Insel, 2003). Maternal behavior is facilitated by oxytocin in several species including rats (Pedersen and Prange, 1979), sheep (Kendrick, Keverne, and Baldwin, 1987), and prairie voles (Bales and Carter, 2003a; Olazabal and Young, 2006).

Under natural conditions, the infant present in the immediate postpartum period would be the female's own offspring, and thus care directed toward that infant would be reproductively adaptive. Male parental behavior is also most likely to be exhibited toward a male's own offspring or at least toward the offspring of his sexual partner. Recent evidence suggests that socially monogamous males, while not experiencing the drastic hormonal changes associated with pregnancy, may experience hormonal changes in association with their mates' pregnancies (Ziegler, Washabaugh, and Snowdon, 2004). Parental behavior in males appears to be regulated somewhat differently from maternal behavior (Wang et al., 1998; Bester-Meredith, Young, and Marler, 1999), and it has been suggested that vasopressin has a primary role in paternal behavior. For example, in prairie voles, injections of vasopressin in the lateral septum increase paternal behavior (Wang, Ferris, and De Vries, 1994). However, oxytocin also may be involved in male parental care (Bales, Kim, et al., 2004). Experiments in prairie voles suggested that it was necessary to block both OTR and V1a receptors to inhibit all aspects of male parental care. In male marmosets (a primate species that displays biparental care), V1a receptor expression dramatically increased in fathers compared to pair-bonded nonfathers; V1aR levels in fathers slowly decreased again with the increasing age of their infants (Kozorovitskiy et al., 2006).

Infant care also may be exhibited by animals other than the biological parents, including older siblings or other members of a family. Animals that care for offspring other than their own may be termed *alloparents* or "helpers at the nest." Alloparental behavior, which is easily measured in the laboratory, is sometimes used as a model for understanding the biological basis of parental behavior in animals that are reproductively naive or in animals that have not been primed by the hormonal events of pregnancy (Carter and Roberts, 1997). The prairie vole model is of interest because reproductively naive males of this species are more reliably parental than naive females (Roberts et al., 1998; Lonstein and De Vries, 1999b; Bales, Pfeifer, and Carter, 2004). Because vasopressin is a peptide hormone and therefore relatively fast acting (at least in comparison to steroid hormones), this peptide may facilitate the very high levels of spontaneous infant care in naive male voles. Adding vasopressin to the endogenous hormonal cocktail for parental behavior might allow males to overcome fear or anxiety and attend to infants even under somewhat stressful conditions. Postpartum females, whose systems are flooded with oxytocin, could be using oxytocin in a similar manner.

Alloparenting behavior appears to be physiologically related to responses of the alloparent in the face of other forms of stressors. Data from prairie voles also support this hypothesis. For example, exposure to a nonsocial stressor, such as swimming, also has sexually dimorphic effects on parental behavior. Following a brief swim, stress parental behavior increased in males but not in females (Bales et al., 2006). The infant also may have the capacity to regulate the emotional responses of its caregiver. Males may use contact with the infant, presumably via hormonal processes, to downregulate HPA responses to a swim stress. Evidence for this hypothesis comes from the finding that corticosterone levels following exposure to the infant were inversely correlated with time spent licking and grooming the infant. Females did not display this inverse correlation (Bales, Kramer, et al., 2006).

The Absence of Social Interaction

Social isolation is a potent stressor, especially for highly social mammals such as prairie voles. In nature, female prairie voles rarely remain in social isolation, although some males may live alone. Social isolation has potent inhibitory effects on various behavioral and autonomic measures, including reductions in sucrose preference (considered a measure of anhedonia), exploration of a novel environment (used to index anxiety and fear), as well as increases in heart rate and inhibition of the vagal (parasympathetic) control of the heart (Grippo, Lamb, Carter, and Porges, 2007; Grippo, Cushing, and Carter, 2007). Females' central vasopressinergic responses, especially during the juvenile period, may be more reactive or vulnerable to the effects of social isolation (Ruscio et al., 2007). Sexually dimorphic responses to stressors also may be due to sex differences in the functions of vasopressin and oxytocin. While oxytocin and vasopressin responses may be especially likely to be sexually dimorphic in adulthood, in juvenile voles, both males and females display isolation-induced increases in HPA axis hormones (CRF and corticosterone) (Ruscio et al., 2007) and an associated reduction in the birth of new cells (including gliogenesis and neurogenesis) (Ruscio et al., 2008). In adulthood, social isolation followed by a social stressor was associated with increased oxytocin as well as vasopressin, CRF, adrenocorticotropic hormone (ACTH), and corticosterone (Grippo, Cushing, and Carter, 2007). It is possible that the release of oxytocin under conditions of isolation / social stress may help to protect animals against the emotional effects of stress. This effect may be of particular importance in females, who may be especially vulnerable to the consequences of isolation.

Developmental Effects of Social Experience

Genetic differences, including species and sex differences, are one source of variance in social behavior (Lim et al., 2004; Hammock and Young, 2005). As described above, the proximate processes necessary for social behavior are sexually dimorphic and also species typical. However, genetic differences (at least those known at present) are not sufficient to explain most individual variations in social behaviors (Carter, 2007). The mechanisms underlying social behaviors also tend to be plastic and within the life span of an individual can be modified by various experiences including those associated with social interactions. Positive or negative social experiences, especially in early life, program the tendency of individuals to show social behaviors and can alter individual patterns of behavioral responses to challenge. Changes in gene expression for hormones and their receptors may permit "retuning" of behavioral responses, in turn helping animals to more effectively adapt to future challenge.

Among the behaviors and neural systems that are modified by early experience are those necessary for pair-bond formation and alloparental behavior (Bales and Carter, 2003a, 2003b; Bales, Abdelnabi, et al., 2004; Bales et al., 2006), as well as other forms of social behavior and the management of stress (Levine, 2001; Weaver, Diorio, et al., 2004; Meaney and Szyf, 2005). Studies in rats have revealed that at least some of the effects of early experience can be reversed by epigenetic manipulations, including drugs capable of altering gene methylation and thus gene expression in later life (Weaver, Meaney, and Szyf, 2006).

As another example of the long-lasting effects of early experience, when prairie vole families are disturbed by brief handling within the first weeks of life, tendencies toward sociality and exploratory behaviors are changed in later life (Boone et al., 2006; Bales, Lewis-Reese, et al., 2007). Prairie voles receiving daily handling in the first week of life also had higher central levels of oxytocin (Carter et al., 2003). These effects are likely to be mediated by changes in maternal stimulation of the young during the postnatal period. Either reduced handling or, alternatively, excess stimulation holds the potential to disrupt or reprogram the development of neuroendocrine systems necessary for the formation of pair-bonds or the ability to show parental behavior (Boone et al., 2006). In one experiment, we found that the effects of reduced stimulation in early life were sexually dimorphic, resulting in reduced alloparenting behavior in males (but not in females) and reduced pair-bond formation in females (but not in males), although measures of anxiety increased in both sexes (Bales, Lewis-Reese, et al., 2007).

In early handling studies, observed effects may be mediated, at least partially, by enhanced parental stimulation of the pups. In female (but not male) rats the oxytocin receptor was upregulated by maternal licking and grooming. In contrast, in male (but not female) rats the effects of maternal stimulation were seen as increases in the expression and binding of vasopressin (V1a) receptor (Champagne et al., 2001; Francis, Young, et al., 2002; Champagne et al., 2003). These data support the general hypothesis that in females oxytocin may be a central component involved in the mediation of the long-lasting effects of early experience, while in males vasopressin may be of particular importance to changes in social behaviors.

Developmental Programming Mediated by Early Peptide Manipulations

Taken together the results from these studies suggest that social experiences are powerful regulators of endogenous peptides, including oxytocin and vasopressin, across the life span. The same peptides, administered during early life, appear to be capable of programming individual differences in sociality. Manipulations of either endogenous or exogenous oxytocin have long-lasting consequences for brain development and social behavior. Again using the prairie vole model, we have observed that a single exogenous oxytocin injection on the first day of life enhanced subsequent social behaviors including the tendency to show alloparental behavior or to form pair-bonds. However, neonatal exposure to either higher doses of oxytocin or blocking of oxytocin through exposure to an antagonist (OTA) may be associated in later life with reductions in social behaviors. One mechanism through which early endocrine experience could have lifelong consequences for behavior is through changes in the expression of neuropeptides and their receptors. For example, exposure to a single treatment with oxytocin on the first day of life upregulated later synthesis of oxytocin in female prairie voles (Yamamoto et al., 2004). This same treatment downregulated the subsequent expression of the vasopressin (V1a) receptor in females (but not males) in several brain regions including the extended amygdala (Bales, Plotsky, et al., 2007). These data are contrary to the assumption that sex differences are based primarily on genetics or genetically regulated gonadal hormones. Rather, they support the hypothesis that early experiences, possibly mediated by changes in central neuropeptides, can play a major role in the later expression of peptide receptors. Changes in receptors in turn could influence the sensitivity of an individual to neuropeptides such as vasopressin across the life of the individual. Because of the central role of vasopressin in both defensive and social behaviors, developmental programming of this system, at the level of production of either the peptide or its receptor, would have broad consequences for later emotional regulation.

Exposure to exogenous oxytocin during neonatal life also can facilitate the subsequent tendency to form pair-bonds (Bales and Carter, 2003b) and may reduce behavioral and neuroendocrine overreactivity to a novel environment (Bales and Carter, n.d.). In addition, females given exogenous oxytocin showed decreases in vasopressin in the hypothalamic supraoptic nucleus in adulthood (Kramer et al., 2006). In males, early oxytocin exposure upregulated V1a receptors in the ventral pallidum, while downregulating receptors in the ventral pallidum in females (Bales, Plotsky, et al., 2007). Increases in V1a receptors in the ventral pallidum in males of nonmonogamous species, through the use of viral vectors, also can increase the tendency to form pair-bonds (Lim, Hammock, and Young, 2004). Oxytocin had similar dimorphic effects in both the lateral septum and posterior cingulate cortex (both areas involved in social behavior), increasing V1a receptors in males and decreasing them in females (Bales, Plotsky, et al., 2007).

In contrast to the effects of oxytocin, even brief neonatal exposure to an oxytocin receptor antagonist (OTA) can disrupt subsequent social behaviors including the tendency to form social bonds, to exhibit parental behaviors, and to manage anxiety or stress, with corresponding effects on peptide receptors and synthesis (Figure 6.1). For example, in males a single neonatal exposure to a dose of OTA resulted in reduced vasopressin (Yamamoto et al., 2004), inhibited alloparenting in males (Bales, Pfeifer, and Carter, 2004), and produced a reduction in vasopressin (V1a) receptor binding in the extended amygdala (Bales, Plotsky, et al.,

· 2007). The androgen dependence of hypothalamic vasopressin and the sexually dimorphic capacity of an OTA to downregulate both vasopressin receptors and vasopressin may help to explain the fact that exposure to an OTA was especially disruptive to male behavior. Although the inhibitory effects of neonatal OTA were most obvious in males, some reductions in V1a receptors also were present in females exposed to early OTA (Bales, Plotsky, et al., 2004). In addition, female prairie voles exposed postnatally to OTA were more reactive than normal in later life to social stimuli from a novel male (Kramer et al., 2006).

These same peptides also play a role in the regulation of the HPA axis and could directly or indirectly influence emotionality or social behaviors. In adulthood, oxytocin in particular is capable of downregulating the HPA axis, and this effect applies to both rodents and humans (Carter, 1998; Heinrichs et al., 2003; Carter, Bales, and Porges, 2005). Manipulations of hormones of the HPA axis also may regulate social behaviors, both in adulthood and during development (Carter and Roberts, 1997). It has been shown in rats that CRF receptors display lifelong

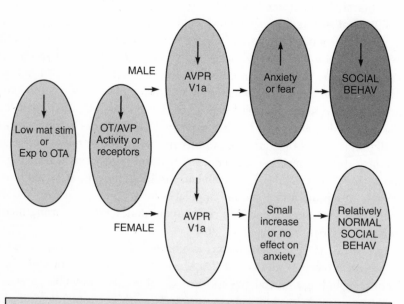

Females may be less reactive to early experience or OTA, possibly because they are less dependent on vasopressin and more dependent on oxytocin than males.

Figure 6.1   The effects of reduced early experience or blocking oxyfocin/anginine vasopressin (OT/AVP) receptors in early life on subsequent anxiety and social behavior are in part sexually dimorphic.

modifications as a function of early experiences, including those associated with maternal licking and grooming (Meaney and Szyf, 2005).
Thus, hormones of the HPA axis have developmental consequences for
various social and reproductive behaviors (Levine, 2001; Pedersen and
Boccia, 2002), possibly interacting with both gonadal steroids and neuropeptides to modulate the expression of the characteristics of social behaviors, as well as reproductive behaviors. The involvement of early
experience and the hormones of the HPA axis in the development of the
central nervous system may be of critical importance to the molding of
both species and individual differences in sociality.

The effects of early experience may also, in later life, be affected by exposure to neuropeptides. For example, developmentally induced deficits,
such as those associated with prenatal stressors including undernutrition
(Olausson, Uvnäs-Moberg, and Sohlstrom, 2003), certain drug treatments (Lee et al., 2005), or blocking the negative effects of postnatal
OTA (Bales and Carter, n.d.), can be reduced or reversed by later treatments with OT. However, the effects of vasopressin manipulations have
received less attention. Prairie vole exposure to exogenous vasopressin in
the first week of life showed higher levels of same-sex aggression in
adulthood (Stribley and Carter, 1999). These effects were seen in both
sexes but most obvious in males. Blocking the vasopressin receptor
neonatally was associated with very low levels of later aggression in both
sexes.

Summary

Taken together, these and other related findings (Carter, 1998, 2003;
Carter and Keverne, 2002) support the general hypothesis that social behavior is regulated in a species- and sex-dependent manner by both oxytocin and vasopressin. The effects of these neuropeptides are seen both in
adulthood and during development. Social experiences, including those
between adults and offspring as well as between two or more adults, are
mediated in part by long-lasting changes in neural systems that incorporate oxytocin and vasopressin.

Vasopressin is associated with increased mobilization and the management of reactivity to challenge. Under conditions of perceived danger, it
may be necessary to increase cardiovascular functions (including heart
rate and blood pressure), and vasopressin might play a role in this autonomic mobilization. Vasopressin is elevated following a social challenge
in male rats that show active, versus passive, patterns of coping (Ebner,
Wotjak, et al., 2005). Vasopressin has the capacity to antagonize the

functions of oxytocin, perhaps through direct effects at the oxytocin receptor as well as through the physiological effects of vasopressin on systems associated with behavioral and physiological mobilization (Engelmann et al., 1996; Pedersen and Boccia, 2006). However, oxytocin also may be released under conditions of perceived threat, especially if immobilization is an adaptive strategy. For example, in rats exposed to a swim stress, oxytocin released into the central nucleus of the amygdala modulates stress-coping behaviors (Ebner, Bosch, et al., 2005). Microinjections of oxytocin into the central nucleus of the amygdala also can inhibit aggressive behavior in rats (Consiglio et al., 2005).

Sex differences are a consistent finding throughout the research on social behavior. The behavioral and endocrine management of stressful experiences also tends to differ between the sexes. Sex differences in neuroendocrine reactivity to early social experiences, manipulations of neuropeptides, and reactions to various kinds of stressors are apparent at least as early as the first week of life. Some components of these sex differences in behavior appear to be based on sex differences in oxytocin and vasopressin. Males may be especially sensitive to developmental perturbations in part because males are more reliant on, or responsive to, vasopressin (Carter, 2007). In males, vasopressin synthesis (Yamamoto et al., 2004) and the expression of V1aRs (Bales, Plotsky, et al., 2007) can be modified by either social or endocrine experiences. Due to its dependence on estrogen, oxytocin may be of particular importance to female development and behavior. Oxytocin synthesis is also sensitive to neonatal peptide manipulations and to early experience. However, oxytocin has functions in both sexes, including the capacity to be neuroprotective and to reduce overreactivity to stressors (Takayanagi et al., 2005). Because oxytocin and vasopressin can bind to each other's receptors, it is likely that there is considerable cross-talk between the functions of these peptides, possibly allowing rapid and dynamic changes in behavior and emotional reactivity.

Although extrapolations to humans and other primates must be made with caution, the neural systems and chemicals responsible for social behavior depend on conserved neuroendocrine systems and thus may be highly relevant to behavior in other species. The capacity of neuropeptide systems, including their receptors, to undergo long-lasting functional modifications presents an epigenetic model that may help to explain the origins of traits and states. Adaptive changes in these systems, especially those mediated at the level of various peptides and relevant receptors, may play a role in individual differences in behavior. Understanding these systems also could offer insights into the development of pathological or maladaptive behaviors—especially those such as autism—that are highly sexually dimorphic (Carter, 2007).

Acknowledgments

We are grateful to our many colleagues who conducted the studies described here. We are also grateful for research support from the National Institutes of Child Health (PO1 38490 to C. S. C.; NRSA 08702 to K. L. B.); the National Institute of Mental Health (RO1 073022 to C. S. C. and K. L. B.; MH73233 NRSA to A. J. G.); the National Alliance for Autism Research (to C. S. C.); and the National Science Foundation (grant #0437523 to K. L. B.).

CHAPTER SEVEN

The Endocrinology of Family Relationships in Biparental Monkeys

Toni E. Ziegler and Charles T. Snowdon

Introduction

Primates are social mammals whose family relationships are highly important. In almost all primate species, group living is the norm. Therefore, primate offspring grow up in rich social environments with plenty of opportunities to develop extensive interactions with other group members and individuals of all ages. Humans also live in social families, but in postindustrial societies, not all offspring have the successful positive experience that our nonhuman primate offspring receive. Social isolation does exist for some offspring, particularly those raised without extended families and by a single parent. As noted in Chapter 5, the fundamental component of the human family is the commitment of parents to feed and nurture their young to independence. Since primates, human and nonhuman, have highly dependent offspring that are slow to mature, the investment is heavy. For those species of primates where the parental investment is shared by both parents, there is an apportioned cost of parenting. With humans having the longest dependency of offspring, the cost of investment is very high. Additionally, our ever-evolving complex world requires a longer time to prepare our young to become independent, and therefore, the demands of proper care of human offspring have increased. Parenting is best achieved when there is social support and no one is left to raise a child on his or her own (Smith, 2005). Additionally, studies on humans in nonindustrial societies indicate that children who reside with nonrelated caretakers or stepparents are more likely to have stress-related conditions to contend with (see Nepomnaschy and Flinn, this volume). The nuclear family provides social support.

While there are data on what constitutes a good mother, father, and family environment in humans, there are limited data on the biological basis of parenting. This chapter will discuss how the social organization of nonhuman primates affects parental involvement, with a particular emphasis on fathers. Several families of New World monkeys are biparental and monogamous, and some are even cooperative breeders. These species are excellent models for understanding the physical and neuroendocrine basis of positive parenting. Our studies on the cotton-top tamarins *(Saguinus oedipus)* and common marmosets *(Callithrix jacchus)* have allowed us to examine how positive parenting is modulated by the hormonal system and experiences that influence successful parenting.

The biological basis of mothering has been studied in rodent species (Bridges, 1996; J. Stern, 1996; Wang and Insel, 1996) and many primate species (Pryce, 1995, 1996; Fleming, Morgan, and Walsh, et al., 1996; Maestripieri and Megna, 2000; Maestripieri, 2001). Nonhuman primate offspring rarely exhibit the neglect or abuse that human offspring too often receive, unless they are isolated or placed in unnatural conditions (Smith, 2005). We have found that fathers and even older offspring exhibit a biological basis for their participation in infant care and survival.

Biparental species, such as the marmosets and tamarins, work in concert to reproduce and rear their offspring. Our work has shown that these primates rely on olfactory communication and other chemical signals, which are effective in optimizing reproductive outcomes. The perception of these reproductive signals allows these species the ability to maintain their strong pair-bond and to ensure that males are available for the demands of infant care. The dynamics of the biparental species ensures that offspring have essential needs met and that they develop in a family environment. The effect of experience and genetics appear to play a role in promoting good parenting skills.

Social Organization and Parental Behavior

Many factors have played a role in the establishment of the social organizations among the various species of nonhuman primates (see Lee, this volume). Population density, predator concerns, competition for essential resources such as food or mates, and the cost and benefits of dispersing to form new groups are all selective pressures on group size and composition (Strier, 2007). Males and females may live alone, in single male-female groups, in multimale and multifemale groups, single male and multifemale groups, polyandrous families, or socially monogamous families. Most nonhuman primate males show only indirect paternal investment in offspring. The females nurse and carry the young, while

Figure 7.1 Cooperative breeding common marmosets (A) and cotton-top tamarins (B) provide communal infant care where fathers invest heavily in their offspring. Photography by Judith Sparkles (marmosets) and Carla Boe (tamarins).

the males are available for protection and sharing resources. In a few species, such as in the Barbary macaques *(Macaca sylvana)*, males assist in several aspects of infant learning (Burton, 1972), and baboon males *(Papio cynocephalus)* are known to protect their genetic offspring from attacks by peers (Buchan et al., 2003). However, mothers provide the majority of the parental investment. Direct paternal care occurs only in the polyandrous and monogamous species. Polyandrous species are rare and may be an alternative reproductive strategy in a small proportion of groups in a primarily monogamous species, such as the mustached tamarins *(Saguinus mystax)* (Huck et al., 2005).

Monogamous pairs occur when males establish long-term bonds with a female. Pair-bonded males have tied their reproductive success to their mate and therefore gain a high degree of paternity certainty (Strier, 2007). Phylogenetic comparisons suggest that pair-bonding precedes infant care by males and that the costs associated with the male's help are offset by the advantages he incurs (Dunbar, 1995). In the marmosets and tamarins, there is a high reproductive rate consisting of twinning and multiple births per year. This reproductive strategy allows the female to raise more off-spring than in other social systems and requires extensive help from males and older offspring. In species such as the owl monkeys *(Aotus)*, titi mon-keys *(Callicebus)*, Goeldi's monkey *(Callimico goeldi)*, marmosets, and tamarins, male investment also involves significant contributions to infant carrying, which improves infant survival (Garber and Leigh, 1997). In ad-dition to infant carrying, the male provides protection and shares food (Tardif et al., 1990; Pryce, 1995; Snowdon, 1996; Nunes et al., 2001). Bi-parental care also exists in some prosimians. There are over 30 pair-bonded species of lemur (Jolly, 1998). However, not all species display direct biparental care where the male actually carries the infant. For ex-ample, the nocturnal fat-tailed dwarf lemur *(Cheirogaleus medius)* will baby-sit and guard the offspring but will not carry and therefore does not display direct biparental care (Fietz and Dausmann, 2003). In contrast, the red-bellied lemur *(Eulemur rubiventer)* shows biparental care, with both the mother and father carrying offspring (Overdorff and Tecot, 2006). This species is also known for its twin births and delayed dispersal of re-productively mature offspring. Further research into the lemur species should provide additional illustrations of cooperative care of infants.

Cooperative Care of Infants

Among nonhuman primates, the marmosets and tamarins are the most notable for their cooperative care of infants (Figure 7.1). However, species of owl and titi monkeys also show infant care by older offspring and therefore should be considered cooperative breeders (Wright, 1984). The cooperative system allows the species to offset the high costs of reproductive output by limiting maternal investment in each offspring and providing extramaternal care for the young (Garber and Leigh, 1997). Both the common marmoset and the cotton-top tamarin have a postpartum ovulation that occurs as early as 10 and 13 days, respectively, following birth (marmosets: Lunn and McNeilly, 1982; tamarins: Ziegler, Bridson, et al., 1987). Conception is high, occurring more than 80% of the time (Ziegler, Bridson, et al., 1987). Therefore, mothers are lactating and pregnant at the same time. This high reproductive rate is energetically costly for mothers, requiring infant care support from the entire family. The ability to simultaneously lactate and conceive, as is seen for the callitrichid monkeys, is due to the lower frequency of suckling because of the lower frequency of nursing bouts allowed by mothers (Ziegler et al., 1990).

Cooperative breeders have delayed dispersal so that offspring of reproductive age forgo their own reproduction to remain in the family and care for younger siblings (Solomon and French, 1997). Cotton-top tamarin daughters are reproductively suppressed while living in such families, with no signs of ovulation even though they have gone through puberty (Ziegler, Savage, et al., 1987). However, in the marmoset species, there appears to be a flexibility of response to fertility control in which some families are associated with complete fertility suppression in daughters, while in other families many eldest daughters, display occasional ovarian cycles (for pygmy marmosets [*Cebuella pygmaea*]: Carlson, Ziegler, and Snowdon, 1997; for common marmosets: Abbott, 1984; Salzmann et al., 1997; Ziegler and Sousa, 2002; for Wied's black tufted-ear [*Callithrix kuhli*]: Smith et al., 1997). While females show suppression of reproductive hormones, males do not. Cotton-top tamarin sons show pubertal hormonal changes while living in the family and even exhibit sexual behaviors but without an appropriate partner (Ginther et al., 2002). A lack of suppression of reproductive hormones in subordinate males has been reported for other species: lion tamarins *(Leontopithicus rosalia)* (French et al., 1999), black tufted-ear marmosets *(Callithrix kuhli)* (French and Schaffner, 1995), and the male offspring of the common marmoset may have a behavioral suppression for sexual behavior (Abbott, 1984).

Cooperative infant care for the cotton-top tamarin allows for a buffering of individual parenting styles. Female tamarins do not provide extensive care to their offspring. A female generally spends only around 20% of her time carrying infants in the first six weeks of their life, and most of that time is spent nursing infants (Ziegler et al., 1990). Fathers will provide for most of the carrying of infants, but as the size of the family increases with more offspring, the amount of a father's carrying time decreases (Achenbach and Snowdon, 2002). Infants receive similar levels of care regardless of group size and parenting experience, but the contribution from individual family members will vary with group size and parental experience (Washabaugh, Snowdon, and Ziegler, 2002). This behavior may also occur for common marmosets (Yamamoto, 1993).

In large families of cotton-top tamarins, the eldest males provide much of the infant carrying (Savage et al., 1996). Thus the number of male offspring will increase the likelihood of infant survivorship (Sussman and Garber, 1987). This effect on survival is found even in captive tamarin groups (Snowdon, 1996). Eldest sons learn tolerance of infants and acquire infant care skills for later reproduction (Snowdon, 1996). Physiological responses to infant contact may prepare males for their continued role as caretaker when their own infants are born.

Prolactin, a hormone involved in milk production and regulation of maternal behavior in mammals (see Wallen and Hassett, this volume), is elevated among fathers and eldest sons during infant interaction (Ziegler, Wegner, and Snowdon, 1996). Postpartum levels of prolactin are higher in fathers and eldest sons than in nonfathers. These levels remain high even after the infants are weaned. In the common marmoset, fathers and eldest sons and daughters show elevated prolactin postpartum, but these levels are highest when the marmosets have been recently carrying infants (da Silva Mota, Franci, and de Sousa, 2006). While it is not known whether other offspring besides the eldest tamarin males have elevated prolactin levels, it appears for the marmosets that the actual contact time carrying infants, and possibly the number of infants, stimulates the release of prolactin (Dixson and George, 1982; Schradin and Anzenberger, 2004; da Silva Mota, Franci, and Sousa, 2006).

Patterns of prolactin changes appear to reflect prolactin's role in male paternal care. Prolactin levels are higher in marmoset fathers than their sons during infant care (Shradin and Anzenberger, 2004). In contrast, father titi monkeys who carry infants about 80% of the time (Mendoza and Mason, 1986) have higher prolactin levels than nonfathers, even when there are no dependent infants (Schradin et al., 2003). In another New World species, the Goeldi's monkey, fathers do not carry infants until the third or fourth week following birth, but prolactin levels become elevated prior to infant care (Schradin and Anzenberger, 2004). Thus prolactin appears to facilitate paternal care rather than be a response to infant carrying. Prolactin may prepare fathers for their role in infant care.

Preparing for Infant Care

In biparental species, males and females have hormonal facilitation of parental responsiveness. The hormonal environment of pregnancy is known to prime females for maternal responsiveness (see Fleming and Gonzalez, this volume; for reviews, see Numan, 1994; Bridges, 1996; Fleming, Morgan, and Walsh, 1996; Pryce, 1996). Prepartum hormones, such as estradiol, progesterone (or the ratio of the two), prolactin, and oxytocin, all participate in initiating maternal behavior (Bridges, 1996; Gonzalez-Mariscal and Rosenblatt, 1996). We are now beginning to see how these hormones are involved in the facilitation of male parental care.

Marmoset and tamarin fathers display physical and hormonal changes in preparation for their role in infant care. The energetic cost for fathers once the infants arrive is high. Twin neonates can weigh up to 20% of a carrier's weight, and a single adult often carries both twins simultaneously (Achenbach and Snowdon, 2002). Father tamarins actually lose weight during the postpartum period, even under laboratory conditions, while carrying infants, but mothers do not (Sanchez et al., 1999; Achenbach and Snowdon, 2002). In cotton-top tamarins, there is a negative correlation between weight loss by males during infant care and number of helpers available to carry (Achenbach and Snowdon, 2002). Therefore, helpers can relieve the father's energetic burden of infant carrying.

Both marmoset and tamarin fathers actually gain weight prior to the birth of their offspring (Ziegler, Prudom, et al., 2006). Expectant marmoset males show a significant weight change across the five months of gestation, with an average 10% increase in weight from the first month to the last month, while expectant cotton-top tamarin fathers showed an average 5% increase in weight across the six months of gestation (Figure 7.2). Males of both species begin to increase their weight prior to their mate's significant weight change. Pregnant females showed their highest increase in weight either during the last month of gestation (marmosets) or during the last two months of gestation (cotton-top tamarins), which is when the greatest fetal growth occurs. This indicates that the males are not following the weight change trajectory of their mates, which might occur if they were eating sympathetically. Human men are the only other species of primates in which weight gain during their mate's pregnancy has been reported. There are no systematic studies in men, although there have been many anecdotal reports. Sympathetic pregnancy symptoms in men are referred to as *couvade* (derived from the French word for "hatch"), indicating that men share some of their mate's pregnancy symptoms (Klein, 1991). Men report symptoms such as weight gain, nausea, headaches, restlessness, and backaches (Clinton, 1986). There are two interesting lines of thought that follow from comparing men to the biparental marmosets and tamarins with respect to weight gain. If men do indeed show a systematic weight gain during pregnancy, then they are most likely preparing for their energetic cost of parenting. If men are preparing for their infant care, then it would be interesting to discern if they are receiving chemical signals from the female that initiates responses in the males. Is there any relationship between males that show increased physical changes and the amount of paternal care they show their offspring?

To date, the most extensive hormonal changes in expectant primate fathers have been documented in the cotton-top tamarins. We have found that experienced fathers have elevated levels of several hormones during their mate's pregnancy (Figure 7.3). Prolactin levels are significantly elevated during the third of the six-month gestation period, and another surge is seen during the last month of gestation (Ziegler, and Wegner, and Snowdon, 1996; Ziegler and Snowdon, 2000; Ziegler et al., 2000; Ziegler, Washabaugh, and Snowdon, 2004). The midgestational elevation was also correlated with the number of surviving infants from the preceding birth and therefore may be related to infant care ($r = 1.0$;

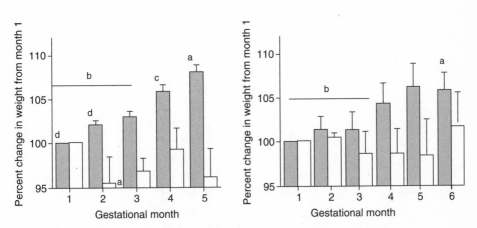

Figure 7.2   Percent change in male and female body weight over the gestational range in common marmosets *(left graph)* and cotton-top tamarins *(right graph)*. *Dark bars* indicate expectant fathers, and *light bars* indicate control males weighed for five consecutive months for the marmosets and six consecutive months for the tamarins. *Letters* indicate months significantly different from each other. Males of both species begin to show weight changes by midgestation. Data presented in Ziegler, Prudom, et al., 2006.

Ziegler and Snowdon, 2000). Prolactin has been implicated in facilitating parental responsiveness in rodent species and birds, where it is elevated prior to birth (R. Brown, 1993; Buntin, 1996).

In summary, prolactin may have two roles in paternal care: (1) for facilitating the onset of paternal care behaviors and (2) for the maintenance of paternal care behaviors in response to infants. First, prolactin may be activated and elevated prior to infant contact, as we see for the cotton-top tamarin prior to the birth of infants (Ziegler, Washabaugh, and Snowdon, 2004) and in the Goeldi's monkey during the first two weeks following birth but prior to any direct infant care in the fathers (Schradin et al., 2003). Second, prolactin appears to be associated with

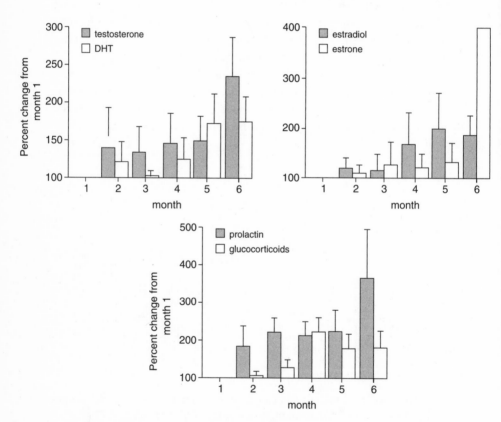

Figure 7.3   Experienced cotton-top tamarin males show changes in their reproductive hormones during their mate's pregnancy. These graphs show the percent change in hormone level from the average first month of the gestation period for: testosterone, DHT, estradiol, estrone, prolactin, and glucocorticoids (cortisol, cortisone, and corticosterone). Data presented in Ziegler, Washabaugh, and Snowdon, 2004.

infant contact and is elevated in the blood immediately following infant carrying (Dixson and George, 1982; Mota and Sousa, 2000; da Silva Mota, Franci, and de Sousa, 2006). Our recent data on experienced tamarin fathers indicate that prolactin elevations are a response to both the pregnancy and the infants (Almond, Ziegler, and Snowdon, 2008). Experienced fathers show both an elevation of prolactin in midgestation of their mate's pregnancy and in the final month before birth (Ziegler and Snowdon, 2004). When these males were removed from the family and paired with an inexperienced female with no offspring present, these males no longer showed the midgestational increase in prolactin but did in the final month of the gestation, as do inexperienced fathers. Therefore, the midgestational rise in prolactin levels was actually due to the previous infants that the males were caring for, whereas the prolactin increase at the end of pregnancy may be due to chemical signals from their mates.

Other reproductive hormones, such as testosterone, DHT (dihydrotestosterone, a potent androgen), estradiol, and estrone were all significantly elevated in the final gestational month of experienced tamarin fathers. Although the findings of elevated testosterone may be surprising, since testosterone is not thought of as a parenting hormone, Trainor and Marler (2002) have shown in the biparental California mouse *(Peromyscus californicus)* that elevated testosterone levels promote paternal behavior by being converted to estrogens. Thus, high levels of both androgens and estrogens found in expectant tamarin fathers may be explained by an increased production of androgens and subsequent conversion into estrogens. Urinary estradiol is known to have a testicular origin in the male cotton-top tamarin (Ziegler et al., 2000), and both urine and blood levels in expectant fathers are higher than peak levels found for nonpregnant tamarin females. These increased estrogens could promote paternal behavior, as in the male California mice, and may prepare fathers for immediate interaction after infants are born. Elevated prepartum estrogens are known to promote maternal behavior in many mammals (Kendrick and Keverne, 1991; Bridges, 1996), and specific studies on female marmosets and tamarins substantiate this (red-bellied tamarins: Pryce et al., 1988; common marmoset: Pryce, 1993; Pryce, Dobeli, and Martin, 1993). Additionally, estradiol is known to stimulate prolactin release from the pituitary (Garas et al., 2006). Thus, the increase in male testosterone levels during pregnancy may ultimately stimulate the production of prolactin.

These studies on the cotton-top tamarin indicate that biparental expectant fathers have both physical and hormonal changes that occur to prepare them for their role in ensuring the survival of their offspring. The physical changes may prepare males for the energetic costs of actively carrying such large infants, while the hormonal changes may be related to activating the motivation to parent, reducing aversion to the novel infant, inducing parental behaviors, or sustaining the pair-bond with the mate to ensure continued investment in these and future offspring. Further studies are needed to understand the mechanisms by which these hormones influence the male's paternal behavior and his pair-bonding to the female.

Experience and Its Influence on Infant Care

All males are not equal when it comes to parenting. Experience plays a role in the facilitation of infant care. Marmosets and tamarins learn how to parent by interacting with the younger siblings in the family (see Snowdon, 1996, for discussion). Mature male and female cotton-top tamarins are unsuccessful in caring for their offspring if they lacked prior experience raising siblings while in their natal family (Smith, 2005). The lack of experience raising siblings has less of an effect on common marmoset males and females, but both species show higher success in rearing offspring after they have experienced at least one birth as a parent (Snowdon, 1996).

The patterns of hormone secretion in male tamarins during his mate's pregnancy are different between experienced and inexperienced males (Ziegler and Snowdon, 2000; Ziegler, Washabaugh, and Snowdon, 2004). Whereas all male tamarins show significant increases in prolactin during the last month prior to the birth of their offspring, only parentally experienced males show significant increases in other reproductive hormones. Levels of testosterone, DHT, estradiol, estrone, and cortisol increase significantly during the final month of the gestation for experienced males, while there are no significant changes at the end of the gestation in the parentally inexperienced males (Figure 7.4). How might parental experience promote these hormonal changes?

These differences could be due to the perception of pregnancy signals sent by the female. Tamarin pairs increase their proximity to each other during the last two months before the birth of their infants (Ziegler, Washabaugh, and Snowdon, 2004). Males groom females more often and approach their mates at a higher rate. Therefore, males appear to be monitoring their mates for scent changes at the end of pregnancy. However, we found that the pair interactions at this time were higher in inexperienced fathers than in the more experienced fathers. One explanation for this occurrence would be that the hormonal changes occurring for

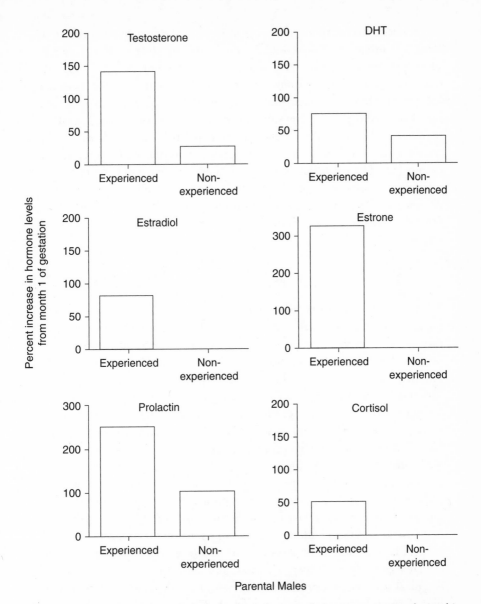

Figure 7.4 Experience plays a role for males in hormonal changes occurring during his mate's gestation period. Experienced males have higher levels of testosterone, DHT, estradiol, estrone, prolactin, and cortisol. Data presented in Ziegler, Washabaugh, and Snowdon, 2004.

males might stimulate an increased interest in the female's reproductive state. Experienced males could be distracted by their busy family life since they are interacting with offspring from previous births. They may not have the opportunity to monitor the pregnant female as closely as the less parentally experienced males with no offspring. Since experienced males showed the most pronounced hormonal changes and the fewest interactions with their mates, experienced males may learn to detect the female's condition through some other means without the need to closely monitor the female. It has been shown that experience in male mice will affect their response to female sex odors (Nyby and Whitney, 1980). Experienced male mice are much more responsive to the chemical signals of females. Detecting and responding to female stimuli may be a learned response.

Communication between the Pair about the Pregnancy and Ovulation

Communication between pair-bonded callithricids may be due to their ability to detect chemical signals. Callitrichids have well-defined olfactory systems. Both the main olfactory system (MOS) and the accessory olfactory system (AOS) are present (Taniguchi et al., 1992; Smith et al., 2003). These species also have well-developed scent glands in their suprapubic and anogenital regions. Callitrichids deposit scent secretions on substrates in their environment and use these extensively in communication of territory, sexual identification, reproductive information, and identification of individuals (Epple, 1986).

Communication of the pregnancy to the expectant father is most likely through either scent secretions or through chemical signaling by the female. Females will scent-mark throughout the pregnancy, and the frequency is higher in the last month before birth (Ziegler, Washabaugh, and Snowdon, 2004). Additionally, females excrete high levels of glucocorticoids into their urine beginning at midpregnancy, and these remain elevated until birth (Ziegler, Scheffler, and Snowdon, 1995). This profile is typical for pregnant primates and is seen in both urine and blood samples. The midgestation rise in glucocorticoids in female primates is attributed to the development of the transitional zone of the fetal adrenal glands where cortisol synthesis occurs (Coulter and Jaffe, 1998). Through the interaction of the fetal-placental-maternal unit, large amounts of cortisol are excreted into the urine. Experienced expectant tamarin fathers also have elevated levels of glucocorticoids during their mate's pregnancy. Fathers show a peak of glucocorticoid excretion generally the week following their mate's onset of cortisol elevation at midpregnancy (Ziegler,

Washabaugh, and Snowdon, 2004; Almond, Ziegler, and Snowdon, 2008). It is possible that this rapid fetal-placental-maternal urinary glucocorticoid increase might act as an endocrine signal that initiates a glucocorticoid response in the expectant father. It has been reported that glucocorticoids regulate pheromones (or chemical signals) in mice, with corticosterone influencing both the release of the signal and the response of the recipient (Marchlewska-Koj and Zacharczuk-Kakietek, 1990; Weidong, Zhongshan, and Novotny, 1998). Further work is planned to elicit the interactions of the chemical signaling.

Male tamarins and marmosets are also responsive to chemical signaling from their mates at the time of ovulation. Although female callitrichids do not have obvious signs of their reproductive phase, such as sexual swellings or colorations, they do communicate their reproductive status by scent secretions. Tamarin males respond to the timing of ovulation in scent marks with sexual arousing behaviors (Ziegler et al., 1993). We presented males who were paired with pregnant females with scents from novel females. Scent marks were collected daily from our donor females over their 23-day ovulatory cycles and presented daily as fresh scents to the males in their home cage. The tamarin males showed their highest responses to the scents during the periovulatory days (the 3 days before and after the female showed a preovulatory luteinizing hormone [LH] peak). They increased rates of mounting to their pregnant mate and increased duration of erections. This indicated that the signal received from the scents was arousing to the males. Males are hormonally responsive to sexual stimuli (see Roney, this volume).

Male marmosets have also been tested with collection of scent secretions from novel females (Ziegler et al., 2005). Males were tested in their home cages with the removal of other marmosets from the cage. Frozen scents from ovulating novel females or a vehicle control were presented on wooden disc to the male. He was given 10 minutes to interact with the scent, and then at 30 minutes a blood sample was taken to obtain circulating levels of testosterone and cortisol. Males had a significant increase in testosterone levels over the vehicle control 30 minutes following the presentation of the scent but no changes in cortisol occurred. Behaviorally, males spent significantly more time sniffing the scent substrate, and the duration of erections were significantly elevated in relation to the vehicle scent. Therefore, the common marmoset male also responds to an arousal signal, as did the tamarin males. When the marmosets' hormonal results were classified by whether they were family males (those who were paired and had young offspring in the cage), paired males (living with a female but with no offspring), or single males (males who had been living without a cage mate for a few days), there was a difference in their response to the novel ovulatory odors. Paired and single males both had a significant testosterone response to the ovulatory odors, compared to the vehicle, whereas the family males did not show an increase (Figure 7.5). It appears that male marmosets have a behavioral and neuroendocrine response to the arousing, sexually relevant scent signal. However, under stable family conditions, there may be an inhibitory process that prevents males from exhibiting a full response, which may be due to their monogamous social system. With the extensive infant care that males must provide, fathers are already invested in their relationship and would have a lot to lose if they were responsive to novel ovulatory females.

The hormonal response to the ovulatory signal from novel females is also associated with changes in the brain. These scents have been shown to activate the medial preoptic area (MPOA) of the hypothalamus and the anterior hypothalamus (AH) during brain imaging of awake marmosets by functional magnet resonance imaging (fMRI) (Ferris et al., 2001). These are areas known to be associated with sexual arousal in

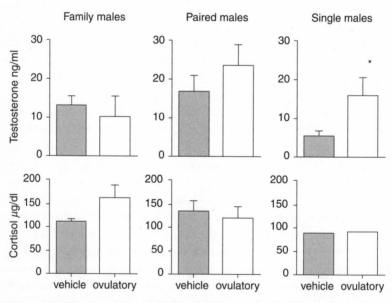

Figure 7.5 Common marmosets have elevated testosterone 30 minutes after receiving scent secretions from a novel ovulating female. Significant differences are seen in single and paired males but not in males who are living with a family where young offspring are present. Family males may not process the novel ovulating scent signals as do other males. Data presented in Ziegler et al., 2005.

male marmosets (Kendrick and Dixson, 1983). Sexual arousal is also associated with activations of brain areas involved in emotional processing and reward (Ferris et al., 2004). These imaging studies on awake marmosets have shown similar neural circuits to humans in sexual arousal and motivation. This circuitry is part of the general appetitive circuit for positive valence in emotional stimuli.

Male cotton-top tamarins also show a hormonal response to their mate's ovulatory signal. We tested male tamarins in their family for changes in urinary steroids—testosterone, DHT, estradiol, and corticosterone—postpartum and through the postpartum ovulation. Whether the female ovulated early (within 15 days as determined by urinary LH peaks) or later (up to 25 days following birth), testosterone levels were elevated during the female's follicular phase of the postpartum ovulation. Additionally, DHT and estradiol were elevated during this time (Ziegler, Jacoris, and Snowdon, 2004). Cotton-top tamarins anticipated their mate's postpartum ovulation with an increase in urinary androgens. Males may increase their reproductive hormones to prepare for the female's fertile phase. This could be a strategy that biparental males use to raise their testosterone levels. In birds, the Wied's marmoset, and even human males, testosterone levels are lower in fathers during times of parenting (Hegner and Wingfield, 1987; Nunes et al., 2001; Gray et al., 2006). However, in species where females are ovulating at the same time that parenting is occurring, the male would need to be at optimal fertility both to guard his mate and to protect his infants. This appears to be what is occurring in the tamarin males. Since fathers are in such close association with mothers and infants during this time, it is not surprising that males receive and respond to fertility signs with increased androgen production. Increased testosterone production in males stimulates spermatid maturation and may influence fertility in males (Williams-Ashman, 1988). The increase in androgens in male cotton-top tamarins may work both to increase a male's fertility and to influence his paternal care.

These studies provide evidence that male callitrichids detect chemical signals from their mates or a novel ovulating female. In a species where the group works together to ensure high fertility and successful rearing of offspring, it is essential that the male is aware of the female's reproductive condition and is ready for his role in conception or parenting.

Neural Basis for Paternal Behavior

Few studies have examined the neural changes associated with parenting that occur in male marmosets and tamarins. This is an exciting new area of research that will soon provide evidence of how the male brain responds to parenting. However, there are numerous brain areas in mothers that have been found to be involved in maternal behavior. For instance, the MPOA of the rodent hypothalamus has cell bodies that are enlarged by the stimulation of estradiol (Sheehan and Numan, 2002), and basic maternal behaviors are eliminated when this area is lesioned (Terkel, Bridges, and Sawyer, 1979). Estradiol stimulation promotes rodent maternal behavior by enhancing pup-stimulated activity in the MPOA and other brain areas such as the ventral bed nucleus of the stria terminalis and the dorsal and intermediate lateral septum (regulation of mood and motivation; Sheehan and Numan, 2002). Pregnancy also stimulates neurogenesis in the forebrain, which is mediated by prolactin (Shingo et al., 2003). Surface areas of the neuronal branches in the hippocampus increase during pregnancy and lactation. The stimulation of the hippocampus (area of learning and memory) during pregnancy is thought to improve offspring recognition. The cingulate cortex (an area that regulates emotions) receives neural input from the thalamus. When the cingulate cortex is damaged, maternal behavior is diminished (Peredery et al., 1992). Other cortical areas include the prefrontal and orbitofrontal cortices. The orbitofrontal cortex is involved in positive emotion and is activated in human mothers while viewing pictures of their infants (Nitschke et al., 2004). The nucleus accumbens, an important area for reinforcement and reward (Lee et al., 1999), has higher neural activity in mother rats when they nurse (Febo, Numan, and Ferris, 2005). Mother rats preferred nursing their pups over a cocaine reward (Febo, Numan, and Ferris, 2005). The nucleus accumbens appears to be involved in maintaining maternal motivation, and electrolytic lesions of the shell significantly disrupt pup retrieval but not other maternal behaviors such as nursing or licking (Lee, Clancy, and Fleming, 2000; Li and Fleming, 2003). Preference for pup stimuli appears to be associated with the lateral and basolateral amygdala (Robbins et al., 1989).

The neuropeptides oxytocin (OT) and arginine vasopressin (AVP) are implicated in parenting behaviors in biparental rodents (see Carter et al., this volume; Carter, Altemus, and Chrousos, 2001). Apparently, common marmosets have the same sex-specific patterns of immunoreactive AVP staining in brain cells as biparental rodents. Immunoreactive AVP staining brain cells located in the paraventricular and supraoptic nuclei, suprachiasmatic nucleus of the hypothalamus, the lateral hypothalamus, and the bed nucleus of the stria terminalis are more numerous in male than female marmosets (Wang et al., 1997). Monogamous species such as the prairie vole, California mouse, and common marmoset express higher levels of V1aR (V1a receptor) in the ventral pallidum region relative to nonrelated monogamous species, suggesting that AVP, acting through the V1aR in the ventral pallidum, may be an important mediator of pair-bond formation (Hammock and Young, 2002). Differences in this area are associated with differences in social behavioral traits relating to pair-bonding and parenting. Additionally, marmoset fathers have an increase in the density of dendritic spines in the prefrontal cortex with the presence of vasopressin in their stomata when compared to marmoset males who have not been fathers, suggesting that fatherhood may enhance dendritic spine density on the same population of neurons that synthesize vasopressin and express the V1aR (Kozorovitskiy et al., 2006).

Oxytocin has been considered to be involved in a number of social effects on behavior and body function (Carter et al., this volume). Studies in the prairie vole have demonstrated the importance of oxytocin on developmental changes in social behavior and aggression (Bales and Carter, 2003a). A recent study in humans showed urinary oxytocin levels in children to be responsive to tactile contact from a known caregiver (mother) but were muted in children who had been institutionalized after birth, with limited contact in their early years (Fries et al., 2005). These studies indicate that social contact may be critical for the oxytocin response to stressful or interactive conditions. In the common marmoset, we have recently shown that peripheral oxytocin enters the urine within 30 to 60 minutes after release and is quantifiable (Seltzer and Ziegler, 2007). Both oxytocin and vasopressin are measurable in the urine and appear to alter with changing social condition. Six male marmosets isolated from their mates for two days showed lower levels of oxytocin and vasopressin during isolation than when the males were exposed to visual and olfactory stimulus from their mates or contact with their mates. Males showed their highest levels of oxytocin when they were in visual contact with their mates rather than tactile contact. This indicates that the type of social contact may be important to the release of these peptides in these species. We found that cortisol levels were high in all males in this study during the isolation period but that there was a significant increase with the visual contact condition. Levels decreased relatively rapidly afterward. It is possible that seeing the mate without the option for tactile contact is a stressful experience for a pair-bonded species. Therefore, oxytocin and vasopressin appear to be responsive to social conditions in these monogamous and biparental species. More research into the conditions affecting oxytocin and vasopressin release are merited.

Genetic Influence on Parenting

While quality care during infancy (Champagne et al., 2001) and parenting experience during development (Snowdon, 1996) can have profound influences on parenting styles as an adult, genetic influences are also likely to be involved. Variability in phenotypic expression of behavior results from variability in specific gene sequences. Studies on the monogamous and biparental prairie vole indicate that there is variability in frequency of the behaviors pup licking and grooming. Males who show a significantly higher frequency of these behaviors also have a longer repetitive DNA sequence (microsatellite) of the arginine-vasopressin receptor 1a *(Avpr1a)* gene (Hammock and Young, 2002). No differences were found in the females harboring the long and short microsatellite allele, indicating that AVP primarily regulates male behavior. Arginine vasopressin has been shown to be involved in paternal care of offspring (Wang, Ferris, and De Vries, 1994). It therefore appears that genetic variance in the specific gene for the vasopressin receptor can influence male paternal care in a biparental vole species. This led us to examine this gene in male marmosets and to determine if there were any variances in the amino acid sequences (polymorphisms). If there are variations, then is there a specific variant, or allele composition, that would be associated with a particular male marmoset parenting behavior?

We collected blood from 13 parentally experienced marmoset fathers and extracted the DNA from leukocytes (Ziegler, Mamanasiri, et al., 2006). The marmoset *Avpr1a* gene was sequenced 56 to 1,857 base pairs upstream from the transcription start site. Although our data failed to reach the putative microsatellite region, we were able to identify polymorphisms. We found six haplotypes that we designated A through F to be common in the marmoset. Therefore, the male common marmoset has the variation in the *Avpr1a* gene, which could activate a specific paternal behavior.

To test whether the genotyped males had any variation in their infant-directed behaviors that would be associated with a particular *Avpr1a* gene haplotype, we developed a method for testing males who have been removed from their home environment (Zahed et al., 2008). Males were placed in a test cage that had access to another cage where infant vocalizations from the male's own infant were playing from a recorder (a digital recording of the infant's vocalizations). Fathers were scored on their frequencies of infant-directed behaviors to the recorder stimulus, including: enter the infant test cage, look at the stimulus, attempt to retrieve the stimulus, search for the infant, and bite the stimulus. We found that the infant-directed behavior, searching for the infant, was significantly higher in frequencies for males who had the B$^+$ form of the *Avpr1a* gene allele. While these data are preliminary, they do suggest that genetic variation in the *Avpr1a* gene may influence the expression of paternal care in the common marmoset.

Data on genetic influences of paternal care in primates will likely be an important source of parental variation. However, the genetic interactions with parental experience and other environmental factors that influence parenting will most likely provide an accurate indication of parenting outcome.

Summary

While there are other species of biparental nonhuman primates that can be studied for proximate mechanisms of parental care, the callitrichids have been studied the most. Additionally, the cooperative care social system provides the most suitable model for studying human parenting. Further research on other species of biparental monkeys would be very informative in their comparisons with the callitrichids. Since fathers in both the owl and titi monkeys are actively involved with infant care but do not have regular twinning and the large family groups, males may exhibit different responsiveness to infants.

As the studies mentioned in this chapter indicate, the social bond within a family is strong. The pair-bond between males and females allows the male to monitor the reproductive state of the female, whether she is pregnant or ovulating. The knowledge of pregnancy allows the male to prepare for his highly energetic cost of parenting. Without this, the female would be unable to ensure the high level of reproductive output typical of callitrichids. While male investment promotes an increased likelihood of the offspring surviving to adulthood, the male is also provided with a high degree of paternity certainty. The ability to identify and prepare for the mate's postpartum ovulation also ensures that the male is at his optimum fertility, and he can monopolize the female during this time.

If the social bond is the potentiator of the physical and hormonal changes occurring in the males during the pregnancy, then these are probably mediated through chemical signaling. Indeed, the callitrichids have both the specialized scent glands and the specialized olfactory systems to make this an ideal method of communication for them. The chemical signaling may provide an important form of family communication and assist in maintaining the presence of the older offspring for helpers.

Acknowledgments

We have had many collaborators on these studies of the marmoset and tamarins. We would like to thank the members of the tamarin colony, especially Kate Washabaugh, Aimee Kurian, Carla Boe, and Anita Ginther. Our studies in the marmosets have been possible through the collaborations with Nancy Schultz-Darken, Shelley Prudom, and Sofia Refetoff Zahed. Samuel Refetoff and S. Mamanasiri are responsible for the genetic analyses. We had many collaborators on the brain imaging work who were led by Craig Ferris. Fritz Wegner, Dan Wittwer, and Steve Jacoris, members of Assay Services of the Wisconsin Primate Research Center, provided hormonal analyses. We acknowledge the grants MH 58700 to C. F. Ferris, MH 35215 to C. T. Snowdon and T. E. Ziegler, and MH 70423 to T. E. Ziegler, from the National Institutes of Health (NIH), and NIH P51 RR000167 to the Wisconsin National Primate Research Center.

CHAPTER EIGHT

Hormonal and Neurochemical Influences on Aggression in Group-Living Monkeys

Lynn A. Fairbanks

Introduction

Nonhuman primate social systems vary widely, but all involve complex social relationships that include cooperation and competition with other group members. Group living provides the benefits of shared vigilance against predators and group defense of a home range that provides food and shelter for all group members. At the same time, some level of within-group conflict over access to resources is inevitable for primates living in multimale, multifemale social groups. The way individuals respond to competitive challenge influences their access to food, social position, and reproductive opportunities. It also influences their risk of ostracism, injury, and even death in aggressive encounters. Successful management of conflict and aggressive interactions has important consequences for individual and group welfare.

Physiological systems that have been implicated most often in the regulation of aggression in nonhuman primates are testosterone and the monoamine neurotransmitters serotonin and dopamine. This chapter will review the literature on these hormonal and neurochemical mediators of aggressive and competitive behavior, with an emphasis on the importance of understanding behavior-physiology relationships in the context of life history and reproductive success. The studies covered here vary in species, age, sex, setting, season, and hormone assessment methods, and often produce conflicting results. The Wingfield challenge hypothesis, originally developed for monogamous birds, has been used to make sense of variation in the testosterone-aggression relationship by considering how the function of aggression varies in different mating systems (Wingfield et al., 1990). The challenge hypothesis proposes that testosterone promotes aggression in situations where aggressive behavior is beneficial for reproduction, such as territory establishment and mate guarding. Elements of the hypothesis are that increases in aggression in response to increasing testosterone are more likely under conditions of social instability and that testosterone interferes with parental behavior, resulting in a trade-off between investment in male-male aggression versus parenting as a reproductive strategy. Because the challenge hypothesis goes beyond a simple linear mechanistic approach and incorporates an understanding of functional outcomes, it has become a valuable organizing principle for interpreting testosterone-aggression results for group-living primates (for example, Cavigelli and Pereira, 2000; Whitten and Turner, 2004; Cristóbal-Azkarate et al., 2006). This type of approach is also used here to interpret the complex relationships between serotonin, dopamine, and aggression in group-living primates.

This chapter will begin with a brief overview of the social organization of the group-living monkeys and the features of their life histories that are most relevant to understanding the context and consequences of aggressive behavior. It will follow with an introduction to the physiological functions of testosterone and the monoamine neurotransmitters dopamine and serotonin. The next section will review the literature on the relationship between testosterone, aggression, and social dominance. And the final section will review the evidence linking serotonin and dopamine to aggressive and risky behavior.

The results indicate that testosterone levels are highly variable from day to day and that high levels of testosterone are associated with aggression during unstable periods of rank challenge and opportunity for males. They provide support for the general hypothesis that testosterone helps prepare males for aggressive response at times when the fitness payoff for aggression is high. In contrast, review of the neurotransmitter literature suggests that baseline levels of serotonin and dopamine metabolism are relatively stable and traitlike and that individual differences are related to alternate life history strategies along a continuum from risk taking to conservative. The consequences of an aggressive and risk-taking strategy would depend on the social and ecological circumstances.

Social Organization of Group-Living Monkeys

The most common form of social organization for group-living monkeys is the multigenerational matrilineal group (Silk, 1987). In this type of system, males and females have very different life histories. Females stay in the natal group with their mothers and female kin for life, while males leave at adolescence and transfer to neighboring groups for breeding.

In stable multimale, multifemale social groups, dominance hierarchies typically form to "institutionalize" priority of access among individuals and avoid the costs of repeated aggressive contests. Noncontact signs and signals such as vocal, postural, and facial threats communicate aggressive intent and are often enough to elicit submissive signals in opponents without resorting to physical contact. The dominance hierarchy is a mechanism that reduces the risks of frequent aggressive behavior for both higher- and lower-ranking animals and is beneficial to all when the fitness disadvantage of being low ranking is less than the potential cost of leaving the group or attempting a rank overthrow. During times of male rank instability, such as when new males enter the group or when the alpha male is sick or injured, these cost/benefit equations change. Adult males need to be able to respond rapidly to defend their position in times of challenge and to take advantage of opportunities to increase in rank.

For males in matrilineal primate societies, emigration from the natal group is a time when the benefits of a successful aggressive campaign can be high. Field reports describing alternate male strategies for entering new groups have shown that males who use an unobtrusive strategy minimize the risk of injury, but they are also less likely to achieve high rank in the new group, compared to males who are more aggressive and confrontational (Henzi and Lucas, 1980; van Noordwijk and van Schaik, 1985; Fedigan and Zohar, 1997). Since high-ranking males generally have greater reproductive success than low-ranking males, the payoff for the more aggressive immigration strategy can be substantial.

For females in matrilineal primate societies, the picture is quite different. Times of acute and intense competition with other group females are relatively rare. Long-term competitive advantages are managed through alliances with kin and dominance hierarchies that can be stable across generations. In some societies, adolescent females have a period of increased aggressiveness as they assume their position in the adult female hierarchy (Pereira, 1995), but the intensity and risk of injury are generally mild compared with male-male aggression at the same life stage. The available data suggest that the effect of female rank and aggression on survival and reproductive success is smaller than the effect of male rank and aggression on the same fitness parameters for matrilineal primate species (Clutton-Brock, 1982).

These differences in life history between males and females, described above, are reflected in hormones that influence aggressive behavior and motivation.

Endocrine and Neurochemical Mediators of Aggression

The hormone that has been implicated most often in modulating aggressive motivation in primates and other animals is testosterone. Testosterone is a steroid hormone that is excreted by the testes of males and, to a much lesser extent, by the ovaries of females (see Chapter 2). During gestation and early in postnatal life, testosterone influences development of male and female genitalia and gender differentiation (Wallen, 2001). At adolescence, an increase in testosterone for males is associated with the achievement of adult primary and secondary sexual characteristics (Bercovitch, 1993). For adult males, testosterone is related to libido; to maintenance of strength, muscle mass, and bone density; and to general energy.

The early studies of behavior-testosterone relationships in group-living monkeys measured circulating testosterone levels in plasma, a sampling method that requires capturing the animals and either using anesthesia or extensive training for awake blood collection. More recent research has taken advantage of new methods for measuring testosterone in fecal samples to allow collection of longitudinal hormonal data in animals without the disruptive effects of repeated capture and anesthesia (Whitten, Brockman, and Stavisky, 1998; Ziegler and Wittwer, 2005). Both methods indicate that daily levels are highly variable and responsive to circumstances. The rapid responsivity of testosterone makes this hormone a good candidate for modulating behavioral changes within short time intervals, particularly for males.

Studies of neurochemical influences on aggressive behavior have focused on the monoamine neurotransmitters serotonin and dopamine. These two neurotransmitters carry signals from one neuron to another in the areas of the brain involved with regulating emotion and motivation, and they function along with other neurotransmitter systems and neuropeptides (for example, oxytocin and vasopressin; see Chapters 6, 7, and 14) to influence mood and behavior. The serotonin system has been implicated in shifting response tendencies from calm, sleepy, and inhibited, at one extreme, to impulsive and aggressive behavior, at the other (Soubrie, 1986). Dopamine is involved in motor activity and in appetite and sexual motivation (Depue and Collins, 1999). Both systems have complex feedback loops that interact with one another to amplify, modulate, or minimize neural signals. After a neurotransmitter is released, its products are transported back into the presynaptic cell and broken down, and the by-products are released into the cerebrospinal fluid (CSF). Measures of the metabolites of serotonin, 5-hydroxyindoleacetic acid (5-HIAA), and dopamine, homovanillic acid (HVA), in CSF are used as indicators of activity in these systems. Even though there are mean changes over the life course, longitudinal research has shown that individual differences in the concentrations of the metabolites of serotonin and dopamine in CSF are relatively stable over time. The stability of these systems makes them good candidates for mediating long-term differences in behavioral and reproductive strategies between individuals.

Testosterone, Dominance, and Aggression

Age and Seasonal Effects

Testosterone varies predictably with age and sex for group-living primates. Circulating levels are relatively low for juvenile males and females and increase markedly at puberty for males. Studies of captive and free-ranging rhesus monkeys have shown that juvenile males begin to increase in testosterone at age three and develop the full adult pattern by age five (Rose et al., 1978; Bernstein et al., 1991; Bercovitch, 1993; Higley, King, et al., 1996; Dixson and Nevison, 1997; Wallen, 2001). Field studies have verified the increase in testosterone during adolescence in a variety of other primate species, including Old and New World monkeys and prosimians (Brockman et al., 1998; Lynch, Ziegler, and Strier, 2002; Gesquiere et al., 2005; Gould and Ziegler, 2007). Increasing testosterone at puberty stimulates the development of the genitals and male secondary sex characteristics. These changes also coincide with a time of life when males prepare for emigration from the natal group and increased competition with other males.

For seasonally breeding primates, testosterone levels are typically higher during the mating season than during the birth season. Seasonal effects on male testosterone have been documented for a number of different species in captivity and in the field (Gordon, Rose, and Bernstein, 1976; Mehlman et al., 1997; Cavigelli and Pereira, 2000; Barrett et al., 2002; Lynch, Ziegler, and Strier, 2002; Ostner, Kappeler, and Heistermann, 2002; Gould and Ziegler, 2007; Muroyama, Shimizu, and Sugiura, 2007). These seasonal increases in testosterone are consistent with the hypothesis that testosterone prepares males for competition in the context of reproduction (Wingfield et al., 1990).

Relationship to Male Dominance and Aggression

The first study to report a relationship between testosterone and dominance in group-living monkeys was conducted at the Yerkes Primate Center (Rose, Holaday, and Bernstein, 1971). The authors placed 34 adult male rhesus monkeys together in a large outdoor corral, and recorded aggressive- and dominance-related behavior for the next four months. The results indicated that males in the upper quartile of the dominance hierarchy had significantly higher testosterone levels than middle- or low-ranking males. This seminal study had a major impact on the field.

Since that time, a large number of studies have looked at individual differences in testosterone, dominance, and aggression for male group-living primates, and the results have indicated that the relationship between testosterone, aggression, and dominance is more complex. A follow-up study at Yerkes by Rose and colleagues (Rose, Bernstein, and Gordon, 1975) suggested that rather than causing higher levels of aggression, changes in testosterone levels may have been a response to the outcome of aggressive competition. When four adult males were introduced to a group of females, testosterone levels of the males who won aggressive contests increased, and the levels of the males who were defeated dropped. The male with an early rise in testosterone following introduction became the dominant male of the new group. When the entire group was merged with a larger mixed-sex group, testosterone levels of all four adult males dropped as they lost rank to the males of the larger group. Bernstein et al. (1979) reported similar results for male pigtail macaques following a merger of two groups. Testosterone levels of the previous alpha and beta males dropped sharply after they were defeated in intense aggressive encounters, and there was an ephemeral increase in testosterone for the alpha male of the victorious group. In a second study involving brief one-week introductions of individual males who were already familiar with one another, the introduced males were not attacked, and even though they became low ranking, their testosterone levels did not decline. These results suggest that inhibition of testosterone release following aggressive defeat is a major factor influencing testosterone differences between high- and low-ranking males.

The importance of social instability in the testosterone-dominance relationship was recognized when longitudinal research on stable mixed-sex groups at Yerkes showed no relationship between plasma testosterone and male dominance rank (Gordon, Rose, and Bernstein, 1976). Differences between males in testosterone levels were not related to dominance status, either for the entire study period or during the mating season.

When testosterone was experimentally increased in four male rhesus monkeys living in a stable all-male group, the dominance hierarchy did not change, and there was no average increase in aggressive behavior for the treated males (Gordon et al., 1979). Several other studies of rhesus and Japanese macaques have also failed to find a relationship between testosterone and dominance for males living in stable social groups. Eaton and Resko (1974) found that plasma testosterone levels varied markedly over time for adult male Japanese macaques living in a large, stable mixed-sex group at the Oregon Primate Center, and individual differences in testosterone were not correlated with rates of aggression or dominance rank during the mating season. Bercovitch (1993) reported no overall relationship between testosterone and dominance rank during the mating season for adult and adolescent male rhesus living in a large, stable captive group. In a study of free-ranging Japanese macaques, Barrett et al. (2002) reported a significant correlation between levels of testosterone and noncontact aggression directed toward females during the mating season in the Arashiyama E population, but there was no relationship between testosterone and male dominance rank, which was stable throughout the study period. Muroyama, Shimizu, and Sugiura (2007) also reported no difference in testosterone levels between high- and low-ranking males in free-ranging Japanese macaques on Yakushima Island during the mating season; and contrary to expectation, high-ranking males had significantly lower levels of testosterone during the nonmating season. These studies suggest that testosterone is not related to dominance rank in stable groups, even during the mating season.

Further research with other primate species is consistent with the idea that the testosterone-aggression relationship is primarily found during periods of rank instability and male-male conflict. Steklis et al. (1985) found no mean differences in testosterone level between dominant and subordinate adult males in a study of captive vervet monkeys, but daily behavioral and serum samples indicated that testosterone level was positively correlated with the rate of aggression initiated the same day. In a study of dyadic relationships of captive pair-housed male cynomolgus macaques by Clarke et al. (1986), no association was found between relative testosterone and dominance rank for male dyads in stable dominance relationships, but the higher-ranking male had significantly higher testosterone levels in contested relationships. Another study of the relationship of testosterone, dominance, and aggression following group formation for male squirrel monkeys found a brief (two-week) elevation in average testosterone (T) levels following introduction, but no relationship between subsequent aggression and testosterone (Lyons, Mendoza, and Mason, 1994). The squirrel monkey males formed clear dominance hierarchies, but high- and low-ranking individuals did not differ in testosterone levels.

Adolescence is a time of increasing testosterone and increasing aggressiveness for group-living males, but within an adolescent age cohort, high testosterone does not necessarily predict higher rates of aggression. In the 1990s a series of studies was published by Mehlman, Higley, and colleagues that described the neurochemical and endocrine correlates of aggressive and impulsive behavior in adolescent rhesus monkey males from a large population of provisioned semi-free-ranging rhesus monkeys on Morgan Island. A number of young males were followed longitudinally from pre- to postadolescence (age three to five), with behavioral, hormonal, and CSF sampling at periodic intervals. At three years of age, testosterone levels for males were not related to rates of aggression or prior wounds in this sample (Mehlman et al., 1994). At age four, free testosterone levels in CSF (but not in plasma) were positively related to aggression initiated, aggression received, and male-male mounting (Higley, Mehlman, et al., 1996). At age six, males were sampled during the mating and nonmating seasons. CSF free testosterone and plasma testosterone increased significantly during the mating season, as did aggressive and sexual behavior, but individual differences in aggression during the mating period were not related to level of testosterone in plasma or CSF (Mehlman et al., 1997). In a study of adolescent development of male rhesus monkeys in another setting, Dixson and Nevison (1997) found that relative testosterone levels for three-year-old natal males were related to dyadic dominance for four of five dyads. These results indicate that adolescence is a time of changing hormonal and behavioral relationships, and the results for any given sample are likely to be related to local circumstances.

Field studies of the testosterone-aggression relationship among adult male baboons indicate that testosterone is not consistently related to stable dominance rank, but it may be related to initiating and winning fights during emigration and times of rank instability. The first major study to collect behavioral and endocrine measures in a wild unprovisioned primate population was conducted by Saplosky and colleagues for anubis baboons in east Africa (Sapolsky and Ray, 1989; Sapolsky, 1991; Virgin and Sapolsky, 1997). The results indicated that, on average, dominant and subordinate males did not differ in serum testosterone levels, but subordinate males who initiated fights had higher basal testosterone levels than subordinate males who did not (Sapolsky, 1991; Virgin and Sapolsky, 1997). Blood samples collected from an anubis baboon population at Ambolesi National Park in Kenya suggested that testosterone may be related to male aggression during group transfer (Alberts, Sapolsky, and Altmann, 1992). An extremely aggressive immigrant male in that population had testosterone levels that were more than double the levels of resident adult males. Among wild chacma baboons in the Okavango Delta, male aggression was positively related to testosterone during periods of rank instability but not during stable periods (Beehner et al., 2006). After controlling for age, there was no significant effect of testosterone on current dominance rank, but males with higher testosterone were more likely to rise in rank in the next year. Testosterone was also higher for males who dispersed during the study than for those who did not. In a related study from this site, Bergman et al. (2006) found that when both adult males had high testosterone levels, levels of dyadic aggression were higher than when one male or neither male had high levels. These results for wild baboons from three different populations support the conclusion that testosterone is related to preparation for, and response to, male-male conflict.

Studies of testosterone-aggression relationships in three different prosimian species produced results similar to those described above for Old World monkeys, for example, that testosterone is related to aggression primarily during periods of conflict. Brockman et al. (1998) used fecal sampling methods to collect a large number of samples per individual for male sifakas in Madagascar. They found that testosterone increased during periods of conflict, which included a male rank challenge, aggression from group females, and aggressive encounters with males from neighboring groups, and declined for two males after loss of rank. During most of the study period, the high-ranking males had higher testosterone levels than lower-ranking males. Ostner, Kappeler, and Heistermann (2002) conducted a similar study of wild redfronted lemurs in Madagascar that included three social groups studied over a 14-month period. They found that male dominance rank was clear during this period, but male rank was not related to individual differences in fecal testosterone levels. Both testosterone and aggression increased during the mating season, but individual differences in aggressive behavior during the mating season were not related to mean levels of testosterone. Testosterone-aggression relationships in ringtailed lemurs have been studied in a semi-free-ranging setting and in the field (Cavigelli and Pereira, 2000; Gould and Ziegler, 2007). Female estrous is highly synchronized in this species, resulting in a high level of scramble competition for mating during the brief mating season. Under these circumstances, both testosterone and male aggression increased during the mating season. During the mating season, individual differences in testosterone and aggression were positively correlated both within and between males in the semi-free-ranging setting but not in the field groups (Cavigelli and Pereira, 2000; Gould and Ziegler, 2007).

Field studies of testosterone-behavior relationships in three New World monkey species that differ in social organization suggest that testosterone is related to aggression only when the potential for male-male conflict is high. Strier, Ziegler, and Wittwer (1999) measured weekly changes in testosterone over a 1.5-year period spanning two mating seasons for wild muriquis, a species where males tend to stay in the natal group, while females transfer to other groups (Strier, 1992). In the muriquis, testosterone levels were not related to measures of mating activity, and no overt signs of aggression or competition over access to mates were observed. Lynch, Ziegler, and Strier (2002) studied tufted capuchins, a New World monkey species that shares a matrilineal and matrifocal social organization with many Old World monkeys. In this species, male testosterone levels increased markedly during the period of adult female mating synchrony, but the increase in testosterone was not accompanied by increases in aggressive behavior, and individual differences in testosterone were not related to dominance rank. A third New World species, the mantled howler monkey, lives in relatively peaceful multimale multifemale groups punctuated by periods of intense competition during group takeovers by solitary males (Clarke, 1983). Cristobal-Azkarate et al. (2006) tested the effects of the presence of solitary males on fecal testosterone levels of males from 10 wild groups of mantled howlers at Los Tuxlas Biosphere Reserve. They found that the best predictors of group male testosterone levels were the number and density of solitary males living in the forest fragment, a result that is consistent with a relationship between testosterone and preparation for male-male conflict.

Relationship to Male Mating Behavior

The Wingfield challenge hypothesis proposes that testosterone is most immediately related to aggression associated with reproduction (Wingfield et al., 1990). Testosterone levels increase during adolescence when males become sexually active (Wallen, 2001) and are generally highest during the season when mating is most common (for example, Lynch, Ziegler, and Strier, 2002; Gesquiere et al., 2005; Muroyama, Shimizu, and Sugiura, 2007). In captive rhesus monkeys, experimental exposure to females causes an increase in male testosterone levels, and suppression of testicular function reduces the rate of male mounting (Gordon, Bernstein, and Rose, 1978; Wallen, 2001). In field studies, changes in testosterone levels over time have been positively correlated with the number of females in estrous and male sexual activity (Lynch, Ziegler, and Strier, 2002; Beehner et al., 2006). These relationships lead to the presumption that testosterone levels are directly related to male sexual motivation and activity. Within the mating season, however, remarkably few studies have demonstrated that males with higher testosterone levels are more sexually active and have more mating success than males with lower testosterone levels. Mehlman et al. (1997) reported that testosterone levels were significantly higher for males observed in consortship with females, compared to males who were not seen consorting during the mating season for semi-free-ranging rhesus monkeys. Bercovitch (1993) reported a single case where the highest-ranking adolescent male had higher testosterone at an earlier age and also engaged in more sexual activity than lower-ranking adolescent males, indicating a pattern of earlier sexual maturation associated with testosterone and rank. In contrast, no significant relationship was found between testosterone and measures of sexual behavior in studies of captive rhesus, free-ranging Japanese macaques, and wild sifakas (Dixson and Nevison, 1997; Brockman et al., 1998; Barrett et al., 2002; Muroyama, Shimizu, and Sugiura, 2007). Results from other studies tend to be complex and only significant when within- and between-subjects samples are combined, confounding seasonal and individual effects (Lynch, Ziegler, and Strier, 2002; Beehner et al., 2006).

Testosterone and Aggression in Females

The few studies that have looked at testosterone-behavior relationships for female group-living monkeys and prosimians have generally found no relationship of individual differences in testosterone to dominance rank or rates of aggression (Altmann, Sapolsky, and Licht, 1995; von Engelhardt, Kappeler, and Heistermann, 2000). The exception is a recent study by Beehner, Phillips-Conroy, and Whitten (2005) for wild baboons in the anubis-hamadryas hybrid zone in Awash National Park, which reported a positive correlation between fecal testosterone, dominance rank, and rates of aggression for females in the study group. Testosterone levels varied significantly by age, season, and reproductive status in this study, and the significant relationship between testosterone and dominance was based on dry season samples of nonpregnant females, excluding the older females. Dominance rank remained stable when individual testosterone levels increased during pregnancy, and older females with low testosterone levels did not drop in rank, suggesting that female rank is only indirectly related to testosterone.

Testosterone: Summary of Results

Reviewing the results for group-living male primates reveals several interesting patterns (Table 8.1). While both aggression and testosterone typically increase during the mating season, individual differences in testosterone between males were not consistently related to aggressive behavior. Approximately half of the studies reporting data on rates of aggression found a positive relationship between testosterone levels and aggressive behavior, and the other half found no significant effect. The expectation that the relationship between testosterone and aggression would be stronger during the mating season was not supported by the data. The testosterone-aggression correlation was not consistently related to season, setting, male age, or taxonomic group. Both positive and negative findings were reported for studies in captivity and in the field— and even for studies of the same population at different time periods. A positive relationship between testosterone and aggression was somewhat more likely to be detected during periods of instability and rank challenge, when rates of aggression were higher, and less likely to be found for species with very low rates of aggressive interaction. The fact that experimentally altering testosterone levels caused an increase in aggressive behavior for only one of four treated males suggests that rising testosterone does not directly trigger aggressive behavior (Gordon et al., 1979).

In contrast to the mixed results for testosterone and aggressive behavior, a summary of Table 8.1 shows a clearer pattern for the relationship of testosterone to male dominance rank. In 12 of 14 studies reporting data on testosterone and dominance for males with stable rank relationships, high-ranking males did not have significantly higher levels of testosterone. This was true across taxonomic groups and for males living in captivity and in the field. In contrast, 7 of 8 studies found significant relationships between testosterone and dominance for males in unstable rank relationships, either from new group formation in captivity or from spontaneous rank challenges in the field; 4 of these reported decreases in testosterone following loss of rank, and 4 reported higher levels of testosterone in males prior to a rise in rank.

The results of these studies suggest that testosterone does not directly trigger aggressive behavior but rather that testosterone prepares the body for aggressive conflict, by increasing male strength and energy, and thereby increases the chances for a successful outcome. Effects of testosterone on behavior are facilitatory, but social factors determine suppression or expression of aggressive behavior (Bercovitch, 1993). Testosterone levels also change in response to conflict, initially rising during the conflict and then

Table 8.1 Summary of studies relating testosterone with aggression and dominance for male group-living monkeys and prosimians.

Study	Age-Sex/Species/n	Measures	Context	Aggression	Dominance	Other
Rose, Holaday, and Bernstein, 1971	Adult male rhesus macaque n=34	Plasma T	Captive new group	High aggression, high T	High dominance, high T	
Rose, Bernstein, and Gordon, 1975	Adult male rhesus macaque n=6	Plasma T	Captive new group, group merge		Rise in rank, increase in T; loss of rank, drop in T	High T predicts dominance
Gordon, Rose, and Bernstein, 1976	Adult male rhesus macaque n=7	Plasma T	Captive stable group		T not related to dominance	
Bernstein et al., 1979	Adult male pigtail macaque n=10	Plasma T	Captive group merge		Loss of rank, drop in T; rise in rank, brief increase in T	No change in T when loss of rank did not involve aggression
Gordon et al., 1979	Adult male rhesus macaque n=7	Inject HCG	Captive stable group	Rise in T, no consistent change in aggression	Rise in T, no change in dominance	
Eaton and Resko, 1974	Adult male Japanese macaque n=21	Plasma T	Captive stable group	T not related to aggression	T not related to dominance	T highly variable within individuals

(continued)

Table 8.1   (continued)

Study	Age-Sex/Species/n	Measures	Context	Aggression	Dominance	Other
Steklis et al., 1985	Adult male vervet monkey n=20	Plasma T	Captive stable group	Within male, high T, more aggression (dominant males only)	T not related to dominance	T highly variable within individuals
Clarke et al., 1986	Adult male cynomolgus macaque n=32	Plasma T	Captive stable/ unstable relationships		Higher T, higher rank (contested dyads only)	
Sapolsky, 1991	Adult male anubis baboon n=20	Plasma T	Field stable groups		T not related to dominance	Dominant males had transient increase in T following capture stress
Alberts, Sapolsky, and Altmann, 1992	Adult male anubis baboon n=1	Plasma T	Field male transfer	Very aggressive immigrant, very high T		
Bercovitch, 1993	Adolescent male rhesus macaque n=9	Plasma T	Captive stable group		T not related to dominance	Highest-ranking adolescent had higher T at a younger age
Lyons, Mendoza, and Mason, 1994	Adult male squirrel monkey n=9	Plasma T	Cative new group		T not related to dominance	New group formation, elevation in T

Reference	Subject	Measure	Condition	Finding	Finding	Finding
Virgin and Sapolsky, 1997	Adult male anubis baboon n=26	Plasma T	Field stable group	High T, more aggression (subordinates only)		
Dixson and Nevison, 1997	Adolescent male rhesus macaque n=8	Plasma T	Captive stable groups		Higher T, higher rank for 4 out of 5 dyads	
Mehlman et al., 1994	Juvenile male rhesus macaque n=26	Plasma T	Semi-free-ranging stable groups	T not related to aggression		
Higley, Mehlman, et al., 1996	Adolescent male rhesus macaque n=24	Plasma T, CSF free T	Semi-free-ranging stable groups	High T, more aggression (CSF, not plasma)		T variable within individuals
Mehlman et al., 1997	Adult male rhesus macaque n=33	Plasma T, CSF free T	Semi-free-ranging stable groups, mating season	T not related to aggression		Plasma T higher in males in consortships with females
Brockman et al., 1998	Juv-adult male sifaka n=7	Fecal T	Field stable/unstable periods	Within male, T higher during periods of conflict	High T, high dominance; loss of rank, drop in T	T variable within individuals
Strier, Ziegler, and Wittwer, 1999	Adult male muriqui n=6	Fecal T	Field stable groups, mating season	T not related to aggression		No mating season increase in T; low male-male aggression

(continued)

Table 8.1 (continued)

Study	Age-Sex/Species/n	Measures	Context	Aggression	Dominance	Other
Cavigelli and Pereira, 2000	Adult male ring-tailed lemur n = 10	Fecal T	Semi-free-ranging stable groups, mating and nonmating season	High T, more aggressive (mating season only)		T increased in mating season
Ostner, Kappeler, and Heistermann, 2002	Adult male red-fronted lemur n = 13	Fecal T	Field stable groups, mating season	T not related to aggression	T not related to dominance	T increased in mating season
Barrett et al., 2002	Adult male Japanese macaque n = 6	Fecal T	Field stable groups, mating season	High T, more aggression (noncontact only)	T not related to dominance	T peaked in early mating season
Lynch, Ziegler, and Strier, 2002	Subadult-adult male tufted capuchin n = 6	Fecal T	Field stable groups, mating and nonmating season	T not related to aggression	T not related to dominance	T elevated during consortships; low male-male aggression
Beehner et al., 2006	Adult male chacma baboon n = 13	Fecal T	Field stable and unstable periods	High T, more aggression (unstable periods only)	High T, high rank (age related); high T predicts future rank rise	T variable within individuals; T higher following dispersal

Study	Subjects	Measure	Setting			
Bergman et al., 2006	Adult male chacma baboon n=11	Fecal T	Field stable group	T not related to aggression	High T predicts future rank rise; T not related to dominance	More dyadic aggression when both males had high T
Cristobal-Azkarate et al., 2006	Adult howler monkey n=17	Fecal T	Field extra-group males			T correlates with number of solitary males in habitat
Muroyama, Shimizu, and Sugiura, 2007	Adolescent and adult Japanese macaque n=10	Fecal T	Free-ranging stable group, mating and nonmating season		High T, low rank in nonmating season; no effect in mating season	T higher in mating season
Gould and Ziegler, 2007	Adolescent and adult ring-tailed lemurs n=13	Fecal T	Field stable groups, mating season	T not related to aggression	High T, high rank for 2 of 3 groups	T and aggression higher during mating season

CFS = cerebrospinal fluid; HCG = human chorionic gonadotropin; T = testosterone.

declining after a serious loss. Dominance hierarchies in multimale groups allow testosterone levels to return to physiologically more conservative levels during periods of stable social relationships.

At this point, results for testosterone-behavior relationships in female group-living primates are too few to draw valid conclusions.

Neurochemistry of Aggression

Age, Sex, and Individual Differences

The measures of serotonin and dopamine functioning most commonly used for primate behavioral studies are the metabolites of neurotransmitter breakdown in CSF, 5-HIAA for serotonin and HVA for dopamine. Longitudinal research with captive and free-ranging group-living monkeys has shown that concentrations of these metabolites decline with age from late infancy to adolescence (Higley, Suomi, and Linnoila, 1992). Levels of the dopamine metabolite HVA continue to decline to old age, while the serotonin metabolite 5-HIAA is relatively stable through adulthood (Shelton et al., 1988; Raleigh et al., 1992; Fairbanks et al., 1999). Interestingly, the ratio of 5-HIAA to HVA is lowest at adolescence (Raleigh et al., 1992; Fairbanks et al., 1999). Sex differences in 5-HIAA appear at puberty, with males having significantly lower levels than females (Raleigh et al., 1992).

Within and across age groups, individual differences in CSF 5-HIAA and HVA are remarkably stable (Higley and Linnoila, 1997). Several different research groups have demonstrated genetic contributions to individual differences in 5-HIAA and HVA for rhesus macaques (Higley et al., 1993; Clarke et al., 1995), vervet monkeys (Raleigh et al., 1992; Freimer et al., 2007), and baboons (Rogers et al., 2004).

Serotonin and the Timing of Emigration

For most species of group-living primates, adolescence is the time of life when males must leave their natal group and work their way into another group. Declining serotonin activity at puberty for males coincides with the typical age of emigration. Two studies support a direct link between individual differences in serotonin functioning and the timing of emigration for rhesus monkeys. Cayo Santiago Island contains a large population of provisioned free-ranging rhesus monkeys, managed for biobehavioral research. The monkeys on the island live in naturally composed social groups with overlapping home ranges. At adolescence, males begin to spend more time away from the natal group, either alone or with other males, and they eventually establish residence in another group (Berard, 1994). As in the wild, this is a time of both risk and opportunity for males. Mortality rates are relatively high, as males leave their matrilineal support system behind and compete with other males for group access and dominance.

Individual animals on Cayo Santiago have been identified, and births, deaths, and group membership have been tracked for many years. All animals are captured for veterinary examination once a year, and samples of blood and cerebrospinal fluid were collected at that time. A study by Kaplan et al. (1995) compared concentrations of the serotonin metabolite 5-HIAA between males who had emigrated prior to five years of age and those who were still living in the natal group at that age. The results showed that the delayed dispersers had significantly higher levels of the serotonin metabolite (Kaplan et al., 1995).

In a similar study on Morgan Island, a cohort of three-year-old males was sampled for cerebrospinal fluid and followed for the next two years (Mehlman et al., 1995). In this study, the age of emigration for each male was significantly correlated with his serotonin metabolite level at age three. Males with higher levels of 5-HIAA spent more time grooming and more time in close proximity to other group members as juveniles, while males with relatively low levels of 5-HIAA were more likely to emigrate from the group at a younger age. Age at emigration was also related to physical risk taking measured by the rate of long leaps between tree limbs.

Results from these two studies are consistent with the hypothesis that decreases in serotonin at puberty increase risk taking for males preparing to emigrate and compete for dominance in a new group.

Serotonin, Dopamine, Aggression, and Impulsivity in Males

The relationship of individual differences in serotonin and dopamine functioning to aggressive and risky behavior in males has been studied in a number of different species and settings. These studies are summarized in Table 8.2.

A series of longitudinal studies of rhesus monkeys on Morgan Island have provided evidence that low indices of serotonin function are directly related to impulsivity and "escalated" aggression in this semi-free-ranging setting. The cohort of males described above was followed from pre- to postadolescence, with annual sampling of CSF to measure 5-HIAA and other monoamine metabolites. These males were then radio-tracked for behavioral observations of aggression, risk taking, and affiliative social

Table 8.2 Summary of relationship of serotonin and dopamine measures with aggression and dominance for male group-living monkeys.

Study	Age-Sex/Species/n	Neuro-chemical Measures	Context	Aggression	Dominance	Other
Botchin et al., 1993	Adult male cynomolgus macaque n=75	Prolactin response to fenfluramine	Captive social group	Low prolactin responders, higher % escalated aggression	Prolactin response not related to dominance	High responders, more social proximity
Mehlman et al., 1994	Adolescent male rhesus macaque n=26	CSF 5-HIAA, HVA	Semi-free-ranging	Low 5-HIAA, HVA, more wounds; low 5-HIAA, higher % escalated aggression		Low 5-HIAA, higher % long leaps
Kaplan et al., 1995	Adolescent male rhesus macaque n=38	CSF 5-HIAA, HVA	Semi-free-ranging			Low 5-HIAA, early emigration
Mehlman et al., 1995	Adolescent male rhesus macaque n=26	CSF 5-HIAA, HVA	Semi-free-ranging			Low 5-HIAA, early emigration; high 5-HIAA, more social proximity, grooming
Keyes et al., 1995	Adult male cynomolgus macaque n=39	Prolactin response to fenfluramine	Slide test	Low prolactin response, more aggression		

Higley, Mehlman, et al., 1996	Adolescent male rhesus macaque n=24	CSF 5-HIAA, HVA	Semi-free-ranging	Low 5-HIAA, higher % escalated aggression	Low 5-HIAA, higher % long leaps; stable individual differences in 5-HIAA and HVA
Mehlman et al., 1997	Adult male rhesus macaque n=33	CSF 5-HIAA, HVA	Semi-free-ranging	5-HIAA and HVA not related to aggression (mating season)	High 5-HIAA and HVA, more mating; 5-HIAA increased in mating season; stable individual differences in 5-HIAA
Kaplan et al., 1999	Adult male anubis-hamadryas hybrids n=22	CSF 5-HIAA, HVA	Field		Lower 5-HIAA in anubis males compared to hamadryas-anubis hybrids
Fairbanks et al., 2001	Adolescent-adult male vervet n=138	CSF 5-HIAA, HVA	Intruder challenge test	Low 5-HIAA, more aggression to intruder	Low 5-HIAA, more impulsive approach to intruder

(continued)

Table 8.2 *(continued)*

Study	Age-Sex/Species/n	Neurochemical Measures	Context	Aggression	Dominance	Other
Gerald et al., 2002	Adult male rhesus macaque n=15	CSF 5-HIAA, paternity	Captive social group			Higher 5-HIAA, more mating success; low 5-HIAA and young, more mating success
Kaplan et al., 2002	Adult male cynomolgus macaque n=25	CSF 5-HIAA, HVA	Captive social group		High HVA, low 5-HIAA, high dominance	
Morgan et al., 2002	Adult male cynomolgus macaque n=20	PET dopamine D2 receptor binding	Captive social group		High D2 receptor binding, high dominance	
Fairbanks, Jorgensen, et al., 2004	Adolescent-adult male vervet n=36	CSF 5-HIAA, HVA	Captive social group		Low adolescent 5-HIAA, high adult dominance	More impulsive as adolescent; high adult dominance
Howell et al., 2007	Juvenile-adult male rhesus macaque n=24	CSF 5-HIAA, HVA	Semi-free-ranging	Low juvenile 5-HIAA, HVA, more adult high-intensity aggression	Low juvenile 5-HIAA, high adult dominance	Low juvenile 5-HIAA, higher mortality

CSF=cerebrospinal fluid; 5-HIAA=5-hydroxyindoleacetic acid; HVA=homovanillic acid; PET=positive emission tomography.

behaviors. Measures that indicated impulsivity or lack of restraint were calculated from the percentage of all aggressive acts that were high intensity (chases, physical assaults), termed % *escalated aggression,* and the percentage of leaps that were longer than three meters *(% long leaps).* Presence of bite wounds during the physical exam was also recorded. For males sampled at age three, there were significant negative correlations between level of 5-HIAA and wounding, % escalated aggression, and % long leaps but not for total aggression or for total leaps (Mehlman et al., 1994). The concentrations of 5-HIAA and HVA, the metabolite of dopamine, were positively correlated, and both were significantly inversely related to the number of fight wounds found during the examinations. This further supports the conclusion that males with low serotonin functioning were more likely to engage in unrestrained forms of aggression. The following year at age four, CSF 5-HIAA was again negatively correlated with escalated aggression and % long leaps (Higley, King, et al., 1996).

Research at Wake Forest University confirmed the principal findings from the Morgan Island project in another species, setting, and method of assessing serotonin functioning. Aggression, affiliative behavior, and dominance rank were assessed in small unisex groups of cynomolgus macaques (Botchin et al., 1993). Serotonin functioning was measured by the increase in prolactin following an injection of fluoxetine, reflecting the responsiveness of the serotonin system to challenge. Consistent with results from Morgan Island, the low responders had a higher ratio of escalated to total aggression. They also spent more time alone and less time in affiliative contact with other males. When placed in a novel enclosure and presented with a series of slide images, males in the low prolactin group displayed more aggression toward a threatening image of a human (Keyes et al., 1995). These behavioral differences are consistent with the conclusion that males with low indices of serotonin functioning are more likely to respond in an impulsive and unrestrained fashion, while males with higher serotonin functioning are more likely to be restrained. This variation in behavioral style did not influence dominance rank, however. There were no differences in dominance status between high and low responders (Botchin et al., 1993). A later study that summarized dominance data across several groups found that HVA, the metabolite of dopamine, was positively related to dominance rank (Kaplan et al., 2002). Males in the upper section of the hierarchy had significantly higher HVA and relatively lower 5-HIAA (after controlling HVA), compared to subordinate males. Further confirmation of the role of dopamine in social dominance in this setting was provided by Morgan et al. (2002), who found significantly higher dopamine activity, as measured by positron emission tomography (PET), for dominant compared to subordinate males in cynomolgus monkeys housed in small unisex groups. Under these circumstances, neurochemical characteristics that promote aggressive and impulsive behavior were positively related to dominance attainment.

A similar set of findings has been reported for adolescent and adult male vervets living in large matrilineal captive groups at the University of California at Los Angeles Veterans Affairs (UCLA-VA) Vervet Research Colony (Fairbanks et al., 2001). The intruder challenge test was designed to provide a standardized measure of individual differences in response to a conspecific intruder. The presence of an adult male stranger at the edge of the home group evokes intense interest and excitement for all animals, but some will immediately rush over and challenge the stranger, while others are more cautious and measured in their responses. Behavioral scores on this test have been compiled into an index of social impulsivity that combines speed of approach with the extent of risky and aggressive behavior displayed toward the intruder. As would be expected, male vervets have higher scores on the impulsivity index than females do, and adolescents score higher than juveniles or adults (Fairbanks, 2001). When the scores of adolescent and adult males on the impulsivity index were compared with levels of the serotonin metabolite 5-HIAA in cerebrospinal fluid, the results showed that males with lower levels of the serotonin metabolite were higher in impulsivity than males with higher levels (Fairbanks et al., 2001). This is consistent with the results described above for semi-free-ranging male rhesus monkeys.

Pharmacological manipulation of serotonin activity has confirmed the role of serotonin in regulating impulsive behavior in male vervet monkeys (Fairbanks et al., 2001). In this study, the alpha male was removed, and one of the two remaining subordinate males in each of six groups was treated for nine weeks with fluoxetine, a serotonin reuptake inhibitor that increases transmission in serotonin neurons. At the end of the treatment period, the fluoxetine-treated males had significantly lower impulsivity scores in the intruder challenge test compared to control males (Fairbanks et al., 2001).

Serotonin and Male Sexual Behavior

Only a few studies have looked at serotonin activity in relationship to mating success. In the study of semi-free-ranging rhesus monkeys on Morgan Island described above, young adult males were sampled during the nonmating and mating seasons, and CSF measures were related to aggressive, affiliative, and sexual behavior (Mehlman et al., 1997). There were significant increases in 5-HIAA from the nonmating to the mating season, but individual differences in level were preserved across the two time periods. Aggressive, sexual, and grooming behavior also increased during the mating season. Unlike the prior studies with adolescent males, low 5-HIAA did not predict higher levels of escalated aggression during the mating season, but there was a significant positive relationship between 5-HIAA and heterosexual behavior (Mehlman et al., 1997). Young adult males with high CSF 5-HIAA had more sexual consorts and more matings than males with low 5-HIAA. This is consistent with the authors' conclusion that males with low CSF 5-HIAA are less socially competent than males with higher 5-HIAA. Gerald et al. (2002) followed up this finding with a study of paternity success in captive multimale breeding groups of rhesus monkeys. They found no difference in mean level of 5-HIAA for males who sired infants and those who did not, but within the social groups, the male with relatively higher levels of 5-HIAA was more likely to sire offspring. In the cases where the male with lower 5-HIAA was the father, he was usually the younger of the two. These two studies suggest that there is a mating disadvantage for rhesus monkey males with low levels of CSF 5-HIAA under some, but not all, circumstances.

Adaptive or Maladaptive Traits

Much of the research described above interpreted low levels of 5-HIAA and related high levels of risky and aggressive behavior as dysfunctional or maladaptive traits. The vervet study, in contrast, suggested that moderate levels of impulsivity were favored over the highest and lowest values. Alpha males were significantly more likely to score in the moderate range, while lower-ranking males were more likely to be extremely impulsive or extremely inhibited (Fairbanks, 2001).

Further evidence that variation in serotonin metabolism may underlie normal, adaptive differences in behavior comes from a study of wild baboons in Ethiopia. Awash National Park contains a hybrid zone where hamadryas and anubis baboons meet and interbreed (Phillips-Conroy and Jolly, 1986). Anubis baboons are a typical savannah baboon species where females remain in the natal group, males emigrate at puberty, and access to mating is controlled by male dominance hierarchies. Hamadryas baboons have a different social organization. In this species, males "kidnap" females from neighboring groups and form one-male, multifemale units in the male's home range. While many one-male units will share the same sleeping cliffs, the primary social bonds are between adult males and females. In 1995, males from the hybrid zone were captured for morphometric assessment and blood and CSF collection (Kaplan et al., 1999). A comparison of the serotonin metabolite for pure anubis males with anubis-hamadryas hybrids revealed that the pure anubis had significantly lower levels of the serotonin metabolite in CSF. This suggests that the species' differences in the tendency to emigrate and compete for dominance may be related to central serotonin activity.

Potential benefits of an impulsive-aggressive behavior style for adolescent males were demonstrated in a longitudinal study that followed males from adolescence to adulthood in a controlled captive setting (Fairbanks, Jorgensen, et al., 2004). In this study, 36 male vervets were first tested for social impulsivity on the intruder challenge test when they were adolescents, still living in the natal group. The males were then removed from the natal group, placed in all-male groups, and later dispersed into nine matrilineal breeding groups, to simulate the natural processes of emigration and immigration. Prior to introduction to the breeding groups, all males were sampled for levels of the metabolites of serotonin and dopamine in CSF. Additional information on the males included mother's dominance rank during development and body weight at the time of immigration. The males' behavior was observed for a full year following introduction to the adult breeding groups, and their dominance rank and impulsivity scores were determined at the end of the year. All of these factors were used to predict dominance rank at the end of the first year in the new groups. The alpha position is particularly important for vervet males, as the highest-ranking males typically father more than half of the offspring born during their tenure (Manson et al., in prep.). Based on our prior research showing that alpha males tended to be moderate in impulsivity, we predicted that adolescent males scoring in the moderate range in both impulsivity and 5-HIAA would be more likely to achieve high rank in the matrilineal breeding groups. Contrary to our expectation, males who were the most impulsive as adolescents and with the lowest levels of 5-HIAA were most likely to rise to the alpha position in their new groups. The males who achieved the highest rank could be predicted with a high degree of certainty by a combination of above-average body weight, high impulsivity, and low 5-HIAA and HVA prior to introduction (Fairbanks, Jorgensen, et al., 2004). A recent report of a longitudinal follow-up of male rhesus macaques from the Morgan Island study also found that males with low levels of 5-HIAA as juveniles were more likely to achieve high dominance rank as adults (Howell et al., 2007). The juvenile males with low 5-HIAA had a higher mortality rate and were more aggressive as adults, but those that survived became high ranking.

These results suggest that neurochemical mechanisms that promote impulsive, aggressive behavior can be beneficial under circumstances when a high-risk strategy can lead to a high gain in fitness. Adolescence is a time when physiological and neurochemical changes prepare males for emigration from the natal group and the challenges of competition with other males for dominance. Field studies describing the behavior of immigrant males have demonstrated the effectiveness of an impulsive, aggressive entry into a group, compared to a more cautious approach (Fedigan and Zohar, 1997; Silk, 2002). The aggressive strategy may involve a greater risk of injury, but the unobtrusive strategy almost never results in high rank attainment (van Noordwijk and van Schaik, 1985). It appears that these individual and age-related differences in male risk taking and aggression may be mediated by the serotonin and dopamine systems in the brain. Once dominance is established, the males' behavior and physiology return to the moderate range.

Serotonin, Dominance, and Aggression in Females

While most of the research on the serotonin, dopamine, and aggression for group-living monkeys focuses on males, there are several studies that report similar relationships in females, summarized in Table 8.3. Males are typically more impulsive and more aggressive than females, and they also have lower levels of 5-HIAA compared to females of the same species (Raleigh et al., 1992).

In one of the first studies of the serotonin—aggression relationship in females, Higley, King, et al. (1996) measured aggression, dominance rank, and wounding in two newly formed captive groups of adult female rhesus monkeys at the National Institutes of Health (NIH) Animal Center in Poolesville. As with the males described above, they found that individual differences in serotonin and dopamine metabolites were stable over time. Noninjurious aggression used to maintain social dominance was not related to the serotonin metabolite, but females with relatively low 5-HIAA had a higher incidence of wounding. This pattern is similar to that reported for semi-free-ranging male rhesus on Morgan Island, described above. Unlike the males, relatively high rates of 5-HIAA predicted high rank attainment for females in the new group.

The relationship of serotonin and aggression in females has also been reported in a series of studies of female rhesus and pigtail macaques by Westergaard and colleagues in captive social groups. In the initial study,

Table 8.3 Summary of relationship of serotonin and dopamine measures with aggression and dominance for female group-living monkeys.

Study	Age-Sex/Species/n	Neuro-chemical Measures	Context	Aggression	Dominance	Other
Higley, Mehlman, et al., 1996	Adult female rhesus macaque n = 16	CSF 5-HIAA	Captive social group	Low 5-HIAA, higher rates of wounding; 5-HIAA not related to total aggression	High 5-HIAA, high dominance rank in new group	5-HIAA, HVA increased following group formation; individual differences in 5-HIAA consistent tent over time
Westergaard, Suomi, and Higley, 1999	Juvenile female rhesus n = 31 Pigtail macaque n = 30	CSF 5-HIAA, HVA	Captive social group	Low 5-HIAA, higher % escalated aggression	High 5-HIAA, high dominance rank in new group	Rhesus more escalated aggression than pigtails; rhesus lower 5-HIAA and higher HVA than pigtails
Kaplan et al., 2002	Adult female cynomolgus macaque n = 21	CSF 5-HIAA, HVA	Captive social group		High HVA, high dominance	
Westergaard, Cleveland, et al., 2003	Juvenile-adult female rhesus macaque n = 45	CSF 5-HIAA	Captive social group	Low 5-HIAA, more aggression received; 5-HIAA not related to total aggression	5-HIAA not related to dominance	Low 5-HIAA, higher mortality; low 5-HIAA, more illness.

Study	Subject	Measure	Condition	Finding 1	Finding 2	Finding 3
Westergaard, Suomi, et al., 2003	Juvenile female rhesus macaque n=41	CSF 5-HIAA, HVA	Semi-free-ranging	Low 5-HIAA, HVA, more total aggression		Low 5-HIAA, higher % long leaps
Manuck et al., 2003	Adult female cynomolgus macaque n=25	Prolactin response to fenfluramine	Intruder challenge test	Prolactin response not related to aggression to intruder	Prolactin response not related to dominance	Low prolactin response, more impulsive approach to intruder
Shively et al., 2006	Adult female cynomolgus macaque n=17	PET serotonin 1A receptor binding	Captive social group	Low serotonin 1A binding, less aggression, more submission		
Fairbanks, nid.	Adult female Vervet monkey n=58	CSF 5-HIAA, HVA	Intruder challenge test	5-HIAA not related to aggression to intruder		

CSF=cerebrospinal fluid; 5-HIAA=5-hydroxyindoleacetic acid; HVA=homovanillic acid; PET=positive emission tomography.

juvenile female rhesus macaques had lower initial levels of 5-HIAA in CSF and higher rates of high-intensity aggression and wounding following group formation, compared with pigtail macaque females (Westergaard et al., 1999). Within each species, lower 5-HIAA predicted higher rates of escalated aggression and lower dominance rank in small unisex groups. In a longitudinal extension of this study for the rhesus females, low levels of 5-HIAA were related to higher mortality and higher levels of aggression received following transfer to mixed-sex breeding groups as young adults, but the relationship with dominance rank was not significant (Westergaard, Cleveland, et al., 2003). In a related study of semi-free-ranging female rhesus macaques on Morgan Island, low levels of both 5-HIAA and HVA were associated with high rates of aggression and also with the long leap ratio, a measure of impulsivity that was also found to relate to the serotonin metabolite for males in this population (Westergaard, Suomi, et al., 2003). These results suggest that low 5-HIAA is a risk factor for aggression and impulsivity for females as well as males for rhesus macaques.

Studies of the relationship between measures of serotonin functioning, aggression, and dominance at Wake Forest University have produced somewhat different results for female cynomolgus monkeys living in captive unisex social groups. Shively et al. (2006) used PET to measure serotonin 1A receptor binding in selected regions of the brains of female cynomolgus monkeys and related the results to measures of aggressive behavior and social dominance in the social group. The results showed that females with reduced serotonin binding in the amygdala initiated less aggression (not more) and were more submissive than females with higher serotonin 1A binding. Another study of cynomolgus females found a relationship between HVA, the dopamine metabolite, and dominance rank. Females living in four unisex groups were followed for 10 months, and CSF samples were taken during the luteal and follicular phases of the menstrual cycle (Kaplan et al., 2002). Higher-ranking females had consistently higher levels of HVA throughout the menstrual cycle, compared with lower-ranking females. There was no relationship between dominance and 5-HIAA, the measure of serotonin functioning, in this study.

The intruder challenge test has been used to measure impulsivity and aggressiveness in a large sample of female vervets living in stable matrilineal breeding groups (Fairbanks, Newman, et al., 2004). Vervet females are territorial in the wild, and females play a major role in defending the group territory from incursions by neighboring groups. In the challenge test in the captive groups, females responded to a stranger on the periph-

ery of the home cage with interest and excitement. They were less aggressive than males, but, like males, test-retest performance showed consistent genetically based individual differences in their response style.
Some females rapidly approached and challenged the stranger, while others hung back and watched from a distance. Unlike the males, however,
individual differences in aggressiveness were not related to the level of
the serotonin metabolite 5-HIAA. Figure 8.1 shows the level of aggressive behavior by quartile of 5-HIAA, calculated separately for males and
females. This analysis replicates the significant relationship between aggressiveness and 5-HIAA for males that was found in the earlier study
described above (Fairbanks et al., 2001). In contrast, there was no effect
of 5-HIAA level on female aggressiveness in this test. Female dominance
rank in this population is socially inherited and tends to be stable across
generations. Under these circumstances, there is no relationship between

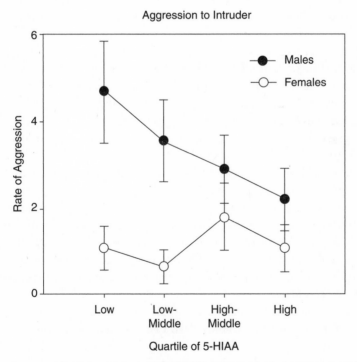

Figure 8.1 Mean (± s.e.) rate of aggression toward the intruder in the intruder
challenge test for adult male and female vervets, by quartile of the serotonin
metabolite 5-HIAA. Quartiles were calculated separately for males (n = 129) and
females (n = 58). Test procedures and scoring described in Fairbanks et al., 2003.

female dominance rank and aggressiveness in the intruder challenge test or in mean levels of 5-HIAA or HVA (Fairbanks, n.d.).

Similar results have been reported using the intruder challenge test to measure impulsivity and aggression in female cynomolgus macaques at Wake Forest (Manuck et al., 2003). Low serotonin function, measured by the prolactin response to fenfluramine, was related to higher levels of impulsive approach to the intruder but not to aggression toward the intruder. This study found no relationship between serotonin function and dominance rank. These results from cynomolgus and vervet females suggest that the serotonin-impulsivity relationship in female monkeys is similar to that found for males, but levels of aggression toward strangers in this context are relatively low and are not related to measures of serotonin functioning.

Serotonin and Dopamine: Summary of Results

Current results on the neurochemistry of aggression in group-living primates are limited by the number of species and research groups that have studied this topic. Nevertheless, a summary of the results reveals some interesting trends.

First, every study that included longitudinal measures of 5-HIAA found that individual differences in this measure were relatively consistent over time. The evidence for genetic heritability of this measure from numerous studies also supports the idea that there are traitlike differences in some aspects of serotonin neurobiology.

The results from Table 8.2 support a relationship between low levels of serotonin turnover and high levels of aggressive behavior for male macaques and vervet monkeys. In two studies, indicators of low serotonin functioning were associated with the overall rate of aggression and in three studies with the quality of aggressive interaction, as measured by the percentage of aggressive encounters that escalated to more serious chases or fights. Similar results were found for subjects in small unisex groups and semi-free-ranging naturally composed groups and for measures of spontaneous aggression and aggressive response to standardized challenge conditions.

The serotonin system has been associated with male emigration from the natal group in three studies. Two of these found that low serotonin metabolism promotes early emigration from the natal group, and the third found lower levels in the type of males who are more likely to emigrate in a baboon hybrid zone.

The results of all of these studies are consistent with the hypothesis that genetically based variation in serotonin functioning is a component of a life history strategy that varies from risk taking to conservative. Low serotonin activity is related to a risk-taking strategy for young males that increases the risk of injury and even death during emigration but may also increase their chances of successful emigration and attainment of high rank following immigration to a new group. The high serotonin males are more conservative, more likely to remain in the home group as young adults, and less likely to initiate aggressive challenges against unfamiliar adult males. Mating success for these two types varied according to age and circumstances. The lack of consistent findings on the relationship between serotonin and dominance rank also fits this picture. On average, the highest-ranking males are probably closer to the population mean than males at either extreme.

The results for females are more limited than those for males and may depend on the species and context. Rhesus monkey females are typically more aggressive than females from other macaque species, and the relationship of serotonin to aggression in this species was similar to that found for males in two studies. In two other studies, rhesus females with low 5-HIAA were more likely to be the victim than the initiator of aggression. Low serotonin functioning was not related to aggression initiated toward an intruder for cynomolgus and vervet monkeys. Research using PET scans on cynomolgus females found that females with low serotonin binding were less aggressive and more submissive compared to females with higher indices of serotonin activity. Taken together, these results suggest that low serotonin functioning may be a result, rather than a cause, of increased aggressive interactions under these circumstances.

As with the male macaque studies, there was no consistent relationship between measures of serotonin functioning and dominance rank for females. High 5-HIAA was related to high rank in two studies involving newly formed groups and not in two others of females in relatively stable groups. Other indices of serotonin function were also not related to female rank in female cynomolgus monkeys.

The relative influence of serotonin and dopamine on aggressive behavior in male and female group-living primates is difficult to calculate because of the high degree of correlation between the two metabolites (5-HIAA and HVA) that are used as measures of functioning for these two systems. Low levels of HVA, the dopamine metabolite, were related to higher levels of aggressive behavior and a higher incidence of wounding in two studies. High-ranking males and females had higher levels of HVA than lower-ranking animals in a study of cynomolgus macaques, while low levels of HVA predicted future rank attainment for vervet males. More research is needed with different measures of serotonin and dopamine functioning to sort out these differences.

Discussion

As this review demonstrates, there has been a lot of attention paid to physiological contributions to aggression in group-living monkeys. While there is a preponderance of evidence that both testosterone and serotonin are involved in regulating aggression, there is also considerable variation across studies in the nature of the relationship.

There are many reasons why one study may find a relationship that is not seen in another. Although they are socially significant, aggressive interactions can be relatively rare events and difficult to measure reliably. The behavior sampling system must be intensive to capture enough occurrences to measure individual rates with sufficient precision to track changes in aggression over time and according to circumstances. Testosterone levels also vary markedly within individuals, from season to season and from day to day. Because of this temporal variability, the number of monkeys sampled and the number of samples per individual constitute an important factor in the generalizability of the research results. Methodological factors such as small sample size or unreliable recording methods can lead to false negatives.

Lack of replication between studies can also be due to true differences related to species, season, age, sex, or circumstances. The Wingfield challenge hypothesis was originally developed to explain species differences in the temporal pattern of testosterone secretion in relationship to breeding strategies (Wingfield et al., 1990). One prediction from the general hypothesis is that testosterone should promote aggression in situations where aggression is beneficial for reproduction. For birds, this includes territory establishment and mate guarding, but for group-living monkeys, the analogous contexts are emigration from the natal group and establishment and defense of a position in the dominance hierarchy. The theory also predicts that increases in aggression in response to increasing testosterone are more likely under conditions of social instability when the costs and benefits of aggression favor an active response.

Summary of the testosterone-aggression results underscored the high degree of variability in testosterone levels from day to day and season to season and supported the hypothesis that fluctuating testosterone levels were highly responsive to season and circumstances related to mating opportunities. Because males play little or no role in infant care in the species reviewed here, the reduction in testosterone levels outside of the mating season cannot be explained by a trade-off with parental care. Seasonal variability in group-living primates is probably related to the physiological costs of maintaining chronically high circulating testosterone levels when there is no counteracting benefit to the individual (Wingfield et al., 1990).

The testosterone results are also consistent with the hypothesis that testosterone and aggression are linked during periods of instability, a time when the opportunity and potential payoff for a successful aggressive response are likely to be greatest. During periods of stability, rising testosterone levels may increase vigor and aggressive motivation, but the risks associated with initiating aggressive behavior and the high probability of losing an aggressive challenge are greater than the expected benefits. Under these circumstances, social and cognitive factors outweigh the motivational effects of testosterone, and seasonal increases in testosterone are not accompanied by changes in the rank structure. This basic set of conclusions appeared to hold across taxa and setting. The importance of stability and opportunity in predicting the testosterone-aggression relationship was found for animals in undisturbed groups in the field and also in captivity. No consistent differences were noted for Old World versus New World monkeys or monkeys versus prosimian primates.

The prediction that the relationship between testosterone and aggression would be more evident during the mating season was not borne out by the data. Numerous studies found seasonal increases in both testosterone and aggression, but males with higher levels of testosterone during the mating season were not consistently more aggressive and did not appear to have a reproductive advantage over males with lower levels of testosterone, after age was taken into account. Although the studies reviewed here did not include verification of paternity, results for many group-living primate species have found a significant reproductive advantage for high-ranking males (de Ruiter and van Hooff, 1993). The effects of testosterone during periods of group instability and rank challenge may serve the reproductive function proposed by the Wingfield challenge hypothesis, if increasing testosterone helps to prepare a male for a successful response to an opportunity to increase in rank or to defend a high-ranking position. Once established, high rank tends to be relatively stable across seasons and is likely to enhance a male's future reproductive success.

Compared to testosterone, which varies widely from day to day, individual differences in measures of serotonin and dopamine metabolites appear to be relatively stable over time. There are age-related changes in these measures, but within age groups, individual differences tend to be relatively consistent. Because of this, the serotonin metabolite has been treated as a measure of traitlike differences in serotonin system functioning. The results of the studies presented here are consistent with the hypothesis that serotonin activity is related to alternative life history strategies. Males with higher indices of serotonin metabolism appear to adopt a conservative life history strategy. They are less aggressive and take fewer risks than males with lower indices of serotonin metabolism. Males with consistently lower levels of serotonin activity are more likely to adopt a high-risk–high-gain strategy that may lead to early death but that may also lead to higher rank attainment and reproductive success at a young age.

The subjects in the serotonin-behavior studies reported here are heavily weighted toward the adolescent and young adult period. This is a time of life when impulsive and aggressive behavior is at its peak for human and nonhuman primates, particularly for males. Declining serotonin and dopamine activity is likely to play a role in preparing adolescent males for emigration from the natal troop. Changing thresholds for risky behavior and aggressive response to social confrontations coincide with the time of greatest challenge and opportunity for emigrating males. The consistency of behavioral and neurochemical changes at this time of life suggests that adolescent impulsivity is an evolved feature of male life history strategies.

The number and variety of species with data on serotonin-behavior relationships are too small to draw valid conclusions about the socioecology of serotonin and life history, but we would expect the benefits of a high or low serotonin strategy to vary according to the social system. For group-living monkeys with male emigration, the way a male handles joining a new group has important consequences for rank attainment and reproductive success. Low serotonin males might be favored when mating is restricted to a few individuals and to brief periods of time. None of the studies reported here were conducted in species—such as the Hanuman langur—that live in one-male groups punctuated by relatively violent overthrows of the resident male by a challenger (Rajpurohit and Sommer, 1993). In winner-take-all systems like this one, a certain degree of risk taking is necessary for a male to father any offspring. In contrast, the high serotonin strategy might be favored when male life expectancy is long, relationships are stable, and males have multiple mating opportunities. Data on neurochemistry and behavior of species with different social systems are needed to shed more light on these interesting questions.

More data on the neurochemistry of aggressive and risky behavior for females living is different contexts are also needed. Primate species differ in the average amount of aggression among females, depending on the level of resource competition and the stability of social relationships. For most group-living primates, dominance rank is buffered by the female kin group, and the effect of dominance on reproductive success is considerably less for females than for males. Females are also less likely than males to respond aggressively to social challenges. For these reasons, the link between neurochemistry and aggression for female group-living primates may not be as clear as it is for males. More research using both male and female subjects will help to reveal the degree to which there are true gender differences in endocrine and neurochemical influences on behavior.

CHAPTER NINE

The Endocrinology of Intersexual Relationships in the Apes

Melissa Emery Thompson

Introduction

As a taxonomic group, the hominoids share many anatomical and behavioral features. Despite this, the 18 species and 23 subspecies in this group display a remarkably diverse range of social structures and mating systems (Table 9.1). Accordingly, there is variation in the pressures and constraints on male-female sexual relationships across these species. In this chapter, I focus on the diversity of social and sexual relationships between males and females in ape societies, with emphasis on how hormonal information has helped make sense of this variation. I will describe how hormones influence the strategies and reproductive success of each sex and what could potentially be learned by extending our studies of wild populations. By looking at how endocrine responses vary with mating system, we can discover which patterns represent generalized physiological relationships and which may be specific to particular aspects of bonding, competition, or reproductive strategy. These examinations may ultimately help elucidate which patterns in the behavioral endocrinology of humans originate from our ape ancestry and which are derived features relating to human socioecology.

Sexual interactions between males and females ultimately represent a combination—or compromise—of male and female strategies and incorporate the constraints of competition and sexual coercion. I will first explore the relationship between hormonally mediated sexual signals, female sexual behavior, and male response in ape species; variation in these relationships provides important clues to the nature and efficacy of

Table 9.1 Ape social systems and intersexual relationships.

	Social Organization	Mating System	Sexual Dimorphism[a]	Dispersing Sex	Dominant Sex	Central Adult Social Bonds (→ direction)	Copulation Initiation (> slight bias; >> strong bias)
Hylobatids (*Bunopithecus hoolock, Hylobates spp., Nomascus spp., Symphalangus syndactylus*)	Bonded pair with dependent offspring	Largely monogamous with covert extra-pair mating	Monomorphic (0.96–1.15)	Both	Codominant (or slightly female biased)	M → F	M >> F
Orangutans (*Pongo abelii, P. pygmaeus*)	Dispersed with aspects of fission-fusion	Polygynandry	Large (1.86–2.25 flanged; ~1.0 unflanged)	Males (?)	Flanged males	None?	F>Flanged M; unflanged M >> F
Gorillas (*Gorilla beringei, G. gorilla*)	Unimale (multimale also common in *G. beringei*)	Harem polygyny (polygynandry in multimale groups)	Large (1.63–1.72)	Both	Males	F → M	F>M
Bonobos (*Pan paniscus*)	Multimale, multifemale with fission-fusion	Polygynandry	Moderate (1.19–1.47)	Females	Individualistic	M → F, FF	M >> F
Chimpanzees (*Pan troglodytes*)	Multimale, multifemale with fission-fusion	Polygynandry	Moderate (1.19–1.41)	Females	Males	MM	M>F

[a]Body size dimorphism figures in parentheses represent the ratio of male to female body weight, as reported in Plavcan & van Schaik, 1997.
Sources: hylobatids: Palombit, 1994a, 1996; Reichard, 1995, pers. comm.; Fuentes, 2002; orangutans: Rijksen, 1978; Galdikas, 1981, 1985; Mitani, 1985; Schürmann and van Hooff, 1986; Fox, 2002; gorillas: Watts, 1991; Sicotte, 1994; Robbins, 1999; Kano, 1992; chimpanzees: Goodall, 1986; Boesch and Boesch-Achermann, 2000; Hohmann et al., 1999; bonobos: Hashimoto and Furuichi, 2006.

female reproductive strategies. Next, I will examine the hormonal correlates of male competitive relationships, particularly as they relate to mating success. Finally, I will focus on the more direct manifestation of conflict of interests between male and female reproductive strategies: sexual coercion.

Female Reproductive Cycles and Sexual Cycles

Ape ovarian cycles are similar to human cycles and can be distinguished from most other primates by associated peaks of estrogen and progesterone in the luteal (postovulatory) phase (Czekala, Shideler, and Lasley, 1988). Aligned relative to ovulation, ovarian steroid profiles are broadly similar across the apes, though the length of the follicular phase, and the resulting total cycle length, varies (Table 9.2); note, however, that intraspecies variation in cycle length exceeds any systematic interspecies variation. Prolonged estrogen production, unopposed by progesterone, may contribute to broader behavioral receptivity of some apes and allows for extended morphological displays of receptive status (Czekala, Shideler, and Lasley, 1988). Thus, the species with prominent sexual swellings, chimpanzees and bonobos, have longer follicular phases (when the swelling occurs) and longer total cycle lengths than other apes.

Unlike many mammals and indeed many other primates, the sexual behavior of female apes is not dependent on female hormonal status (Dixson, 1998). Mating is possible and does occur on any day of the cycle, though both males and females often act to restrict matings to more defined periods. While hormones do not determine female sexual behaviors, they have indirect and mediating effects. Late follicular estrogen activity correlates with increases in female sexual behaviors, as in most other species. It may also have more subtle and interesting effects on sexual relationships, including mate selectivity and attractiveness to males.

Mammalian female sexuality is composed of three interrelated components, each of which can be influenced by hormones: (1) receptivity, the willingness (or in some cases, physical capacity) to mate; (2) proceptivity, the active solicitation or initiation of mating; and (3) attractiveness, the stimulus value of the female to the male (Dixson et al., 1973; Beach, 1976). It can be difficult to identify each of these components independently of the others. In laboratory settings, the use of restricted access tests to allow females to govern sexual access by their much larger mates has helped to decouple proceptive behaviors from receptive or coerced sexual behaviors (Nadler, 1995). Similarly, careful attention to patterns of solicitation, aggression, and copulation approach can help discrimi-

Table 9.2 Sexual behavior and ovarian cycle characteristics of apes.

Species	Context	Cycle Length	Swelling Duration (days/cycle)	Swelling Size (dorsal area, cm²)	Copulation Rate of Receptive Females (cops/hr)	Copulation Duration (s)	References
Hylobates lar, H. agilis, H. muelleri	Captive	19–24	5–7	9.6			1–3
Nomascus leucogenys	Captive	22				34	4
Hylobates lar	Wild	21–29	9			15	3, 5, 6
Symphalangus syndactylus	Wild		Present				7
Pongo pygmaeus/abelii	Captive	26–31	None (4–6 proceptive)	None		840–900	8–11
P. pygmaeus	Wild	22–51	None (1–6 proceptive)	None	<0.2	618	12, 13
P. abelii	Wild	30	None	None	<0.3	300–900	14, 15
Gorilla gorilla gorilla	Captive	30–33	2	3.6		53	16, 17
G. gorilla beringei	Wild	28–29	2		0.3–0.5	80–97	18–20
Pan troglodytes	Captive	33–38	11–18	297	0.8	8	21–24
P. troglodytes verus	Wild	40	11	187		8	25, 26
P. troglodytes schweinfurthii	Wild	35–36	10–13	120	0.4–5.0	7	27–31
P. paniscus	Captive	34–49	13–22		0.3	20	32–34
P. paniscus	Wild	42			0.3–0.4	12–15	34–36

Note: In cases where multiple studies are cited, ranges represent the range of averages from all studies rather than the range of individual variation.

[1]Dahl and Nadler, 1992; [2]Cheyne and Chivers, 2006; [3]Barelli and Heistermann, 2007; [4]Lukas et al., 2002; [5]Ellefson, 1974; [6]Barelli et al., in press; [7]Chivers, 1974; [8]Nadler, 1977; [9]Nadler, 1982; [10]Nadler, 1988; [11]Masters and Markham, 1991; [12]Galdikas, 1981; [13]Mitani, 1985; [14]Schürmann, 1982; [15]Wich et al., 2006; [16]Nadler, 1975; [17]Nadler et al., 1983; [18]Harcourt et al., 1980; [19]Harcourt, Stewart, and Fossey, 1981; [20]Watts, 1991; [21]Yerkes, 1939; [22]Dahl, Nadler, and Collins, 1991; [23]Emery and Whitten, 2003; [24]Shimizu et al., 2003; [25]Deschner et al., 2003; [26]Stumpf and Boesch, 2006; [27]Goodall, 1986; [28]Tutin and McGinnis, 1981; [29]Emery Thompson, 2005; [30]Watts, 2007; [31]Emery Thompson, n.d.; [32]Heistermann et al., 1996; [33]Reichert et al., 2002; [34]Paoli et al., 2006; [35]Furuichi, 1987; [36]Kano, 1992.

nate male and female mating preferences in natural settings. Ultimately, sexual behaviors occur in a complex hormonal environment; peak estrogen, peak androgen, and peak gonadotropin production are temporally associated in the periovulatory period of apes (Shideler and Lasley, 1982; Nadler et al., 1985; Graham, 1988). While sexual behaviors have been studied following ovariectomy, adrenalectomy, and other hormonal manipulations in some monkey species, such experiments have not been performed in apes, and mechanisms revealed in these experiments are not necessarily generalizable across taxa (Dixson, 1998).

Though hormones are frequently discussed in relation to female *behavior* (receptivity and proceptivity), they are less frequently discussed in relation to female attractiveness independent of overt behavioral cues. These mechanisms are still being investigated, though it appears likely that males do not mate with females indiscriminately, nor do they respond uniformly to the display of a signal of female receptivity; instead, males likely use both direct and indirect cues of female ovarian function in order to evaluate mating opportunities. Female attractiveness is an important—though perhaps involuntary—component of her reproductive strategy as it can come with significant costs as will be shown later in this chapter.

A key component of attractiveness in most apes is the physical display of sexual swellings. Where it occurs, swelling (or tumescence) of the anogenital tissue takes place in conjunction with increasing estrogen levels during the follicular phase of the cycle (chimpanzees: Graham et al., 1977; gorillas: Nadler et al., 1979; bonobos: Dahl, Nadler, and Collins, 1991; gibbons: Cheyne and Chivers, 2006). Estrogen receptors promote fluid retention in those tissues; progesterone has an antagonistic effect, leading to rapid detumescence (Ozasa and Gould, 1982). Because ovulation itself is associated with the transition from peak estrogen production by the preovulatory follicle to progesterone production by the recently vacated follicle (the corpus luteum), it can be expected that swellings will be reliable in identifying ovulatory timing and may even have evolved to signal this information to potential mates (Hamilton, 1984). However, the strength of this relationship, and the size and duration of swelling, varies (Table 9.2).

Hylobatids.   Sexual swellings in hylobatids are relatively small, being confined to reddening and tumescence of the vulval and clitoral tissues. While not considered "exaggerated" in terms of size (Nunn, 1999), they are displayed for a substantial portion of the menstrual cycle (Barelli et al., 2007). In wild *Hylobates lar,* approximately 80% of ovulations occur during maximal sexual swelling, and approximately 50% occur three to four days after the onset of swelling; however, ovulation is not strictly tied to swelling development and can occur considerably before or after tumescence (Barelli et al., 2007). In contrast, the vast majority of copulations in *H. agilis* and *H. muelleri* occur during only two days of the swelling period, just before detumescence but after the peak in sexual swelling size, suggesting that ovulation might be more restricted in these species (Cheyne and Chivers, 2006).

Swellings are commonly associated with multimale mating (Nunn, van Schaik, and Zinner, 2001) and concealed ovulation with monogamy (Sillen-Tullberg and Møller, 1993). It may therefore seem surprising to find an extended swelling period in hylobatids, which characteristically form socially and sexually bonded pairs (Fuentes, 2002). While hylobatids are frequently classified as monogamous, considerable field evidence indicates that the true mating system includes extrapair copulations and sequential or simultaneous polygyny and polyandry (Palombit, 1994a, 1994b; Reichard, 1995, pers. comm.; Fuentes, 2000). Male partners invest more in maintaining the pair-bond than females, suggesting that an exclusive relationship may be of greater advantage to males than females (Palombit, 1996, 1999; Fuentes, 2002). The display of sexual swellings by female gibbons suggests that females may have reproductive interests in attracting copulations from outside the pair-bond.

Gorillas. Gorilla sexual swellings are similar in size and conformation to those of gibbons but are more restricted in duration (Dahl and Nadler, 1992). In the context of the predominantly unimale mating system, silverbacks face little competition for immediate mating opportunities. Females have a vested interest in assuring the paternity certainty of the silverback because they depend on this male's protection from infanticide by outside males (Watts, 1989). We should expect a brief period of sexual behavior within the menstrual cycle, as well as female proceptive behavior directed at the silverback. The first expectation is supported by relatively precise sexual advertisement. Gorillas largely restrict copulations to a short period (one to four days) near the time of maximal tumescence (Nadler, 1976; Harcourt and Stewart, 1978). Peaks in sexual behavior are associated with peak follicular estrogen and testosterone production by the female (Czekala et al., 1983; Nadler et al., 1983; Mitchell et al., 1985). Female sexual behavior does not, however, show a dose-response relationship with estrogen, suggesting that this may be a threshold effect that is promoted by the subsequent androgen peak (Nadler et al., 1983; Mitchell et al., 1985). However, this pattern may also represent differing male and female responses to ovarian cycle variation. Female attractiveness, as suggested by male solicitation, increases as midcycle estrogen increases; female proceptivity is more closely correlated with the androgen peak that follows the estrogen peak (Nadler et al., 1983). Females are physically receptive throughout the cycle, though assessment of behavioral receptivity is of little value for this species where sexual dimorphism limits female sexual resistance.

Orangutans.   Orangutans show no outward signal of sexual receptivity, though they display small labial swellings during pregnancy (Galdikas, 1981). In addition to the lack of sexual advertisement, frequent forced copulation and long durations of intercourse (Table 9.2) distinguish orangutan sexual behavior from the other apes. Extreme sexual dimorphism and the propensity for forcible copulations with resistant females may seem to negate the importance of receptive behavior in orangutan females. However, female proceptive behavior in these species is quite pronounced. Females frequently approach males for copulation, facilitate intromission, perform pelvic thrusting, and even orally stimulate males. Even copulations initiated by male force or containing elements of male-female aggression may also include active female sexual behaviors (Fox, 1998).

The lack of an overt ovulatory signal has hindered the assessment of hormonal influence on sexual behavior in orangutans. Due to low population densities, males might benefit from and face few costs by attempting to mate whenever they encounter a female. Matings do occur throughout the cycle and even during pregnancy and lactation (Schürmann and van Hooff, 1986; Fox, 1998). Females, on the other hand, may benefit by showing no conspicuous indication of cycling or ovulatory status. The pressure to confuse paternity is likely to be less than in other apes: infanticide is thus far unknown in orangutans (but see van Schaik, van Noordwijk, and Nunn, 1999), and males do not provide paternal care or infant protection. Furthermore, the dispersed social system makes a visual indication of ovulatory status of limited usefulness to attract mates. Without a sexual advertisement, females may be able to exercise mate choice while avoiding increased feeding competition and sexual coercion. Physical concealment of ovulation is supported by Fox's (1998) observation of a flanged male responding possessively to a female during one periovulatory period but failing to mate even while associating with the same female during another. In the first instance, the female displayed proceptive behaviors; in the second, she did not. This has led researchers to conclude that female proceptive behavior may be the strongest indication of periovulatory period and the strongest determinant of male mating effort (van der Werff ten Bosch, 1982; Fox, 1998). Female proceptive behavior is strongly biased toward dominant flanged males. Females almost exclusively approach the long calls of the dominant resident male (who calls most frequently) and may subsequently stay in consort and initiate copulation with him (Fox, 2002; Delgado, 2006). While females do resist copulations with both flanged and unflanged males, cooperation is most likely with the most dominant of the resident flanged males (Mitani, 1985).

By soliciting dominant flanged males, females may gain protection from harassment or sexual coercion by unflanged males (Schürmann and van Hooff, 1986; Fox, 2002). The dominant male can maintain priority of access to a proceptive female, chasing away other males attempting to copulate. While unflanged males are not often the direct recipients of proceptive behavior, they aggregate around consortships between proceptive females and flanged males (Galdikas, 1985; Fox, 2002). When unflanged males are nearby, females tighten their spatial proximity with associated flanged males, supporting the hypothesis that females are actively using flanged males as protection (Fox, 2002). Thus, females are largely responsible for initiating copulations with flanged males, while unflanged males are almost exclusively responsible for initiating their copulations, frequently with considerable female resistance (Galdikas, 1985; Schürmann and van Hooff, 1986).

These patterns from wild populations strongly suggest that while ovulation is physically concealed, females selectively display behavior that indicates periovulatory status to males, though hormonal information is still lacking to confirm this. Preliminary data from Borneo indicate that female ovulatory status and male reproductive development produce interacting effects on female proceptivity and resistance (Knott, Emery Thompson, and Stumpf, 2007). Captive studies find that proceptivity, as indicated by initiating access to a confined male, is cyclical in nature and restricted to the periovulatory period (Nadler, 1988; Masters and Markham, 1991). These studies also suggest that males lack other cues of ovulatory status, as they initiate mating over the majority of the female's cycle when given free access (Nadler, 1988). It is unclear whether the midcycle peak in estrogens or in androgens might facilitate female sexual behavior, though the luteal increase in progesterone has been linked with abrupt postovulatory cessation of proceptivity (Graham, 1988; Nadler, 1988).

Chimpanzees.   While promiscuous mating appears to be a feature of mating systems in all ape species, it is particularly pronounced among the multimale, multifemale communities of chimpanzees and bonobos. In these species, females mate hundreds of times per conception, often with all available mates (Wrangham, 2002). Accordingly, sexual swellings are substantially larger (incorporating the circumanal as well as labial tissues) and longer in duration than in gorillas or gibbons.

In chimpanzees, sexual activity is primarily confined to the period of maximal tumescence, approximately 10–12 days of a 35-day cycle (Wallis, 1997; Emery and Whitten, 2003). Midcycle peaks in testosterone in female chimpanzees are larger and more prolonged than those of gorillas or orangutans, suggesting a role in the extension of sexual receptivity relative to these other species (Nadler et al., 1985). Ovulation occurs reliably within the period of maximal swelling, but its timing is variable (Deschner et al., 2003; Emery Thompson, 2005). Peak estrogen secretion occurs most frequently between days D-5 and D-2 (D0=first day of detumescence). Incorporating mammalian gamete survival times (Johnson and Everitt, 1988), this predicts the maximum probability of fertile mating between days D-7 and D-3, earlier than is generally assumed (Deschner et al., 2003; Emery Thompson, 2005). For male chimpanzees, the sexual swelling may be a mixed signal, suggesting an increased probability of reproductive success but with a degree of imprecision. This imprecision is compounded because the duration of maximal swelling varies substantially, and fecundability appears to be best predicted retrospectively from swelling cessation. Photographic evidence from West African chimpanzees indicates that there are subtle changes in the size of sexual swellings (~70–110% of average size) that are considered maximally tumescent to human eyes and that the largest swelling occurs during the periovulatory period (Deschner et al., 2004). It is difficult to know whether this change is perceptible to male chimpanzees—or if they might need to compete for close access to females to evaluate these changes.

Whether or not this information is directly conveyed via sexual swelling changes (olfactory cues have also been implicated: Matsumoto-Oda et al., 2002), mating interest intensifies during the fertile window. Increased attractiveness during the fertile window is suggested by a greater proportion of male-initiated copulation, greater proportion of adult versus subadult mates, high copulation rates by dominant males, more males in association, and higher rates of male interference in copulations (Deschner et al., 2004; Emery Thompson, 2005; Emery Thompson and Wrangham, in press).

In the ape species discussed previously, female behavior typically focuses copulations on the most fertile times of the cycle. In chimpanzees, however, females initiate copulations at higher rates during the earlier, nonfertile days of the maximal swelling period (Stumpf and Boesch, 2005; Emery Thompson and Wrangham, in press). Available evidence suggests that proceptive behavior in female chimpanzees functions to increase the number of mating partners and obfuscate paternity (Hrdy, 1981; Watts, 2007). This paradoxically can result in increased female proceptivity toward nonpreferred partners during nonfertile days; during the fertile window, females increase their resistance to these same partners (Stumpf and Boesch, 2006). Maximal sexual swelling is attained well before estrogen levels reach their peak (Emery and Whitten, 2003); this threshold effect of estrogens may underlie the correlated increase in female proceptive behavior at this time. On the other hand, increases in mating activity during the periovulatory period are chiefly the result of male initiative (Emery Thompson and Wrangham, in press), and the mating preferences of females may be overshadowed by male-male competitive outcomes (Matsumoto-Oda, 1999), as well as male possessive (Tutin and McGinnis, 1981) and coercive behaviors (Muller, Kahlenberg, et al., 2007). Prolonged display of large sexual swellings in chimpanzees may facilitate a mixed-mating strategy with the goal of giving all males a probability of paternity while inciting male competition to "select" the highest-quality sire (Nunn, 1999). This type of strategy may have led to increased sensitivity of estrogen receptors on the anogenital tissues, such that tumescence occurs with relatively low estrogen levels; it may even have led to selection on ovarian cycles for a relatively extended follicular phase to allow for extended swelling and receptivity.

Bonobos. Like chimpanzees, bonobos display exaggerated sexual swellings for an extended portion of the menstrual cycle. A few studies have found significantly longer cycle lengths and swelling duration in bonobos relative to chimpanzees, proposing that selection has promoted longer swelling duration to facilitate the use of sex in social mediation (Furuichi, 1987). However, data inferred from interswelling interval (as opposed to intermenstrual interval or hormonal monitoring) may exaggerate cycle length by including very long inferred cycles (Paoli et al., 2006); these "cycles" of greater than 60 days are more likely to comprise multiple cycles or periods of ovarian quiescence. In addition, while chimpanzee swellings change rapidly from maximal to minimum tumescence after ovulation, the bonobo perineum can remain partially enlarged for much of the cycle after maximal tumescence, complicating direct species comparisons (Heistermann et al., 1996; Furuichi and Hashimoto, 2004). While bonobo cycle lengths are highly variable, there is not consistent evidence to support a marked divergence from chimpanzees in cycle length or sexual swelling duration (Stanford, 1998; Paoli et al., 2006).

Despite this broad similarity, sexual cycles and behavior differ in interesting ways between chimpanzees and bonobos. Like chimpanzees, ovulation in bonobos is unlikely during the first half of the sexual swelling period; in contrast to chimpanzees, a significant proportion of ovulations occur after swelling detumescence, making swelling in bonobos both an imprecise and inaccurate signal of ovulation (Heistermann et al., 1996; Reichert et al., 2002). Chimpanzee females are relatively active in initiating and resisting copulation attempts, but male aggression (to both receptive females and competing males) seems effective in thwarting female choice (Muller, Kahlenberg, et al., 2007). While bonobo female sexuality is frequently emphasized, females rarely initiate copulations with males; females are successful in exerting choice via resistance to male mating attempts and interference with copulations of other females (Kano, 1989; Furuichi and Hashimoto, 2004; Hashimoto and Furuichi, 2006). Like chimpanzees, male bonobos attempt copulations most frequently within the maximal swelling period (Furuichi and Hashimoto, 2004). However, relative to chimpanzees, bonobo males both attempt and secure a larger proportion of total copulations (approximately one-third) outside of the swelling period. This suggests that in bonobos, attractiveness and receptivity are somewhat independent of the swelling signal and the hormonal correlates of ovulation (Furuichi and Hashimoto, 2004). The sexual dimorphism in testosterone production is lower in bonobos than in chimpanzees, which might be predicted to influence intensity and duration of female sexual behavior (Sannen et al., 2003). However, whereas testosterone levels correlate with the frequency of nonreproductive sexual behavior in bonobos (for example, genital-genital rubbing between females), they do not correlate with heterosexual copulation rates (Jurke et al., 2001; Sannen et al., 2003; Sannen et al., 2005).

Between-Cycle Variation in Female Attractiveness

In addition to within-cycle effects, variation in hormone production between cycles influences both male and female sexual behavior. While many primates breed only during restricted mating seasons or experience one to two cycles before conception, apes typically experience many nonconceptive cycles prior to each pregnancy. Increases in the number of sexually receptive days per conception are expected to influence dynamics of sexual behavior, including the intensity of male-male competition (Mitani, Gros-Louis, and Richard, 1996) and the degree of paternity confusion. This additional dimension of variation in conceptive ability of receptive females increases the costs of copulation and results in a greater diversification of mating strategies of both males and females.

Female apes, including humans, show considerable variation in the proficiency of ovarian cycles, as reflected in basal and peak follicular estrogen concentrations and luteal estrogen and progesterone production (humans: Ellison et al., 1993; Lipson and Ellison, 1996; chimpanzees: Emery and Whitten, 2003; Deschner et al., 2004; gorillas: Nadler and Collins, 1991; orangutans: Knott, 1999). For chimpanzees (Emery Thompson, 2005), gorillas (Nadler and Collins, 1991), and humans (Lipson and Ellison, 1996), relative ovarian hormone production at these key times is a strong predictor of conception success. This has also been suggested for Bornean orangutans, for which periods of high estrogen production coincide with increased conception success (Knott, 1999). In addition, preliminary evidence for chimpanzees (Emery Thompson, 2005) and humans (Venners et al., 2006) suggests that conceptions that do occur when ovarian hormone production is low are unlikely to result in successful birth.

In chimpanzees (Emery Thompson and Wrangham, 2008) and orangutans (Knott, 2001), ovarian function has been positively correlated to the consumption of high-quality or preferred foods, indicating that conceptions are likely to result when mothers have sufficient energy reserves to carry out the long and costly reproductive effort. These species, and perhaps other apes, share this reproductive adaptation with humans (Ellison et al., 1993; Knott, 2001), in contrast to the majority of primate species, which have adaptations that time lactation for the season of highest energy availability (Lindburg, 1987; Ganzhorn et al., 2003).

In orangutans, two lines of evidence suggest that males may respond directly or indirectly to differences in female fecundity across cycle. First, copulation rates during male-female consortships increase during seasons of high fruit availability (Fox, 1998), when conceptions are most likely (Knott, 1999). Second, both flanged and unflanged males fail to mate with females with young infants (Mitani, 1985). As previously described, female proceptive behavior is likely to be the relevant cue to fecundity within cycles; it is unclear whether male or female behavior is responsible for shifts in sexual activity between cycles.

In chimpanzees, copulation rates are significantly higher in conception versus nonconception cycles (Emery Thompson, 2005; Emery Thompson and Wrangham, in press). While one might suggest that conceptions occurred because copulation rates were high, as opposed to vice versa, this is highly unlikely. First, even during nonconception cycles, copulation rates are high and comparable to mean copulation rates of other ape species (Emery Thompson, 2005). Second, chimpanzee testes are particularly large and appear to be adapted for maximum fertilizing potential under circumstances of sperm competition (Harcourt et al., 1981). Thus, it seems likely that sexual behavior is facultatively related to the increased probability of conception in cycles with high ovarian function (Emery Thompson, 2005). Like within-cycle effects on copulation timing, males appear to drive this difference in copulation frequency (Emery Thompson and Wrangham, in press). It is puzzling that in a sexual milieu that appears designed to obscure the exact timing of ovulation, males can nonetheless glean important information about female fecundity. The exaggerated size of sexual swellings in chimpanzees and bonobos relative to other species, like gorillas and gibbons, is costly (Nunn, 1999), and while the presence of swelling may be relevant to ovulation proximity, the size of swelling may contain additional information. Indeed, the peak size of sexual swellings increases as the number of cycles to conception decreases and is significantly correlated with indices of relative ovarian function (Emery and Whitten, 2003; Deschner et al., 2004). Honest indication of fecundity may be a mechanism of subtle female mating competition for access to preferred mates (Pagel, 1994) or may simply be a side effect of the hormonal control of swelling physiology. It is also not yet known whether males attune directly to swelling size differences in assessing mating opportunities, though experimental manipulations in other primate species suggest that this may be the case (Bielert and Anderson, 1985).

Male Reproductive Strategies: Competition and Coercion

Testosterone, Rank, and Aggression

Male reproductive strategies vary both within and between ape species. Ultimately, however, competitive abilities drive male sexual access to females, be it relatively exclusive access to a female or group of females, as in gibbons or gorillas, or the ongoing competition over immediate mating access in polygynandrous chimpanzees and bonobos. Male competitive ability is not absolute but is relative to other males and fluctuates dramatically over their lifetime. Competition has strong potential to increase reproductive success but is also a costly enterprise. Thus, while genetic factors surely influence male competitive ability, hormonal physiology has a critical role to play in the adjustment of reproductive strategies and competitive investment over time. Testosterone, in particular, is a key mediator of male mating investment through many interrelated effects.

Available evidence thus presents a different model of the interaction of reproductive hormones with behavior for male versus female apes (Muller and Wrangham, 2001). Whereas fluctuations in concentrations of the key reproductive hormones are critical players in the ability of female apes to reproduce, normative testosterone variation is relatively inconsequential to reproductive performance (that is, sperm production and erectile function) in male primates (reviewed in Muller and Wrangham, 2001). However, testosterone can influence a number of traits important for male competitive and mating success: anabolism of muscle tissue, reproductive motivation, aggressive propensity, and risk-taking behavior (Bribiescas, 1996; Mazur and Booth, 1998; Nelson, 2005). Unlike female reproductive hormones that are highly sensitive to fluctuations in energy balance, male testosterone levels may be thought of as responsive to socioecological shifts (Muller and Wrangham, 2001, 2005).

The relationship between testosterone and aggression is a complicated one (see Fairbanks, this volume), as is the relationship between aggression and dominance rank. Nonetheless, evidence from primates and other taxa generally support a role for testosterone fluctuations in facilitating male aggressive behavior, particularly in mating contexts (Wingfield et al., 1990; Muller and Wrangham, 2004a). The strength of this relationship may vary depending on mating system (Wingfield et al., 1990; Whitten, 2000), rank stability (Sapolsky, 1993), and the importance of aggressive competition in determining mating success (Strier, Ziegler, and Wittwer, 1999; Muller and Wrangham, 2001).

The nature and intensity of male-male competition are variable across the apes (Table 9.3). The large sexual dimorphism in gorillas predicts particularly intense male-male competition (Mitani, Gros-Louis, and Richard, 1996). For gorillas, size and competitive ability are most relevant for acquiring and maintaining long-term access to females (Watts, 1990). By contrast, the frequency of competitive interactions is most extreme among chimpanzees, both because males are more frequently interacting with other males and because dominance relationships seem to be a source of continual negotiation (Muller, 2002). Perhaps more important, short-term access to attractive estrous females generates periodic bursts of agonism among males within a chimpanzee community (Muller, 2002; Muller, Emery Thompson, and Wrangham, 2006). As suggested by the discussion of female reproductive strategies, the duration and precision of sexual swellings, variation in female attractiveness, and the ability of males to monitor female fecundity are also expected to produce variation in the intensity of aggression, as well as in the degree to which male competitive interactions influence mating success.

Chimpanzees. There is a significant positive relationship between testosterone levels and male dominance rank in wild chimpanzees (Muehlenbein, Watts, and Whitten, 2004; Muller and Wrangham, 2004a).

Table 9.3 Behavioral and endocrine correlates of male competitive behavior.

Species	Context	Testosterone and Male-Male Aggression: Social Context	Testosterone and Male-Male Aggression: Individual Levels	Testosterone and Rank	Rank and Aggression	Rank and Mating Success	Rank and Reproductive Success (paternity)	Rank and Cortisol	References
Gorilla beringei	Wild	No relationship with group composition		Positive trend	Positive correlation	Positive correlation	Positive correlation (but alpha not exclusive father)	No relationship	1–5
Gorilla gorilla	Wild						Positive correlation		6
Gorilla gorilla	Captive	No relationship with group composition						No relationship	7–8
Pongo sp.	Wild			Higher in flanged versus unflanged adults	Positive correlation	Less mating resistance	No relationship		9–12
Pongo sp.	Captive			Higher in flanged versus unflanged adults (T or DHT)				Higher in flanged versus unflanged adults	13–15

Pan troglodytes	Wild	Both higher in mating contexts	No relationship	Positive correlation	Positive correlation	Positive correlation	Positive correlation	Positive correlation	16–23
Pan troglodytes	Captive		No relationship	No relationship					24
Pan paniscus	Wild			Positive correlation		Positive correlation	Positive correlation	Positive correlation	25–28
Pan paniscus	Captive	No relationship with rank stability/instability	Positive correlation with provoked aggression	No relationship	Hypothesized no relationship			Positive correlation	29–30

DHT = dihydrotestosterone; T = testosterone.

[1]Watts, 1991; [2]Robbins and Czekala, 1997; [3]Robbins, 1999; [4]Sicotte, 2002; [5]Bradley et al., 2005; [6]Bradley et al., 2004; [7]Stoinski et al., 2002; [8]Peel et al., 2005; [9]Mitani, 1985; [10]Schürmann and van Hooff, 1986; [11]Delgado, 2006; [12]Crofoot and Knott, in press; [13]Kingsley, 1982; [14]Maggioncalda, Sapolsky, and Czekala, 1999; [15]Maggioncalda, Czekala, and Sapolsky, 2002; [16]Tutin, 1979; [17]Constable et al., 2001; [18]Vigilant et al., 2001; [19]Muller, 2002; [20]Muller and Wrangham, 2004a; [21]Muller and Wrangham, 2002; [22]Muehlenbein, Watts, and Whitten, 2004; [23]Muller, Emery Thompson, and Wrangham, 2006; [24]Klinkova, Heistermann, and Hodges, 2004; [25]Kano, 1996; [26]Gerloff et al., 1999; [27]Hohmann and Fruth, 2003; [28]Marshall and Hohmann, 2005; [29]Sannen et al., 2004b; [30]Sannen et al., 2004a.

These findings are strong evidence for the role of testosterone in mediating mating competition for a number of reasons. First, higher-ranking males consistently display more aggressive behavior than lower-ranking males (Muller, 2002; Muller and Wrangham, 2004a). Second, both male-male aggression and male testosterone increase in reproductive contexts, when attractive, maximally swollen females are present (Muller and Wrangham, 2004a; Muller, Emery Thompson, and Wrangham, 2006); both aggression and testosterone increase relatively more when females are most likely to conceive (Muller and Wrangham, 2004a; Emery Thompson and Wrangham, in press). Third, behavioral observations indicate that male rank increases mating access (Tutin, 1979; Muller, Emery Thompson, and Wrangham, 2006). Finally, genetic studies confirm that high rank confers male chimpanzees a significant advantage in siring offspring (Constable et al., 2001; Vigilant et al., 2001).

Despite these associations between rank, mating competition, and testosterone, the testosterone levels of individuals are not strong predictors of individual rates of aggression (Klinkova et al., 2004; Muller and Wrangham, 2004a; but see Anestis, 2006, for adolescents). This may indicate that males differ in their behavioral sensitivity to testosterone. However, it more strongly suggests that testosterone is related to male motivation to compete rather than being a direct conduit to aggressive behavior (Sapolsky, 1997). Preexisting dominance relationships likely mitigate conflict escalation because past experience influences the perceived costs and benefits of instigating aggression. Furthermore, competitive experience is likely to feed back on testosterone levels and the incentive to compete: this "winner-loser effect," described extensively in humans, is a potential mechanism by which positive outcomes of aggression enhance future tendencies to compete by altering testosterone levels (Mazur and Booth, 1998). In support of this, researchers find that chimpanzee male testosterone levels do not necessarily determine their place in the hierarchy but that gain or loss in rank can influence testosterone production (Muller and Wrangham, 2004a). Higher testosterone levels have also been associated with decreased rates of strong aggression received from other males (Klinkova et al., 2004; Anestis, 2006).

Bonobos. The chimpanzee social and sexual environment would seem to provide an ideal context for the expression of rank-aggression-testosterone relationships. Given a similar social structure and mating system, bonobo males might be expected to have similar socioendocrine mechanisms. However, bonobo sexual dynamics differ in important ways from chimpanzees. Bonobo sexual activity is more widely distributed across female cycles and ovulation less reliably indicated by sexual signals. Coupled with larger party sizes (Kano, 1992) and shorter birth intervals (de Lathouwers and van Elsacker, 2005), this is thought to increase the frequency of male access to reproductive females, thus reducing the intensity of male competition (Stanford, 1998). Furthermore, females figure into the individualistic hierarchy of bonobos and can both influence the relative rank of males (particularly their sons) and form coalitions with other females to dominate males (Ihobe, 1992; Furuichi, 1997; Hohmann et al., 1999). The result is that females exert considerable influence on the success of male mating attempts, mitigating the influence of male-male competition (Kano, 1996; Furuichi and Hashimoto, 2004). Nevertheless, male rank and aggressive behavior influence mating success in the wild (Kano, 1996; Gerloff et al., 1999; Hohmann and Fruth, 2003), and rates of male aggression increase in mating contexts (Hohmann and Fruth, 2003).

How do these similarities and differences impact the interaction of testosterone with dominance rank and aggression? In a study of captive bonobos, no relationship was found between urinary testosterone and dominance rank, nor was a change in rank associated with a correlated change in testosterone (Sannen et al., 2004b). This study reported that neither testosterone nor aggression differed significantly between stable and unstable periods in the social hierarchy, suggesting that aggression may have little effect on rank in this species. Interindividual differences in testosterone were still pronounced in this captive population, however, and though individual levels did not correlate with overall aggression, they did correlate with aggressive reaction to provocation (Sannen et al., 2004a). Results from the captive group may not be representative, however, because of small group size and potential impacts of captivity on social interactions. In a preliminary examination of wild bonobos, testosterone and dominance had a strong positive correlation, consistent with data from chimpanzees (Marshall and Hohmann, 2005).

Gorillas. Gorillas present a potential natural experiment for understanding the relationship between testosterone and short- versus long-term mating competition. As in chimpanzees, dominant males have increased mating success over subordinates (Watts, 1991; Robbins, 1999). In one-male units, silverback gorillas face little active competition for mating access, though the importance of aggression is still paramount in protecting infants from infanticidal attack by outside males (Watts, 1989). By contrast, some groups of mountain gorillas contain several mature males, all of whom may mate with females. We might expect testosterone to be an important correlate of dominant silverback status in both types of group-ings, though perhaps differently manifested. For instance, testosterone might be expected to be relatively higher in subordinate gorillas within multimale, heterosexual groupings versus subordinate gorillas in bache-lor groups (that is, those excluded from the one-male units) because the latter are more activity involved in mating competition. Curiously, how-ever, preliminary investigations indicate no significant difference in the testosterone levels of subordinate mountain gorilla males in bachelor versus multimale, heterosexual groups (Robbins and Czekala, 1997). Sim-ilarly, a captive examination of western lowland gorillas found no signif-icant difference in the testosterone levels of same-aged males living in all-male versus heterosexual groups with one silverback (Stoinski et al., 2002). Not surprisingly, testosterone levels in both species are higher in mature or maturing males than juveniles (Kingsley, 1988; Robbins and Czekala, 1997; Stoinski et al., 2002). In mountain gorillas, dominant males also tend to have higher testosterone levels than subordinate males (Robbins and Czekala, 1997).

Unfortunately, there has been no direct comparison of the relative testosterone levels of silverbacks in one-male versus multimale groups. Another prediction, thus far not tested, is that whereas adults in multi-male groups should exhibit chimpanzee-like increases in testosterone during periods of mating activity, silverbacks in one-male units might in-stead experience increased testosterone (and cortisol) when there are de-pendent infants in their group, a time when aggressive competition should prove most critical to their reproductive success. Among captive lowland gorillas, Stoinski et al. (2002) have documented increased testos-terone levels in young adults (14–20 years) relative to adults (> 20 years) and subadults (10–13 years). This may simply reflect androgen influence on growth and sexual maturation, but it is intriguing that testosterone would be most elevated at the age when a male might be most apt to mount a challenge for control of a group of females.

Orangutans. Among orangutan males, relative competitive status of males is broadly reflected in the size and behavioral differences between flanged and unflanged male morphs. Fully developed (flanged) males grow to approximately twice the size of adult females and develop prominent cheek pads and throat patches (Crofoot and Knott, in press). Flanged males use these impressive morphological characteristics in dominance displays and booming long-call vocalizations that may func-tion in male spacing and social competition and as a means to attract fe-males (Fox, 2002; Delgado, 2006). The unflanged male category includes not just subadults but also sexually mature males that remain in developmental arrest for as long as 20 years after maturity (Kingsley, 1982; Schürmann and van Hooff, 1986; Utami et al., 2002). Given the mostly solitary nature of this species, unflanged males can opportunistically secure copulations with lone females despite the fact that larger flanged males could easily outcompete them in a face-to-face confrontation. Indeed, recent paternity studies indicate that some unflanged males are successful in siring offspring (Utami et al., 2002).

Flange status itself has hormonal correlates. Testosterone increases several months prior to the initiation of flange growth and remains stable as flange size increases, suggesting a threshold effect on male development (Kingsley, 1982). There are significant differences in the androgen levels of flanged and unflanged males in both captive (Kingsley, 1982; Maggioncalda, Sapolsky, and Czekala, 1999) and wild settings (Knott and Emery Thompson, n.d.). Flanged males who begin to lose condition ("past-prime males") have a corresponding decrease in testosterone (Knott and Emery Thompson, n.d.). However, there is also variation in hormonal profiles (Zhou, 2007) and dominance status (Mitani, 1985) among flanged males that warrants further exploration to separate the endocrine correlates of rank from correlates of development or body size.

Costs of Male-Male Competition

Endocrine studies are valuable for assessing the relative costs and benefits of dominance rank for males. Glucocorticoids, key regulators of glucose availability (Genuth, 1993), help modulate an organism's adaptive response to physiological or psychosocial stressors (Sapolsky, 1992). These responses are critical in the short term, but repeated stress can also lead to chronic activation of the hypothalamic-pituitary-adrenal (HPA) axis (Sapolsky, 1992). Thus, measurement of cortisol (the major glucocorticoid in primates) can be used to evaluate short-term reactions to changing social context or the long-term costs of social or ecological stress.

Theoretically, one might expect stress levels, and thus cortisol, to be highest among subordinate animals that tend to lose aggressive contests and reap fewer nutritional or sexual benefits of competition (Creel, 2001). Conversely, dominant animals are expected to enjoy increased predictability and control of social interactions, contexts known to mitigate the glucocorticoid stress response (Sapolsky, 1992). However, among wild primates and social carnivores, the reverse relationship is often found: dominant individuals have higher cortisol than some or all subordinates (Creel, 2001, 2005). It has been proposed that dominant individuals may face higher psychosocial stress only during periods of rank instability, when they face challenges to their own social standing (Sapolsky, 1993). Alternately, individual animals may vary in their "personalities," including their physiological processing of social situations, and this may be in some way predictive of dominance rank (Sapolsky and Ray, 1989). Dominant individuals may face higher stress from increased participation in the behaviors that correlate with dominance, such as higher rates of inter- or intragroup aggression (Creel, Creel, and Monfort, 1996; Muller and Wrangham, 2004b). Muller and Wrangham (2004b) argue that the correlation between rank and cortisol may result not from psychosocial stress but because aggressive conflict and dominance displays are energetically costly. Dominance often confers the nutritional advantages that allow individuals to initiate such expensive behaviors, so one might expect this to lead to a neutral relationship with cortisol. However, rank is of little consequence unless it can be successfully converted to reproductive advantage; males may assume increased metabolic costs for the brief periods that they can maintain the mating advantages of high rank.

Thus, it is unclear by what mechanism male dominance rank might translate into increased cortisol production, making it difficult to predict under what conditions we might expect this pattern in a given species (Creel, 2005). However, among apes, high social status is frequently associated with those behaviors expected to be psychologicatlly or metabolically costly. Aggressive displays, directed at single individuals or for general intimidation, characteristically involve terrestrial charging, sometimes while dragging or throwing objects (Goodall, 1986). High-ranking male chimpanzees and orangutans travel long distances in order to monitor territory or females within their range (te Boekhorst, Schürmann, and Sugardjito, 1990; Watts and Mitani, 2001). In multimale groups, such as in chimpanzees, males may forgo foraging effort while jockeying for an opportunity to mate with an attractive female. Male orangutans trade the potential competitive benefits of large body size with its increased energetic cost and locomotor limitations (Maggion-calda, Czekala, and Sapolsky, 2002). Thus, under Muller and Wrangham's (2004b) hypothesis, we reasonably expect that dominant ape males will have higher cortisol than subordinates. However, demographic factors might also impact this; under some situations, a U-shaped relationship between rank and cortisol would reflect the very poor energetic and social situation of subordinates. Contexts of male dominance in some apes also predict that costs of dominance figure into life history decisions for males, such as when to begin the developmental process in orangutans (Maggioncalda, Czekala, and Sapolsky, 2002) or when to initiate a challenge to take over a harem of female gorillas.

Positive correlation of cortisol and dominance in one wild chimpanzee community is consistent with the hypothesis that chimpanzees experience physiological costs for maintaining high rank (Muller and Wrangham, 2004b). Similarly, flanged male orangutans have higher cortisol levels than unflanged conspecifics in captivity, though this has not yet been validated in the wild nor for differences in rank among flanged males (Maggioncalda, Czekala, and Sapolsky, 2002). Among wild mountain gorillas, no significant difference in cortisol has been found between dominant and subordinate males; immature males have the highest cortisol levels (Robbins and Czekala, 1997). Cortisol also does not vary significantly with social organization in either wild mountain gorillas (Robbins and Czekala, 1997) or captive western lowland gorillas (Stoinski et al., 2002; Peel et al., 2005). Further research on the apes is needed to fully test the hypothesis that high-ranking animals experience physiological costs related to their dominance behavior.

Sexual Coercion

Observed mating dynamics presumably represent some compromise between the reproductive interests of the two sexes, though most understanding of this potential conflict is still theoretical. Direct conflicts of male and female interests are readily observed in intersexual aggression. Among apes, it is becoming increasingly clear that intersexual aggression functions primarily as a means of sexual coercion, the use of force by males to circumvent the mating preferences of females. The use of coercive aggression by male apes varies in form and prevalence, from near absence among monomorphic hylobatids (U. Reichard, pers. comm.) to directly forced copulations in orangutans (MacKinnon, 1979). Endocrine data can potentially assist researchers in understanding the context of sexual coercion and the physiological costs for its victims.

Forced copulations have been described in all populations of orangutans, and aggression plays some role in a considerable proportion of all copulations in these species (MacKinnon, 1979; Mitani, 1985; Schürmann and van Hooff, 1986; Stoinski et al., 2002). Most examinations suggest that forced copulation is primarily a strategy used by unflanged males to counteract female resistance and that females display very little proceptive behavior to these males in comparison to flanged males (Galdikas, 1985; Mitani, 1985; Fox, 2002). Similarly, Rijksen (1978) proposed that forced copulations might serve in establishing intersexual dominance when size differences alone are insufficient. However, flanged males have also been observed forcing copulations with varying frequency across study sites, suggesting that flanged status is not the only variable influencing the use of force by males or resistance by females (Mitani, 1985; Schürmann and van Hooff, 1986; Knott, in press). Even among flanged males, however, decreasing rank appears to be related to increased use of force in copulations (Mitani, 1985).

Laboratory experiments illustrate that females only choose to mate with males during a few days of the cycle but copulate more frequently when males have unrestricted access to them (Nadler, 1982); given the threatening size of males in orangutans, this finding implicates coercion outside of the fertile period regardless of overt female resistance. Recent endocrine research on wild Bornean orangutans expands on this finding by demonstrating that female ovulatory status, in addition to male dominance rank, influences the degree of resistance or proceptivity exhibited by females to flanged or unflanged mates (Knott, in press). This last finding supports the hypothesis that female mating resistance demonstrates preferences for certain types of mates, rather than being a general feature of orangutan sexual interactions. It is still puzzling, however, that resistance is exhibited so often despite its apparent ineffectiveness. More perplexing is the common mixture of prosexual and resistant behavior by females during single copulatory events (Knott and Kahlenberg, 2006).

Although forcible copulations occur rarely in other ape species, aggression can also be used less directly to increase male mating access to females. For instance, in the captive setting, copulations occurring outside of the periovulatory period of gorillas were primarily the result of male initiative (Nadler and Miller, 1982). The more frequently females received aggression, the more frequently they presented to males for copulation, particularly outside of the fertile period (Nadler, 1982; Nadler and Miller, 1982). In wild mountain gorillas, there is not a clear association between male display rate and courtship contexts or mating success (Sicotte, 2002). However, during intergroup encounters, males aggressively herd reproductive females, apparently to coercively maintain sexual access (Sicotte, 1993).

The data available imply that coercion in orangutans and gorillas is commonly used by nonpreferred mates or when a male attempts to secure matings outside of the periovulatory period when females are reluctant to copulate. However, in chimpanzees, female proceptivity is thought to be equally or more actively expressed outside of the periovulatory period (Stumpf and Boesch, 2005, 2006). Recent findings indicate that male aggression toward females in chimpanzees is directed at thwarting female attempts to mate promiscuously. While male aggression toward other males is rank related in chimpanzees (Muller and Wrangham, 2004a), male aggression toward females is not related to rank but can be very intense (Muller, Kahlenberg, et al., 2007). Males of both high and low ranks direct their highest rates of aggression at females during reproductive contexts and secure higher copulation rates with those females that they direct aggression toward (Muller, Kahlenberg, et al., 2007). This study also suggests physiological costs to coercion by demonstrating increased cortisol excretion during maximal swelling in parous females, those that receive significant courtship aggression from males; by contrast, nulliparous females receive little aggression and do not show cortisol elevation during estrus (Muller, Kahlenberg, et al., 2007).

Given the enhanced social status of female bonobos, it is not surprising that intersexual aggression is reduced in this species relative to chimpanzees. However, female bonobos receive aggression from males more often than expected during estrous periods (Hohmann and Fruth, 2003). Unlike in chimpanzees, aggression from male bonobos does not lead to increased probability of copulation with female targets (Hohmann and Fruth, 2003). In fact, female on male aggression in the Lomako study site was reported to be more than twice as frequent as the reverse, suggesting that males' risk of receiving aggression from females and their allies may explain the lack of coercive mating in this species (Hohmann and Fruth, 2003).

Conclusion

It is difficult to make any broad generalizations about the social or sexual relationships of the apes. In contrast to many other mammals, including some primates, it has been emphasized that sexual behavior of apes is largely divorced from the control of hormones. While true, this generalization glosses over the considerable variation in how ape mating patterns do relate to hormones. Similarly, while hormonal preconditions do not appear to be *necessary* for sexual behavior in either sex, this review indicates that endocrine physiology is an important *influence* on intersexual behavior. Ultimately, these relationships are best studied in the field because dynamic association patterns are such critical aspects of sociosexual behavior and because ecological variation is an important influence on sociality and reproductive function. However, isolating specific hormonal effects is challenging due to the many correlated hormonal shifts at critical times such as ovulation and the probability that hormones and behavior feed back on one another in fairly complex ways. Furthermore, the key variables we wish to explore, such as energy balance, health, psychosocial stress, senescence, and social status, are highly dynamic and may have interacting effects and reproductive function and behavior.

It is still unclear to what extent the nature of male-female social relationships is species typical or is flexible based on the mating system. For example, female proceptivity in gorillas seems to act to focus sexual behavior during the periovulatory period, while proceptivity in female chimpanzees is focused outside of the periovulatory period. When the gorilla mating system becomes more chimpanzee-like (for example, in promiscuous multimale groups), does female proceptive behavior change (Robbins and McNeilage, 2003)? Though a gorilla female's mating behavior has been correlated with periovulatory hormonal events, it is not clear that her behavior is determined by these hormones or whether these could be dissociated under certain circumstances.

Across the apes, hormonal fluctuations influence female sexual behavior and attractiveness, though the quality of the correlation between copulatory peaks and midcycle hormone peaks varies. Hormonally directed sexual signals focus copulations in all but orangutans, though this signal varies in its informational content and reliability. The degree to which variation in mating activity is directed by male attraction to females or by female proceptivity also differs among the apes and warrants further study. In chimpanzees, where promiscuity is an important means to avert infanticide, sexual signals are exaggerated in size and duration and, when combined with female behavior, provide a mixed signal to males about their probability of fertilization. By contrast, female mountain gorillas avoid infanticide by assuring the paternity certainty of the dominant male. They display a small but reliable cue and reinforce this with proceptive behavior. Orangutans show no sexual swelling; their best strategy to avoid sexual coercion may be to effectively conceal ovulation while using behavioral means to cue preferred males to their ovulatory status.

The relationship between testosterone, dominance rank, and aggression might be expected to vary with the intensity of competition in the apes. A role for testosterone in mediating male competitive propensities and rank-related aggressive behavior is well supported for wild chimpanzees, for whom short-term competition over mating opportunities is frequent, as well as for mountain gorillas, for whom competition is less frequent but more intense and relevant to long-term mating access. We still lack important data about the rank-testosterone-aggression contrasts in other social contexts, such as in pair-bonded gibbons, between one-male and many-male gorillas groups, and among flanged male orangutans. Direct cross-species comparisons cannot yet be performed; comparative studies applying similar endocrine methods are warranted to more directly examine the variance in testosterone production among males as a function of social organization and the importance of aggression as a reproductive strategy.

Available data indicate that high-ranking male chimpanzees face energetic costs associated with dominance behavior, while similar relationships between rank and cortisol production have not been found in gorillas. Thus, continuing study, particularly incorporating longitudinal data on energy availability, is needed to further test whether high or low rank imposes significant psychosocial and energetic costs to males and whether these costs are balanced by temporary increases in reproductive success.

Sexual coercion is apparently absent in some apes, frequent and direct in others, and more indirect—though apparently effective—in still others. While data from gorillas and orangutans suggest that coercive behavior is likely to occur when females are least fecund or when male aggressors are less preferred, data on chimpanzees suggest that aggression is used widely by both high- and low-ranking males and may focus on females of highest fecundity. Furthermore, while one might expect the most sexually dimorphic species to have increased male control over copulatory initiation, the actual relationship is just the opposite (Table 9.1): increased female proceptivity where size differences are most pronounced. This relationship suggests a broader role for subtle forms of sexual coercion in apes. Because rates of male-female aggression appear to be independent of male-male aggression, further studies should focus on discriminating the role of testosterone in facilitating one or both of these behaviors, the role of female fecundity in initiating intersexual aggression, and the associated psychological or metabolic costs of coercion.

Many endocrine studies to date have focused on describing average patterns of hormonal production, in some cases with reference to individual behaviors. In this chapter, I hope to have directed attention to the behavioral importance of temporal and individual variation in endocrine physiology and its relevance to understanding the dynamic nature of relationships between the sexes. Much of what is known about the endocrinology of apes has originated in laboratory settings, though noninvasive methods of urine and fecal analysis have allowed for endocrine monitoring of free-living populations. Captive conditions benefit from the luxury of daily sampling, the ability to monitor a greater range of hormones and short-term responses via blood sampling, and the potential to manipulate endocrine physiology. Field studies, however, are required to adequately study the dynamics of sexual interactions, as well as to quantify how captive conditions might influence endocrine physiology. As in this review, information from laboratory and field studies can be combined to provide strong inferences about the endocrinology of social relationships in the apes.

Acknowledgments

I thank Ulrich Reichard, Claudia Barelli, and Cheryl Knott for access to unpublished and in press data. I thank Peter Gray and Peter Ellison for inviting me to contribute to this volume; they, Margaret Crofoot, Carole Hooven, Cheryl Knott, Matthew McIntyre, and Martin Muller provided helpful discussions and comments.

PART THREE

Social Relationships among Humans

Human Sex Differences in Social Relationships: Organizational and Activational Effects of Androgens

Matthew H. McIntyre and Carole K. Hooven

I N THIS CHAPTER, we review the development of human sex differences in social relationships and propose a simple framework for understanding the effects of androgens during different periods. We focus on androgens, especially testosterone, and human male development, with some reference to effects of androgens in women. This focus is warranted given the vast literature on sex differentiation in mammals and the key role played by testicular androgens in the prenatal development of anatomical sex differences.

We further organize this broad and diverse literature around traits for which developmental trajectories can be posited and focus on sex differences in personality and their effects on relevant aspects of social relationships. A developmental perspective is essential because sex differences in adult social relationships begin to be established in childhood. Moreover, growing evidence supports the idea that while a minimum level of testosterone in adults may be necessary for normal mood, boldness, and libido, even substantial changes in testosterone have little effect on core personality, including gender identity. This relatively limited role of testosterone appears to be far from sufficient to explain the obvious and substantial sex differences in personality and gender identity. We propose that androgens have direct psychological effects primarily on labile emotional states and temperamental dispositions. In particular, androgens may inhibit the fear response and elevate energy and mood. These simple dispositional effects manifest in adulthood and, more important, also in infancy, setting boys and girls on somewhat different developmental trajectories and ultimately contributing to the complex sex differences in adult personality that are predicted by sexual selection theory.

Sexual selection theory provides a useful framework for making predictions about sex differences in human personality and the psychology of social relationships. As discussed in Chapters 3 and 5 of this volume, sexual selection theory provides an ultimate explanation for the division of an individual's "energy budget" into growth and development, maintenance and repair, and reproduction. The portion of energy spent on reproduction can be further divided into mating and parenting effort. The ways in which individuals bias their energy expenditure toward mating and parenting effort can be referred to as *reproductive strategies*. For instance, mating effort may involve male-male competition over access to females, or it may involve the expression of adaptations associated with stealth and speed when females are distributed over the landscape or are otherwise difficult to monopolize. Parenting effort may consist mainly of producing many offspring or of intensely nurturing a smaller number. In some cases, different strategies can coexist in a single species and in a single individual, depending on the ecological conditions.

In most mammals, mating and parenting effort are unambiguously associated with the male and female sexes. In humans, men also practice substantial parental investment, and psychological sex differences may therefore be less pronounced. Nevertheless, to the extent that consistently observed differences in personality and social relationships between men and women exist and were sexually selected, mating and parenting should remain salient functional domains. In particular, dominance and status seeking, and the related practices of aggression and affiliation, on the one hand, and proceptive heterosexuality, on the other hand, are predicted to characterize men.

Sex differences in fitness distributions—for example, distribution of lifetime number of surviving children—lead to predictions about sex differences in the psychology underlying reproductive behaviors. Among men in many societies, variance in fitness is relatively high and right skewed—median fitness is low, but some males may have many children. Among women, variance in fitness is lower and less skewed (see Figure 10.1). Given these sex differences in the shape of fitness distributions, men should have been selected to tolerate higher risk of injury and even death because the potential payoffs in terms of increased reproductive success are relatively high. Conversely, women should be relatively risk averse, because gains in reproductive success depend on health and longevity. Women should also benefit more from empathy and nurturing behaviors to the extent that they care for infants and young children,

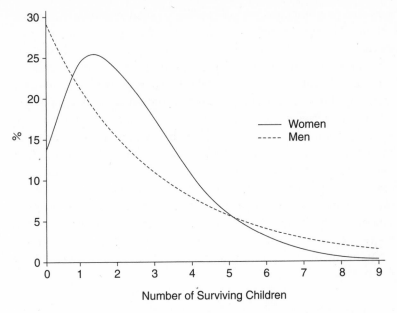

Figure 10.1   Hypothetical distributions of number of surviving children for men and women with moderate polygyny and low population growth.

whereas in the context of mate seeking, men may be better served by a heightened motivation for power and dominance.

What are the psychological dimensions on which these reproductive behaviors depend? How are they predicted to differ according to sex? Researchers have recently suggested that interpersonal dominance orientation and empathy are opposite manifestations of an evolved spectrum of traits (Montanes-Rada, Ramirez, and Taracena, 2006). According to this formulation, interpersonal dominance orientation is characterized by narcissism, instrumental motivations, novelty seeking, impulsiveness, and boldness. Anne Campbell (2006) has further traced the development and expression of this trait complex to infant sex differences in the fear response. The psychological dimensions associated with high dominance orientation are in turn closely associated with the types of aggressive and antisocial behaviors that have been characterized as both "dominant" and testosterone dependent (Mazur and Booth, 1998). In particular, high dominance orientation (and low empathy), combined with threats to status, can result in impulsive aggression. In contrast, prosocial and nurturing behaviors depend on high empathy and anxiety about the well-being of others.

Following on these ideas, we will outline a possible developmental trajectory of "dominance orientation–empathy" writ broadly, show how it is associated with other important domains of sex difference, and describe the changing roles of androgens in elaborating dominance orientation from infancy through adulthood. While testicular-derived testosterone is the primary androgen regulating sex differentiation, we will use the more general term *androgens* when testosterone is not specifically implicated. Most of the findings we discuss pertain either within males or within females or both. Because the sexes differ in many respects, this focus on associations within sex prevents likely confounding owing to extensive sex differences in androgen exposure and psychobehavioral factors that are unrelated to one another. We recognize that in order to take the next step and show that androgens "explain" sex differences, one must consider binary sex and hormone levels simultaneously. However, we do not take this step, owing to an insufficient development in the scientific literature on this point.

Male Androgen Production over the Life Course

In male fetuses, testosterone production rises to a peak between the tenth and eighteenth weeks of gestation, then declines, probably for a combination of reasons including an increase in hypothalamopituitary sensitivity to negative feedback, rising concentrations of placental steroids, and rising prolactin production (Forest, 1990). Sexual differentiation of the genitalia occurs during this period and is normally complete by 18 weeks. Fetal steroid profiles in the weeks immediately preceding birth have not been well characterized, but testosterone levels are high in the few hours after birth (Forest, 1990). Testosterone levels decline rapidly in the first week after birth, but parturition events simultaneously cause a surge in luteinizing hormone (LH), which stimulates the testes to again produce testosterone beginning in the first week after birth and peaking within a few months (Andersson et al., 1998; Bergadá et al., 2006), before falling to the low levels observed in children by one year of age. The significance of this early rise in testosterone has not been clearly established.

Given the timing and level of testosterone activity during fetal and newborn life, the entire perinatal period must be considered a period in which biological and psychological sex differentiation may occur. But studying fetal androgens is difficult because peripheral blood cannot be drawn from fetuses in utero. In an effort to learn more about fetal androgen exposure, researchers have developed techniques to assess prenatal androgens indirectly (Cohen-Bendahan, van de Beek, and Berenbaum, 2005). The indirect measure that most closely approximates fetal serum levels of testosterone is concentration in amniotic fluid, but these studies have also been limited to opportunistic clinical samples. Two commonly used alternatives, umbilical cord venous blood and gravid maternal blood, have been shown to be poor indicators of fetal production, relative to amniotic fluid testosterone concentration (van de Beek et al., 2004). Therefore, in our review, we place greater weight on results obtained using amniotic testosterone measures, while also including studies that used umbilical cord blood and gravid maternal blood to assess fetal exposure to androgens.

As a result of the difficulty associated with more direct methods of assessing prenatal androgen exposure, the standard method of studying its effects has been the study of children with clinical syndromes that specifically influence androgen production in prenatal life, especially congenital adrenal hyperplasia (CAH). CAH cases serve as excellent natural experiments because the disorder causes excess adrenal androgen production without other serious health problems and can be effectively treated after birth. Treated CAH serves as a model specifically for the effects of excess prenatal androgen exposure; but studies must be careful to control for the higher levels of clinical intervention experienced by children with CAH and other differences in their experiences, including the possibility of differences in parental attitudes about their gender (Cohen-Bendahan, van de Beek, and Berenbaum, 2005).

We also consider the ratio of the length of index to ring finger (2D:4D) as a potential marker of perinatal androgen activity (Manning et al., 1998; McIntyre, 2006). This measure is lower in boys than in girls, a sex difference that persists into adulthood. One small study investigated the association of 2D:4D with amniotic testosterone in two-year-old children and found that 2D:4D was negatively associated with the testosterone-to-estradiol ratio but not testosterone alone (Lutchmaya et al., 2004).

Further biological sex differentiation is low during childhood, before resuming again at puberty. During puberty, male testosterone levels rise to reach lifetime peak levels by the end of adolescence. During adolescence, along with the surge in androgens, dominance-seeking, rebellious, aggressive, and sexual behaviors increase (Spear, 2000). It may be tempting to infer that the increase in androgen levels *causes* this behavioral shift, but the factors that lead to behavioral changes in adolescence are likely complex and, at present, are poorly understood. One concern with the interpretation of the associations among androgen, behavior, and personality in adolescents is the strong longitudinal association between androgen production and physical maturation. In any sample of adolescent boys, those who are in a more advanced stage of puberty will have higher testosterone levels and will be taller, more muscular, and otherwise more physically developed and more masculine in their secondary sexual traits. These outward expressions of maturity could indirectly lead to behavioral changes through changes in social responses. Despite these concerns, the few studies investigating the role of androgens during adolescence have been careful to incorporate controls for spurious associations of this kind.

Ecological conditions also affect androgen levels. The peak levels of testosterone production attained in young adulthood are sensitive to factors such as energetic stress and may also determine the pattern of decline associated with senescence (Ellison et al., 2002). While less of a concern than that of adolescent maturation, confounding by developmental history and patterns of aging have not always been carefully considered. That is, just as adolescent boys who are more physically developed will also have both higher testosterone and greater psychological maturity (without any necessary causal association between the latter two), adult men who are older, or perhaps more physically aged, will have both lower testosterone and changes in other psychological dimensions, such as energy level (again, without any necessary causal association between the latter two).

Androgens and the Early Origins of Sex Differences in Personality

Sex differences in behavior begin to emerge prior to puberty. Beginning by three to five years of age, children play increasingly in same-sex groups (Whiting and Edwards, 1973; Munroe and Romney, 2006). Groups of boys are more likely to engage in a type of social-physical play called "rough-and-tumble," a sex difference that is also common in other young mammals.

Eleanor Maccoby suggests that early "gender segregation" is the strongest influence in the development of adult psychological sex differences. She argues that gender segregation may buttress children's emerging understanding of normative gender role and identity communicated by adults and that this segregation may highlight and reinforce relatively small biological sex differences (Maccoby, 1998). The context of all-male play groups further amplifies the tendency to play rough-and-tumble games, apparently through a process of peer socialization (Maccoby, 2000, 2002; Pellegrini, 2004). Therefore, even small early influences, be they biological or social, may initiate a cascade of more complex socialization processes yielding later sex differences. While Pellegrini and Smith

(1998) have suggested that rough-and-tumble play contributes directly to the development of dominance orientation in adult men, this relationship has not been empirically supported. However, some evidence indicates that boys who avoid rough-and-tumble play are perceived as feminine by their parents, especially if their fathers are relatively masculine (Roberts et al., 1987). Avoidance of rough-and-tumble play and male playmates in childhood also predicts adult nonheterosexual orientation (Whitam, 1977; Green, 1985; Whitam and Mathy, 1986; Green et al., 1987; Udry and Chantala, 2006). Among men who self-identify as gay, childhood avoidance of rough-and-tumble play predicts adult preference for a passive over active role in anal intercourse (Weinrich et al., 1992; McIntyre, 2003). Although erotic role is not central to modern definitions of sexual orientation, it does define, along with gender role, male sexual identity in most societies that have multiple sexual categories for men. An important question is whether the apparent association between childhood play and adult personality is causal or simply reflects multiple manifestations of a biologically masculinized brain.

It has become increasingly clear that if androgens have any important effects on personality, they must persist for long periods. Androgen levels in childhood are low in boys and girls. Any biological significance of these low levels has only begun to be elucidated, owing to the development of more sensitive radioimmunoassay techniques (Granger et al., 1999). Nevertheless, given the low levels, if androgens play an important role in the development of childhood sex differences, effects are most likely vestiges of androgen exposure during the perinatal period.

As rough-and-tumble play is so widely observed in young male mammals, experimental research is particularly relevant. For example, male rats show higher rates of play fighting than female rats. Female rats implanted with testosterone capsules in the amygdala during the neonatal period develop the male-typical pattern of play fighting (Meaney and McEwen, 1986), suggesting that rough-and-tumble play observed in humans may also be influenced by early testosterone. More relevant to human play are studies of testosterone and antiandrogen administration in rhesus macaques. This area of research has also shown that androgens promote the development of rough-and-tumble play but that effects are moderated by social factors and highly sensitive to the period and duration of exposure (Wallen, 1996, 2005). In particular, the key period for behavioral effects appears to be later in gestation than that for anatomical effects.

Human studies investigating the effects of excess testosterone exposure in girls with CAH have shown at least modest, positive associations with preference for male-typed toys and playmates in girls 12 months or older (Hines, 2003; Servin et al., 2003; Hines, Brook, and Conway, 2004; Meyer-Bahlburg et al., 2004), but none of these studies specifically focused on rough-and-tumble play. One study of a non-CAH sample of boys and girls failed to find an association between amniotic testosterone and rough-and-tumble play (Knickmeyer et al., 2005) within either sex. Likewise, another study did not find an association between rough-and-tumble play and 2D:4D in a small sample of either boys or girls (G. Alexander, 2006). We suggest that a possible explanation for the apparent weakness of the observed relationship between perinatal androgen exposure and play behavior is that androgens act indirectly through effects on infant temperament, which disposes children to prefer sex-typed play patterns under certain conditions.

Infant temperament varies on a relatively small number of dimensions that are observed through crying, gaze, eating, sleeping, and movement. Two dimensions show notable sex differences. The first dimension can be characterized as fearfulness or aversion to novelty, which is higher in girls (Martin et al., 1997; Else-Quest et al., 2006). Fear (also called shyness or inhibition) is primarily assessed through observations of crying in response to new or uncomfortable situations and is the most clearly validated dimension of infant temperament (Majdandzic and van den Boom, 2007). The second dimension is motor activity level, which is higher in boys (Campbell and Eaton, 1999; Else-Quest et al., 2006). Unlike fear, motor activity can be studied even during gestation. Sex differences in gestational activity level have also been reported (Almli, Ball, and Wheeler, 2001), and higher male activity persists from gestation through infancy (Eaton and Saudino, 1992; Groome et al., 1999). At least part of the sex difference in activity level may be attributed to the female advancement in development, both during the perinatal period (Almli, Ball, and Wheeler, 2001) and later in childhood (Eaton and Yu, 1989). Given that the development of cognitive inhibition enhances motor control (A. Diamond, 1985, 1988; Diamond and Gilbert, 1989), higher fear *and* lower activity level might result from developmental advancement in girls. However, confounding bidirectional effects should also be considered, as motor activity may directly contribute to inhibitory development by exposing the child to more novel experiences (Calkins, Fox, and Marshall, 1996; Campbell, Eaton, and McKeen, 2002). Supporting this hypothesis, higher fetal activity predicts reduced crying in response to frustration at one years old and behavioral inhibition at two years old (DiPietro et al., 2002). In addition, other mechanisms unrelated to inhibition could underlie sex differences in motor activity or fear, especially the function of the amygdala (Wood, 1996; van Honk et al., 2004), development of which is restricted by prenatal androgens (Giedd et al., 2006).

Whatever the neural mechanisms underlying sex differences in infant temperament, it seems likely that androgens play some role in their development. Another hypothesis about the origins of infant sex differences is that mothers reinforce gender-appropriate behavior in infants. But maternal influence is unlikely as a primary explanation, given that the maternal response to infant signals is probably not strongly influenced by the infant's gender per se. Rather, mothers tend to respond to infant behaviors, which are presumably influenced by infant temperament and not strongly influenced by maternal gender ideals or gender-role teaching objectives (Maccoby, Snow, and Jacklin, 1984). Moreover, twin studies show that nearly all of the variation in infant activity level, and distress to limitations and novelty, can be explained by genetic factors (Goldsmith et al., 1999). These results indicate that socialization alone is an unlikely explanation for objectively observed sex differences at this young age. More directly, timidity on the part of 6- to 18-month-old children, defined as moving toward or away from a novel toy, has been associated positively with testosterone and negatively with estrogen and progesterone levels in umbilical cord blood (Jacklin, Maccoby, and Doering, 1983), further supporting the claim that infant behavioral sex differences result from variation in perinatal androgen exposure.

Sex differences in behavior, observed later in childhood, appear to be strongly associated with infant and early childhood sex differences. A relatively high fear response early in infancy predicts the development of both empathy and social anxiety (Spinrad and Stifter, 2006), which reach higher levels in girls than in boys. Infant temperament and interests are commonly assessed through looking-time experiments: the longer an infant looks at a stimulus, the more interested in or attracted the infant is to the stimulus. Even in infancy, girls, relative to boys, show preferences for social over nonsocial stimuli (Lutchmaya and Baron-Cohen, 2002). In contrast, autism, a disorder strongly characterized by lack of empathy and relative attentional bias away from people and toward things, is substantially more common in boys than in girls. Autism itself (Manning et al., 2001; de Bruin et al., 2006) and related traits like low empathy (Knickmeyer, Baron-Cohen, Raggatt, et al., 2006), lack of social attention (Lutchmaya, Baron-Cohen, and Raggatt, 2002), low social skill level (Knickmeyer, Baron-Cohen, Fane, et al., 2006), and low prosociality (Fink et al., 2007) have been associated with higher perinatal androgen exposure (Baron-Cohen, 2002; Baron-Cohen, Knickmeyer, and

Belmonte, 2005). In humans, low 2D:4D in both men and women predicts unconscious attentional bias to masculine, relative to feminine, toys (G. Alexander, 2006). Masculine toy preferences have been interpreted as a manifestation of attention toward things over people (G. Alexander, 2003). Even vervet monkeys show sex differences in preferences for trucks and dolls (Alexander and Hines, 2002), which implies that cultural gender ideals are not the only underlying cause of observed sex differences in preferences for sex-typed toys.

Another childhood disorder more common in boys than in girls is attention-deficit hyperactivity disorder, or ADHD (Nøvik et al., 2006). ADHD symptoms reflect many of the masculine traits already discussed, including reduced social attention, hyperactivity, and lack of inhibitory control. Perhaps unsurprisingly, recent evidence suggests that ADHD is also positively associated with perinatal androgens, assessed via 2D:4D (de Bruin et al., 2006; Stevenson et al., 2007). Nonclinical associations of conduct and hyperactivity problems among boys with low 2D:4D have also been observed (Fink et al., 2007).

We have so far provided evidence that at least some sex differences in infant temperament can be attributed to the early action of androgens. But do established sex differences in infant temperament or toy preferences explain the emergence of typical adult sex differences that have direct relevance to sexual selection theory, perhaps via the phenomenon of play group gender segregation? Or are these observations interesting only insofar as they establish the plausibility of androgenic influences on other behaviors, which are programmed by some more direct pathway?

A stronger argument can be made based on existing evidence that infant and childhood temperament, and their effects on social experiences, shape adult sex differences in social relationships. Under this view, we would predict that infants who are relatively fearful, inhibited, and less active would engage in female-typed play styles as children and that infants who are less inhibited and fearful, and more active, should engage in male-typed play styles, including rough-and-tumble play. But few studies have directly tested this hypothesis, and existing evidence linking infant or early childhood temperament with play preferences has been somewhat inconsistent. For example, Hoffmann and Powlishta (2001) failed to detect sex differences in objectively scored play styles in a sample of 39 two- to five-year-old children. But they found that, in both boys and girls, more active and aggressive play styles were associated with *less* gender segregation. However, Moller and Serbin (1996) found that while gender-typed toy preferences and gender awareness did predict gender segregation in a sample of 42 one- to four-year-old children, gender segregation was higher in more active boys and more socially sensitive girls. Alexander and Hines (1994) showed in a sample of four- to eight-year-old children that while boys prefer other boys as playmates in a simulated design, they prefer girls with masculine play styles over boys with feminine play styles. Together, these results support the idea that rather than segregating strictly according to gender, children segregate based on a number of factors, including common, sex-typical interests and play styles.

Alternative hypotheses about the factors that lead to childhood gender segregation have also failed to find consistent support. Rather than suggesting that differences in temperament lead to gender segregation, these hypotheses argue that it is the emerging assimilation and understanding of cultural gender roles (Powlishta, Serbin, and Moller, 1993; Serbin et al., 2001) and parental attitudes (Jacklin, DiPietro, and Maccoby, 1984) that influence children to segregate on the basis of gender. A heritability study found that about a third of the variability in gender-typed attitudes among adolescents could be explained by genetic factors and nearly all the rest of the variation by environmental factors not shared among siblings, such as peer groups (Cleveland, Udry, and Chantala, 2001). In line with other studies, these findings leave little room for an important gender-socializing role of parents but a potentially large role for peers. More research should definitively resolve the question of whether and how peers socialize one another into sex-typical patterns of behavior and also more clearly elucidate the root causes of same-sex peer group formation.

Androgens in Adolescence

By late childhood, social pressures to conform to normative gender roles may minimize the effects of androgens. While, in principle, the pubertal surge in testosterone levels provides a new opportunity for androgens to refine and organize sex differences in behavior, these and later adult effects are strongly modified by prior developmental experiences and social context (Booth et al., 2006; van Bokhoven et al., 2006). These multiple and interrelated factors present additional challenges to researchers who endeavor to understand the relationship between androgens and behavior. Particularly in adolescence androgens directly affect, and are indirectly associated with, physical traits that may alter the individual's social environment. As already discussed, overall physical development, body size, and visible secondary sexual characteristics may all affect behavior in addition to, or instead of, direct effects of androgens on the brain. One can easily imagine that peers and authority figures may respond quite differently to more and less physically developed boys. There is evidence to suggest, however, that even independent of physical growth, testosterone levels do predict the onset of sexual activity in boys (Halpern et al., 1994; Halpern, Udry, and Suchindran, 1998). However, given the important role of sexual teasing in the maintenance of childhood gender segregation (Thorne and Luria, 1986), it is reasonable to posit that rather than acting simply to increase sexual desire through a direct neural route, testosterone may enhance boldness, which would be an important trait for boys initiating sexual activity prior to their peers. That is, bolder boys may be more willing to risk teasing from their peers in order to interact more with girls. In fact, boys with higher testosterone in adolescence are less inhibited and more spontaneous than their peers (Udry and Talbert, 1988; Granger et al., 2003), which could partly explain earlier onset of sexual activity.

In other social primates, both male lack of inhibition (Fairbanks, Jorgensen, et al., 2004) and higher testosterone levels (Muehlenbein, Watts, and Whitten, 2004; Muller and Wrangham, 2004a) in adolescence predict later dominance. Similarly, adolescent boys in North America with higher testosterone are rated as being more dominant, that is, as being leaders and as being tough (Tremblay et al., 1998). In human societies, the life course trajectory of dominance is likely more complex and variable than in other species, especially as children come into more contact with social institutions.

We might summarize the effects on personality and behavior of early testosterone exposure as reducing inhibition and social anxiety, reducing empathy, and elevating self-esteem/self-regard. Together this suite of dominance traits characterizes what are referred to as "externalizing" symptoms, which are higher in men than in women (De Clercq et al., 2006). While people with these personality types may win a certain respect and admiration, they can also alienate potential cooperative partners and have trouble maintaining close relationships (Baumeister et al., 2003). When these dominant, externalizing adolescents come into contact with restrictive social institutions, one possible outcome is delinquency or later criminality, which have also been positively associated with adolescent testosterone levels (Rowe et al., 2004; van Bokhoven et al., 2006). Nevertheless, it appears that on balance and in North American settings the positive social effects of a dominant personality, including control over social situations and overall popularity, outweigh the negative effects among peers, particularly if dominance is also moderated by affiliative traits (Pellegrini and Bartini, 2001; Hawley, 2003). Even though dominant adolescents may gain advantages among peers, perhaps unsurprisingly, teachers and other adults may give preference to more prosocial or empathetic adolescents (Hawley, 2003).

In adolescence, as in infancy, high androgen levels appear to masculinize behavior. But questions remain regarding the extent to which behavioral changes that are contemporaneous with exposure are under the direct, higher neurological influence of androgens or are primarily mediated by secondary pathways such as physical growth or basic temperament. In terms of how androgens in adolescence may help to shape adult personality in men, one possibility is that the adolescent's overall experience with peers and with authority figures or social institutions influences the trajectory followed into adulthood. Societies and social classes vary substantially in the degree to which dominant personalities are rewarded with high status. The North American context, in which dominant people (who manage to avoid excessive entanglement with the justice system) tend to be rewarded, might be exceptional with respect to human history and prehistory (Boehm, 1999).

Effects of Exogenously Administered Androgens in Adults

Studies of exogenous testosterone administration allow investigators to test hypotheses about causal relationships between hormones and behavior that are not possible without hormonal manipulation. Although these studies may be informative, investigators are limited in their ability to generalize results to healthy, normal populations. These limitations arise for several reasons, the most important being the clinical nature of most samples (subjects may be seeking medical treatment for various physical or behavioral problems) and a lack of consistency in administering testosterone and measuring its effects, including differences in dose and, more critically, in the period of treatment and latency to measuring outcomes. Given these considerations, we draw only tentative conclusions from these results, considering them indicative of areas worthy of future study.

Some of the psychological effects of exogenous testosterone in adults appear to be similar in some respects to the ways in which male temperament differs from female temperament in infants. Van Honk and colleagues have shown through a series of interesting experiments that a single, sublingual administration of testosterone to young women results in notable physiological and behavioral effects after only four to six hours. They found that increasing testosterone mutes the fear-potentiated startle response (Hermans et al., 2006), reduces unconscious but not conscious fear (van Honk, Peper, and Schutter, 2005), and reduces unconscious empathy in the form of tendency to mimic the facial expressions of others (Hermans, Putman, and van Honk, 2006).

If androgens affect activity level in infants, they may have similar effects in adults. Mood, particularly when taken to include energy level,

might be a good candidate. Findings have been varied. Effects of exogenous testosterone on mood have included effects only on energy level
components (Snyder et al., 2000; Pope et al., 2003; O'Connor, Archer,
and Wu, 2004), only on mania components (Pope, Kouri, and Hudson,
2000), or no effects at all (Anderson, Bancroft, and Wu, 1992; Wang
et al., 1996; Gray et al., 2005). Stronger effects of exogenous testosterone on mood have been observed in women and hypogonadal men
(Wang et al., 1996; Alexander et al., 1997; O'Connor et al., 2002), and
affective responses to testosterone administration can vary substantially
even in eugonadal men (Pope, Kouri, and Hudson, 2000).

A key problem in interpreting studies of exogenous testosterone administration is that different outcomes have been assessed in different populations, owing to varying clinical objectives. Studies of supplementation in
older men and women, or in hypogonadal men, often focus on depression
and overall quality of life (Lu et al., 2006). Studies in anabolic steroid users
and young men often focus on anger-hostility and aggression (Pagonis et
al., "Psychiatric and hostility factors," 2006; Pagonis et al., "Psychiatric
side effects," 2006). A better theoretical formulation of the underlying psychological dimension of interest might someday aid in the integration of
these qualitatively different findings. For example, energy level components of mood are captured differently, or not at all, by different depression rating scales (Shafer, 2006). Moreover, other components of affect
may be associated with emotions like anger and fear in complex ways and
in ways that differ between men and women (Verona and Curtin, 2006).
The effects of short-term changes in testosterone on this system of emotions may take on apparent complexity against the background of individual differences in personality, despite acting through simple and highly
labile psychological or nonpsychological mechanisms, including heart rate
response to stimuli (van Honk et al., 2004).

Cross-Sectional Studies of Androgens and Adult Personality

In the case of results already presented in the chapter, we could more or
less localize the exposure to androgens and the development of a psychological outcome to a particular developmental period. That is, the direct
effects of perinatal androgens are on early development, effects of pubertal androgens are in puberty (though in principle they may be correlated
with perinatal androgens), and the effects of exogenous androgens are
sometime between administration and measurement of effects. A separate body of evidence links current endogenous androgen levels, or
markers of previous exposure like 2D:4D, to personality traits and social

relationships in adults. How can we fit these cross-sectional observations into a developmental account of the sex differences in adult social relationships? The answer to this question depends on whether androgen exposure at different ages is correlated, how the adult trait of interest develops, and also whether direction of causation can be inferred. Although researchers may infer causality from contemporaneous associations between androgen levels and specific behaviors or personality traits, the relationship between androgens and behavior is usually more difficult to interpret in the context of development. While the evidence certainly indicates that testosterone affects labile states in adults, in the case of more complex features of personality, it is difficult to establish the relevant period of androgen exposure or when the psychological outcome being measured was established. Clearly a cross-sectional association between testosterone and a complex personality trait or behavior in adults probably *does not* imply that testosterone is currently acting to maintain said trait. If so, exogenous testosterone should be able to alter personality traits, which it cannot. Sex differences in personality are well established prior to adulthood, and the administration of sex hormones to adults does not cause psychological transgendering, nor do transgendered people show substantial differences in endogenous sex hormone concentrations (Meyer-Bahlburg, 1977; Singh et al., 1999).

Researchers sometimes assume, perhaps for the sake of simplicity, that testosterone levels are a stable individual trait. While some evidence suggests that this may be true over the course of several years for adult men and women (Missmer et al., 2006; Mohr et al., 2006), we should be more skeptical about whether adult male testosterone levels are informative about, for example, pubertal exposure (van Bokhoven et al., 2006), let alone perinatal exposure. Evidence establishing a link between markers of perinatal exposure, such as 2D:4D, and adult levels of testosterone is lacking.

As our understanding of the development of sex differences in behavior increases, cross-sectional evidence becomes increasingly interpretable in retrospect. In particular, the significance of associations between adult personality and 2D:4D is substantially benefited by an understanding of the early antecedents of personality that were developing closer to the time of the presumed androgen exposure. A simple example is sensation seeking, which is higher in men and associated with low 2D:4D (Austin et al., 2002; Fink et al., 2006). One could reasonably hypothesize that infant sex differences in activity level and fear, by causing children to choose same-sex playmates and styles of play, ultimately contribute to adult sex differences in sensation seeking. This explanation is offered as an alternative to the more dominant hypothesis that traits like sensation seeking are under more direct control of androgen exposure. Both hypotheses explain sex differences in traits like sensation seeking as the result of differences in androgen exposure, and both can be accounted for by a history of sexual selection, provided that children had sufficient opportunities for play in relevant evolutionary conditions.

Narcissism appears to be a particularly important component of dominance, but little is understood about its development. While moderate narcissism may increase personal happiness and be perceived as charming in brief interactions, pathological narcissism characterizes the most extreme forms of psychopathy (Ruiz, Smith, and Rhodewalt, 2001; Brown and Zeigler-Hill, 2004; Lee and Ashton, 2005; Jakobwitz and Egan, 2006). The developmental precursors of adult narcissism have been little studied outside of psychoanalysis, but narcissistic personality disorder is sometimes diagnosed in children and is associated with low empathy and poor inhibition of impulses (Weise and Tuber, 2004). Narcissism is higher in men and has been associated with low 2D:4D (McIntyre et al., 2007). The pattern of associations between 2D:4D and the widely studied Big-Five personality dimensions (extraversion, agreeableness, conscientiousness, neuroticism, openness to experience) may also be partly explained by an underlying association with narcissism. While associations between 2D:4D and Big-Five personality dimensions in adults have been weak and inconsistent (Austin et al., 2002; Fink, Manning, and Neave, 2004; Luxen and Buunk, 2005; Lippa, 2006), the general pattern of findings is compatible with an underlying association between narcissism or dominance and 2D:4D. That is, lower 2D:4D, or higher perinatal androgen exposure, would predict higher levels of narcissism and dominance (Neave et al., 2003) and also correlated personality traits. Individuals higher in narcissism also score lower on the agreeableness scale and higher on the extraversion and openness scales of the Big-Five (Paulhus and Williams, 2002; Lee and Ashton, 2005). In a very large sample, Lippa (2006) found weak but somewhat contradictory results—a negative association between 2D:4D and openness (suggesting perinatal testosterone increases openness) but also a positive association between 2D:4D and extraversion (suggesting perinatal testosterone reduces extraversion). Luxen and Buunk (2005) found moderate but more consistent results in small samples of both men and women— positive associations between 2D:4D and agreeableness (perinatal testosterone reduces agreeableness). However, Fink, Manning, and Neave (2004) found a *negative* association between 2D:4D and agreeableness, but only among women. The key to clarifying associations between adult personality, including various formulations of dominance, and 2D:4D will be to identify clear developmental trajectories linking androgen-mediated sex differences in infant temperament with the measures of adult personality most implicated in theoretically relevant sex differences.

As already discussed, mood contains components, such as energy level, that may be relevant to sexually selected effects of androgens. However, contrary to narcissism and sensation seeking, depressed mood is not generally associated with 2D:4D (Martin, Manning, and Dowrick, 1999; Austin et al., 2002; Bailey and Hurd, 2005) despite a large sex difference. The lack of an association between 2D:4D and depression may, in part, be due to difficulties already discussed in the context of effects of exogenous androgens on mood, including the fact that these studies all assessed depression using different instruments, which may not strongly correlate with underlying dimensions that are most relevant to normal social relationships. Alternatively, sex differences in depression might be more strongly influenced by differences in the actual experience of social stress by men and women, in addition to the emotional reaction to the stress (Shih et al., 2006), and be relatively unrelated to core sex differences in personality. The lack of clear association between depression and 2D:4D is perhaps somewhat less surprising in light of the uncertain association between exogenous androgens and mood.

Even more difficult to interpret than associations with 2D:4D are associations of adult personality and social behavior with current testosterone levels. One problem of interpretation is that intraindividual variation in testosterone levels may result *from* social interactions (Archer, 2006). For example, numerous studies show that men's testosterone levels, but not women's, tend to respond to the anticipation and outcome of competitive encounters, generally rising in anticipation, and staying high or rising upon winning, and falling upon losing (Dabbs and Dabbs, 2000). Nevertheless, aggression was an early area of research in the testosterone-behavior area, which has accumulated substantial evidence for *at most* a small overall association, even without accounting for bidirectional effects (Campbell, Muncer, and Odber, 1997; Archer, Birring, and Wu, 1998; Mazur and Booth, 1998; Book, Starzyk, and Quinsey, 2001).

In contrast to the weak associations between aggression and testosterone levels, adult salivary testosterone may be correlated with unconscious features of dominance orientation, including reduced smiling (Dabbs, 1997), similar to the previously discussed effects of exogenous testosterone administration (Hermans, Putman, and van Honk, 2006). In this case, it seems likely that the association reflects current effects of testosterone, given the demonstrated effects of exogenous testosterone.

Even associations between testosterone and cognitive tasks known to favor men might be explicable at least partly by noncognitive factors (Hooven et al., 2004), including dominance orientation (Newman, Sellers, and Josephs, 2005).

In a classic study of over 4,000 men, Jim Dabbs showed that serum testosterone level predicts employment in a low-status occupation and that this effect was mediated by educational attainment and antisocial behavior (Dabbs, 1992). In this case, given the large sample size, the relationship might well reflect small individual differences persisting from earlier periods of development (such as adolescence), when dominant personalities of boys with higher testosterone first came into conflict with teachers and society at large. As androgens modulate observable personal qualities like appearance (Penton-Voak and Chen, 2004) and voice pitch (Dabbs and Mallinger, 1999), it would be interesting to know whether these qualities mediate and/or confound effects of testosterone on social status and employment, as might be expected from prior evidence (Puts, Gaulin, and Verdolini, 2006). In particular, the negative effects of androgens on status attainment acting through antisociality may be even stronger if effects are masked by the positive treatment that more masculinized men enjoyed from others, especially peers.

Contrary to expectations, sexual orientation and gender identity appear to be at most weakly related to either adult or prenatal androgens. It has been known for some time that hormones administered to adults have no effect on gender identity or sexual orientation, which are undeniably central features of "sex differences in social relationships." While girls with CAH are more likely to develop homosexual orientation than controls, the effects are small, and no effect is observed on gender identity (Hines, Brook, and Conway, 2004; Meyer-Bahlburg et al., 2004). Studies comparing 2D:4D in heterosexual and homosexual samples were initially inconsistent with respect to both female and male homosexuality (Robinson and Manning, 2000; Williams et al., 2000; Rahman and Wilson, 2003; van Anders and Hampson, 2005; Voracek, Manning, and Ponocny, 2005; Kraemer et al., 2006), but two large Internet-based studies have both found that homosexual men have slightly higher 2D:4D, suggesting lower perinatal testosterone (T) exposure, but no effect in women (Lippa, 2003; Manning, Churchill, and Peters, 2007). One study has also found that male-to-female transsexuals have higher 2D:4D than control men (Schneider, Pickel, and Stalla, 2006). But the paucity of evidence supporting an important relationship between androgens and core features of gender identity and sexual preferences is notable.

Conclusion

We offer several conclusions about how changing life course exposure to androgens influences sex differences in personality and social relationships. First, at all ages, androgens act primarily on emotional states. Second, to the extent that androgens influence the development of core personality traits, it is only a secondary effect of emotional states experienced during formative periods of development. Third, sex differences in social experiences and emotional reactions to those experiences together shape sex differences in personality, including dominance orientation. And fourth, the psychological effects of testosterone in humans, unlike other species, may hinder the attainment of power, particularly in settings strongly regulated by social institutions. These points are not entirely new and draw extensively from related arguments (Wallen, 1996; Campbell, Muncer, and Odber, 1997; A. Campbell, 2006). Figure 10.2 summarizes some of the differences between the biosocial framework that is supported by current evidence and the neuroendocrine model that is commonly articulated. While we relegate adult androgen to a role in promoting labile emotional states (rather than personality), this is not to imply that prolonged declines in adult androgen exposure, resulting, in men, for example, from romantic relationships, parenting, or the aging process, cannot conversely influence social relationships through labile emotional states. Rather, we suggest only that those processes, which are carefully discussed in this volume by Gray and Campbell, do not explain broad sex differences in personality.

The implications of this indirect pathway of androgenic influences on social relationships are not revolutionary but worth considering. In particular, the biological effects of androgens on sex differences in social relationships may be modified by the social environment. We would argue that even if the role of androgens is developmentally complex, sexual selection theory remains a useful tool for understanding their effects on social relationships. However, an indirect, biosocial developmental pathway in conjunction with sociocultural change can dramatically reduce the utility of sexual selection theory for understanding observed patterns of adult behavior, as opposed to developmental systems. This results not only from the movement of the environmental adaptive space away from the organism (*mismatch* in the usual evolutionary psychology sense of the word) but also from changes in the development of the organism itself.

While the direct effects of androgens on temperament seem to be highly conserved, the social conditions under which human sex differences develop and operate have changed substantially, including the whole

Models of the Development of Sex Differences in Social Relationships

More Biosocial Model

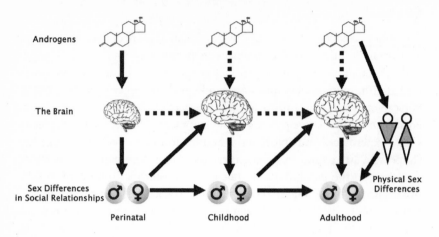

More Neuroendocrine Model

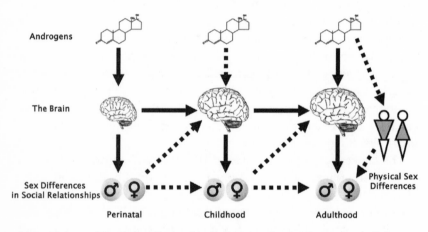

Figure 10.2 Two models of the effects of androgens on the development of adult sex differences in social relationships. *Solid arrows* show stronger effects, and *dashed arrows* show weaker effects. We argue that existing evidence supports a more biosocial model *(top)* focusing on the role of androgens in shaping early social interactions, which, in turn, are critical to later development, rather than a model in which androgens directly and continuously induce neural sex differentiation *(bottom)*, thereby causing sex differences in social relationships.

system of male-male competition. Humans and perhaps chimpanzees differ from other primates and mammals in that a pattern of between-group aggression and within-group cooperation has partly replaced or augmented classical forms of male-male competition (Wilson and Wrangham, 2003). While one might imagine that the psychological effects of testosterone that we reviewed contribute to intergroup competition in some ways, they seem likely to interfere with intragroup cooperation. The psychometric construct that most closely approximates attitudes toward intergroup conflict is social dominance orientation (SDO). SDO is higher in men than women and shares some features with interpersonal dominance, including low empathy (Pratto et al., 1994), but it does not appear to be associated with testosterone or 2D:4D (Johnson et al., 2006; McIntyre et al., 2007). An obvious difference between interpersonal and intergroup dominance orientation is that individuals with high SDO support social hierarchy, whereas individuals with high interpersonal dominance orientation support social hierarchy too, provided that they are on top. Personal glory certainly remains an important motivation in the context of small-scale intergroup raiding and intragroup social mobility. However, in egalitarian or rigidly stratified societies, and in societies that conduct true warfare, motivation for personal glory may not lead so clearly to reproductive success. Recall that even among the ancient Greeks, who certainly valued personal glory, Achilles died a violent, childless death, while Odysseus ultimately returned home to die quietly in the bosom of his family.

Acknowledgments

We thank Peter Gray and Peter Ellison for inviting us to contribute to this book. We thank Ellison in particular for his comprehensive perspective and work on the relationship between ecology and endocrinology and for the many helpful conversations that informed our thinking and writing about sex differences and endocrinology. We would also like to thank the Anthropology Department at Harvard University.

CHAPTER ELEVEN

The Role of Sex Hormones in the Initiation of Human Mating Relationships

James R. Roney

A LARGE BODY of research has demonstrated that sex hormones play important roles in mate choice and mate attraction across a broad range of nonhuman vertebrate species (for reviews, see Becker, Breedlove, and Crews, 1992; Nelson, 2005). The general story that emerges from this literature is that sex hormones increase when conditions are advantageous for mating, and elevated hormone concentrations in turn enhance attractiveness, cause attraction to opposite-sex conspecifics, and promote behavioral and physiological investment in mate choice and mate attraction. Although the effects of sex hormones on human fertility and sexual physiology have been well-investigated (for reviews, see Carter, 1992a; Ellison, 2001), relatively little research has investigated the role of sex hormones in human mate choice and mate attraction. The goal of this chapter is to review existing evidence regarding the possible importance of endocrine variables in human relationship initiation. Using the nonhuman literature as a guide, the chapter will focus on three specific research questions: (1) whether sex hormones regulate perception of others' mate attractiveness, (2) whether sex hormones predict individuals' attractiveness to others, and (3) whether sex hormones increase in reaction to interactions with potential mates. Preliminary findings on these questions seem promising enough to suggest that further incorporation of endocrine variables into human research may significantly enhance our basic understanding of human mating psychology.

Elucidation of hormonal mechanisms may mesh particularly well with evolutionary psychological approaches to understanding human mating psychology. Evolutionary psychology posits that the human mind is composed of a set of domain-specific psychological mechanisms that were naturally selected to address specific adaptive problems, such as mate choice, food choice, social exchange, and parenting (Tooby and Cosmides, 1992). An interesting consequence of any such modular organization of functionally distinct brain mechanisms is the need for some type of signaling process that couples the activation of specific mechanisms to the circumstances in which this activation is functional, especially if, as seems likely, some mechanisms produce contradictory or otherwise functionally incompatible outcomes (for example, aggressiveness promoted by mate competition mechanisms versus gentleness promoted by parenting mechanisms). Sex hormones may have an interesting role to play in any such signaling process, given that the information they carry is diffusely broadcast throughout the brain and body (recent evidence, for instance, supports a broad distribution of sex hormone receptors throughout both subcortical and cortical brain regions; see DonCarlos et al., 2006) and is therefore positioned to broadly adjust physiological and psychological parameter settings associated with functionally distinct mechanisms. As discussed below, sex hormone concentrations in women may largely index nutritional status and other energetic factors in such a way that specialized mating mechanisms are up- or downregulated depending on the degree to which current physiological conditions would likely support a successful pregnancy. In men, on the other hand, sex hormone concentrations may be largely responsive to the presence and potential availability of fertile mates, in addition to being calibrated by energetic considerations associated with functional resource allocation to mating effort versus maintenance of physical health. In both sexes, then, sex hormones may turn on and off the domain-specific mating mechanisms envisioned by evolutionary psychology, with sex hormones in turn calibrated to condition-dependent circumstances that index the current functional utility of energy investment in mating effort. If sex hormones are in fact priming mating mechanisms in a condition-dependent way, then members of the opposite sex could also use observable cues of sex hormone concentrations as indicators of condition that could inform mate preferences. As such, sex hormones may calibrate mate choice mechanisms, respond to cues of mating opportunities, and in part determine attractiveness to members of the opposite sex.

Sex Hormones and Perceptions of Mate Attractiveness

Females' Perception of Males

Individual conspecifics can be evaluated along any number of functional social dimensions—such as potential mates, kin, competitors, and social exchange partners—and sex hormones may promote attention to, and evaluation of, stimuli relevant specifically to judgments of mate attractiveness. Among a range of nonhuman vertebrate females, for instance, ovarian hormones tend to promote attraction to androgen-dependent cues that may mark conspecific males as potential mates. Ovariectomized female rodents exhibit no preferences for associating with males over females or intact over castrated males, but ovariectomized females administered estradiol and progesterone show clear preferences for gonadally intact males over both females and castrated males (Edwards and Pfeifle, 1983; Xiao, Kondo, and Sakuma, 2004). Natural variations in ovarian hormones appear to produce similar preference shifts, with stronger preferences for male stimuli among females tested during fertile versus nonfertile phases of the estrous cycle (for example, Beach et al., 1976; Johnston, 1979; Clark et al., 2004) and during the breeding versus non-breeeding season (for example, Ferkin and Zucker, 1991; Michael and Zumpe, 1993). These differential reactions to male stimuli may be hormonally modulated even at the level of perceptual inputs, as estrogen-treated ovariectomized mice exhibit greater fos-immunoreactivity in the vomeronasal organ in reaction to soiled bedding from intact males than do ovariectomized females without hormone replacement (Halem, Cherry, and Baum, 1999). In sum, a large body of evidence in nonhuman species suggests that ovarian hormones modulate attraction to androgen-dependent male stimuli.

Ovarian hormone concentrations are in turn calibrated to physiological conditions that support the functional allocation of energy to reproductive efforts. In seasonally reproducing species, for instance, it is widely accepted that gonadal quiescence during the nonbreeding season functions to conserve energy under conditions in which food scarcity or thermoregulatory challenges reduce energy availability beneath levels necessary to support both basic physiological processes and reproductive activities. Seasonal inhibition of ovarian activity represents only the extreme end of a continuum, though, in which energetic stress can inhibit ovarian activity. Food deprivation or increased energy expenditure has been shown to rapidly reduce ovarian hormone concentrations, fertility, and sexual receptivity across a wide range of mammalian species, with the key variable appearing to be the net availability of oxidizable fuels (for example, Schneider and Wade, 1990; Williams et al., 2001; for a review, see Wade and Jones, 2004). The hormonal calibration of attraction to androgen-dependent male stimuli may therefore function to couple the activation of mate preference mechanisms to the conditions under which mating effort is most adaptive: during conditions of suppressed fertility, attention and behavioral effort are probably better allocated to problem domains such as foraging, thermoregulation, predator avoidance, immune function, or maternal investment in offspring.

An intriguing recent line of research suggests that ovarian hormones may also calibrate attraction to androgen-dependent stimuli in humans. Growing evidence suggests shifts in women's evaluations of men across distinct phases of the menstrual cycle, with the time near ovulation associated with stronger preferences for facial masculinity (Penton-Voak et al., 1999; Penton-Voak and Perrett, 2000; Johnston et al., 2001), deeper voice pitch (Puts, 2005; Feinberg et al., 2006), olfactory cues associated with body symmetry (Gangestad and Thornhill, 1998; Thornhill and Gangestad, 1999), and more forward and aggressive behavioral displays (Gangestad et al., 2004). The traits that are more preferred near ovulation may be unified by their associations with higher androgen concentrations, as faces perceived as more masculine tend to belong to men with higher testosterone concentrations (Penton-Voak and Chen, 2004; Roney et al., 2006), androgens are known to promote deeper voice pitch (reviewed in Feinberg et al., 2006), and various lines of evidence suggest positive correlations between testosterone and aggressive or competitive behaviors (for reviews, see Mazur and Booth, 1998; Archer, 2006). On the perceiver end, preferences for the above traits tend to increase from approximately the midfollicular phase until ovulation, a time period when estradiol concentrations are also rising. In addition, a recent study reported that women with higher estradiol concentrations had stronger preferences for the faces of men with higher testosterone concentrations and that women's testosterone preference and estradiol curves tracked one another across days of the cycle (Roney and Simmons, 2008; compare Welling et al., 2007). The human data thus exhibit patterns that are quite similar to those seen in nonhuman species in that studies converge in finding that women are more attracted to masculine, androgen-dependent traits during high estrogen times of the cycle, and preliminary evidence suggests that estradiol itself may be the physiological signal that regulates these effects.

The most prominent proposed explanation for cycle phase shifts in women's attractiveness judgments is that they are products of an adaptation designed to facilitate extra-pair copulations under circumstances in which women may obtain higher-quality genes than may be available from their primary partners (for example, Penton-Voak et al., 1999; Gangestad, Thornhill, and Garver-Apgar, 2006). This argument is predicated on the assumption that cues to androgen concentrations may serve as heritable fitness indicators (see discussion below) combined with a cost-benefit analysis that posits that the potential costs of infidelity are constant across the menstrual cycle (such costs could include abandonment or violence if infidelity is discovered), but the potential genetic benefits are realizable only near ovulation when conception is actually possible. As such, ancestral women may have maximized their reproductive fitness via a mixed strategy in which they sought material support from high-investing long-term partners but also opportunistically sought better-quality genes from higher androgen men during extra-pair copulations that occurred during fertile portions of the menstrual cycle. Proponents of this position argue that cycle phase shifts in preferences for masculinity appear inexplicable otherwise, as there would be no functional reason to downregulate interest in masculine features at nonfertile times of the cycle, absent the potential costs of a discovered infidelity (for example, Gangestad, 2000).

The broader context provided by the nonhuman literature suggests possible alternative explanations for menstrual phase shifts in women's attractiveness judgments. As reviewed above, ovarian hormones promote attraction to androgen-dependent male stimuli in species such as rats and mice that do not form pair-bonds, thus excluding extra-pair mating as a possible cause of estrous cycle shifts in such attraction. Rather, sex hormones appear to modulate the categorization of conspecifics along sexual versus nonsexual dimensions, depending on the current relevance of mating as an adaptive problem domain, with sex hormones in turn indexing changes in fertility associated with estrous cycle physiology and energetic conditions. Insofar as humans have inherited similar mechanisms that calibrate ovarian hormones and fertility to energetic conditions, the menstrual cycle shifts in women's mate preferences may have less to do with special design for infidelity and more to do with adaptive allocation of attention to mate evaluation during stretches of the life cycle in which women experience higher fertility.

A large body of research demonstrates that women's fertility and ovarian hormones are in fact positively associated with energetic conditions that are favorable for sustaining successful pregnancies and ensuing lactation (for reviews, see Ellison, 2001; also see Ellison, this volume). It is well established that fertility and ovarian hormone concentrations are suppressed during lactation, for instance, and a number of studies have demonstrated that energy balance is the key variable regulating the degree of suppression (for example, Lunn et al., 1981; Ellison and Valeggia, 2003; Valeggia and Ellison, 2004). Likewise, in industrialized societies, even moderate exercise (for example, Bullen et al., 1985; Ellison and Lager, 1985; Schweiger et al., 1988) and weight loss (for example, Pirke et al., 1985; Lager and Ellison, 1990) have been associated with reduced estradiol or progesterone concentrations, and similar effects of both calorie intake restrictions (for example, Ellison, Peacock, and Lager, 1989; Panter-Brick, Lotstein, and Ellison, 1993; Bentley, Harrigan, and Ellison, 1998; Vitzthum et al., 2002) and elevated energy expenditure (for example, Panter-Brick and Ellison, 1994; Jasienska and Ellison, 1998) have been shown in natural fertility populations. Recent evidence also implicates stress as a possible suppressor of ovarian hormones (Nepomnaschy et al., 2004), which may complement mechanisms sensitive to energetics in delaying reproduction when aversive circumstances reduce the prospects for a successful pregnancy.

The sources of ovarian suppression discussed above would likely have entailed that women in ancestral environments usually experienced suppressed fertility and only in rare months between births would have experienced fertile cycles with elevated ovarian hormone concentrations (see Strassmann, 1997). This may have created circumstances loosely analogous to breeding seasons in nonhuman species in which evaluation of male sexual attractiveness is downregulated during stretches of suppressed fertility (when other adaptive problems are more pressing) but then upregulated when fertility returns. I propose that women may exhibit temporal shifts in their mate preferences due to the operation of a mechanism that uses elevated estradiol as a signal that women are currently experiencing high fertility cycles and thus upregulates evaluation of men's sexual attractiveness across cycle days in these more fertile months; since estradiol also peaks near ovulation within individual cycles, though, this mechanism might generate within-cycle preference shifts even if it were primarily designed to shift psychology across different cycles.

A full discussion of empirical tests of this between-cycle alternative to the extra-pair mating theory is beyond the scope of this chapter. Briefly, though, the extra-pair mating explanation predicts that preference shifts should be tightly coupled to the "fertile window" (the portion of an ovulatory cycle when conception is possible), while the between-cycle alternative explanation predicts that preferences for androgen-dependent traits should correlate with ovarian hormone concentrations more broadly at various times in the cycle. The luteal phase of the cycle is especially interesting since conception is not possible at this time, but estradiol is still likely to be higher in the luteal phases of more versus less fertile cycles: the extra-pair mating position thus predicts that attraction to masculine traits should be uniformly downregulated during the luteal phase in order to avoid the costs of infidelity when genetic benefits are impossible to obtain, while the between-cycle position predicts that higher luteal phase estradiol should predict stronger attraction to masculine traits since elevated estradiol signals a more fertile cycle. Figure 11.1 presents data from Roney and Simmons (2008) demonstrating that women's luteal phase estradiol predicted the strength of their preference for the faces of higher testosterone men. Although certainly not definitive, this result provides at least preliminary evidence for the proposed between-cycle position.

In summary, the human and nonhuman data present intriguing parallels with respect to the role of sex hormones in females' perceptions of male attractiveness. Attraction to androgen-dependent cues is enhanced when fertility is elevated, and both experimental manipulations in nonhumans and correlational work in humans suggest that estradiol may be the physiological signal that regulates these effects. The precise functional explanation for these temporal shifts in women's mate preferences is still a matter of some debate and may provide interesting directions for future empirical research.

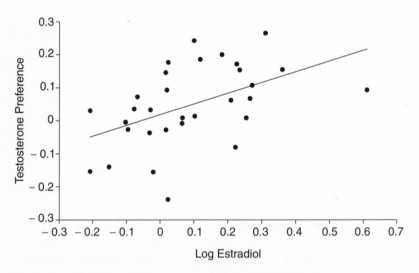

Figure 11.1   Relationship between luteal phase estradiol levels in women and the strength of their preference for the faces of higher testosterone men. From Roney and Simmons, 2008.

Males' Perception of Females

Sex hormones also modulate males' preferences for cues of fertility in females. Male rodents' preference for odors from estrous versus ovariectomized females is eliminated by castration (Xiao, Kondo, and Sakuma, 2004), for instance, and preferences for estrous over nonestrous odors can be restored after gonadectomy via testosterone replacement (for a review, see Gandelman, 1983). As in females, seasonal fluctuations in male preferences for opposite-sex stimuli are largely mediated by concomitant shifts in sex hormones (for example, Ferkin and Gorman, 1992) and may function to distribute attention, energy, and motivation away from mating effort when the absence of fertile females reduces or eliminates reproductive return on such effort.

There do not appear to be any extant studies testing whether fluctuations in men's hormone concentrations are associated with shifts in attractiveness judgments analogous to those documented among women across different phases of the menstrual cycle. Experimental induction of hypogonadism via gonadotropin-releasing hormone (GnRH) antagonists produces reductions in sexual desires and behaviors that can be reversed with testosterone supplementation, but the replacement concentrations necessary to restore normal sexual function are well below the normal baseline concentrations of circulating testosterone (Bagatell, Heiman, Matsumoto, et al., 1994; for a general review, see Bancroft, 2005). Such results suggest that male sexual interest may require a fairly low threshold concentration of androgen, as opposed to exhibiting a more continuous dose-response relationship across the normal range. If such a pattern extends to mate choice, it may turn out that variations in men's sex hormone concentrations do not generally predict preferences for particular features in potential mates. This is an empirical question, though, and future research could examine whether higher testosterone men exhibit stronger preferences for, say, more feminized facial or bodily features that may index higher fertility in women (see below). The relatively low cost of sperm production and the potentially large reproductive benefits from copulation with fertile mates lead one to expect that men's attraction to cues of women's health and fertility will be robust across most conditions. Rather than modulating libido and mate preferences, then, men's sex hormone concentrations may primarily modulate whether energy is invested in mate competition and courtship effort versus being invested in the maintenance of physical health (Ellison, 2001).

Sex Hormones and Attractiveness to Others

Female Attractiveness

Ovarian hormones appear to index fertility and energetic conditions across mammalian species, and, as such, males could use observable cues of sex hormone concentrations in mate choice decisions. At least two somewhat related types of information might be provided by such cues. First, hormone-dependent cues could mark the window of fertilizability within ovulatory estrous cycles. The well-established finding that gonadally intact males exhibit stronger preferences for cues from estrous versus nonestrous or ovariectomized females (see partial review above) provides strong evidence for this function. A second possible type of information provided by such cues might pertain to distinctions in female quality when position in the estrous cycle is held constant: if two females are both in a fertile portion of an ovulatory estrous cycle, might cues associated with differing sex hormone concentrations make one female more attractive to males than the other? In general, there is very little evidence that mammalian females produce ornamentlike cues that may indicate health or fecundity independent of indicating position in the estrous cycle (Andersson, 1994). A possible exception may be sexual swellings produced by some nonhuman primates. Emery and Whitten (2003) reported that chimpanzees with larger sexual swellings exhibited greater estradiol concentrations across equivalent portions of the cycle than did females with smaller sexual swellings. They also reviewed evidence demonstrating that cycles with greater estradiol are associated with higher conception probabilities and that male primates tend to exhibit stronger attraction to larger sexual swellings. The tendency for females to advertise and males to pay attention to more continuous hormonal signals of fertility may be associated across species with slower rates of reproduction and greater energetic costs of offspring production, as such conditions reduce the number of highly fertile mating opportunities and thus limit the potential payoffs of male strategies that involve relatively indiscriminate mate choice in the service of maximizing the quantity of copulations. The pattern of large energetic investment in a small number of offspring characterizes humans, of course, and leads to the expectation that observable cues of sex hormone concentrations might in part determine women's attractiveness to men.

Human menstrual cycles that result in successful conceptions are characterized by higher ovarian hormone concentrations than are ovulatory nonconception cycles that were similarly exposed to unprotected intercourse (Lipson and Ellison, 1996; Venners et al., 2006). This pattern suggests adaptive advantages for male mate choice mechanisms that could identify not only external signs of ovulatory versus anovulatory cycles but also any available cues of continuous gradations in women's ovarian hormone concentrations. Elements of body shape may provide just such cues, as women with larger breasts and smaller waist-to-hip ratios were found to have higher salivary estradiol and progesterone concentrations than other women when compared on equivalent days across the menstrual cycle (Jasienska et al., 2004). Other research has provided evidence that these body dimensions are in fact attractive to men (Singh, 1993). Likewise, women's late follicular estradiol concentrations were significantly correlated with attractiveness ratings of their faces, at least among women who were not wearing makeup (Law-Smith et al., 2006). The faces of higher estradiol women were also subjectively rated as more feminine, and a large body of research consistently demonstrates that feminized facial features are attractive in women (for a review, see Rhodes, 2006). These studies provide the most direct evidence that men's attractiveness judgments are informed by morphological cues of women's hormone concentrations, though indirect evidence supports this possibility as well. Some evidence suggests that higher estradiol may be associated with higher voice pitch (Abitbol, Abitbol, and Abitbol, 1999), for instance, and men rate higher frequency voices more attractive in women (Collins and Missing, 2003). Other research suggests that women's facial attractiveness correlates with both their vocal attractiveness (Collins and Missing, 2003; Feinberg et al., 2005) and their body attractiveness (Thornhill and Grammer, 1999), which implies that underlying hormonal condition may have feminizing effects on multiple phenotypic cues that men's mate choice mechanisms could in turn use as indices of likely fertility.

The studies by Jasienska et al. (2004) and Law Smith et al. (2006) suggest that facial and bodily cues of women's attractiveness may indicate relative differences in fertility across women when time in the menstrual cycle is held constant. Other research has examined whether phenotypic cues may indicate the window of fertilizability within given menstrual cycles. When presented with photographs of the same women taken during the late follicular and luteal phases, raters judged the follicular phase photos more attractive at rates that were slightly above chance (Roberts et al., 2004). This effect could be a product of ovarian hormone fluctuations altering soft tissue traits. A number of other studies have provided evidence that women's scent is rated more attractive by men when odors are collected near ovulation as opposed to at other times in the cycle (Singh and Bronstad, 2001; Thornhill et al., 2003; Kuukasjarvi et al.,

2004; Havlíček et al., 2006; cf. Thornhill and Gangestad, 1999), and such studies have been interpreted as further evidence that ovulatory timing is not entirely concealed.

Although the above studies are presented as supporting the existence of choice mechanisms in men that are attuned to cues of women's ovulatory timing, the results are equally consistent with mechanisms designed to promote greater attraction to women with chronically higher estradiol concentrations. The Havlicek et al. (2006) scent study, for example, found that between women variance in odor attractiveness was far greater than within-women variance associated with phase of the menstrual cycle. This means that the odors of some women outside the fertile window were judged more attractive than scents of other women near ovulation, which is a pattern inconsistent with a mechanism designed to identify the timing of ovulation in order to target short-term mating effort toward currently fertilizable women. This study also found that scents collected during the luteal phase were judged more attractive than scents collected during menstruation: estradiol concentrations are higher in the luteal phase versus at menstruation, but fertilizability is zero in both cases. These patterns are consistent with a mechanism designed to track cues of estradiol across the cycle in order to promote attraction to more fertile women, with greater attraction to ovulatory scents simply following from the presence of higher estradiol concentrations at midcycle.

Alternatively, a true ovulation-detection mechanism might have evolved in order to target men's mate guarding to fertile portions of their partners' cycles. Some research indicates greater mate guarding near ovulation (Gangestad, Thornhill, and Garver, 2002), but another study found that this pattern held only for relatively less attractive women and that men paired with more attractive women showed no shift and, if anything, engaged in greater guarding during the luteal phase (Haselton and Gangestad, 2006). Given that more attractive women appear to exhibit higher estradiol concentrations than other women throughout most of the cycle (for example, Jasienska et al., 2004), this pattern is again consistent with a mechanism that responds to cues of absolute estradiol concentrations rather than being targeted specifically to ovulation. In sum, the state of the evidence in humans appears to be the reverse of that found in nonhuman species: the existence of hormone-dependent cues that differentiate women's fertility at equivalent points in the cycle is more strongly supported by evidence than is the existence of cues that reliably indicate the window of fertilizability within specific cycles.

Although initial evidence suggests that observable cues of sex hormone concentrations in part determine women's attractiveness, research in this area is just beginning, and there are many unanswered questions. It is unclear, for example, to what extent facial and body attractiveness are statelike cues that may change within women with temporary changes in circumstances as opposed to being more stable traitlike cues that may better index lifetime fertility than cycle-to-cycle fluctuations. Do variables like facial femininity and waist-to-hip ratio change concurrent with drops in ovarian hormone concentrations that have been documented in response to seasonal increases in energetic stress within subsistence societies (for example, Ellison, 1994)? Such traits do change with age (for a review, see Singh, 1993) and parity (Lassek and Gaulin, 2006), but degree of short-term change within age and parity classes appears unknown. It is also not entirely clear what circumstances might ensure that external cues of sex hormone concentrations are in fact honest signals of health or reproductive condition. Some authors have suggested that only healthier women can afford higher estrogen due to the possibility that high estrogen may handicap the immune system (for example, Thornhill and Grammer, 1999), but evidence for immunosuppressive effects of estrogen is mixed at best, and many components of the immune response appear to be improved under conditions of high ovarian hormones (for reviews, see Beagley and Gockel, 2003; Bouman, Heineman, and Faas, 2005). Body shape dimensions that involve specialized fat storage may provide signals of reproductive condition that would be especially difficult to fake under conditions of either current or developmentally experienced energetic stress. Facial femininity, on the other hand, does not intuitively appear to have high production costs, and it thus seems uncertain why the development of feminine features could not be decoupled from hormonal influences as a means of dishonestly signaling a fertile hormonal profile. Theoretical arguments related to the selection pressures that may have promoted men's attraction to phenotypic cues of ovarian hormones are thus needed in addition to further empirical work linking variability in hormone profiles to variability in attractiveness.

A final avenue for future research concerns cross-cultural work on possible hormonal correlates of women's attractiveness. The extant studies that have examined hormonal correlates of facial and body attractiveness have been performed in industrialized countries, and similar findings in subsistence societies could substantially bolster the case for species-typical psychological adaptations in men that are attuned to phenotypic cues of women's hormonal status. Consistent with the existence of such adaptations are findings showing fairly strong agreement in judgments of facial attractiveness among raters from diverse cultures (for example, Cunningham, 1986), including the finding that neotenous and feminized facial features that may plausibly reflect estrogen exposure tend to be preferred cross-culturally (for example, Jones, 1995).

The case for species-typical standards for women's bodily attractiveness, on the other hand, is complicated by studies showing that men in some cultures seem to prefer higher overall body fat and not lower waist-to-hip ratio (for example, Yu and Shepard, 1998; Marlowe and Westman, 2001). Although these findings could be artifacts of a failure to use a range of body weights typical of the local population (see Sugiyama, 2004, 2005), another intriguing possibility is that the morphological cues of ovarian hormones and thus fertility differ according to food availability in the local environment. Jasienska et al. (2004) found that women with low waist-to-hip ratio and large breasts actually had lower body mass index (BMI) than women with other body shapes but nonetheless had higher estradiol concentrations. An interesting question posed by Jasienska et al. is whether under conditions of food shortage BMI itself would be a better predictor of ovarian hormones and fertility than would waist-to-hip ratio. Cross-cultural research on the endocrine correlates of body shape and attractiveness would be necessary to test this idea and thus determine whether men's preference mechanisms are designed to facultatively track different phenotypic cues depending on the nature of local conditions.

Male Attractiveness

The role of androgens in signaling male attractiveness has played a prominent role in tests of handicap models of sexual selection within evolutionary biology. Folstad and Karter (1992) proposed that the immunosuppressive effects of androgens allow androgen-dependent traits to honestly signal immunocompetence. There has been confusion in the ensuing literature regarding whether such a model should predict high or low correlations between measured androgens and parasite loads, however, and empirical studies have reported conflicting findings (for a review, see Muehlenbein and Bribiescas, 2005). This debate may be based on misconceptions, as there seems no clear reason to expect correlations in either direction. Trade-offs between investment in androgen-dependent traits and investment in immune function are presumably enforced by finite energy availability. Assume that males in a population vary in their energy budgets due to some underlying, at least partly heritable aspects of their physiology. Males with larger budgets can, other things equal, afford larger investments in both secondary sex traits and immune function. Depending on factors such as the intensity of mate competition and the rate of future discounting (for example, the number of remaining breeding seasons), males with larger budgets may sometimes need to invest so much energy in courtship signaling to outcompete other large-budget rivals that their immune function is compromised to the level of smaller-budget males who have equivalent immune functioning but could not also afford to grow larger ornaments. In other conditions in which the intensity of mate competition is more relaxed, larger-budget males may be able to afford both better immune functioning and larger ornaments. In all circumstances, though, ornament size should correlate positively with size of energy budget, and females' offspring should always benefit from inheriting larger budgets (assuming that energy can be allocated to survivorship or mating effort in a plastic manner). Thus, an androgen-mediated energy trade-off model of this sort makes no predictions about relationships between ornament size and parasite load but clearly entails two other predictions: (1) that androgen concentrations will correlate positively with ornament size and (2) that females will prefer larger ornaments. These two predictions are in fact fairly well supported in the nonhuman literature (for reviews, see Folstad and Karter, 1992; Andersson, 1994; Parker, Knapp, and Rosenfield, 2002).

The handicap model implies that circulating androgen concentrations should directly correlate with male attractiveness, though the size of the correlation should be attenuated by imperfect relationships between androgens and secondary sex traits and between secondary sex traits and female judgments of attractiveness. Female preferences for intact versus castrated males (see partial review above) supports this correlation at the extreme end of the distribution but does not demonstrate that continuous variation in androgens covaries with degree of mate attractiveness. Ferkin et al. (1994) may have provided the strongest evidence for such covariation among mammals by demonstrating a dose-response relationship between the amount of testosterone injected into castrated male meadow voles and the attractiveness of their odors to females. Other research suggests that androgen doses may in fact signal immune status. The odors of male mice experimentally infected with sheep red blood cells are less attractive to females than the odors of control males, for example, but testosterone administration in conjunction with red blood cell infection can restore odor attractiveness (Litvinova et al., 2005). Although immune responses to testosterone manipulations are not always consistent across species or across studies within species (Muehlenbein and Bribiescas, 2005), and although much more research is needed on androgen mediation of overall energy distribution, the results in the nonhuman literature on the whole suggest that females may often use androgen-dependent phenotypic cues as signals of good condition in males. This renders plausible the possibility that androgens also play a role in determining human male attractiveness.

There is fairly strong evidence that men's androgen concentrations at least coarsely index their general condition. Bribiescas (2001a) has proposed an energy allocation model in which higher circulating androgens are associated with greater energy investment in the construction of men's skeletal muscle mass, as opposed to investment in either energy storage in adipose tissue or energy allocation to immune function. He reviews evidence in humans showing that chronic or acute food deprivation is associated with decreases in testosterone and muscle mass, as are certain illnesses and physical traumas (see also Muehlenbein and Bribiescas, 2005). Testosterone administration, on the other hand, stimulates fat catabolism and has been associated with very large and rapid increases in muscle mass and measures of physical strength (Bhasin et al., 2001). Likewise, in vitro studies in human cell cultures have generally shown that testosterone inhibits components of immune responses such as immunoglobulin production (for example, Kanda, Tsuchida, and Tamaki, 1996), though in vivo studies have been less conclusive (Muehlenbein and Bribiescas, 2005). Taken together, these findings imply that women could use phenotypic cues of high androgen concentrations as signals that men are in good enough condition to afford allocating energy away from functions such as fat storage and immune responses.

Despite common appeals to the idea that men's testosterone concentrations may correlate with their heritable fitness (for example, Penton-Voak et al., 1999), there is surprisingly little research that has directly examined correlations between men's circulating testosterone and ratings of their attractiveness. Neave et al. (2003) reported that attractiveness ratings of men's face photographs were uncorrelated with men's salivary testosterone concentrations. Another study found a small but significant tendency for raters to choose composite faces constructed from the upper half of the testosterone distribution as more attractive than composites constructed from the lower half of the distribution when judgments were made in forced-choice tests; similar forced-choice tests performed on the original faces, however, revealed no significant effect of testosterone (Penton-Voak and Chen, 2004). Roney et al. (2006) reported a small but significant relationship between men's salivary testosterone and women's ratings of the men's faces for short-term mate attractiveness; for subsequent samples of faces collected in my lab, though, we have not always found positive correlations between facial attractiveness and circulating testosterone (unpublished data). With respect to men's facial attractiveness, then, there appears at best a weak positive influence of circulating testosterone concentrations.

Facial architecture may be relatively stable after puberty, though, and so may not provide the most sensitive index of condition-dependent fluctuations in androgen concentrations. Androgen-dependent changes in odor attractivity similar to those seen among nonhuman mammals could provide a more transient signal of condition, and researchers have suggested that more attractive scents among men with lower fluctuating asymmetry could be mediated by higher androgen concentrations in such men (for example, Gangestad and Thornhill, 1998). Nonetheless, in the only study that seems to have directly measured the relationship between men's testosterone concentrations and their odor attractiveness, there was no correlation between women's ratings of men's T-shirt odors and the men's salivary testosterone measures (Rantala et al., 2006). Although research has found that women express attraction to deeper voice pitch (Collins, 2000), muscularity (Maisey et al., 1999; Dixson et al., 2003), and some types of behavioral displays (Gangestad et al., 2004), there do not appear to be any studies that have directly assessed whether attraction to these traits classifies men according to their circulating testosterone concentrations.

The relationship between androgens and attractiveness may be complicated in humans by women's mate preferences for signs of paternal quality that could either conflict with, or at least be independent of, androgen-mediated cues of physical condition. Some research suggests that women may infer poor paternal potential from highly masculinized faces (Perrett et al., 1998; Johnston et al., 2001), and similar inferences for other androgen-derived cues could attenuate relationships between testosterone concentrations and rated attractiveness. Future research might test whether cues of higher androgen concentrations become more attractive under conditions in which women are provided information that assures them that the men in question are relatively high in kindness, interest in children, and so on.

A potentially more difficult problem in testing the role of sex hormones in men's attractiveness relates to the fact that androgen-mediated energy allocation might produce downregulation of testosterone production during circumstances in which elevated concentrations are not necessary for courtship or intrasexual competition. In olive baboons, for instance, testosterone concentrations do not correlate with dominance under stable social conditions but correlate positively during conditions of social instability in which dominance is being challenged (Sapolsky, 1983). Perhaps analogously, female mice exhibited no preferences between scent marks of male mice that were or were not infected with foreign antigens unless the males were first exposed to female urine before marking (such exposure causes transient increases in testosterone; see below), in which case the females preferred marks from the uninfected males (Zala, Potts, and Penn, 2004). In both cases, basal testosterone concentrations may not distinguish males based on their condition—perhaps because the higher-condition males gain energy savings by reducing testosterone when it is not necessary for competition or signaling—but the males in better condition are better able to upregulate concentrations during challenge. The expression of similar energy allocation mechanisms in humans would imply that the relationship between men's androgen concentrations and their attractiveness might be revealed most clearly under conditions of energetic, immune, or social challenge. Although empirical tests of this idea would likely be challenging, the successful demonstration of positive correlations between androgens and attractiveness during conditions of challenge could represent an important advance in our understanding of the physiological determinants of men's attractiveness.

Effects of Mating Stimuli on Sex Hormones

Male Reactions to Female Stimuli

The energy allocation model reviewed above is predicated on the presumption that androgen-mediated investments in muscle development, competitive behavior, and ornament growth are functional only if the mating benefits of such investments outweigh the costs of diverting energy away from survival-related processes such as fat storage and immune function. This implies that testosterone concentrations should be sensitive to cues of mating opportunities. Androgen suppression during the nonbreeding season among seasonally breeding species supports this general position, as it strongly suggests that the benefits of testosterone elevation are coupled specifically to mate competition (see Daly and Wilson, 1983). Breeding season reactivation of gonadal function may be triggered in part by cues from fertile females, furthermore, as male macaques exhibited testosterone elevations after exposure to females who were artificially brought into estrous during the nonbreeding season (reviewed in Michael and Zumpe, 1993), and male Siberian hamsters failed to exhibit breeding season restoration of luteinizing hormone pulses under conditions of long-day photoperiod unless they were also exposed to females (Anand et al., 2002; see also Hegstrom and Breedlove, 1999). Even within breeding seasons or among continuously breeding species, though, mating opportunities vary over time, and we might therefore expect to see androgen-mediated variations in energy allocation over much shorter time scales. Among continuously breeding chimpanzees, for instance, males exhibited both higher testosterone concentrations and more aggressive behaviors on days when parous females displayed maximal sexual swellings (Muller and Wrangham, 2004a). This demonstration potentially extends to primates evidence for the "challenge hypothesis" developed for avian species in which testosterone elevations above those necessary for sperm production promote mate competition behaviors during periods of pair formation (Wingfield et al., 1990). These studies together suggest that male androgen production and its sequelae are functions not only of male condition but also of cues to mating opportunities that can alter the cost-benefit structure of energy investment in mating effort versus survival.

Male hormonal reactions to conspecific females have been demonstrated even more conclusively at shorter time scales than those addressed in the above studies. Across various species, testosterone increases are typically reported within about 20 minutes of onset of exposure to females placed behind transparent barriers (Purvis and Haynes, 1974; Batty, 1978; Popova and Amstislavskaya, 2002; Amstislavskaya and Popova, 2004; Bonilla-Jaime et al., 2006) or to female chemosensory stimuli such as urine or vaginal secretions (for example, Macrides et al., 1974; Pfeiffer and Johnston, 1994; Richardson et al., 2004; for demonstrations specifically in primates, see Ziegler et al., 2005; Cerda-Molina et al., 2006). Testosterone increases are not reported after comparable nontactile exposure to other males (Macrides, Bartke, and Dalterio 1975; Pfeiffer and Johnston, 1992; Amstislavskaya and Popova, 2004), which suggests mechanisms that are designed to respond more specifically to cues of mating opportunities. Other research suggests that common brain mechanisms may regulate behavioral and hormonal reactions to females. Courtship behaviors emitted during exposure to female stimuli have in some cases been found to correlate positively with the magnitude of testosterone responses at about 30 minutes after exposure onset (Pinxten, Ridder, and Eens, 2003; James, Nyby, and Saviolakis, 2006), and lesions to subcortical forebrain structures such as the medial preoptic area disrupt both courtship behaviors (for example, Lloyd and Dixson, 1988; McGinnis and Kahn, 1997; Riters and Ball, 1999) and testosterone reactions to females (Kamel and Frankel, 1978). Castration is also known to reduce or abolish behaviors such as ultrasonic courtship vocalizations in response to female urine (Nyby et al., 1977), but, interestingly, testosterone implantation localized to only the medial preoptic area can restore these reactive vocalizations to normal levels (Matochik et al., 1994; Sipos and Nyby, 1996). A limbic-hypothalamic pathway including structures such as the medial preoptic area may therefore act as a gating mechanism that determines the extent to which cues from females trigger behavioral courtship effort and the upregulation of testosterone production (see also Wood, 1997). Since the hypothalamus processes numerous sources of information regarding male condition (for example, stress hormone concentrations, energetic status, immune activation), furthermore, this pathway may integrate information about mating opportunities with signals of male condition to generate decision rules regarding current investment in mating effort.

The function of short-term testosterone increases is less clear than changes over time scales of, say, breeding seasons, as transient elevations might not substantially adjust energy allocation processes. Transient elevations could play a role in priming longer-term shifts in androgen production and energy allocation if multiple exposures to mating cues have positive feedback effects on neuroendocrine mechanisms—perhaps by upregulating receptors in the limbic-hypothalamic circuit—though I am not aware of systematic tests of this idea. Short-term testosterone spikes might also play a role in regulating behavioral response strategies on short time scales. In particular, cues to mating opportunities may dramatically alter the momentary potential benefits of risky and competitive behaviors. Consistent with a risk-modulation function, Aikey et al. (2002) demonstrated in mice that exposure to mating stimuli, injections of testosterone, and injections of metabolites of testosterone known to act as gamma-aminobutyric (GABA) agonists (such as 3-alpha-androstanediol) each independently produced anxiolytic effects on males within 30 minutes of the respective manipulations. Similar manipulations have also been shown to reduce mount latencies in tests with receptive females (James and Nyby, 2002). Other research has shown that testosterone administration can enhance muscle metabolism within minutes in vitro (Tsai and Sapolsky, 1996), and such effects could plausibly support mate-seeking behaviors. Although rapid adjustments of risk aversion and competitiveness might be accomplished via neurotransmitters without the need for changes in hormone concentrations, it may be functional to maintain some hormonal modulation if androgen production is in fact coupled to general condition. A male in poor condition due to pathogen-based or energetic stress, for instance, may have less functional reason to reduce risk aversion than a male in better condition, and coupling behavioral adjustments to condition-dependent testosterone responses may therefore allow for a better integration of cues to condition with cues to mating opportunities.

The evidence supporting the condition dependence of androgen production in men (see partial review above) supports the functionality of human males also using cues to mating opportunities as moderators of testosterone concentrations. Rapid testosterone responses to cues from potential mates are amenable to experimental tests and could suggest that homologous brain mechanisms produce reactive hormone increases across human and nonhuman mammals. Some studies have provided evidence that exposure to sexually explicit films can trigger testosterone increases within about 20 minutes of exposure onset (Hellhammer, Hubert, and Schurmeyer, 1985; Stoleru et al., 1993; Redoute et al., 2000; but for negative results, see also Carani et al., 1990; Kruger et al., 1998), though the ecological validity of such manipulations seems uncertain. Roney, Mahler, and Maestripieri (2003) reported that men's salivary testosterone increased significantly over baseline concentrations 20 minutes after the onset of a conversation with a woman confederate. The magnitude of testosterone change was also positively correlated with the women confederates' ratings of the degree to which the subjects tried to impress them, which is consistent with the possibility that common mechanisms regulate the behavioral and hormonal responses. Subsequent research has provided further evidence for rapid testosterone increases in men after social interactions with women (Roney, Lukaszewski, and Simmons, 2007). Although the amount of human data is fairly limited, the studies to date present interesting parallels with the nonhuman literature both in terms of the time course of testosterone increases and the correlations between courtshiplike behaviors and time-lagged hormonal responses.

No studies directly address the functions of men's short-term testosterone reactions to mating stimuli. Rapid anxiolytic effects as in nonhuman species are a possibility, and studies that have administered testosterone to subjects have found some evidence for reduced fear responses within a few hours (for example, van Honk, Peper, and Schutter, 2005; Hermans et al., 2006). Although rapid adjustments in risk aversion may seem unnecessary in modern environments in which adults may meet opposite-sex strangers on a fairly regular basis, it is important to keep in mind that encounters with novel young adult women may have been fairly rare events in the small group environments commonly thought to characterize ancestral living arrangements. A face-to-face conversation with an unattached, unfamiliar young woman may have signaled a new mating opportunity that could have rapidly altered the cost-benefit structure of investment in mating effort. If so, transient elevations after initial encounters may play a role in priming longer-term testosterone increases (and hence adjustments in energy allocation) if subsequent encounters suggest the possibility of relationship initiation. Perhaps consistent with this is the finding that although men in relationships typically have lower testosterone than unattached men, men within the first few months of a new relationship had higher testosterone concentrations than even single men (Gray, Chapman, et al., 2004)—such men may be exhibiting the culmination of an energy reallocation process that began at the initial encounter. Reduced risk aversion and rapid adjustments in courtship motivation within hours of meeting a potential mate may also be crucial in modulating mating success if initial impressions play an important role in women's mate choice. Coupling such adjustments to condition-dependent androgen production could in turn ensure that mating effort is calibrated to potential costs as well as benefits. All of this is speculation at this point but could be tested in research designs that look at the effects of testosterone increases in reaction to encounters with potential mates: greater testosterone increases could predict outcomes such as lower risk aversion, greater willingness to engage in competition, or greater willingness to approach women when such outcomes are measured not during the encounters with potential mates but at least 30 minutes after such encounters when testosterone increases have been detected in past studies.

Other evidence supports the idea that men's testosterone concentrations are calibrated to levels of mating effort over longer time scales. Perhaps the strongest support for this possibility comes from studies that have shown that testosterone is lower among men who are in committed relationships or who are fathers (see Gray and Campbell, this volume), as the costs of androgen-mediated energy allocation away from fat storage and immune function may not be functional in men who are not actively seeking mating opportunities. Consistent with the idea that degree of mating effort may be an important variable here, McIntyre et al. (2006) reported that men in relationships who expressed interest in extra-pair sex maintained higher levels of testosterone. Regardless of motivation for pair formation, though, androgen production should also be reduced over longer time periods in the absence of available mating opportunities. Anecdotal evidence from men performing field research away from women suggests that testosterone may fall during such times and then rise again when men are reunited with women (Anonymous,

1970; Bribiescas, 2001a). Over more intermediate time scales, a correlational study found that degree of self-reported, psychological sexual stimulation (estimated by amount of exposure to stimuli such as reading materials, photos, or actual women) predicted serum testosterone concentrations measured the following day (Knussman, Christiansen, and Couwenbergs, 1986). Various lines of evidence are thus consistent with the possibility that motivation for pair formation and/or exposure to stimuli from potential mates can modulate men's sex hormones over a range of different time scales.

The evidence that higher testosterone concentrations actually do improve men's ability to compete for mates is entirely circumstantial. A couple of studies have reported positive correlations between either circulating testosterone (Bogaert and Fisher, 1995) or dihydrotestosterone (Mantzoros, Georgiadis, and Trichopoulos, 1995) and self-reported number of sex partners, though direction of causality is ambiguous. Correlational studies have also reported positive associations between testosterone and personality characteristics that would logically facilitate courtship and intrasexual competition, such as disinhibition, self-confidence, assertiveness, extraversion, and sensation seeking (Daitzman et al., 1978; Daitzman and Zuckerman, 1980; Dabbs, Hopper, and Jurkovic, 1991; Gerra et al., 1999; Aluja and Torrubia, 2004). Likewise, various lines of evidence suggest that higher testosterone individuals are generally more focused on maintaining personal power (Schultheiss, Campbell, and McClelland, 1999), status (Josephs et al., 2003), and dominance (for reviews, see Mazur and Booth, 1998; Archer, 2006), and other research suggests that such characteristics may enhance men's mate attractiveness (for example, Sadalla, Kenrick, and Vershure, 1987). Unclear in all of this correlational research is the extent to which these traits may change over time within individuals concurrent with changes in testosterone concentrations—if men becoming fathers experience drops in testosterone, for instance, do they also exhibit reduced disinhibition or sensation seeking? An androgen-mediated energy allocation model clearly predicts such within-individual changes, but direct empirical evidence will require longitudinal research designs that may be especially challenging to implement. Also needed are empirical studies that can more directly assess whether higher testosterone concentrations predict greater courtship effort in ecologically realistic circumstances. In sum, the available evidence suggests that mating stimuli can affect men's sex hormone concentrations across a range of time scales, though more evidence is needed regarding the functional effects of these physiological adjustments on the initiation of mating relationships.

Female Reactions to Male Stimuli

Interactions with males trigger GnRH release in species with induced-ovulation such as rabbits, ferrets, and cats, but even among spontaneous ovulators there is evidence in various species that exposure to chemosensory or tactile stimuli from males can in some cases advance the preovulatory luteinizing hormone (LH) surge and thus the timing of ovulation (for a review, see Bakker and Baum, 2000). The functionality of any such responses in humans seems uncertain, as it would appear that selection pressures for calibration of fertility to energetic conditions sufficient for successful pregnancy would be far greater than selection pressures for increasing fertility on perception of available mating opportunities—mating opportunities are essentially irrelevant if a woman does not have sufficient energy resources for pregnancy. Nonetheless, there could be marginal functional benefits to advancing ovulation as a means of cryptic female choice in the context of multiple mating or longer-term benefits in terms of energy savings associated with the calibration of ovarian hormones to social circumstances such as relationship status.

Some correlational research has suggested that women who self-report sexual activity or otherwise greater contact with men tend to have shorter and more regular cycles than women who are less exposed to men (for example, McClintock, 1971; Veith et al., 1983), though the direction of causality in these studies seems uncertain. Cutler et al. (1986) experimentally manipulated exposure to male axillary secretions but reported only nonsignificant trends toward shorter or "less aberrant" cycle lengths; furthermore, this manipulation had no effect on ovarian hormone concentrations. Another study that also exposed women to men's axillary secretions found that the time to next LH pulse was reduced after axillary exposure relative to control exposure, but the physiological significance of this effect was uncertain, given that the manipulation had no effect on number of LH pulses, pulse amplitude, or total secretion of LH over the course of the testing session (Preti et al., 2003). It thus appears that there is not yet any direct evidence that exposure to stimuli from men can affect sex hormone concentrations in women.

Conclusion

The energetic costliness of internal gestation and lactation in mammalian reproduction may have selected for females designed to convert energy into offspring and males designed to convert energy into mating opportunities with females (Daly and Wilson, 1983). Evidence for the role of sex hormones in human mating psychology—although very preliminary—appears generally consistent with this position. Women's ovarian hormone concentrations are highly sensitive to variables like energy balance (Ellison, 2001; also see Ellison, this volume), and cycle phase shifts in women's mate preferences suggest that hormone concentrations may in turn calibrate attraction to androgen-dependent traits in men. Ovarian hormones may thus function to couple women's fertility and mate preferences to propitious reproductive conditions. Men's sex hormone production is not as finely sensitive to energetic conditions as in women (Bribiescas, 2001a) but does appear to respond to cues of mating opportunities in ways that have not been demonstrated in women. This suggests a chain of causality in which favorable energetics upregulate women's ovarian hormones (and hence fertility), and men's perception of cues to enhanced fertility in turn upregulates men's androgen production. If sex hormone allocation of energy to fertility in women or courtship effort in men is essentially condition dependent because of finite energy availability, furthermore, then cues of elevated hormone concentrations may enhance attractiveness by honestly indicating good condition. Some direct evidence now indicates that women's physical attractiveness may correlate with their estradiol and progesterone concentrations. The evidence that men's attractiveness correlates with their testosterone production is more indirect, but this relationship may be complicated by the possibility that men may reduce androgen production when it is not needed for mate competition such that correlations with attractiveness may be revealed consistently only under conditions of challenge.

Greater understanding of the role of sex hormones in human mating psychology may substantially advance an area of research that has heretofore been dominated by self-report survey methodologies. Not only can hormonal investigations potentially integrate psychological measures with neurobiological mechanisms and phylogenetic patterns, but the linkages between sex hormones and systems regulating immune function and energy storage suggest that research on sex hormones may be crucial to understanding how mating mechanisms function within the context of multiple systems that have been designed to interact in adaptive ways. This chapter was intended to lay out what little is already known about the role of sex hormones in human mate choice and mate attraction, but clearly this is an area with broad opportunities for further investigation.

CHAPTER TWELVE

Human Male Testosterone, Pair-Bonding, and Fatherhood

Peter B. Gray and Benjamin C. Campbell

Introduction

One of the fundamental life history challenges facing human males is the allocation of reproductive effort to its primary components: mating and parenting effort (Trivers, 1972; Low, 2000; Kaplan and Lancaster, 2003; Lee, this volume). Investment in male-male competition and mate seeking (the main components of mating effort) may be incompatible with the maintenance of affiliative pair-bonds with a long-term mate or involvement in paternal care. There may be insufficient time and energy to engage in all of these facets of reproductive effort. Physiological mechanisms promoting one component of reproductive effort may interfere with another aspect of it.

Researchers have highlighted a variety of physiological mechanisms thought to adaptively regulate male mating and parenting effort (Carter, 1998; Dixson, 1998; Ziegler, 2000; Adkins-Regan, 2005; Nelson, 2005), including cortisol, prolactin, vasopressin, and oxytocin. Importantly for our purposes, the steroid hormone testosterone has been implicated in various taxa (see Chapters 3, 4, 8, and 9). This chapter focuses on the role of human male testosterone in contexts of pair-bonding and parenting relationships. To the degree that male testosterone levels are important in the modulation of reproductive effort, we would expect them to be elevated in contexts of human mating effort and lowered in contexts of affiliative pair-bonding and paternal care.

Testosterone and Human Male Mating Effort

Male-male aggression is a central aspect of male mating effort. As such, we expect to find positive links between human male-male aggression and testosterone. In fact, a recent meta-analysis observed a significant, but weak, positive correlation between human male testosterone levels and aggression (Book and Quinsey, 2005). The types of studies contained in this meta-analysis included ones showing elevated testosterone levels among prisoners convicted of violent crimes, compared to those convicted of nonviolent crimes (Dabbs et al., 1995). Furthermore, in some competitive athletic paradigms, male testosterone levels were elevated among participants actively engaged in the contests, especially winners (reviewed in Salvador, 2005; Archer, 2006). Moderating factors in this meta-analysis included the age of subjects (observing higher correlations between testosterone and aggression among males aged 13–34 than younger or older males) and time of sample collection (higher correlations between 12:00 P.M. and 5:00 P.M., compared to mornings or evenings).

Testosterone may be more closely linked to dominance-related behavior than physical aggression. Short-term changes in male testosterone levels associated with winner-loser effects have been observed in chess matches, in soccer fans (see Mazur and Booth, 1998), and in a pilot study of domino players (Wagner, Flinn, and England, 2002). As an egalitarian species (Boehm, 1999) in which most male-male competition does not entail physical aggression, we might expect social status–related behavior to be most closely linked to testosterone, especially among young adult males. Within a small community of lowland Ecuadorian Amazonians, male social status, controlled for age, is positively correlated with salivary testosterone levels (John Q. Patton, pers. comm.).

The other central element of mating effort, mate seeking, has also been positively associated with testosterone. In Chapter 11, Roney reviews the links between elevated male testosterone and mate seeking. If male testosterone levels are elevated during mating effort, are they lower in contexts of pair-bonding and paternal care?

Male Testosterone, Pair-Bonding, and Parenting in North America

Since Booth and Dabbs (1993) first demonstrated a relationship between male testosterone levels and marital status, research on male testosterone and relationships has surged. We summarize results of the relevant North American studies in Tables 12.1 and 12.2. Table 12.1 presents *between-group* data—that is, results of North American studies that have compared

Table 12.1 North American and European between-group data on male testosterone, pair-bonding, and paternal care.

Study Population	Research Design	Results	Reference
U.S. Army veterans	4,462 men (mean age = 37 years) providing one morning serum sample	Married men had lower T; T was also positively associated with divorce and extramarital sex	Booth and Dabbs, 1993
U.S. Air Force veterans	1,881 men aged 32–68 providing 1–4 morning serum samples over a 10-year period	Married men had ~9% lower T	Mazur and Michalek, 1998
Boston-area men	584–1,662 men aged 40–70 at baseline providing morning serum samples in longitudinal study	Men whose wife died experienced decreases in T	Travison et al., 2007
Boston-area men	58 men aged 21–40 providing 2 morning and 2 evening saliva samples	No morning differences; in evenings, married nonfathers (19%) and fathers (32%) had lower T	Gray et al., 2002
Harvard Business School students	122 men aged 23–34 providing 1 midmorning saliva sample	Pair-bonded men had 21% lower T	Burnham et al., 2003
U.S. men	65 men aged 28–40 providing 4 morning, 4 afternoon, and 4 evening saliva samples	In multivariate analyses, married men had only lower evening T	Gray, Campbell, et al., 2004
Harvard undergraduates	107 men aged 17–26 providing 1 saliva sample	Pair-bonded men had 49% lower T only at times later in day	Gray, Chapman, et al., 2004

New Mexico undergraduates	Pair-bonded men had 17% lower T	74 men (mean age=20 years) providing 2 morning saliva samples	McIntyre et al., 2006
Canadian fathers	Fathers had 33% lower T 3 weeks after birth of baby compared to 3 weeks before birth of baby	34 men providing venous blood samples around the time of their wife giving birth	Storey et al., 2000
Canadian fathers	Fathers had lower evening, but not morning, T compared with controls	23 fathers and 13 nonfather controls	Berg and Wynne-Edwards, 2001
Canadian men	Paired heterosexual, but not nonheterosexual, men had lower T levels than unpaired men	76 heterosexual men and 47 nonheterosexual men provided afternoon saliva samples	Van Anders and Watson, 2006a

T = testosterone.

Table 12.2   North American within-group data on male testosterone, pair-bonding, and paternal care.

Study Population	Research Design	Results	Reference
Ohio men with professional occupations	39 middle-aged men aged 39–50 providing 1 morning serum sample	Marital satisfaction and T were negatively correlated	Julian and McKenry, 1989
Boston-area men	58 men aged 21–40 providing 2 morning and 2 evening saliva samples	Among married fathers, spousal investment and time spent with spouse negatively correlated with evening T	Gray et al., 2002
New fathers and controls from Ontario, Canada	43 fathers and 24 controls collected saliva samples between 11:00 A.M. and 2:00 P.M.	Fathers' T inversely correlated with sympathy and need to respond to infant cries	Fleming et al., 2002
Recently married Pennsylvanians	92 newlywed men (mean age = 27 years) collected 2 morning saliva samples	Interaction effects: greater male support provision and better problem solving when wives and husbands had low T	Cohan, Booth, and Granger, 2003
Harvard undergraduates	107 men aged 17–26 providing 1 saliva sample	Lower T among "inexperienced" men; highest T among men in new relationships	Gray, Chapman, et al., 2004
Pennsylvanians	307 married men (mean age = 42.1 years) providing 2 morning saliva samples in successive years	No main effects of T and marital quality; interactions observed between marital quality and role overload on T	Booth, Johnson, and Granger, 2005
New Mexico undergraduates	74 men (mean age = 20 years) providing 2 morning saliva samples	Interaction: among paired men, higher T if interested in extra-pair sex	McIntyre et al., 2006

T = testosterone.

the testosterone levels of men differing in relationship status. Table 12.2 presents results of *within-group* analyses; these studies examined whether, within a group of men (for example, married fathers), variation in male testosterone levels was related to variation in some measure of social behavior.

The studies reviewed here display a variety of methodological differences. These differences include controlling or not for potential confounding variables like body mass index (BMI) and time of day for sample collection. A few of these studies involved clinically based sample collection, but others a variety of convenience sampling methods that gave rise to the variation in sample collection times. The measurement of relationship categories also differed across studies, with some studies distinguishing fathers from nonfathers and some studies differentiating pair-bonded unmarried men from unpaired, unmarried men, but others not. Such methodological variation precludes a formal meta-analysis. Instead, we present elements of the studies in tabular summary and describe patterns emerging from them.

One of the primary patterns emerging from the North American data is that pair-bonding status served as a consistent predictor of variation in male testosterone levels. In all but one study where males were distinguished by their involvement (or not) in a long-term committed relationship, they displayed lower testosterone levels, compared with their unpaired counterparts. The one exception was a sample of men comprising the Massachusetts Male Aging Study (Travison et al., 2007); this exception appears due to rapid health deterioration of these older men following the death of a spouse rather than alterations in mating effort. However, there were no significant differences in the total or bioavailable testosterone levels of men in this Massachusetts study according to whether men were married, single, or divorced (Travison, pers. comm.). In three of the studies where testosterone levels differed by relationship status, this was true whether or not these pair-bonded men were married.

Two of the studies found even lower testosterone levels among fathers compared with unmarried men, but the differences compared with pair-bonded nonfathers were not statistically significant (Gray et al., 2002; Burnham et al., 2003). In two other studies that compared fathers with nonfathers, fathers exhibited lower testosterone levels (Berg and Wynne-Edwards, 2001; Fleming et al., 2002).

Several variables appeared to have moderating effects on the links between male testosterone levels and relationship status. The time at which biological specimens were collected in some, but not all, cases affected the results observed. In four studies in which only morning samples were collected, differences in testosterone levels according to relationship status appeared (Booth and Dabbs, 1993; Mazur and Michalek, 1998; Burnham et al., 2003; McIntyre et al., 2006). In four other studies in which testosterone levels were measured at various times of day (for example, mornings and evenings), significant differences in testosterone levels according to relationship status were only observed in afternoon or evening samples (Berg and Wynne-Edwards, 2001; Gray et al., 2002; Gray, Campbell, et al., 2004; Gray, Chapman, et al., 2004). In no North American studies were significant differences in testosterone levels according to relationship status observed in mornings but not afternoons or evenings. These patterns suggest that significant testosterone–relationship status effects may be more likely to emerge among samples collected later in the day. If so, this is an important empirical observation, given that most clinical studies standardize their measurement of male testosterone by morning, fasting samples and may thus underestimate the links between male testosterone and social variables.

In addition, while all of the studies in Table 12.1 are from North America, there is substantial heterogeneity among study populations. Some early commentators criticized the relevance of military samples (for example, the Booth and Dabbs [1993] sample of army veterans) to the civilian population. However, similar findings have since been observed in samples differing in measures of socioeconomic status like educational attainment (for example, Gray, Campbell, et al., 2004). Thus differences in testosterone levels according to relationship status appear fairly robust across socioeconomic strata, at least in North America. Also important to the interpretation of these North American data is that cultural differences have not themselves been the subject of study.

Age distributions of samples may also be relevant. It is feasible that hormone–social behavior relationships will be stronger among young adults; social inertia, experience, and alternative behavioral tactics may serve as more important correlates of behavior as adults age (Adkins-Regan, 2005). With the exception of the Massachusetts Male Aging Study, all of the studies reviewed here focus on men aged 20–50. This older sample of Massachusetts men was also the only North American sample for which male testosterone levels did not differ according to relationship status (apart from the presumed health-related decline in men's testosterone after a spouse's death). It may be that differences in testosterone levels according to relationship status are most pronounced among men in their late teens to thirties and least likely to appear among the oldest men.

Turning from the *between-group* to the *within-group* analyses in North America, each of the published studies shown in Table 12.2 observed a significant association between variation in male testosterone levels within a group of men (for example, fathers) and either a marital or parenting variable. In six of these studies, negative relationships between male testosterone levels and some aspect of pair-bonding was observed. Among unpaired Harvard undergraduates, those with sexual experience had higher testosterone levels than inexperienced men (Gray, Chapman, et al., 2004). Among paired University of New Mexico undergraduates, males reporting more interest in extra-pair mating had higher testosterone levels. In a sample of newly married central Pennsylvanians, married men with similar testosterone levels as their wives scored higher on marital interactions. In two other studies—drawn from Boston and Ohio samples—the quality of marital relationships was negatively associated with testosterone levels among married men. Among a different Pennsylvanian sample, marital quality and testosterone were only inversely related among those men reporting a high "role overload" (for example, having too much work to do).

Relationships between male parenting outcomes and testosterone were less commonly addressed within samples of fathers. Still, one study found that men with lower testosterone levels reported greater emotional responsiveness to infant stimuli such as tape-recorded cries (Fleming et al., 2002). No differences in paternal care were linked to testosterone among Boston fathers (Gray et al., 2002).

The within-group analyses appeared less robust and exhibited more interaction effects compared with between-group analyses. In large part, this may occur because the between-group contrasts offer more dramatic differences in mating and parenting effort across groups of men; accordingly, it is perhaps not surprising they would be more likely to show differences in testosterone levels. At the same time, these groups are clearly not homogenous categories—they encompass variation in reproductive effort. While conceptually, then, we would expect to find differences in male testosterone levels associated with variation in reproductive effort within groups of men (for example, within married fathers), it may be more difficult to accurately measure these indices of reproductive effort. Across the relevant studies, within-group differences in social behavior were typically assessed by means of paper-and-pencil questionnaires or, to a lesser degree, behavioral observations in laboratory settings.

The North American studies on which both between-group and within-group inferences are based are subject to several limitations. With the exception of the two large military studies, sample sizes for these studies were small by epidemiological standards, although moderate to large for nonhuman behavioral endocrinology studies. The total number of subjects surveyed in these North American studies is about 7,000, a small fraction of the several hundred million people currently living in the United States and Canada.

Furthermore, with three exceptions, all of the studies reviewed in this section were cross-sectional. These cross-sectional designs do not enable determining how variation in male testosterone and relationship status are causally linked. Only the Mazur and Michalek (1998), study, the Canadian study of fatherhood (Berg and Wynne-Edwards, 2001), and the Canadian study of relationships (van Anders and Watson, 2006a) employed longitudinal designs. The first two of these studies found effects of changes in social variables (for example, divorce, becoming a father) on male testosterone levels, showing that social behavior can lead to changes in male testosterone levels.

Finally, there have been no studies of experimental testosterone interventions and pair-bonding or parenting behavior. Thus, it is not known how responsive, if at all, male psychology and behavior in the context of pair-bonding and paternal care are to experimental alterations in testosterone levels. At best, clinically based human testosterone intervention studies provide a few indirect insights into whether increasing or decreasing male testosterone levels affect relationship dynamics. In a short-term study of the effects of supraphysiological testosterone, Bagatell, Heiman, Matsumoto, et al. (1994) reported no effect of testosterone administration on pair-bonding or parenting based on two to three unspecified items from a larger questionnaire on male-female dyads. However, this study was flawed for use of only a few questions (rather than a complete, validated questionnaire) and low statistical power. Choi and Pope (1994) observed increased physical abuse toward partners among men on cycles of anabolic steroids compared to these same men when off cycles of steroids or control subjects, offering one inference that elevations in testosterone levels may affect relationship dynamics.

One other limitation of this review is potential publication biases. We are unaware of negative, unpublished findings within this research domain despite familiarity with many of the researchers conducting this type of research. Several large-scale epidemiological studies (for example, Baltimore male aging study) presumably have collected information on male relationship status (at least marital status) and so could enable tests of the prediction that married men will have lower testosterone levels than unmarried men. We suspect findings, negative or positive, from these biomedical databases have not been published because they were not conceived of as important dimensions of male physiology and health.

Lastly, although the association between relationship status and lower male testosterone levels is consistent, the size of the effect as judged by % differences in testosterone levels, correlation coefficients, and p values is modest. We cannot well predict an individual's reproductive effort by measuring his testosterone level. Nonetheless, the patterns are sufficiently robust to conclude that, at least for North America, these findings offer general support for the view that variation in male testosterone levels in part reflects variation in reproductive effort. A practical implication is that measures of relationship status should be included as predictors or covariates of male testosterone levels, along with variables such as BMI, disease states, age, and so forth, in multivariate models.

Male Testosterone, Pair-Bonding, and Parenting outside North America

All of the studies reviewed above are based on samples from North America. We assume that, consistent with analyses reviewed above, male testosterone levels are also modestly, but significantly, associated with male reproductive effort in non-Western societies. We expect these relationships play out adaptively in local sociocultural contexts (see Kaplan and Lancaster, 2003).

To address the expectation that elevated male testosterone levels are associated with male-male competition and mate seeking (mating effort) in non-Western societies too, few data are available to directly examine this proposition. In a study of the !Kung San, former foragers of southern Africa, Christiansen and Winkler (1992) found no differences in testosterone levels between men having previously been involved in violence, compared with those without a history of violence. However, among those men with a history of violence, the greater rate of violence was positively correlated with testosterone. Relying on paper-and-pencil questionnaire measures of aggression, male testosterone levels in Beijing, China (Yang et al., 2008) and South Africa (Gray et al., 2006) were not related to male aggression. A pilot study of Dominican male checker players showed greater testosterone elevations when facing opponents from a different village (Wagner, Flinn, and England, 2002). We know of no non-Western studies that have assessed links between human male mate-seeking behavior and testosterone. The overall mixed support for links between elevated testosterone and mating effort appears to reflect, in part, measurement issues since the two studies that relied on behavioral indices found significant differences, whereas those relying on paper-and-pencil questionnaires did not.

Table 12.3 summarizes the studies on human male testosterone, pair-bonding, and paternal care conducted outside North America. One of these studies contained preliminary data from a sample of 30 Dominican men described in Flinn et al. (1998), in which there were no differences in salivary testosterone levels between unmarried and married men. However, these data did not include ages, times of saliva collection, anthropometrics, and divorce or parental status, making interpretation of them difficult. They have been followed up by more readily accessible data presented at a conference (Gangestad et al., 2005). These follow-up Dominica data revealed that, controlling for age, fathers had lower testosterone levels than nonfathers.

Additional Caribbean data from urban Jamaica also showed that fathers had lower testosterone levels than single men (Gray, Parkin, and Samms-Vaughan, 2007). Jamaican fathers in this sample were of two types: so-called coresidential fathers, who live with their partner (mate) and youngest child, and visiting fathers, who live apart from their partner and youngest child. Approximately half of Jamaican children are born into these types of visiting parental relationships. If separating this Jamaican sample into three groups (single men, coresidential fathers, visiting fathers) rather than two (single men, fathers), significant group differences remained, but the significant contrast lay with visiting fathers having lower testosterone levels than single men.

Several Asian and African studies have been conducted on male testosterone, pair-bonding, and fatherhood. Fathers in Beijing, China, had lower testosterone levels than either married nonfathers or unmarried men (Gray, Yang, and Pope, 2006). However, Bangladeshi fathers did not have lower testosterone levels than either married men without children or unmarried men (Magid et al., 2006). Among Kenyan Swahili men, there was a statistical trend whereby fathers of younger children tended to have lower evening testosterone levels than fathers of older children. Among neighboring populations in rural Tanzania, fathers among the hunting-and-gathering Hadza had lower testosterone levels than single men, but there was no difference in the testosterone levels of fathers and single men among the pastoralist, and more polygynous, Datoga (Muller, Marlowe, et al., 2007). Collectively, then, these data on testosterone and fatherhood from outside North America lend mixed support to the idea that fathers have lower testosterone levels than other men.

When investigating links between male testosterone and pair-bonding status, patterns in these data outside North America are less clear. Married men in Beijing did not have significantly different testosterone levels

Table 12.3 Cross-cultural data on male testosterone, pair-bonding, and paternal care.

Study Population	Research Design	Results	Reference
Dominica	30 men of unspecified age, body composition, and time of day	No differences in T levels between unmarried and married men	Flinn et al., 1998
Dominica	46 men aged 18–53 provided multiple saliva samples at different times of day	Controlling for age, fathers had lower T levels than nonfathers	Gangestad et al., 2005
Kingston, Jamaica, community sample	43 men aged 18–38 provided saliva samples before and after a 20-minute behavioral interaction	Fathers (particularly "visiting fathers") had lower T levels than single men	Gray, Parkin, and Samms-Vaughan 2007
Kenyan Swahili	88 men aged 29–52 provided 3 A.M. and 3 P.M. saliva samples	No differences in T levels between unmarried and monogamously married men; polygynously married men had higher T levels than other men; trend toward fathers of younger children having lower P.M. T levels	Gray, 2003
Ariaal pastoralists, Kenya	203 men aged 20 and older provided 1 A.M. and 1 P.M. saliva sample	Monogamously married men had lower T levels than unmarried men aged 20–39; no differences in T levels between monogamously and polygynously married men	Gray, Ellison, and Campbell, 2007
Hadza hunter-gatherers of Tanzania	31 men aged 17–72 provided A.M. and P.M. saliva samples	Fathers had approximately 50% lower T than single men	Muller, Marlowe, et al., 2007

(continued)

Table 12.3 *(continued)*

Study Population	Research Design	Results	Reference
Datoga pastoralists of Tanzania	85 men aged 18–68 provided A.M. and P.M. saliva samples	T levels did not differ between married men or fathers and single men	Muller, Marlowe, et al., 2007
Urban Bangladesh	52 men aged 18–64 provided 2 waking (A.M.) and 2 before-bed (P.M.) saliva samples	No differences in T levels associated with marriage or fatherhood	Magid et al., 2006
Beijing, China, university community	126 men aged 21–38 provided 1 A.M. and 1 P.M. saliva sample	Fathers had lower T levels than married nonfathers and unmarried men	Gray, Yang, and Pope, 2006
Japanese university community	87 men aged 18–35 provided 2 A.M. and 2 P.M. saliva samples	Pair-bonded men tended to have lower P.M. but not A.M. T levels—but perhaps sleep rather than behaviorally related	Sakaguchi et al., 2006
Japanese sample from university and biomedical settings	44 men aged 20–66 provided A.M., early P.M. and later P.M. saliva samples	AM but not PM levels tended to be lower among paired men, but sexual activity appeared to account for this effect	Sakaguchi et al., 2007

T = testosterone.

than unmarried men, although there was a trend by which married men showed inverse correlations between marital quality and testosterone levels. Paired men from a Japanese university setting exhibited a trend toward having lower evening, but not morning, testosterone levels than unpaired men. In a smaller, different Japanese sample, paired men exhibited a trend toward having lower morning testosterone levels than unpaired men. Paired status was not a significant predictor in Dominica; a contrast between unmarried and monogamously married men was significant in one but not the other Kenyan sample and not significant among Bangladeshi men. In one, but not the other, sample of Kenyan males, polygynously married men had higher testosterone levels than other men.

The time of day at which testosterone levels were measured was not clearly linked with significant findings in these non–North American data. Contrary to North American findings, when some significant links between male testosterone and relationship status were more likely observed at later times in the day, this was not the case outside North America. Two trends appeared in evening samples (Kenyan Swahili and one Japanese study) that did not occur in morning samples, but in one Japanese study, a trend appeared only in morning, but not evening, samples. All other studies showed similar findings for morning and afternoon or evening samples.

There may be several reasons why general patterns outside North America are less clear compared with those from North America. One reason is that North American families sampled in the testosterone and relationships research tend to be minimalist by comparison with families around the globe. That is, the North American samples tend to consist of nuclear households of a husband, wife, and children but few extended kin or other individuals also residing with them (Van Den Berghe, 1979; Roopnarine and Gielen, 2005). By contrast, cross-cultural patterns of families encompass greater social complexities; these include the presence within the household or nearby of extended kin like grandparents, siblings, aunts, and uncles and the presence of biological and stepchildren. In these more socially complex families, men may be exposed more to children, including young siblings, nephews, and nieces, and may have greater social responsibility to other kin like grandparents. It may be the case that transitions from mating to parenting effort occur more starkly among males living in North American societies than in these other types of families, in turn accounting for the more consistent findings of testosterone and relationships in North America than outside the region.

A second reason why these findings may be less consistent outside North America is that relationship categories serve as poorer proxies of mating and parenting effort outside North America than within it. While we may feel confident that pair-bonded men in the United States engage in less mating effort than their unpaired counterparts, with some independent evidence corroborating such views (for example, Daly and Wilson, 1999), we may be less justified making this assumption in other social contexts. Indeed, Magid et al. (2006) suggest that traditions of arranged marriages in Bangladesh may help account for null findings there. In societies allowing polygynous marriage or requiring minimal direct paternal involvement, is it necessarily the case that men's mating effort decreases once marrying a wife? Or if men commonly spend time around young children and other kin, they may experience less dramatic behavioral shifts associated with pair-bonding and fatherhood.

A third reason is that this "outside North America" category itself encompasses considerable cross-cultural variation. The sample of men from Dominica is a multiethnic population for which marital relationships are fairly dynamic, expressed as comparatively high rates of divorce and mixed family status. The two Kenyan samples include an agropastoral population—the Ariaal—as well as a coastal community engaged in fishing, tourism, mango farming, and various other activities (the Swahili). The two Tanzanian populations include both hunter-gatherers and pastoralists. University samples from Japan and China are also considered, with a second older sample of Japanese represented by both members of a university community and a biotechnology company. Patterns of mating and parenting effort differ across these samples, meaning that expectations about how male testosterone levels are associated with pair-bonding and parental status may also vary. Indeed, Muller, Marlowe, and colleagues' (2007) contrasting findings for neighboring Tanzanian populations were *predicted* based on recognized differences in paternal involvement in these groups. Because of such cross-cultural complexities, we elaborate on several aspects of these studies for which we have had involvement—those in Kenya and China.

In China, the sample of young and well-educated Beijing men reported high levels of extra-pair matings, consistent with independently conducted research suggesting a recent rise and high prevalence of extra-pair matings in urban China (see Gray, Yang, and Pope, 2006). In these Chinese data, men engaged in extra-pair sex did not have higher testosterone levels. This contrasts with findings by Booth and Dabbs that among U.S. military veterans who have generally lower socioeconomic status (SES), those with higher testosterone levels were more likely to engage in extra-marital sex (Booth and Dabbs, 1993).

However, the cultural and psychological process of engaging in extra-pair sex in China could be very different from that in the United States.

Though we do not have detailed information about extra-pair sex partners in the Chinese sample, other literature (for example, Suiming, 2004) suggests that some of these partners may have been prostitutes or mistresses. If extramarital relationships among these Chinese men arose because they were targeted by mistresses (because of their relatively high status) or because these men had sex with prostitutes, the psychological processes linking testosterone and extra-pair sex could be quite different than those arising in, say, existing North American samples. Thus, interpretation of these discrepant results could be contingent on the social context of the behavior.

In comparing the two Kenyan studies, several other cultural differences emerge as potentially mediating the relationship between testosterone levels and marriage. Monogamously married Ariaal men had lower testosterone levels than their unmarried counterparts; this was not true among the Swahili. Several factors may help to explain this difference. First, among the Ariaal men, unmarried men were those who had never married. In contrast, some of the unmarried Swahili men were previously married and divorced, exposing them to longer-term effects of pair-bonding–related experiences (and fatherhood for several men). Moreover, the Ariaal sample included younger men (in their twenties) than the Swahili, potentially important if hormone-behavior correlations are stronger among younger adults. We suspect that the "cleaner" or more homogeneous a sample with respect to age, marital, and reproductive history, the more likely differences in testosterone levels associated with different reproductive effort will be likely to emerge. The Ariaal, in this respect, may have enabled a cleaner comparison of unmarried and monogamously married men's testosterone levels.

Among Kenyan Swahili men but not Ariaal men, polygynously married men had higher testosterone levels. In first reporting this result among the Swahili (Gray, 2003), it was noted that the proximate cause of these elevated testosterone levels could not be convincingly attributed to nutritional/energetic factors, elevations in mate guarding, or libido. Still, it is notable that this finding, in the only other study for which it has been investigated, was not replicated among the Ariaal. If long-standing energetic effects (for example, linked to developmental set points) were responsible, they might have been more likely to have emerged among the more socioeconomically variable Swahili than the Ariaal. On the other hand, the strong age-related increase in polygyny among the Ariaal may be due to political, social, and economic factors unrelated to testosterone-mediated behavior. Since the Ariaal sample contained more older men, this may help make sense of the null finding there but not in the slightly younger (ages 29–52) Swahili sample.

Finally, energetic constraints may be important contexts of male gonadal function in many non-Western societies (for example, Campbell
and Leslie, 1995; Ellison, 2001; Bribiescas, 2006; Gray et al., 2006; Ellison, this volume). The role of chronic energetic factors (intake, expenditure, disease burden) probably accounts for the lower testosterone
levels consistently observed in subsistence populations. These energetic
constraints are also more apt to represent the evolutionary context of
male social behavior. High testosterone levels among men in Western
populations, unhinged from long-standing energetic constraints, appear
to represent an evolutionary novelty (Bribiescas, 2006). Apart from this
evolutionary perspective, energetic factors may also need to be controlled for in non-Western populations before examining human male
hormone-behavior relationships. Such considerations also mean that we
should not expect differences in male testosterone levels across populations to reflect differences in behavioral reproductive effort; rather,
analyses should be conducted within populations, as true of all studies
reviewed here.

These evolutionary and energetic considerations are also relevant to
rapid shifts in human ecology. As human dietary intakes, energy expenditures, and disease burdens change rapidly, these factors can be viewed
as unpredictable environmental factors in the framework presented by
Wingfield (this volume). In the face of such rapid shifts, we find that urban Aymara in Bolivia have higher testosterone levels than their rural
counterparts (Beall et al., 1992) and that across a rural-urban gradient in
South Africa urban professionals have the highest testosterone levels
(Gray et al., 2006).

The Hadza findings warrant further attention because they are the
only ones obtained from a hunter-gatherer population. Evolutionary sce-
narios of human behavior privilege hunter-gatherers because they live in
socioecological contexts more similar to those in which our species' behavior evolved than, say, Las Vegas or Tokyo and thus speak more
closely to the "human adaptive complex" alluded to by Lancaster and
Kaplan (this volume). Although testosterone research had been published
on several human hunter-gatherer populations—the Efe of the former
Zaire (Bentley, Harrigan, and Campbell, 1993) and the !Kung of southern Africa (Christiansen, 2002)—researchers had not addressed whether
male testosterone levels in these societies differed according to indices of
mating and parenting effort until these Hadza data were reported.

The Hadza data are consistent with the view that fathers exhibit
considerable paternal investment and reduced mating effort compared to
their single counterparts. Much of this paternal investment may be in the

form of indirect care (for example, food provisioning), but the degree of direct paternal care (for example, holding and carrying offspring) tends to be relatively large compared to cross-cultural patterns of male care (Marlowe, 1999, 2003), even if it pales in comparison to the direct care given by older siblings (Kramer, 2006) and adult female relatives (Hrdy, 1999). And because these Hadza data are embedded in the extended kin networks and face-to-face social context of a hunter-gatherer society, they speak most directly to male testosterone and relationship patterns in an evolutionarily relevant context, in turn suggesting that male variation in testosterone according to variable investment in mating and parenting effort likely has a long evolutionary history.

Neuroendocrine Mechanisms and Human Male Behavioral Reproductive Effort

The causal basis for lower testosterone levels among pair-bonded men and fathers is not entirely clear. Are men with lower testosterone more likely to become pair-bonded? Or do men's testosterone levels drop as they become pair-bonded and fathers? Current evidence suggests support for both views.

Two of the three human longitudinal studies reviewed above showed that changes in relationships resulted in changes in testosterone. At the same time, a man's testosterone level is highly correlated across days, weeks, or even years (Dabbs, 1990b; Dabbs and Dabbs, 2000), suggesting stable individual differences in testosterone levels somewhat predictive of different reproductive effort. Moreover, the fact that raising or lowering male testosterone levels (especially into the "hypogonadal" or "supra-physiological" range) can affect human cognition and behavior (for example, Pope, Kovri, and Hudson, 2000; Gray et al., 2005) also suggests that testosterone can effect changes in behavior. Regardless of the direction of causality, however, the association of testosterone, pair-bonding, and fatherhood may be mediated by the same neuroendocrine processes.

Current functional magnetic resonance imaging (fMRI) studies of attachment have implicated a number of different brain areas associated with romantic love (which we take to be commonly associated with the early stages of pair-bonding), including regions of the striatum, the dentate gyrus of the hippocampus, the dorsal region of the anterior cingulate cortex, and the ventral tegmental area (VTA) (Bartels and Zeki, 2004; Aron et al., 2005). Among these areas, the striatum and the VTA are of particular interest. The striatum is associated with reward, suggesting that the feelings associated with romantic love are experienced as rewarding (see also Jankowiak, 1995). In the rat, areas of the striatum exhibit androgen receptors (Creutz and Kritzer, 2004), suggesting that changes in testosterone could influence pair-bonding through effects on reward and motivation.

The VTA is of particular interest because of its association with vasopressin and oxytocin (Depue and Morrone-Strupinsky, 2005), hormones implicated in pairbonding and parental care (see Carter et al., this volume; Sanchez et al., this volume). Oxytocin has been linked to human trust (Kosfeld et al., 2005; Zak, Kurzban, and Matzner, 2005), suggesting that it may play a crucial role in attachment associated with pair-bonding. Furthermore, oxytocin has been related to reduction of fear and social recognition (Kirsch et al., 2005), suggesting a potential relationship between the reduction of amygdala activity and increased attention to a particular individual as part of human pair-bonding (Fisher, Aron, and Brown, 2005). Again, in the rat, the VTA exhibits androgen receptors (Kritzer, 1997), suggesting that this area may act as a node linking testosterone and oxytocin.

Prolactin has been associated with paternal behavior in animals, including callitrichids (see Ziegler and Snowdon, this volume). Two studies in humans, both conducted in Canada, have found that elevated prolactin was associated with paternal reponses to infants in humans (Storey et al., 2000; Fleming et al., 2002). Jamaican fathers experienced flat prolactin profiles during a 20-minute behavioral interaction with a partner (mate) and youngest child, whereas prolactin levels of single men sitting along, reading a newspaper, declined during this same time. These Jamaican results suggest that prolactin levels were maintained (rather than increased) by paternal stimuli.

In addition, prolactin could mediate effects of sexual activity on testosterone levels. Sexual activity, or more specifically orgasm, is associated with increased prolactin levels (Kruger et al., 2003; Bancroft, 2005). It has long been known that prolactin has inhibitory effects on the hypothalamic-pituitary-gonad al (HPG) axis (Carani et al., 1996). Therefore, it is possible that increased sexual activity among married men is associated with elevated prolactin, which in term leads to lower testosterone levels. However, the inhibitory effects of prolactin on the HPG axis are most readily observed in cases of hyperprolactinemia (Sobrinho, 2003), suggesting that elevated prolactin levels due to paternal or sexual stimuli are unlikely to account for lower testosterone levels among pair-bonded or paternal men. Oxytocin also increases at orgasm (Kruger et al., 2003). Yet increases appear to be related directly to ejaculation in males (Filippi et al., 2003; Vignozzi et al., 2004) and are unlikely to mediate effects of sexual activity on decreased testosterone levels.

Could activation of the stress response, as indicated by cortisol release, account for patterns of male testosterone, pair-bonding, and paternal care? Several United States studies have investigated husbands' and wives' stress responses, including adrenocorticotropic hormone (ACTH) and cortisol release, during brief discussions designed to induce marital discord (Kiecolt-Glaser et al., 1997; Robles et al., 2006). In these studies of both newlywed and older couples, women showed greater stress reactivity than men. The stress buffering ability of these men and the fact that it takes enormous effort for cortisol release to suppress the HPG axis (for example, when administering supraphysiological doses of synthetic glucocorticoids or especially acute stressors) suggest that cortisol release in contexts of family stress is unlikely to account for the observed patterns of male testosterone, pair-bonding, and paternal care. Indeed, cortisol levels may be lower (married men compared with single men: Mazur and Michalek, 1998; fathers compared with nonfathers: Berg and Wynne-Edwards, 2001) or comparable (married Kenyan Swahili men: Gray 2003; Jamaican fathers compared with single men: Gray, Parkin, and Samms-Vaughan, 2007b) between men in relationships compared with control men.

In addition, other behavioral changes associated with pair-bonding and paternal care might have indirect effects on testosterone levels. At least two studies found that married men exhibit higher BMI than unmarried men (Jeffery et al. 1989; Lipowicz, Gronkiewicz, and Malina, 2002), and BMI has been inversely associated with testosterone levels (for example, Allen et al., 2002; Gray et al., 2006). Because many of the human studies reviewed above adjusted for BMI, however, it is unlikely to have accounted for associations between male testosterone levels and relationship status. Furthermore, testosterone shows a marked circadian rhythm with levels elevated at the time of waking (Dabbs, 1990a; Diver et al., 2003), thanks to sleep-entrained HPG activity. If testosterone levels measured later in the day (for example, afternoons or evenings) better index effects of social interactions, this might help to account for the fact that, in some studies reviewed above, relationship status was more likely to be related to testosterone levels measured later in the day, rather than in the morning.

Perhaps more important, sleep disruption has been shown to have a substantial impact on testosterone levels throughout the day (Schiavi, White, and Mandeli, 1992; Luboshitzky et al., 2002; Axelesson et al., 2005). Though sleep disruption may not be associated with pair-bonding, fatherhood commonly is (especially with young children). As one illustration, Jamaican coresidential fathers self-reported 5.7 hours of sleep the previous night, compared with 7.7 hours self-reported by single men (Gray, Parker, and Samms-Vaughan, 2007). Thus, sleep disruption associated with having a young child may contribute to lower testosterone levels among fathers, and it could be particularly interesting to look at this issue with respect to co-sleeping with an infant (see Worthman and Melby, 2002).

While this chapter has focused on testosterone levels, it is important to note that variation in the androgen receptor that binds testosterone may have consequences for androgen action. Functional imaging technologies are, at this time, unable to discern between-individual differences in androgen receptor localization and density among living humans. However, species and population differences in the androgen receptor have been observed, most notably in the number of CAG (cytosine-adenosine-guanine) repeats in the promoter region. Fewer repeats appear to be associated with greater "androgenicity," including from in vivo experiments (La Spada et al., 1991; Choong, Kemppainen, and Wilson, 1998). Across primates, a trend toward longer repeats has been observed; a basic local alignment search tool (BLAST) search conducted by Luz Pfister in March 2006 (personal communication) showed that lemurs have 4 CAG repeats; macaques, 4; gibbons and siamangs, 8; orangutans, 12; gorillas, 8–17; bonobos, 13–19; chimpanzees, 9–23; and humans, 11–31. The overall trend, then, appears to be reduced androgenicity, perhaps consistent with the modest effect sizes and increasing cortical control of behavior. Decreased effectiveness of the androgen receptor in humans may reduce the role of testosterone in promoting sexual and aggressive behavior, thus facilitating the effects of nonhormonal mechanisms and other hormonal mechanisms such as oxytocin that may facilitate pairbonding.

CAG repeats show population and individual differences that may have small yet continuous effects on androgen action, including subtle differences in signal transduction of a given testosterone concentration (Zitzmann and Nieschlag, 2003). However, the effects of these differences seem unlikely to play an important role in population differences in patterns of pair-bonding compared with other factors. Among a polygynous, east African pastoralist population (the Ariaal), a higher average number of CAG repeats was observed than among West African and African American populations and at the upper end of more global population variation, thus arguing against selection for greater androgenicity in the face of polygyny and relatively little direct paternal care (Campbell et al., 2008).

Species differences in the orbital frontal cortex may also shed light on the evolution of pair-bonding in humans. The orbital frontal cortex is associated with inhibition of emotional impulses, including aggression (Kolb, Pellis, and Robinson, 2004) and sexual motivation (Stoleru et al., 2003). Orangutans, known for forced copulations, exhibit a smaller orbital frontal cortex compared to other hominoids (Schenker, Desgouttes, and Semendeferi, 2005), while humans and bonobos exhibit a larger orbital frontal cortex (Semendeferi et al., 1998). Again, increased inhibition of sexual motivation or aggression by the orbital frontal cortex in humans might result in increased impact of other neural mechanisms that act to promote pair-bonding.

Finally, it is interesting to note that age-related variation in mating and parenting effort follow a pattern similar to age-related changes in testosterone (Daly and Wilson, 1988), with young adult males showing high levels of testosterone and low levels of parenting. This suggests a potential trade-off between male testosterone and dominance, on the one hand, and pair-bonding and paternal care, on the other, that may be related to declining physical capacity and energy with age. This could also be linked with the view that testosterone-related male behavior is more important among younger, rather than older, adults: as men transition from investment in mating effort to parenting effort typically at young adult ages, the differences in testosterone levels associated with relationship status are most likely to emerge. At older ages, social inertia and socioeconomic status may be more closely tied to relationship status than testosterone.

Conclusion

This chapter began with the theoretical proposition that variation in male testosterone levels may correspond with differential reproductive effort. This theoretical framework suggests that higher testosterone levels may be associated with male-male competition and mate seeking, and lower testosterone levels may be associated with affiliative pair-bonding and paternal care. In reviewing the human data, consistent differences in male testosterone levels associated with pair-bonding status and paternal care have been observed in North America. Some 10 of the 11 relevant studies indicated lower male testosterone levels associated with either pair-bonding or fatherhood, and the exception was a study containing the oldest sample of men. Moreover, 6 North American studies showed differences in testosterone levels associated with either pair-bonding or paternal variables (for example, testosterone levels inversely linked to marital quality among married men). The time of day of sample collection moderated some, but not all, of the findings of these studies. Outside of North America, 4 studies found lower testosterone levels among fathers (Beijing, China; urban Jamaica; a rural village in Dominica; Hadza in Tanzania), but relationships between male testosterone levels and marital status were less consistent than North American data. Reasons for the more variable results outside North America include the heterogeneous social contexts in which pair-bonding and paternal care emerge.

Summarizing this human literature, the links between male testosterone levels, pair-bonding, and paternal care appear relatively consistent, particularly in North America. Several limitations should be considered, however, when reviewing these data. Almost all of the available data have been obtained from cross-sectional research designs, although evidence suggests that bidirectional relationships characterize the links between male testosterone levels and relationship variables. Sample sizes of these studies tend to be small by epidemiological standards, though large by field or lab animal behavior standards. The effect size of the links between male testosterone levels, pair-bonding, and paternal care is modest. We thus urge the appropriate caution when extrapolating from these findings to implications.

Future research in elaborating the effects of testosterone on pair-bonding and parenting might fruitfully concentrate on three avenues: cross-cultural research, ongoing clinical studies involving testosterone administration, and use of testosterone outside clinical settings. Cross-cultural variation in human pair-bonding and parenting behaviors provides a range of "natural experiments" to investigate the role of testosterone in reproductive effort across different socioeconomic and cultural contexts (see Lee, this volume). For instance, current results on testosterone and fatherhood include Canadian samples of men highly committed to child care. They are actively involved in prenatal classes, in birth, and in unusually high levels of direct paternal care of neonates. Other cultural contexts in which men are not as actively involved or committed to pair-bonding or child care—such as arranged marriages (Low, 2000), adoptions (Rogoff, 2003), and partible paternity (Beckerman and Valentine, 2002)—provide important comparisons.

In addition, the evidence for a link between relationship status and testosterone reviewed here suggests that the incorporation of indices of relationship quality and parental care into studies of clinical human testosterone interventions is warranted. Hundreds of thousands of men have had their testosterone levels suppressed for the treatment of prostate cancer. Such dramatic alterations in testosterone have been shown to change mood, sexual function, and a variety of other outcomes (for example, Wang et al., 2000; Gray et al., 2005), but the potential effect on pair-bonding and parenting dynamics remains unexplored. In particular, the use of implicit (subconscious) or behavioral assessments of pair-bonding and parenting dynamics would enable consideration of the effects of clinically based testosterone interventions (for example, van Honk et al., 1999).

Furthermore, the explosion of testosterone use and abuse in Western countries, among otherwise normal men (Pope, Phillips, and Olivardia, 2000), suggests that the effects of testosterone supplementation on pair-bonding and parenting in the larger population should also be investigated. Many men are now using androgens, whether to bolster athletic performance, for aesthetic reasons, or to increase mood and sexuality. What are the effects of taking exogenous testosterone on male pair-bonding and paternal care? If the men involved in these interventions begin with depression, lethargy, low libido, and so forth, they (and the relationship in which they are involved) may benefit. However, if these men gain a heightened sense of self-worth and elevated responsiveness to social challenges and, put simply, engage in elevated mating effort, their relationships may suffer.

In addition to potential psychological and behavioral effects, the association of lower testosterone and relationship status suggests one possible pathway by which marriage is associated with lowered morbidity and mortality (Hu and Goldman, 1990; Johnson et al., 2000). Reduced testosterone levels associated with affiliative pair-bonding and paternal care may be associated with better immune function and cardiovascular health among married men (see Robles and Kiecolt-Glaser, 2003). Indeed, one of the strongest effects on male longevity yet observed was acute testosterone suppression (via castration) of human males (Hamilton and Mestler, 1969), and similar effects have been observed in other animals (Austad, 1997).

CHAPTER THIRTEEN

Neurobiology of Human Maternal Care

Alison S. Fleming and Andrea Gonzalez

Conceptualization and Measurement of Maternal Behavior

In this chapter, we describe maternal behavior in humans from a psychobiological perspective. Grounded in a foundation of animal behavior and behavioral endocrinology, we view and analyze reproductive behavior in humans in a comparative fashion across species, drawing on our own research and that of others. In doing so, we consider this functional behavior system in a species-characteristic context, where components or features of the behavior have a certain degree of stereotypy. However, the existence of stereotypes in no way implies that the behaviors themselves, the motivation to express them, and the timing of their appearance are not molded by contextual features of the social and physical environment and by earlier experiences during individual development. Hence, we view behaviors as emerging from interactions between environmental and genetic influences across development and at the same time as being under the proximal control of sensory, contextual, and hormonal factors acting on a complicated neural structure.

Defining Measures of Maternal Behavior: Observable *"Consummatory" Behavior*

Western art has elevated maternal nurturance to a holy state; in fact, the image of the young mother Mary cradling her newborn son in her arms is one illustration that may come to mind when conjuring up an image of maternal nurturance. This is an image that is repeated in the works of

Renoir, Cassatt, Monet, Klimt, and many others. The standard represen-tation depicts a mother cradling the infant on her front and making di-rect eye contact as she rocks the infant back and forth and talks soothingly. If the new analysis of the Mona Lisa, uncovering a veil on her dress typical of new of expectant mothers has any credence, her smile re-flects the inner feelings of a pregnant or recently parturient mother.

To what extent do such "beatific" images reflect the reality of mother-hood? Is it a universal one? And does it depend on common universal bi-ological mechanisms? The behavior of the new mother toward her infant shows both large variations and striking similarities across individuals within a culture and across cultures (Hrdy, 2000; Bornstein, 2002; Hark-ness and Super, 2002; Cote and Bornstein, 2005). The similarities tend to reflect the constraints of our biological underpinnings—our inclusion in the class Mammalia—and they are surprisingly stereotyped in form. It is true that most mothers in most cultures nurse their newborn infants, and they do so by cradling the infant on the ventrum during nursing (Corter and Fleming, 2002). Some mothers in some cultures do not nurse. Instead, they bottle-feed or arrange for other women or relatives to either nurse or bottle-feed (Clarke-Stewart and Allhusen, 2002; Moore and Brooks-Gunn, 2002). Regardless of who fulfills the nurturing role, the infant is typically in face-to-face orientation with the caregiver and is usually embraced and stimulated tactilely by the caregiver (Brazel-ton, 1977; Konner, 1977; Leiderman and Leiderman, 1977). Given that mothers lactate during the early weeks and months of the infant's life, the mother is usually the primary caregiver (Leiderman and Leiderman, 1977), although in humans, as in some other animals, fathers and other alloparents may also care for the young (see Carter et al., this volume; Ziegler and Snowdon, this volume; Fleming et al., 2002).

Associated with cradling and nursing, mothers and infants stimulate one another through a variety of sensory systems. Mothers typically pro-vide the infant with various forms of tactile stimulation and are in turn stimulated by them. Over time a synchrony usually develops between in-fant and mother (D. Stern, 1977; Winberg, 2005; Forcada-Guex et al., 2006). Visual stimulation is represented in many cultures via eye-to-eye contact between mother and infant during nursing, activating the visual systems of each member of the pair (Lier, 1988; Papoušek and Papoušek, 1995). In terms of the auditory system, mothers often respond to infant cries by picking up the infant (Bell and Ainsworth, 1972) and to infant noncry vocalizations by vocalizing themselves (Fleming et al., 1988). Mothers are drawn to approach crying infants, even from a distance, and attempt to soothe infants through rocking, stroking, or adjusting the infants' position, providing somatosensory input (Barr and Elias, 1988; Lier, 1988; Ludington-Hoe, Cong, and Hashemi, 2002). In addition, as a precursor to verbal communication, soon after birth, mothers and infants develop a dialogue in which both members engage in "mutual visual attention and turn-taking between listening and vocalizing" (Papoušek and Papoušek, 1995, 127). These interactions have been described in most cultures where they have been studied as examples of "intuitive parenting" (Papousek and Papousek, 1995, 2002). Mothers and infants are usually in close proximal contact with one another for some period of the day and have the opportunity to smell one another's body odors (Schaal and Porter, 1991; Porter and Winberg, 1999) and be affected by them, thereby activating olfactory systems in the dyad (Fleming et al., 1993; Fleming, Steiner, and Corter, 1997; see Corter and Fleming, 2002). Despite these similarities in sensory activations, the variations in mothering style and cognitions within and across cultures are evident, deriving both from individual differences in motivation and attitudes toward mothering and, more plainly, from variations in cultural norms, expectations, and social and economic structures (see Corter and Fleming, 1995, 2002; Harkness and Super, 2002; Bornstein and Cote, 2004). Not surprisingly, differences across cultures in caregiving behaviors become increasingly pronounced as the infant grows older and into childhood and when sensory processes diverge from more cognitive and affective systems that regulate parenting (Bornstein and Cote, 2004).

Behavioral Indices of "Motivation" to Mother

In our studies on mothering in new mothers from North America, we use two approaches to measure maternal behavior. One approach constitutes measurement of discrete behaviors on a microanalytic level and complex analyses involving contingencies, while the other approach is on a more global, contextual scale. We videotape mothers interacting with their two-day- to six-month-old infants over a 15 to 30-minute period during a nonfeeding interaction and code the videotapes in terms of a variety of maternal and infant behaviors, including maternal touch and stroking, affectionate kisses and hugs, vocalizations, visual inspection, rocking, singing, instrumental caregiving, and toy playing, as well as infant smiles, arm waves, and vocalizations. These behaviors are coded in terms of the duration and frequency of behaviors (Fleming et al., 1988; Fleming, Steiner, and Corter, 1997). The global index of maternal responsiveness can involve either videotaped interactions or home observations in which mothers' sensitivity to their infants is coded using the

Maternal Behavior Q-Sort (MBQS; Pederson et al., 1990) or Ainsworth scales (see Ainsworth, 1979).

Using these coding schemes for maternal responsiveness/sensitivity, we have shown reduced affectionate touching and/or "sensitivity" and, often, higher levels of caretaking/instrumental behaviors as a result of low levels of prior mothering experience (Fleming et al., 1988) in depressed mothers (Gonzalez, Steiner, and Fleming, in prep-a), in adolescent mothers (Krpan et al., 2005; Giardino et al., 2008), and in mothers who experienced disruptions in their families of origin (Krpan et al., 2005; Gonzalez, Steiner, and Fleming, in prep-b).

Self-Reported Experiences, Beliefs, and Desires

Mothers demonstrate nurturance not only through direct interaction with their infants, as described above, but also through self-report measures. During pregnancy, they engage in indirect preparatory behaviors; they prepare the infants' room, clean the house, change their diets, become less social and more introspective, have more infant-related dreams, and prepare for the birth experience (for review, see Corter and Fleming, 2002; also see Fleming et al., 1995).

Like other mammals, for many mothers, "maternal responsiveness" or the motivation to mother or to engage in the behaviors described above increases throughout pregnancy (Corter and Fleming, 2002; Numan, Fleming, and Levy, 2006). These attitudes and self-reported feelings highlight the issue of motivation, indicating that well before the birth of the baby and before actual maternal behaviors are performed, mothers-to-be experience changes in their feelings and cognitions relating to mothering. These changes can be viewed as precursors to behavior and may function to prepare the mother psychologically for mothering. They are, in some sense, "appetitive behaviors" at a temporal distance.

In one Canadian study, first-time, middle-class mothers were assessed longitudinally on a maternal attitude questionnaire at seven time points; they were administered the questionnaire five times across pregnancy, on day two postpartum, and at three months postpartum. Although there were clear individual differences in the maternal feelings across gestation, as a group, mothers often reported elevated feelings of attachment to the fetus starting at the time of "quickening" at about 20 weeks and again after the birth of the infant, when mothers' nurturance ratings tended to reach the upper asymptote (Fleming et al., 1997). More sensitive indices of nurturant feelings were obtained using a structured interview, which indicated that feelings of attachment continued to grow across the first postpartum year, as mothers spent significantly more of their interview time devoting their thoughts and descriptions to their infants, shifting the focus from their partners (Fleming et al., 1988). In terms of changes in other attitudes, across pregnancy mothers reported more positive feelings about caretaking activities and preparations for the baby, but they reported no changes in their feelings about their partners, their parents, or other children (Fleming et al., 1997). In fact, Deutsch and colleagues (1988) found that the "work" of pregnancy in terms of the issues attended to by prospective mothers seems mostly related to the pregnancy itself and to the upcoming birth, rather than to the actual baby or postpartum-related issues. It is only with the birth of their infant that issues specifically pertaining to mothering and child development become salient.

Physiological Correlates of Nurturance

Across pregnancy and the postpartum period, mothers also experience physiological changes that likely reflect their underlying feelings and, in some cases, may affect those feelings, as shown below. For instance, consistent with what we know based on other mammals, there is now considerable evidence that new mothers have "affective" and autonomic responses to infants and to infant cues that may be consistent with mothers' reported feelings and behaviors but may at times be discrepant with these measures of responsiveness (Giardino et al., 2008; see below).

In a study comparing mothers, nonmothers, and pregnant women with a variety of prior experiences, Bleichfeld and Moely (1984) found that hearing unfamiliar infant crying elicited heart rate accelerations among newly parturient women, indicative of states of arousal, interest, and motivation (Wisenfeld and Klorman, 1978; Leavitt and Donovan, 1979; Furedy et al., 1989), and decelerations among nonpregnant women, where decelerations are interpreted as orienting responses, without the arousal component (Wisenfeld and Klorman, 1978; Leavitt and Donovan, 1979; Furedy et al., 1989). In contrast, pregnant women demonstrated both accelerations and decelerations dependent on their parity and prior experience with infants: accelerations among multiparous pregnant mothers and decelerations among inexperienced women (Bleichfeld and Moely, 1984). These data suggest that the salience of infant cues varies as a function of a woman's hormonal state and prior maternal experience.

The relations between physiological and attitudinal responsiveness seem linked in a logical fashion. However, discrepancies in patterns of responding are evident when multiple measures of nurturance or responsiveness are taken. As an example of such a discrepancy, teenage and adult mothers respond quite similarly to infant cues, demonstrating the same affect profile to recorded infant cries—that is, the same level of expressed sympathy, distress, and alertness (Giardino et al., 2008). Teen and adult mothers also do not differ in their self-reported attachment to their new infants or in their subjective reports about motherhood. Nevertheless, when one explores their endocrine and autonomic responses to infant cries, the patterns of response in the two populations are quite different, with adult mothers showing elevations from baseline levels in both cortisol and heart rate in response to cries and teen mothers showing minimal changes (see Figure 13.1). It is important, therefore, to explore maternal motivation using multiple direct and indirect measures of mothering and associated behavioral and physiological indices of that motivation.

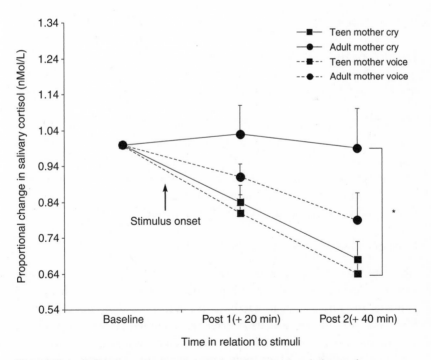

Figure 13.1 Salivary cortisol concentrations (proportional change from baseline + SEM) in response to cries and control stimuli in teen mothers and adult mothers. The bracket represents a Group × Cortisol interaction (p < 0.017). Adapted from Giardino et al., 2008.

In all these studies of affect and autonomic function, the effectiveness of infant stimuli depends on a multitude of factors including mothers' parity, intensity of the stimulus, origin of infant cues (own infant versus another infant), and complex attributes of the infants such as temperament, sex, illness, and prematurity, all of which have also been shown to influence maternal responsiveness and may well affect autonomic responsivity (Corter and Fleming, 2002).

Effects of Psychological State

Although we know a great deal about the importance of processes around the time of parturition and hormones for the onset of maternal responsiveness in new mothers, as will be documented later, precisely how these processes initiate a change in the mothers' behavior is less clear. Based on the nonhuman literature, we know that changes in physiology associated with parturition produce changes in mothers' perception, affect, attention, and plasticity and that these changes enhance the likelihood that the mother will approach, rather than avoid, the infant; that she will experience a positive affect in response to infant cues and pay attention to those cues as necessary; and that she will learn from that experience. These changes clearly affect the likelihood of occurrence and quality of the new mothers' maternal behavior. There is now a substantial animal literature documenting each of these effects (see Numan, Fleming, and Levy, 2006).

In order to determine whether the same processes exist for human mothers, below we outline, first, the relation between mothers' maternal responsiveness and changes in their perception, affect, and attention. Later we explore the role of hormones and brain systems presumed to be involved in these behavioral characteristics and in maternal behavior (see Corter and Fleming, 2002; Numan, Fleming, and Levy, 2006).

Perception and Maternal Behavior

Since human mothering involves responding to multiple sensory inputs and is under multimodal control, it is not surprising that mothers have affective and physiological reactions to infant-related stimuli that differ from responses of nonmothers (Corter and Fleming, 2002). For instance, a mother shows a strong response to the infant's odor; mothers are more attracted to newborn infant body odors than are nonmothers (Fleming et al., 1993) and are able to recognize their infants based on their odors following very little experience with them (Porter, Cernoch, and McLaughlin, 1983; Schaal and Porter, 1991; Fleming et al., 1995). The sources of the relevant odors are likely sebaceous and epocrine glands on the baby's body and head (Schaal and Porter, 1991). Somatosensory experiences in the form of skin-to-skin and nipple contact are also important for mothering feelings, behavior, and physiology as well as for physiological and behavioral adjustment in the neonate (Widstrom et al., 1990; Matthiesen et al., 2001; Mizuno et al., 2004; Porter, 2004).

In addition to olfactory experience, tactile interactions with the infant both affect the mother's nursing behavior and competence and also permits the mother to become familiar with the infant's unique characteristics and thereby enhance feelings of attachment to the baby. With very little experience with their newborns, soon after parturition mothers are able to discriminate their own infants from other infants based solely on the quality of the skin surface and touch characteristics of the dorsal surface on the infant's hand (Kaitz et al., 1992). Less surprising perhaps is that mothers also show increases in infant face recognition and recognition of their own newborn infants' pictures soon after birth (Kaitz, Rokem, and Eidelman, 1988).

The study of mothers' responses to infant cries is equally instructive, given that infants' cries signal distress, a need for interaction, feeding, or physical contact and are signals that mothers have difficulty not responding to through approach behaviors to the infant (Frodi and Lamb, 1978a, 1978b; Frodi, 1980; Kaitz et al., 2000; Stallings et al., 2001; Ludington-Hoe, Cong, and Hashemi, 2002; Soltis, 2004). It is clear that mothers have the ability to recognize infant cries and respond to them based on their acoustic properties, which are well characterized (Petrovich-Bartell, Cowan, and Morse, 1982; Runefors et al., 2000; Ludington-Hoe, Cong, and Hashemi, 2002; Michelsson et al., 2002; Bellieni et al., 2004; LaGasse, Neal, and Lester, 2005). Moreover, a majority of adult new mothers are able to recognize the cry of their own infants (Formby, 1967) and, when compared to adult nonmothers, report greater sympathy in response to cries (Stallings et al., 2001).

In a recent series of studies, we explored the subjective self-report, autonomic, and endocrine responses to recorded "pain" and "hunger" cries of unfamiliar infants in new mothers versus nonmothers. These cries were recorded from newborns immediately before a scheduled feed (hunger cry) and during a circumcision (pain cry). As shown in Figure 13.2, we found that first-time primiparous mothers are affectively more sympathetic and alert to the more intense pain cries of infants (but not to hunger cries) when compared to nonmothers (nulliparous women), indicating that with the onset of motherhood there occurs a change in the valence of infant signals and, further, that in comparison to the less sym-

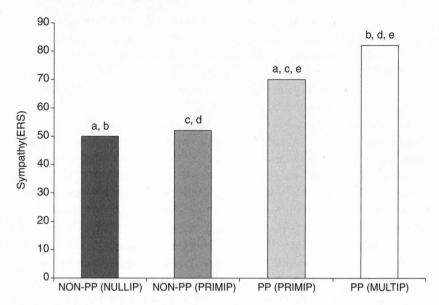

Figure 13.2   Effects of postpartum status and parity on sympathy (emotional reaction score [ERS]) to infant hunger and pain cries in women. *PP,* postpartum (days 2–4 pp); *NULLIP,* nulliparous women (no prior children); *PRIMIP,* primiparous women (first birth and infant); *MULTIP,* multiparous women (more than one prior infant); histograms sharing *letters* are significantly different, $p < 0.05$. Adapted from Stallings et al., 2001.

pathetic mothers, the more sympathetic mothers show elevated baseline heart rates and salivary stress hormones prior to and while listening to recorded cries (Stallings et al., 2001). The mother-nonmother parity difference likely reflects a combination of differences in endocrine state and prior experience with infants (Fleming, Steiner, and Corter, 1997; Fleming et al., 1997; Stallings et al., 2001).

Affect and Maternal Reactions to Infant Smells,
Expressions, and Vocalizations

In parallel with perceptual changes, mothers often experience intense emotions after birth and when first interacting with their infants. New mothers are known to exhibit "emotional" heightened feelings of well-being, anxiety, and sometimes negative affect. The intensity of the feelings is the most noteworthy characteristic. Varied affect states translate into differential interactions with the infant. There is ample evidence that feelings of well-being are significantly associated with more positive attitudes about the infant and motherhood in general (Fleming et al., 1988; Fleming et al., 1997) and are associated with higher levels of positive and "sensitive" responding to the infant (Cohn et al., 1990; Field et al., 1990; Righetti-Veltema, Bousquet, and Manzano, 2003). In fact, it is likely that because of these positive feelings, mothers are able to withstand considerable duress and sleep deprivation to remain attentive to, and in close proximity to, their new infants. However, as indicated above, these positive feelings are neither uniform within mothers, who often experience postpartum lability, nor across mothers, where the predominant feelings may initially be intense anxiety and sometimes depression (Fleming et al., 1988; Righetti-Veltema, Bousquet, and Manzano, 2003).

Despite the fact that the majority of mothers experience positive affect with the birth of a new baby, the relation between maternal behavior and affect in humans has primarily been studied in the context of postpartum depression. Epidemiologic studies, primarily from North America, have found that between 10% and 15% of women exhibit depressive symptoms in the first weeks following delivery (Beck, 2006), which can persist for up to two years but which usually resolves spontaneously within three to six months (Cooper and Murray, 1995). However, several researchers argue that the prevalence of postpartum depression varies across cultures, ranging from a low reported prevalence of 0% to up to 60% (Halbreich and Karkun, 2006); others suggest that a "morbid state of unhappiness" following delivery is universal across cultures (Oates et al., 2004; Goldbort, 2006). In one recent meta-analysis of 140 studies on prevalence rates of postpartum depression (PPD) across cultures, Halbreich and Karkun (2006) argue that in interpreting the wide variability in reported prevalence rate, it is important to take into account cross-cultural differences in symptoms and syndromes, culturally specific reporting styles, and absence of use of culturally relevant assessment instruments. In Western cultures, the symptom profile of PPD resembles that of a major depressive episode experienced at other times in life, but it is unique in its timing and it always involves at least the mother-baby dyad and in most cases an entire family unit. Cross-culturally, risk factors share similar themes including previous episodes of depression, unplanned/unwanted pregnancies, significant life stressors within the last year, child-care stress, low social support, and fatigue (Oates et al., 2004; Goldbort, 2006). The main difference across cultures lies in whether the "morbid unhappiness" postpartum is facilitated through interventions and requires or is referred to health-care professionals (Oates et al., 2004).

Comparisons between depressed and nondepressed mothers indicate that depressed mothers are irritated, respond less sensitively and less contingently, and are either withdrawn or intrusive during interaction with their infants compared to nondepressed mothers (Fleming et al., 1988; Cohn et al., 1990; Field et al., 1990; Righetti-Veltema et al., 2002). When observed later in the postpartum period, depressed dyads exhibited reduced mutual attentiveness, vocal and visual communications, touching interactions, or smiling, compared to nondepressed dyads (Fleming et al., 1988; Field et al., 1990; Righetti-Veltema et al., 2002; Gonzalez, Steiner, and Fleming, in prep-a). Although many of the negative effects of postpartum depression on maternal behavior usually disappear by six months (Fleming, Klein, and Corter, 1992), there is growing evidence that difficulties in maternal responses to the infant may have implications for the longer-term social and emotional development of the child even after the mother's depression remits, although evidence on this is mixed (Field, et al., 1988; Isabella and Belsky, 1991). Infants of depressed mothers exhibit a distinct physiological profile marked by greater relative right frontal electroencephalogram (EEG) activation in withdrawn mothers, greater left frontal EEG activation with intrusive mothers, delayed heart-rate deceleration to music presentation, and higher cortisol and lower serotonin and dopamine levels as neonates (Diego et al., 2004; Diego et al., 2006; Field et al., 2006). In general, there seem to long lasting adverse effects on children's cognitive and emotional development (Beck, 1998).

The effects of postpartum depression on mothers' interactive behaviors are also reflected in depressed mothers' responses to infant cries. We find that in comparison to nondepressed mothers, depressed mothers were particularly prone to feeling anxious when listening to cries and differentially responded more negatively to the pain as opposed to the hunger cries (Gonzalez, Steiner, and Fleming, in prep-a), indicating that the problematic interactions seen in depressed mothers may also reflect a more general emotional reaction to infant cues.

Attention and Maternal Behavior

Dyadic synchrony between a mother and her infant consists of a variety of necessary components, including shared focus of attention, temporal coordination, and contingency, which are all achieved primarily by the caregiver (Isabella and Belsky, 1991). We believe that a mother's maternal responsiveness is related to other behavioral systems, in this case, related to how focused the mother is on the infant, how easily distracted she is, and how consistent she is in her caregiving. This relation has been well established in the animal literature in which rat mothers who show reduced prepulse inhibition (PPI), a measure of the reduction of the startle response seen after presentation of a weaker stimulus and attention set–shifting performance (analogous to the Wisconsin Card Sorting Task), are more easily distracted in the maternal context and spend less time in contact with, and actively licking, their young (Lovic and Fleming, 2004). Among human mothers, we see a similar relation between attention and maternal behavior. Gonzalez, Steiner, and Fleming (in prep-b) found that mothers who respond with fewer errors on the extradimensional shifting (ED) task at two to six months postpartum, indicating better cognitive flexibility, are also more sensitive in their interactions with their infants. Based on the Ainsworth attachment scales, they show more contingent responding to infant cues and less time looking away from their infants during a 20-minute interaction (Gonzalez, Steiner, and Fleming, in prep-b).

Physiological Mechanisms: Hormonal Causes and Correlates of Behavior

Hormones: Estrogen, Progesterone, and Glucocorticoids

The observation that mothers exhibit physiological changes associated with exposure to infants and their cues is not informative when attempting to elucidate the physiological underpinning of maternal behavior or the direction of effects between physiology and behavior. Although it is difficult to ascribe causal mechanisms in humans where studies are usually correlational, nonhuman animal literature provides some comprehensible predictions regarding the pattern of changes that would be expected if changes in mothers' physiology in fact affected their nurturant feelings and behavior. There is, for instance, considerable cross-species similarity in the neuroendocrine events associated with pregnancy and with the onset of parturition; and for many of the nonhuman mammals that have been studied, these hormonal events contribute to the expression of maternal behavior after parturition (Numan, Fleming, and Levy, 2006). Differences in regulatory mechanisms likely emerge after birth when the socialization of the young commences, and, not surprisingly, these psychological mediators are most influential among the primates and humans.

Hormones and Maternal Behavior in Representative Nonhuman Mammals

In many mammals that have been studied, pregnancy is characterized by elevations in progesterone and prolactinlike lactogens, followed by a decline in progesterone toward the end of pregnancy and in elevations in estrogen, prolactin, and oxytocin (Numan, Fleming, and Levy, 2006). After birth the glucocorticoid corticosterone may also enhance mothers' earliest responses to their offspring by increasing the new mothers' attention and arousal state, necessary for optimal responding to the young (Rees et al., 2004, 2006; Graham et al., 2006). There are, of course, species differences in the timing of these endocrine changes within pregnancy and their importance for the onset of maternal behavior close to the time of parturition (see Numan, Fleming, and Levy, 2006).

Hormones and Maternal Behavior in Humans

In humans, there is some evidence that, as in other mammals, hormones may also exert important influences on mothers' early interactions with their infants; however, there is considerable variability among mothers in the pattern of their maternal feelings across pregnancy and the postpartum period. As indicated earlier, for many mothers there is an increase in attachment-related feelings during the second trimester and once again after the birth of the baby (Fleming et al., 1997). The role of steroids in this elevated responsiveness shown by new mothers has received some attention. In one study, Fleming et al. (1997) assessed mothers prospectively in a longitudinal study as well as cross-sectionally, across pregnancy and the early postpartum period, measuring circulating hormones and self-report feelings of attachment. Although few hormone-maternal attitude correlations were found during pregnancy, mothers who exhibited a smaller early to late pregnancy decline in the estradiol/progesterone ratio expressed higher feelings of attachment to their infants during the puerperium.

However, when behavioral, as opposed to attitudinal, measures of responsiveness are considered, adrenal hormones rather than ovarian or placental steroids seem to be important for maternal responsiveness (Fleming et al., 1993; Fleming, Steiner, and Corter, 1997; Corter and Fleming, 2002). Consistent with a role of the glucocorticoids in the modulation of maternal behavior during the early postpartum period in mother rats (Rees et al., 2004, 2006), Fleming and colleagues (Fleming, Steiner, and Anderson, 1987) found in human mothers immediately after parturition a significant positive correlation between postpartum plasma levels of cortisol and the amount of contact and "affectionate" approach responses shown by mothers toward their nursing infants. However, hormonal effects work in concert with mothers' feelings and attitudes and are affected by mothers' parity. When hormones and affect are combined in an analysis, levels of day-two postpartum cortisol and expressed positive maternal attitudes during pregnancy accounted for 46% of the variance in postpartum maternal behavior (Fleming, Steiner, and Anderson, 1987). This finding has since been replicated in two separate samples of similarly aged primiparous mothers but not among older or multiparous mothers (Fleming, Steiner, and Corter, 1997; Krpan et al., 2005). In addition, in primiparous mothers (but not multiparous mothers), day-two postpartum concentrations of cortisol were associated with mothers' greater attraction to the odors of their own infants' T-shirt, other infants' T-shirt, and infant urine but not to other odor stimuli (Fleming, Steiner, and Corter, 1997). These results indicate that hormones may enhance maternal affect and responsiveness in inexperienced mothers; however, in multiparous mothers, prior experience itself enhances maternal responsiveness by reducing mothers' overall anxiety and enhancing their mood state, thereby negating any additional effect of the hormones. This is similar to what has been observed in other species where hormonal effects on maternal behavior are seen most clearly in the first-time mother; multiparous females do not require hormones to respond to young (see Numan, Fleming, and Levy, 2006).

In terms of accuracy discriminating own and other infants' T-shirt odors, the cortisol-parity effects are, however, reversed. Multiparous mothers with higher cortisol levels showed significantly higher accuracy scores, whereas among primiparous mothers there was no relation between accuracy scores and cortisol levels (Fleming, Steiner, and Corter, 1997). It may be that in the instance of individual recognition, where experience effects are acting diffusely on affect but instead relate to learning about specific features, multiparity provides the mother with the learned information regarding what type of infant cues to attend to, and cortisol then enhances that focused attention. Finally, in a study of three- to six-months to postpartum mothers of mixed parity, Stallings et al. (2001) reported that mothers experiencing more sympathy in response to infant cries had higher salivary cortisol levels. Finally, mothers with higher levels of cortisol also showed a greater difference in their affective responses to infant pain and hunger cries than did women with lower postpartum cortisol levels (Stallings et al., 2001).

These studies, although only showing correlations (not causal relations) between hormonal levels and behavior, suggest that in humans the adrenal system may play a role in mothers' initial responses to their infants. There is evidence that the adrenal corticoids are activated during high arousal states and that they enhance attention and acuity in a variety of sensory systems and a variety of behavioral contexts (Rubinow et al., 1984; Schaal and Porter, 1991; Henry, 1992; Spangler et al., 1994; Spangler and

Schieche, 1998; Lupien, Gillin, and Hauger, 1999; Stallings et al., 2001; Erickson, Drevets, and Schulkin, 2003). The enhancement induced by the adrenal system may influence maternal responsiveness by enhancing the salience of infant stimulation and attention to it. However, these relations are only maintained during the immediate postpartum period and under particular circumstances. During later postpartum periods, after endogenous cortisol levels have declined by three to six months postpartum, and especially in women who experience postpartum depression (Gonzalez, Steiner, and Fleming, in prep-a) or who experienced instability in the family of origin (Krpan et al., 2005; Gonzalez, Steiner, and Fleming, in prep-b), high cortisol levels are associated with reduced well-being, greater depression, and more negative interactions with infants. These bimodal cortisol relations will be discussed in greater detail below.

Early Experiences and Hormonal Effects on Mothering

Evidence from other animals suggests that early life experiences modulate the development of a number of endocrine systems, including the hypothalamic-pituitary-adrenal (HPA) axis, known to affect the expression of maternal behavior (Champagne and Meaney, 2001). Recent work shows that newly parturient adult rats raised postnatally without their mothers show elevations in postpartum glucocorticoids and that, under these conditions, dams licked their pups less (Gonzalez et al., 2001; Burton et al., 2007). Results with human mothers map onto these HPA findings quite directly. In a series of recent studies, we have found that mothers who were either at risk teens (Krpan et al., 2005) or clinically depressed (Gonzalez, Steiner, and Fleming, in prep-a) were more likely to show less affectionate and disrupted interactions with their infants and had elevated basal cortisol levels. Teen mothers with this endocrine and behavioral profile were also more likely to experience inconsistent care and to have experienced multiple and changing caregivers (Krpan et al., 2005). Adult mothers who were exposed to early adversity (inconsistent care and/or maltreatment) had higher levels of diurnal cortisol (measured from an area under the curve over two consecutive days) and were less sensitive when interacting with their infants (Gonzalez, Steiner, and Fleming, in prep-b). What is not known, of course, is whether the elevated cortisol levels in our high-risk mothers are restricted to the postpartum period or whether elevated baseline and/or stress cortisol levels were also present and/or produced in childhood (Spangler et al., 1994; Spangler and Schieche, 1998). Based on the literature and our preliminary work, we now know postpartum mothers experiencing early adversity, established retrospectively using the Life History Inventory and the Childhood Maltreatment Questionnaire, are associated with elevated cortisol and likely with dysregulated HPA axis and autonomic responses in adulthood. These relations are dually related to mothers' affect, attention, maternal sensitivity, present contextual factors, and hormonal responses (Krpan et al., 2005; Gonzalez, Steiner, and Fleming, in prep-b) and to infant attachment status in the Strange Situation and measures of infant emotional and social development (Madigan et al., 2006). Taken together, this suggests a possible mechanism for the transgenerational transmission of maternal style.

Hormones and Psychological Processes

Hormones are not only related directly to mothering behavior; they are also related to the psychological processes that affect mothering. These relations are outlined below.

HORMONES AND AFFECT

Although it is generally believed that postpartum depression is based on hormonal dysregulation after birth, few consistent relations have been found between levels of hormones during either the prepartum or the postpartum periods and postpartum affect or depression. In one study, we (Fleming et al., 1997) found that while there were no relations between mothers' pregnancy hormone levels and any of the maternal attitudes during pregnancy, mothers with high free levels of estradiol or with a high estradiol-to-progesterone ratio during the second or third trimester experienced elevated negative affect during both pregnancy periods as well as during the postpartum. In terms of a role for the HPA axis in mothers with postpartum affective dysphoria, results are mixed. Increased cortisol plasma levels correlated with PPD in some studies (Handley et al., 1977), including higher cortisol levels in response to a stressor (Nierop et al., 2006) and general HPA disregulation (Jolley et al., 2007), but not in others (Handley et al., 1980; Smith et al., 1990; O'Hara et al., 1991). Conversely, higher cortisol has also been associated with positive affect (Harris et al., 1994a, 1994b; Fleming, Steiner, and Corter, 1997; Stallings et al., 2001). Krpan et al. (2005); and Gonzalez, Steiner, and Fleming (in prep-a) find that cortisol is most consistently associated with mothers' positive and negative affect; however, whether the relation between cortisol levels and affect is positive or negative depends on multiple and interacting factors, including the following:

1. *Maternal age.* For teen mothers at six weeks postpartum, high cortisol is associated with more energy and increased well-being, whereas for older mothers it is associated with negative affect and fatigue (Krpan et al., 2005).
2. *Pregnancy and postpartum period.* In older mothers, cortisol has an inverse relation to dysphoria during pregnancy (Shea et al., 2007) and at two to four days postpartum (Fleming et al., 1988; Fleming et al., 1997) and a positive relation to caretaking but negative relation to attention to the infant and maternal sensitivity at three to six months postpartum (Gonzalez, Steiner, and Fleming, in prep-a, in prep-b).
3. *The mother's early life experiences.* If associated with prior early life stressors, cortisol is associated with negative mood (Krpan et al., 2005; Gonzalez, Steiner, and Fleming, in prep-a).

The situation becomes yet more complicated when one considers other "stress" hormones (parasympathetic and sympathetic hormones/neurochemicals) that are likely also released along with cortisol; the profile of these may contribute to differences in mood state. For instance, in one recent study, Shea et al. (2007) found that depression, anxiety, or recent stress during pregnancy was inversely related to morning cortisol levels but positively associated with salivary alpha-amylase and heart rate variability, two measures of sympathetic tone that were also related inversely to fetal development and neonatal head circumference (Shea et al., 2007; Gonzalez, Steiner, and Fleming, in prep-a, in prep-b). Other researchers have found that depressed mothers' prenatal biochemical profiles— higher cortisol and norephinephrine and lower dopamine and serotonin— were mimicked by their newborn infants (Lundy et al., 1999; Field et al., 2004). Collectively, these studies suggest that physiological and biochemical processes influence behavior and affect differently, depending on reproductive stage, ages, and parity, and presumably act against a milieu of varying experiential and genetic dynamics.

HORMONES: PERCEPTION AND REACTION TO STRESSORS

To illustrate the importance of evaluating the pattern of physiological responses to derive any meaning from these measures, in a recent study of PPD and responsiveness to infants and to infant cries, we found that in terms of physiology, depressed mothers had higher overall levels of cortisol than did the nondepressed mothers; however, the nondepressed mothers showed a brief cortisol elevation to the cries, whereas the depressed mothers did not (Gonzalez, Steiner, and Fleming, in prep-a).

Moreover, the different physiological measures were associated with a different pattern of responses. In contrast to the cortisol results, when examining autonomic responses to the cries, there were no differences between depressed and nondepressed mothers in their heart rate responses either at baseline or in response to cries. In fact, both groups showed elevated responses to the cries and differed in their responses to cry versus voice stimuli (Gonzalez, Steiner, and Fleming, in prep-a). In addition, among nondepressed, primiparous mothers, cortisol has been positively associated both with the degree of expressed attachment to their infants and to their positive ratings of infant odors (Fleming et al., 1993; Fleming, Steiner, and Corter, 1997).

Cortisol has also been associated with stress reactivity in lactating, non-depressed women (Altemus et al., 1995; Uvnas-Moberg, 1997). There is evidence that lactation in non-depressed mothers is associated with decreases in general fearfulness and stress reactivity, which may well enhance positive approach responses in mothers (Altemus et al., 1995; Uvnas-Moberg, 1997). The effects of breastfeeding interact with parity, with primiparous mothers not exhibiting variations in cortisol reactivity in response to a psychological stressor, while among multiparous mothers, breastfeeding was associated with reduced responsiveness to the Trier Social Stress Test and a child-related stressor (Tu, Lupien, and Walker, 2006a). Differences in parity and feeding status were also observed in cortisol diurnal rhythms, with multiparous, bottle-feeding mothers showing higher levels of cortisol at awakening and in the afternoon, whereas no differences were found in primiparous mothers based on feeding status (Tu, Lupien, and Walker, 2006b). Taken together, these studies suggest that affect, parity, and lactation differentially modulate HPA function, which in turn underlies divergent physiological responses to stressful infant cues (cries and videos) and to more generalized stressors.

HORMONES AND ATTENTION

Cortisol has been associated not only with mood, fear, and anxiety but also with sustained attention and effort (Erickson, Drevets, and Schulkin, 2003). The effects of cortisol appear to act in an "inverted U-shaped" or curvilinear manner on many of these systems such that moderate levels are optimal, while extremely high or low concentrations have adverse behavioral and cognitive outcomes (Erickson, Drevets, and Schulkin, 2003). For example, chronic cortisol elevations have detrimental effects on attention, whereas short-term moderate or high concentrations appear to have no negative effects on sustained or selective attention in humans (van Londen et al., 1998; Lupien, Gillin, and Hauger, 1999; Baghai et al., 2002).

However, depending on a host of factors, cortisol may be positively related to "arousal," "attention," perceptual function, and heightened "vigilance" (as occurs early in the postpartum period) (Henry, 1992) or negatively related to these end points (see also Mason, 1968; Goodspeed et al., 1986; Henry, 1992). In one recent study, in postpartum mothers, higher diurnal levels of cortisol were associated with more errors on an extradimensional shifting task, indicating that cortisol had a negative impact on cognitive flexibility (Gonzalez, Steiner, and Fleming, in prep-b). Increased levels of diurnal cortisol and number of errors on the ED setshifting task were both negatively associated with maternal sensitivity (Gonzalez, Steiner, and Fleming, in prep-b). This is the first study to date linking neuropsychological and endocrinological correlates together with maternal behavior.

Neural Mechanisms Regulating Maternal Behavior

Numan has provided us with an exquisite description and analysis of the functional neuroanatomy of maternal behavior in the rat (see Numan and Insel, 2003; Numan, Fleming, and Levy, 2006). The circuit involves both excitatory and inhibitory systems. However, most work in the area has focused on the final common path for the expression of the behavior, the medial preoptic area/ventral bed nucleus of the stria terminalis (MPOA/vBNST) and its downstream projections into the midbrain (ventral tegmental area [VTA]) and hindbrain (periaqueductal gray [PAG]) and sensory, limbic, and cortical systems that project into the MPOA/vBNST. The MPOA contains receptors for all the hormones involved in the activation of maternal behavior, including receptors for estradiol (E), progesterone, prolactin, oxytocin, vasopressin, and opioids (see Numan, Fleming, and Levy, 2006). As shown in Figure 13.3, neurons projecting into the MPOA are involved in other behavioral changes, including changes in mothers' affect (amygdala, orbitofrontal cortex), stimulus salience (amygdala and striatum/nucleus accumbens [NAC]), attention (NAC and medial prefrontal cortex [mPFC]), and memory (NAC, mPFC). Some of these sites also contain hormone receptors (amygdala, mPFC) and may be the sites where the periparturitional hormones likely act to change behavior at the time of parturition (Numan, Fleming, and Levy, 2006). The relatively complicated neuroanatomy of maternal behavior is based predominantly on work with rats, voles, sheep, and primates (summarized in Numan, Fleming, and Levy, 2006). Taken together, these cross-species studies indicate a striking similarity in the neuroanatomy that underlies mothering.

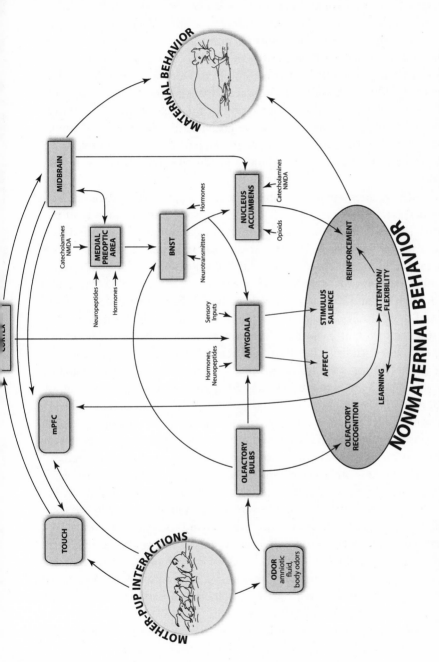

Figure 13.3 Functional neuroanatomy mediating maternal and related behaviors in mammals. Neuroanatomical structures include olfactory bulbs, amygdala, nucleus accumbens, bed nucleus of the stria terminalis (BNST), medial preoptic area, midbrain, parietal cortex, and medial prefrontal cortex (mPFC). Adapted from Fleming, Kraemer, and O'Day, 1999.

Work on neural bases of maternal behavior in humans is derived primarily from functional magnetic resonance imaging (fMRI) studies where mothers, nonmothers, and sometimes fathers are presented with either pictures of their own infants or same-aged unfamiliar infants (Bartels and Zeki, 2004; Leibenluft et al., 2004; Nitschke et al., 2004), recorded infant cries (Seiftriz et al., 2003), or videotapes of infants (Ranote et al., 2004). These studies in general focus on the effects of these infant signals or cues on brain activation in the regions described above, which are known to mediate positive and negative emotions and stimulus salience and are involved in reinforcement, cognitive and executive function, memory, experience, and reproductive and affiliative behaviors.

Presently, there are 8 to 10 studies that explore mothers' responses to infants' cries and/or pictures. All studies demonstrate that many of the same hypothalamic, limbic, and cortical sites important for emotional or social (face) processing (Seifritz et al., 2003; Bartels and Zeki, 2004; Leibenluft et al., 2004; Nitschke et al., 2004) or for regulation of maternal behavior in other mammals (Lorberbaum et al., 2002; Seifritz et al., 2003; Bartels and Zeki, 2004; Swain et al., 2007) are implicated. A number of these studies also show that mothers exhibit a different pattern of responses to infant cues than do nonmothers or fathers (Seifritz et al., 2003). Other studies have illustrated that brain systems underlying maternal motivation include systems that are specific to maternal behavior, recruited in a variety of motivational contexts, and involve rewarding social stimuli. For instance, two studies by Bartels and Zeki (2000, 2004) presented either infant-related pictures or pictures of romantic partners to women and found that some brain regions were activated by both sets of stimuli (especially the striatum, middle insula, and anterior cingulate cortex), whereas other brain regions were more specialized and were activated only by infant stimuli (orbital frontal cortex [OFC] and the PAG) or only by romantic stimuli (dentate gyrus, hippocampus, hypothalamus). In general, "both studies on maternal and romantic attachment revealed activity that was not only overlapping to a large extent with each other, but also with the reward circuitry of the human brain" (Bartels and Zeki, 2004, 1162) and with sites rich in receptors for two neurohormones that have been associated with social affiliation, oxytocin and vasopressin (Bartels and Zeki, 2004).

To determine brain activation patterns in mothers-at-risk for parenting problems, a recent fMRI study (Popeski et al., in prep) presented 8 to 10-months postpartum mothers with pictures of their own infants or unfamiliar infants in either a happy face or a neutral face. Mothers had either experienced adversity during their pregnancies or were healthy controls—a difference in experience that was also reflected in differences in state anxiety and cortisol levels (high pregnancy adversity was associated with high postpartum anxiety and high cortisol levels). Similar to the Bartels and Zeki (2004) studies, blood-oxygen-level-dependent (BOLD) responses were obtained from a number of regions within the limbic and cortical regions known to be important for emotion, for stimulus salience, or for maternal behavior. Analysis from the fMRI sites revealed significant activation in the middle insular cortex and posterior cingulate region in the Low Maternal Adversity group when viewing pictures of their own infant, in comparison to the unfamiliar infant—a pattern that was not present in the High Maternal Adversity group. These regions are important for the mediation of positive emotion. In contrast, mothers in the High Maternal Adversity group showed less activation within the nucleus accumbens when viewing pictures of their own infants than when viewing other infants, a pattern not seen among low adversity control mothers. These results demonstrate that mothers who experience more stress show reduced activity in the dopaminergically rich nucleus accumbens regions, suggesting that the image of their own infants did not elicit the usual positive incentive motivation that characterizes the healthy mother.

Although promising, these fMRI studies are still few, often with small sample sizes, and in their formative stages in conceptualizations of brain function. However, they demonstrate that in humans, as in other animals, brain regions that are involved in processing emotion, reinforcement, and stimulus salience and are not specific to maternal behavior but are necessary for its occurrence are also activated by infant cues; moreover, they show that there are different patterns of activation in nonmothers, in mothers and fathers, and in mothers with different experiences and stress levels. Further, these studies illustrate that different stimuli with a common emotional valence can activate the same brain regions, whereas the same stimuli can produce different activation patterns in different kinds of subjects. In order to more fully understand the complete circuitry that may be activated when a mother interacts with (or thinks about) her baby in humans, as we have in animals, and to understand its general properties as well as its specificity, it will be necessary to use a combination of imaging and electrophysiological (EEG and early reception potential [ERP]) techniques that both permits imaging of nuclei in deep brain structures and provides information on the extent of and time course of activation in mothers, fathers, and nonparents in response to a variety of infant and noninfant stimuli. Given advances in technologies, the next few years are likely to produce exciting discoveries about the neuroanatomy of human maternal behavior.

Final Integration

The maternal role is vitally important to ensure the safety, survival, and well-being of the infant. In order to adapt to the maternal role, mothers undergo a number of changes across various systems. Mothering involves multiple facets, including perception, affect, and cognition/attention. Each of these components is associated with underlying biological and endocrinological systems that contribute to how behaviors are expressed and experienced. Expression depends on prior experience, in terms of both a mother's early experience in her family of origin and her previous mothering experience. Numerous hormones have been implicated in maternal behavior. Our research has focused on the role of glucocorticoids—in particular, cortisol—in maternal responsiveness and sensitivity. There are two functional benefits to focusing on the HPA axis and its end point in humans: First, cortisol is easily measured through salivary samples and is noninvasive. Second, research exists on the HPA axis and its involvement in maternal behavior (Rees et al., 2004) and the transmission of maternal behavior (Champagne and Meaney, 2001).

In our research, we find cortisol is positively related to maternal arousal and perception of infant cues early in the postpartum period but also negatively related to these same cues in high-risk populations; it also has negative effects on maternal sensitivity and cognitive flexibility later. Through a series of studies involving normative samples, we have found that higher maternal cortisol is related to increased attraction to own infant odor and better recognition of own infant odor. This suggests that newly parturient mothers are learning about specific characteristics of their infants, and this learning experience may be facilitated through glucocorticoid and, possibly, neuropeptide functions (see Numan, Fleming, and Levy, 2006). Glucocorticoids also play a role in maternal responsiveness to infant cries. Mothers who respond with more sympathy in response to listening to infant cries have higher salivary cortisol levels and also show more discriminative responding with respect to their affective responses to infant pain and hunger cries when compared to mothers with lower levels of cortisol (Stallings et al., 2001). In the studies outlined above, higher levels of cortisol are positively associated with maternal responsiveness. In our research examining maternal responsiveness in high-risk populations, we find that this relation does not always hold. Mothers with postpartum depression and adolescent mothers exhibit different endocrine profiles with respect to maternal responsiveness to infant cues. It is likely that cortisol in new mothers may be related to higher arousal states, such that higher cortisol facilitates learning and responding to infant cues and may also serve to orient attention and acuity so that the mother can attend to her infant across a variety of environmental contexts and competing demands. However, we also believe that cortisol operates in a U-shaped fashion, such that abnormally high levels of cortisol may reverse this facilitation and instead prevent mothers from functioning "normally."

Research outside the mother context supports our hypotheses. Glucocorticoids also play a role in learning and memory outside the maternal context (see Erickson, Drevets, and Schulkin, 2003; Roozendaal and de Quefvain, 2005). Recent studies have shown that glucocorticoids enhance consolidation of long-term memories of emotionally arousing experiences (Het, Ramlow, and Wolf, 2005) but impair memory retrieval and produce no facilitation of memory consolidation in nonarousing situations (Kuhlmann and Wolf, 2006). As outlined above, in our research we find that cortisol is related to learning about infant cues. It can be inferred that infant cues are emotionally arousing stimuli; and in this context, glucocorticoids facilitate learning about the infant. Similar to our research with our high-risk populations, other studies have suggested that heightened glucocorticoid levels selectively impair cognitive functions (Brunner et al., 2006). It appears that there are optimal levels of glucorticoids that facilitate cognitive functions, particularly learning about emotionally arousing experiences, but that once a certain threshold is passed and levels are "too" high, there are impairments in functioning.

In addition to learning about perceptual infant cues, mothers must use higher-level cognitive processes when interacting with their infants. *Maternal responsiveness-sensitivity* refers to maternal behaviors that are contingent on the infant's prior behavior and are timely and appropriate (Ainsworth, 1979). We know that components of executive functioning, flexibility, and modifying behaviors are all incorporated within the construct of sensitivity. In a recent study, we examined the relations between diurnal cortisol, cognitive flexibility, and maternal sensitivity in mothers. We found that mothers with higher levels of diurnal cortisol showed reduced maternal sensitivity and more errors on an attentional set–shifting task; mothers showing reduced attention were also less sensitive with their infants. Hence, HPA functioning is both directly and indirectly related to maternal sensitivity.

Taken together, these studies indicate that HPA function is related to behavioral systems whose modulation is essential for the optimal expression of maternal behavior: these include mothers' mood, perception, learning, and attention. The glucocorticoid-behavior relations are mediated through hormonal actions on specific brain and neurotransmitter systems including the hypothalamus, the amygdala, the striatum, and the cingulate and prefrontal cortices (example, Mizoguchi et al., 2004; Seamans and Yang, 2004; Cerqueira et al., 2005). When activated, these "generalist" systems integrate with the activity of the final common pathway for the expression of maternal behavior at the biologically relevant time.

CHAPTER FOURTEEN

Oxytocin, Vasopressin, and Human Social Behavior

Roxanne Sanchez, Jeffrey C. Parkin,
Jennie Y. Chen, and Peter B. Gray

Introduction

Research on nonhuman animal models of social behavior informs our views of human social behavior. This has been especially true of the exciting, and rapidly growing, body of research implicating the peptides oxytocin and vasopressin in social behavior. Based on findings inspired by original nonhuman animal work, others have hypothesized that these neuropeptides also play central roles in human social behavior. For example, Taylor's (Taylor et al., 2000; Taylor, 2002) concept of females seeking social partners when facing social stressors ("tend and befriend") emphasizes the role of oxytocin in facilitating human affiliative behaviors.

In the present chapter, we review the evidence linking oxytocin and vasopressin to human social behavior. Based on both nonhuman and human research, we present a conceptual model highlighting the basic functions of oxytocin and vasopressin as well as cognitive, mood, and ultimately behavioral effects associated with oxytocin and vasopressin. We review both human observational and experimental studies, including several studies that we have conducted. Finally, we conclude with a critical evaluation of the human literature, offering directions for future research.

Functional Roles of Oxytocin and Vasopressin

Oxytocin and vasopressin are closely related hormones belonging to a larger family known as the nonapeptides (see Wallen and Hassett, this volume; Carter et al., this volume). Oxytocin plays a classically recognized role in parturition by stimulating smooth muscle contractions in the uterus during labor (Greenspan and Gardner, 2001). For this reason a synthetic oxytocin agonist (for example, pitocin) is often used in labor induction (Ramsey, Ramin, and Ramin, 2000). Oxytocin also plays a classically recognized role in milk ejection during lactation (Greenspan and Gardner, 2001). Beyond these classic functions, oxytocin has recently been linked with a number of other physiological effects. It is associated with lower blood pressure (Uvnäs-Moberg, 1998; Grewen et al., 2005) and slower heart rate in response to stress (Weisenfeld et al., 1985). Oxytocin also suppresses cortisol, thereby attenuating the stress response (Heinrichs et al., 2003). By suppressing cortisol, oxytocin has ancillary effects such as the facilitation of wound healing (Detillion et al., 2004) by its anti-inflammatory properties (Işeri et al., 2005).

Vasopressin has long been implicated in two classical allostatic responses. Initially, vasopressin was identified as a potent vasoconstrictor. Later, its important effects on water reabsorption and blood osmolality were recognized (Mutlu and Factor, 2004). Either directly (vasoconstriction) or indirectly (increasing fluid volume), vasopressin increases arterial blood pressure and plays a role in the cardiovascular stress response.

Oxytocin and vasopressin display various species-specific and sex-specific roles. For example, we will emphasize the primary relevance of oxytocin to female behavior and vasopressin to male behavior throughout this chapter. Nonetheless, oxytocin and vasopressin also serve consistent functions across various taxa. They play consistent roles with respect to homeostasis: oxytocin helps maintain homeostasis, while vasopressin facilitates allostatic responses. Both hormones have been implicated in reproduction, with additional roles in sociosexual behavior among social animals.

Oxytocin and vasopressin facilitate mood and exert cognitive and behavioral effects that build on their classically recognized physiological functions (see Figure 14.1). Perhaps the most important mood effects involve anxiety regulation. Oxytocin tends to decrease anxiety (is anxiolytic), whereas vasopressin promotes anxiety. These hormones also exhibit different effects on memory: oxytocin commonly inhibits memory, whereas vasopressin enhances it. Moreover, these hormones have been implicated in various behavioral functions. Oxytocin has been linked with parental behavior, pair-bonding, and affiliative interaction with non-kin. Some of the social effects of oxytocin apply both to males and females, but the links appear more pronounced among females. The overarching behavioral effects of oxytocin promote "calm and connection"

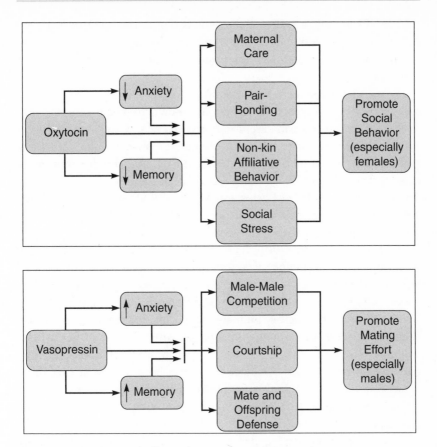

Figure 14.1   Behavioral effects of oxytocin and vasopressin.

(Uvnäs-Moberg , 2003) or "tending and befriending" (Taylor, 2002). In contrast, vasopressin has a role in courtship and male-male competition, the latter particularly when embedded in mate and partner defense (Winslow et al., 1993). Vasopressin has also been implicated in aggression, social recognition, and male pair-bonding.

The Release of Oxytocin and Vasopressin: Rhythms and Stimuli

In the brain, oxytocin and vasopressin are released from all sections of the neuronal membrane, particularly in the dendrites of the hypothalamic nuclei and axon terminals in some limbic brain areas. Oxytocin and vasopressin are synthesized in the supraoptic nucleus (SON) and the paraventricular nucleus (PVN) of the hypothalamus (see Gimpl and Fahrenholz, 2001; Greenspan and Gardner, 2001; Wallen and Hassett, this volume). Especially in males, vasopressin is synthesized in other brain areas, including the bed nucleus of the stria terminalis. Oxytocin and vasopressin are then released from the posterior pituitary into the peripheral circulation.

The effects of oxytocin and vasopressin occur through activation of their respective receptors. There is one known oxytocin receptor and three distinct vasopressin receptor subtypes, V1a, V1b, and V2. Oxytocin and vasopressin receptors are found in various limbic structures and the cortex (Gimpl and Fahrenholz, 2001), and the distribution of these receptors is highly variable across species and among individuals (Insel and Shapiro, 1992; Phelps and Young, 2003). Species and individual variation in oxytocin and vasopressin receptors and their distributions may be linked with differences in social behavior. For example, Hammock and Young (2005) have used rodent transgenic experiments to show that microsatellite instability in the promoter region of the V1a receptor is capable of modulating diverse patterns of social behavior. Specifically, they suggest that genotype differences in V1a binding in the lateral septum and olfactory bulb are responsible for the divergent social behavior between the monogamous prairie vole and the rather asocial montane vole. While the microsatellite's role in the regulation of the V1a gene in humans is poorly understood, it has been suggested that these variations have functional significance including behavioral effects. Human behaviors that have been linked to variations in the V1a gene include autism (Kim et al., 2002), eating disorders (Bachner-Melman et al., 2004), and even creative endeavors such as dance (Bachner-Melman et al., 2005).

Recent work has contributed to the relatively few human data studies on circannual, circadian, and other rhythmic aspects of oxytocin. Central levels of oxytocin demonstrate circadian variation, as levels in cerebrospinal fluid (CSF) peak midday (McCarthy and Altemus, 1997). In a small clinical study of 10 men, Kostoglou-Athanassiou et al. (1998) observed subtle but nonsignificant peripheral circadian variation in plasma oxytocin and vasopressin. The authors state that the rise of oxytocin and vasopressin levels during the night cannot be considered true circadian rhythms, given that factors such as hydration and estrous cycle status can alter hormone concentrations. Similarly, Forsling (2000) has reported increased oxytocin secretion during the sleep of rats and humans. Oxytocin also demonstrates rhythms associated with estrogen in women: levels increase with pregnancy, with use of oral contraceptives, and during midcycle and decline with menopause (McCarthy and Altemus, 1997; Salonia et al., 2005; Taylor et al., 2006). Moreover, external factors influence the concentrations of oxytocin. Alcohol decreases oxytocin measured in the plasma of lactating women (Mennella, Pepino, and Teff, 2005), whereas modest intakes of alcohol in nonpregnant women and men may increase oxytocin (Uvnäs-Moberg, 2003). Uvnäs-Moberg (2003) also suggests that cigarette smoking increases oxytocin levels and may be related to the sucking and social contact involved with this activity.

Research on vasopressin has also revealed important information on circadian and circannual rhythms. Vasopressin is subject to circadian variation, as central levels increase during sleep to reach their peak around the time of waking (Forsling, 2000). In addition to circadian variations, there is a notable sex difference in vasopressin levels, as the hormone is potentiated by testosterone (Share, Crofton, and Ouchi, 1988). In support of these findings, it has been demonstrated that castration reduces testosterone and vasopressin in males. External factors, such as pathologies and intake of various substances, also influence hormone concentrations. For example, damage to the pituitary gland can affect vasopressin levels in various ways, depending on the location of the damage (Makara et al., 1995). In the case of diabetes insipidus, vasopressin levels are lowered, leading to frequent urination and persistent thirst. With syndrome of inappropriate antidiuretic hormone (SIADH), excessive production of vasopressin leads to overconcentrated urine and is commonly associated with central nervous system dysregulation (for example, hyperanxiety) (Saito et al., 1997). Smoking can increase vasopressin (Uvnäs-Moberg, 2003), whereas alcohol and caffeine, known as antidiuretics, inhibit the hormone. Harding et al. (1996) suggest that alcohol's inhibition of vasopressin via neuronal degradation of the PVN and SON is dose dependent and requires approximately 100 grams of alcohol consumption per day for 20 to 25 years. It has also been shown that exercise increases vasopressin, while lying prone reduces hormone levels (Forsling, 2000).

Oxytocin is released in response to a range of stimuli. Physical contact (Uvnäs-Moberg et al., 1996), nipple stimulation (Christensson et al., 1989; Zinaman et al., 1992), or simply being in the presence of conspecifics (Detillion et al., 2004) can increase oxytocin levels. Oxytocin also increases during vaginal-cervical stimulation and orgasm in men and women (Todd and Lightman, 1986; Murphy et al., 1987; Carter, 1992b; Kruger et al., 2003). The release of oxytocin during social interactions may positively reinforce social experiences including reduced anxiety and activation of oxytocin-facilitated pathways, which makes sense for species that display even a limited degree of conspecific interaction. Oxytocin also has analgesic properties, and it is noted that the hormone can raise the pain threshold (Crowley, Rodriguez-Sierra, and Komisaruk, 1977; Brown and Perkowski, 1998).

Vasopressin tends to be associated with allostatic responses, especially those involving courtship and male-male aggression (Goodson and Bass, 2001). Vasopressin increases with sexual arousal and typically drops to low levels with orgasm (Murphy et al., 1987; Kruger et al., 2003). It is modulated by testosterone and may be associated with increased aggression and repetitive behaviors (Delville, Mansour, and Ferris, 1996). Increased levels of vasopressin have also been associated with stressful and defensive circumstances (Carter, 2007).

Role of Oxytocin and Vasopressin in Nonhuman Animal Behavior

The behavioral effects of oxytocin and vasopressin have been elucidated through well-controlled, experimental research in nonhuman mammals (see Carter et al., this volume). Both oxytocin and vasopressin play instrumental roles in the formation of partner preferences in socially monogamous voles and several other species (for example, Winslow et al., 1993). Although both sexes are capable of responding to both peptides, when administered exogenously, it seems likely that endogenous oxytocin is of greater importance in females and vasopressin in males (Carter, 2007). Similarly, oxytocin is fundamental to the development of the mother-infant interaction (Keverne, Nevison, and Martel, 1997), and vasopressin plays an important part in paternal behavior (in those species where the father invests in the offspring) (for example, Bamshad, Novak, and De Vries, 1994). These social attachment effects are likely the result of connections between oxytocin and vasopressin and dopamine-mediated memory and reward (Bielsky and Young, 2004; Young and Wang, 2004). While oxytocin may inhibit general memory, both oxytocin and vasopressin appear to enhance social memory (Carter, 1998).

Oxytocin and vasopressin, through their effects on social memory formation, lead to the formation of a partner preference or pair-bond. The release of oxytocin and vasopressin associated with sexual activity is one way these partner preferences develop. In males, shortly after copulation, vasopressin leads to an increase in aggression against intruders, which if successful would guard the mate from mating with other males (Winslow et al., 1993; Insel, Preston, and Winslow, 1995). With the birth of any resultant offspring, oxytocin in females and vasopressin in males facilitate parent-offspring interaction and a concomitant increase in intruder aggression in both sexes, thereby further protecting the offspring. Further, a recent study suggests that vasopressin or oxytocin can stimulate parenting in male prairie voles (Bales, Abdelnabi, et al., 2004).

The relationship of oxytocin and parental behaviors has been demonstrated in a number of mammalian species including sheep, rats, prairie voles, and nonhuman primates. For example, sheep direct maternal behavior toward their young via oxytocin release from vaginal-cervical stimulation and suckling (Keverne, Nevison, and Martel, 1997). Experimental studies have shown that under normal conditions unfamiliar lambs are rejected; however, oxytocin injections can facilitate ewes to become attached to an unfamiliar lamb (Keverne, Nevison, and Martel, 1997). In a small but important study of primates and maternal behavior, Holman and Goy (1995) injected oxytocin in two nulliparous female rhesus monkeys. Following the injection, the monkeys demonstrated an increase in touching and watching of infants, with a subsequent decline in aggressive facial expressions toward observers.

Role of Oxytocin and Vasopressin in Human Behavior

Due to a number of obstacles, both ethical and methodological, we know more about the roles of oxytocin and vasopressin in vole pair-bonding and parental care than we do in humans. However, research on human oxytocin and vasopressin is growing rapidly. The foundation for this human research rests with the type of nonhuman research summarized above and in Carter et al. (this volume). We begin with the assumption that the physiological and behavioral functions of oxytocin and vasopressin are conserved in humans. Although humans (and monkeys and apes) exhibit a greater degree of behavioral freedom from hormonal control compared with smaller-brained vertebrates (Keverne, 2005), we anticipate that the social contexts regulated by oxytocin and vasopressin are largely consistent with those of other species. This leads to the expectation that the conceptual model of oxytocin and vasopressin outlined in Figure 14.1 also applies to humans. We now review the relevant human literature. We first survey the effects of oxytocin and vasopressin on mood and cognition, then turn to behavioral human studies.

Cognitive and Mood Research into Oxytocin and Vasopressin in Humans

Several reviews highlight the effects of oxytocin and vasopressin on cognition and mood in humans (Meisenberg and Simmons, 1983b; Fehm-Wolfsdorf and Born, 1991; Heinrichs et al., 2004; Landgraf and Holsboer,

2005). Many of these human studies have relied on intranasal or peripheral administration of oxytocin or vasopressin. Rarely have human intervention studies used oxytocin or vasopressin antagonists or receptor antagonists. A central concern surrounding the use of intranasal sprays has been whether any observed effects were due to peripheral or central hormone concentrations, with convincing evidence since showing that intranasal sprays can have central effects (Born et al., 2002). The available human intervention studies have the advantage of demonstrating causal, rather than correlational, effects of peptides on human behavior. Despite showing patterned responses, the effects of oxytocin and vasopressin on cognition and mood appear to be moderate in humans (Fehm-Wolfsdorf and Born, 1991).

Oxytocin has known anxiolytic effects. For example, 15 male Vietnam veterans with posttraumatic stress disorder experienced reduced reactions to aversive combat imagery when given intranasal oxytocin (Pitman, Orr, and Lasko, 1993). In an experimental economic investment game, intranasal oxytocin spray increased interpersonal trust behavior in college-aged men (Kosfeld et al., 2005). Further analysis showed that it was not simply risk-taking behavior that was increased by oxytocin administration; rather, social risks were affected. The results of this intervention corroborate similar cross-sectional findings of a positive correlation between oxytocin and trust behavior in an economic game (Zak, Kurzban, and Matzner, 2004). In a different approach, Bell et al. (2006) noted that in a sample of depressed male (n = 23) and female (n = 37) subjects, endogenous oxytocin levels were positively correlated with reward dependence, a theory particularly relevant to social affiliation. These studies suggest that, in the face of potentially anxious social contexts, oxytocin promotes social affiliation.

As further evidence of oxytocin's powerful anxiety-reducing effects, Kirsch et al. (2005) used functional magnetic resonance imaging (fMRI) to demonstrate that intranasal oxytocin spray decreased amygdala activation in response to potentially anxiety-inducing stimuli. Though using a modest sample size of 15 male subjects, their findings speak to the calming effect of oxytocin and provide an important pathway by which oxytocin can facilitate social interaction. Similarly, Heinrichs et al. (2003) discussed the interplay of oxytocin and social support in a sample of 37 male subjects. Intranasal oxytocin combined with the social support of a male subject's best friend was found to be the strongest suppressor of salivary cortisol after a stressful event, which consisted of a public speaking task and public performance of mental arithmetic. Together, these studies clearly demonstrate the conservation of oxytocin's ability to decrease conspecific anxiety.

Vasopressin, on the other hand, has been positively linked with anxiety, including clinically significant anxiety disorders (Landgraf and Holsboer, 2005). For example, van Londen et al. (1997) reported elevated vasopressin levels in 57 clinically depressed male (n=22) and female (n=30) subjects. It has also been suggested that administration of vasopressin up-regulates general arousal, which may be associated with behavioral effects related to anxiety (Landgraf and Holsboer, 2005). To investigate the effect of intranasal vasopressin spray on social communication, Thompson et al. (2004) examined how 14 college-aged males responded to emotional facial expressions after receiving treatment with vasopressin. While the cognitive response remained the same, unconscious reactions by the facial muscles were noted among the subjects who had observed emotionally neutral facial expressions. This response was similar to the 13 control subjects who had viewed angry facial expressions, which suggests that vasopressin may enhance aggression in humans by influencing individuals to react aggressively to emotionally neutral stimuli. More recently, Thompson et al. (2006) investigated the effects of vasopressin on both male and female social cognition. Here, they showed an increase in state anxiety in men (n=20) and women (n=20) receiving intranasal vasopressin. However, women channeled that anxiety into socially affiliative responses, which was indicated by the unconscious activation of facial muscles suggestive of prosocial projections onto strangers. In contrast, males were more primed toward agonistic responses and showed signs of unconscious aggression to neutral faces.

Generally, oxytocin diminishes short-term memory. Eleven males given intranasal oxytocin showed compromised short-term memory and a decrease in vigor (Bruins, Hijman, and Van Ree, 1992). The authors note that the reduction in vigor may have contributed to impaired initial storage and a decline in the rate of storage of information. However, other researchers have supported the link between oxytocin and diminished memory. Heinrichs et al. (2004) administered intranasal oxytocin to 19 males in order to examine the effects on memory formation. They found that oxytocin inhibited memory formation and recall for a word task utilizing "reproduction-related" words. While some enhancement of memory in reproductive contexts might have been expected (akin to promoting social memory specifically), the effects of this intervention are more consistent with the general inhibitory effects on memory observed in previous research relying on oxytocin sprays. Indeed, this is unsurprising since use of reproduction-related words bears little ecological validity to reproductive behavior.

A number of studies suggest that vasopressin facilitates short- and long-term memory (Bruins, Hijman, and Van Ree, 1992). Further, the benefits of vasopressin treatment have been studied and are thought to be most useful for individuals with mild cognitive disorders (Bruins, Hijman, and Van Ree, 1992). In a small study of cognitively sound women, a single dose of vasopressin was found to enhance memory in 9 females (Bruins, Hijman, and Van Ree, 1992). The authors suggest that these effects were related to the retrieval of information, as no effects were seen 75 minutes after administration of the treatment. Others postulate that the enhanced memory associated with vasopressin may be ancillary to increased arousal or heightened attention (Fehm-Wolfsdorf et al., 1984). For example, Fehm-Wolfsdorf and Born (1991) speculate that a 1984 study by Beckwith, Till, and Schneider, which showed improved memory in males (n=64) who received a vasopressin analogue, was due to changes in subject motivation. However, a subsequent study by Beckwith et al. (1987) demonstrated improved recall of "idea units" of high and medium importance by 20 males who had received a vasopressin analogue.

Behavioral Research into Oxytocin and Vasopressin in Humans

Much of the human behavioral research involving oxytocin and vasopressin comes from observational studies. Observational studies do not directly involve hormone or hormone receptor manipulations but, rather, attempt to correlate endogenous hormone concentrations with behavior in either clinical or naturalistic settings.

Many of the observational studies in humans have focused on the connection between oxytocin and social support. Knox and Uvnäs-Moberg (1998), for example, examined the role that oxytocin plays in preventing cardiovascular disease. These authors emphasized that the presence of social support appeared to benefit cardiovascular health and that this social support was likely mediated by oxytocin. Several observations are consistent with this view. Women report greater positive affect, which includes calmness and tolerance of monotony, after a breastfeeding rather than bottle-feeding bout, possibly due to increases in oxytocin concentrations associated with nursing (Carter, Altemus, and Chrousos, 2001). Moreover, breastfeeding mothers (n=10) show reduced stress responsiveness, including hypothalamic-pituitary-adrenal (HPA) activity, when subject to an exercise stressor, compared with bottle-feeding mothers (n=10) (Altemus et al., 1995).

A related line of research focuses on oxytocin and behavior in new mothers. Women who deliver vaginally (n=20) have higher oxytocin levels two days postpartum compared with women delivering by emergency cesarean section (n=17) (Nissen et al., 1996). Moreover, in this same sample of women, oxytocin levels were inversely related to anxiety among mothers delivering by cesarean section. Oxytocin pulses were positively associated with social desirability both in women delivering by emergency cesarean section (n = 17) and those delivering vaginally (n = 20) (Nissen et al., 1998). Matthiesen et al. (2001) showed that oxytocin levels in mothers (n = 10) increased in response to body contact with infants. However, mothers' oxytocin levels can also increase in response to an infant cry or other stimuli before a lactation bout begins (n = 10) (McNeilly et al., 1983). Light et al. (2000) studied new mothers, divided into breastfeeding (n = 14) and bottle-feeding (n = 12) groups, and their reactions to a speech stressor. They observed a link between oxytocin increase and blood pressure decrease in humans. Not surprisingly, women that had a higher oxytocin level and lower blood pressure throughout the study were breast-feeding at home. In a later project, Light et al. (2004) also studied the effects of a speech stressor on mothers (who were primarily bottle-feeding) that either had (n = 10) or had not (n = 25) used cocaine during their pregnancy. They found that cocaine use had, effectively, short-circuited the oxytocin pathways that provide the stress-buffering benefits to the "no drug" control group. Control mothers displayed all of the general stress-reducing effects of oxytocin, such as lower blood pressure and less self-reported hostility and depression. Interestingly, in their study, the cocaine group demonstrated blood pressure reductions similar to control mothers when they held their infant in a laboratory setting, suggesting that cocaine used during pregnancy led to a problem with the attachment-forming effects of oxytocin.

Two recent studies highlight a different type of social support but find the same stress-mediating effects of oxytocin. These studies examined the interplay of oxytocin, stress, and partner support in a romantic relationship. In one such study, researchers evaluated the importance of warm partner contact in the form of hugs (Light, Grewen, and Amico, 2005). They found that among a sample of 59 premenopausal women, those that reported more frequent hugs had higher oxytocin levels. These higher levels were associated, as expected, with lower blood pressure as well as a lower heart rate. Another study on partner interactions (n = 38) considered the effects of oxytocin on men and women (Grewen et al., 2005). Both men and women reporting greater partner support exhibited higher oxytocin levels. However, only women that reported greater partner support demonstrated the possible cardioprotective effects of oxytocin, which include lower blood pressure and cortisol levels. This suggests that the benefits of oxytocin, particularly those related to health, may be more apparent in females.

Recent research by Taylor et al. (2006) and Turner et al. (1999) suggest a different role for oxytocin in the absence of interpersonal support. Turner et al. (1999) investigated oxytocin, mood, and interpersonal support among 25 women. They reported that "relaxation massage" increased plasma oxytocin concentrations and that negative emotion, particularly sadness, seemed to decrease it, although neither result was significant. They found that a greater increase in plasma oxytocin was associated with being in a romantic relationship. In a follow-up analysis, they also found that women recounting relationship experiences displayed a positive correlation between increases in oxytocin levels and affiliation cues (for example, in facial expressions) but not with sexual cues (Gonzaga et al., 2006). Surprisingly, though, Turner et al. (1999) observed that single women had higher basal plasma oxytocin concentrations and were more likely to demonstrate unhealthy interpersonal traits such as "intrusiveness, anxiety, and coldness." Similarly, Taylor et al.'s (2006) study of 73 postmenopausal women reported higher basal plasma oxytocin concentrations in association with low social support. Authors of both papers suggest different reasons for these unexpected findings. Turner et al. (1999) suggest this correlation may be "indicative of poor oxytocin regulation in response to social stimuli" and not causal in nature. Taylor et al. (2006) propose that the increased oxytocin levels could provide an unconscious drive to make affiliative contact with a conspecific.

The social benefits related to oxytocin are not restricted to human-human interactions. Odendaal and Meintjes (2003) have found that oxytocin increases in both adult male humans (n = 18) and dogs (n = 10) after brief, affiliative interaction between the two species. Further, in humans, but not dogs, an oxytocin-related reduction in cortisol was observed. In addition to this possible cardioprotective effect of oxytocin, humans were found to have lower blood pressure after brief contact with the dogs. This decrease in blood pressure from social affiliation is in contrast to other findings that have suggested that oxytocin's cardioprotective effects are greatest in women (Grewen et al., 2005). Thus, the health effects of oxytocin, particularly those related to the cardiovascular system, warrant further investigation.

Few human studies have investigated the links between vasopressin and behavior. In one such study, Coccaro et al. (1998) observed a link between cerebrospinal fluid concentrations of vasopressin and a history of aggression. In a sample of 26 subjects including males (n = 18) and females (n = 8) diagnosed with personality disorders, they found positive associations between vasopressin levels and aggression (Coccaro et al.,

1998). The authors suggest that the aggression may be particularly targeted at other people. In keeping with these findings, Landgraf and Holsboer (2005) suggest that while a continuous release of vasopressin is beneficial during times of stress, long-term exposure may increase susceptibility to anxiety and depressive disorders.

In addition to the individual observational studies on oxytocin and vasopressin, there have been recent investigations of the combined effects of these hormones. For example, Fries et al. (2005) have addressed the developmental basis of human oxytocin and vasopressin function in a recent study. These researchers studied social and endocrine function of adopted children initially raised in socially deprived orphanages who were subsequently adopted into warm households. They observed lower baseline urinary vasopressin and lower reactive oxytocin levels in response to social interaction among adopted children (n=18) compared with controls (n=21). These findings demonstrate the plasticity of the human oxytocin and vasopressin systems and the role that early social experience can have in canalizing subsequent behavior.

As a complement to the existing observational studies of human oxytocin, vasopressin, and behavior, we summarize results of several studies that the authors have recently conducted.

Oxytocin and Female Attachment Study

Since previous studies have linked oxytocin with social support, it is conceivable that support during a social interaction may be related to higher levels of oxytocin. We build on adult attachment research, which focuses on the amount and type of support given by partners, to determine whether female oxytocin is associated with human romantic relationships. Adult attachment research has grown from the original theory of attachment proposed by John Bowlby (1969), which identifies a romantic partner as the attachment figure (Chisholm, 1999). Similarly, adult attachment scales assess the degree of emotional attachment one has in romantic relationships. Securely attached individuals trust their partners and are comfortable becoming emotionally close. Avoidantly attached individuals are not comfortable trusting others. Anxious individuals tend to seek emotional closeness but also worry that their partners do not really love them. Securely attached individuals do give more support, while avoidantly attached individuals do not seem to want or need support. These facets of human adult attachment have yet to be examined in relation to oxytocin.

Chen, Simpson, and Rholes (in prep) conducted a study examining the relationship between social support and oxytocin: 30 couples were recruited to participate in a laboratory experiment. Mean age for the participants was 19.4 (SD = 1.2). Mean length of relationship was 21.5 months (SD = 17.2). Mean score for relationship quality (Fletcher, Simpson, and Thomas, 2000) across satisfaction, commitment, intimacy, trust, passion, and love was 18.5 (SD = 2.7). After completing a battery of inventories about relationships and attachment styles, the woman of the couple was then primed into a support-seeking situation. A female researcher talked with the woman about personal problems the woman wanted to change in her life. The function of this paradigm is to put the woman in a situation where she would be emotionally stressed and also elicit social support from her partner. Then the couples were instructed to discuss an issue that the woman wanted to change about herself. During this videotaped interaction, blood samples were drawn from the woman every 90 seconds via a butterfly line, beginning with a prediscussion baseline and ending with a postdiscussion draw. This allowed the couple to engage in the interaction with no interruption or physical contact from the researchers during the interaction. After the interaction, the couple viewed the interaction privately and documented at which time they thought the man was being supportive of his partner. Independent coders also coded the videotapes for supportive behavior of the men. These ratings were not influenced by the men's or women's preconceived notions and perceptions of the man's behavior and thus provide a more objective rating of the man's behavior.

Results showed that there was a significant positive correlation between an increase in women's oxytocin levels and the amount of support given by the men as rated by researchers ($r = .36$; $p = 0.05$). The correlation between rated support and oxytocin release in the female remained significant ($r = .40$; $p = 0.04$) even after controlling for relationship quality factors (intimacy, trust, love, passion, and commitment). This supports previous research that social support and bonding can increase oxytocin levels and also shows that it is not mediated by relationship quality. More specifically, individuals in high-quality relationships (high on passion, commitment, trust, love, and intimacy) did not differ in oxytocin release from individuals in low-quality relationships. Stated differently, individuals can be in a high-quality relationship and yet not receive support and thus not release oxytocin. Individuals can also be in low-quality relationships, receive support, and experience oxytocin release. Additionally, results showed a strong negative correlation between oxytocin levels and anxious attachment style. Specifically, women who had anxious attachment styles had lower levels of oxytocin ($r = -.57$; $p = 0.001$) and also released less oxytocin than women who were not anxiously attached ($r = -.43$; $p = 0.01$).

These results indicate that oxytocin levels are affected by current partner support, while overall relationship quality is not related to the release. Specifically, the findings of the present study suggest that anxiously attached women have increased oxytocin levels. Marazziti et al. (2006) have found the opposite effect in romantic attachment styles, which is consistent with Taylor et al. (2006) and Turner et al. (1999), who reported elevated baseline oxytocin among women lacking interpersonal support. A difference in subject populations, current cognition, social environment, and the distinction between baseline versus reactive oxytocin levels likely account for the disparity between our results and Marazziti's. As the present study indicated anxiously attached women had lower levels of oxytocin, it can be theorized that these anxious subjects do not perceive nor get the same benefits from social support received by other people. Anxious people may have hypersensitive HPA axes that stay aroused and possibly inhibit the release of oxytocin. Avoidantly attached individuals perhaps have other mechanisms of self-soothing during stressful situations that do not directly involve oxytocin.

Oxytocin, Vasopressin, and Human Paternal Care

Because of the nonhuman links between oxytocin and vasopressin and pair-bonding and parental care, we investigated whether human males involved in long-term pair-bonds (such as marriage) and paternal care would display predicted peptide differences. In light of sex-specific functions of vasopressin, we anticipated that links would be more apparent in male relationships to vasopressin than to oxytocin. To facilitate subject recruitment, we relied on urinary measures of oxytocin and vasopressin, in turn subject to the caveats described by Wallen and Hassett (this volume) concerning peripheral versus central measures of these peptides.

To compare hormone levels of pair-bonded men with unpaired controls, Parkin (n.d.) recruited 45 men aged 18 to 42 in Las Vegas. The study relied on baseline measurement of urinary oxytocin and vasopressin levels standardized by creatinine. Questionnaire responses were used to establish relationship status (paired/unpaired) and other facets related to male social behavior (for example, aggression, paired men's scores on a "love" scale). While one previous study by Grewen et al. (2005) suggested that pair-bonded men might display oxytocin increases associated with brief periods of affiliative bonding, it was unclear whether we would find baseline differences in men's oxytocin linked to pair-bonding. No previous studies had compared vasopressin levels by human male relationship status. Results revealed no differences in oxytocin or vasopressin levels among men according to whether or not they were paired. Further, no links between the hormones and self-evaluated measures of social behavior (mate seeking, love and attachment, aggression, and paternal care/investment) were observed.

To expand the cross-cultural scope of human peptide and relationships research, we investigated the roles of oxytocin and vasopressin among men in urban Jamaica (Gray, Parkin, and Samms-Vaughan, 2007). We recruited 43 men aged 18 to 38 to participate: 15 single men, 16 "coresidential" fathers (fathers who live with their mate and youngest offspring), and 12 "visiting" fathers (fathers who live apart from their mate and youngest offspring). Since half of all newborns in Jamaica are born into these "visiting" households, this household structure represents an interesting facet of both adult male relationships and children's early social environments.

We enlisted men to report to the University Hospital of the West Indies, where they completed informed consent before subject to a 20-minute "cool-down" phase. Subsequently, single men were asked to sit quietly alone, reading a newspaper, while fathers spent this same length of time interacting with their mate and youngest offspring. Blood pressure was obtained before and after this 20-minute behavioral session, and a urine sample was collected after the behavioral session from which oxytocin and vasopressin levels were measured.

Results revealed no differences in oxytocin or vasopressin levels according to male relationship status (single/father, whether visiting or coresidential). However, in this sample of fathers of very young children (newborns to a $3\frac{1}{2}$-year-old), age of a father's (n=28) youngest child was significantly, and negatively, correlated with his urinary vasopressin levels (r=−.431; p=0.022). In this same sample, vasopressin levels were also positively correlated with systolic and diastolic blood pressure. Thus, we suspect the correlation between age of child and vasopressin represents some effect of anxious arousal (faced with unpredictable infant stimuli to which a father responds) and physical exertion (holding a young child). We speculate that male increases in vasopressin during early interactions could enhance reward and reinforce social memory and thus aid in paternal attachment to a child. Interestingly, we did not find any relationships between fathers' oxytocin levels and age of their youngest child or to blood pressure or salivary cortisol levels, suggesting that paternal interactions in human males could be more closely tied to vasopressin than oxytocin. Our findings, albeit with a peripheral measure of vasopressin levels, resonate with the observation that in a sample of captive common marmoset fathers age of youngest offspring and fathers' anginine vasopressin (AVP) 1a receptor density in the prefrontal cortex were negatively correlated (Kozorovitiskiy et al., 2006). The contrasts between these preliminary Jamaican and U.S. male vasopressin and social relationships findings also suggests the importance of social interactions preceding urine sample collection since significant findings emerged among dads after interactions with their kids in Jamaica but not among Las Vegas males.

Synthesis and Conclusions

As shown in the literature reviewed above, oxytocin and vasopressin have been implicated in human social behavior. The links between oxytocin and vasopressin observed in humans appear relatively consistent with more controlled, experimental research on nonhuman animals. The conceptual model implicating social roles of oxytocin and vasopressin depicted in Figure 14.1 largely holds for humans, too.

In broad terms, oxytocin suppresses the anxiety reaction to conspecifics, helping make us the social animals we are. It is also possible that oxytocin has direct effects on social motivation or on the capacity of a social interaction to be rewarding. As in nonhuman animals, the release of oxytocin and vasopressin during sexual activity in humans may help condition a preference for the partner. Oxytocin's role in human parturition and postpartum interactions such as breastfeeding reinforce the important bond between mother and infant. Increases in oxytocin associated with touch and other positive interactions may cement a variety of social relationships, from adult male-female pair-bonds, friendships, and even relationships between people and their beloved pets. Consistent with nonhuman research and what little research exists on humans, vasopressin appears to promote anxiety and enhance memory. Both effects may be functionally advantageous during male-male competitive behavior, courtship, and mate and offspring defense (Storm and Tecott, 2005). In addition to the social effects of these neuropeptides, work by Fries et al. (2005) demonstrates the plasticity of the human oxytocin and vasopressin systems and the role that early childhood experience may have in shaping later reproductive and affiliative behavior.

In surveying the available human data and considering pathways to future research, we should highlight some of the challenges posed by human research. What are the primary considerations for both interpreting the available findings and guiding development of future research designs? What are some of the appropriate caveats to bear in mind in reviewing the human data and planning for the next wave of research on oxytocin, vasopressin, and social behavior?

One consideration is the subject pool. The bulk of both observational and experimental human studies have involved small sample sizes with typical numbers of 20 or 30. Such small samples are vulnerable to both Type I and Type II statistical errors, while also providing the impetus for assuming that such effects may generalize to millions of other people. Given the sex-specific effects typically observed (oxytocin in females, vasopressin in males), it seems curious that so many of the human intervention studies have involved males. A likely reason is that fluctuations in oxytocin across the menstrual cycle or reproductive state can be avoided as a confounding influence in males (Taylor et al., 2000); another reason may be that administration of exogenous oxytocin to women could affect birth outcomes (Carter, pers. comm.). However, we view females as more relevant subjects in such interventions, with complex interaction effects (for example, depending on menstrual cycle phase) appearing in such interventions.

The dearth of human male vasopressin-behavior studies is remarkable. Both community-based samples and specific nonclinical subpopulations (for example, competitive athletes—a human model of stereotyped aggression—used frequently in testosterone and aggression research: Salvador, 2005) are almost entirely lacking despite the strong rationale expected for human vasopressin-behavior relationships to appear. Also, with the exception of the Jamaican findings reported above, no studies have been conducted on human oxytocin, vasopressin, and social behavior outside North America or Europe. Given variable human social environments around the world, we are nonetheless left to guess how findings from some corners of the world can be extrapolated throughout it. Furthermore, most of the research has focused on adult populations aged 20 to 40. Research into the developmental roles of these hormones in young children is only addressed by Fries et al. (2005). It is also likely that older adults, particularly postmenopausal females, will yield important information about the effects of oxytocin and vasopressin across the life cycle.

Also, assay methods have traditionally relied on the measurement of serum or plasma, presenting the challenge of interpreting the biological significance of peripheral hormone concentrations. As ethics prevent the measurement of central hormones in living humans, researchers have questioned the interpretation of peripheral hormone concentrations as reflections of central levels. However, studies from the central nervous system and peripheral circulation suggest a positive correlation between central and peripheral oxytocin and vasopressin levels in some cases (Landgraf and Neumann, 2004), thereby suggesting peripheral measurements can be used with the appropriate cautions to understand the central functions of oxytocin and vasopressin (see also Wallen and Hassett, this volume). In addition, it is likely that peripheral levels are well correlated with central levels in dynamic settings (for example, responding to a social stressor). This may be one reason why peripheral hormone data on oxytocin and vasopressin responses to social stimuli may be more telling than baseline levels. Further, a convergence of human functional imaging, human postmortem autoradiographic receptor localization study, and nonhuman research implicates the central effects of fluctuating hormone concentrations even when these central effects cannot be measured directly.

One of the recurring problems with human research has been the measurement of oxytocin and vasopressin. These hormones occur at low concentrations and are biologically unstable, making the development of precise and sensitive assays a challenge. Improved antibodies and assay techniques have lessened some of these problems, but it remains more technically challenging to measure these hormones than, say, testosterone in human males. Measurement of oxytocin and vasopressin in cerebrospinal fluid provides the best index of central hormone levels, but at the disadvantage of such an invasive technique (for example, lumbar puncture). Plasma and serum, which require a less invasive blood draw, can also provide a tool for measuring oxytocin and vasopressin. The development of urinary oxytocin and vasopressin assays may provide yet a better methodological tool in human studies: urine samples typically index hormone secretion over hours, providing a longer time integration of these hormones than observed in cerebral spinal fluid, serum, or plasma. Salivary measurements provide greater logistical flexibility in studies of hormones and human behavior. Recent work by Carter et al. (2007) demonstrates that saliva may be a valid biomarker for the measurement of oxytocin. Carter et al. (2007) note that a similar study by Horvat-Gordon et al. (2005), which reported that oxytocin could not be measured in saliva, had several methodological differences including extraction and did not concentrate samples.

Despite some of the methodological challenges posed by research with humans, the prospects for a greater understanding of the roles of oxytocin and vasopressin in human social behavior are fantastic. The number of open research questions seems boundless. Just a few of these unanswered questions include: Is the mechanism for partner preference identical in heterosexual and same-sex couples? What roles do oxytocin and vasopressin play in grandparenting? What role do these hormones play in sibling interactions? In light of cross-cultural variation in human social relationships (see Lancaster and Kaplan, this volume), additional questions arise concerning the roles of these hormones in polygynous marriages, in kin adoption, and in marital relationships of varying degrees of affiliation.

Other lines of human research will likely contribute to a better understanding of the roles of oxytocin and vasopressin in social behavior. For example, work in functional imaging is further clarifying the pathways through which oxytocin and vasopressin may facilitate intimate human social relationships. Guided by the neural localization of oxytocin and vasopressin receptors, Bartels and Zeki (2004) showed that the brain regions rich in these receptors were also activated when viewing photos of a mate and one's child when subject to fMRI. Ongoing study of the human oxytocin and vasopressin receptor genes may expand the range of receptor variation by including a wide sample of human societies (Pfeifer and Tishkoff, 2007).

A better understanding of the psychological and behavioral effects of oxytocin and vasopressin in humans may have clinical implications. For example, clinicians, epidemiologists, and social scientists alike have increasingly recognized the health benefits of social relationships (for example, House, Landis, and Umberson, 1988; Hollander et al., 2003). The possible clinical and mental health implications gained from a more thorough understanding of hormones and human social behavior are important outcomes of this type of research. The evidence reviewed here suggests that oxytocin may facilitate some of these adaptive benefits of human social relationships, particularly those related to cardiovascular and mental health. Vasopressin's role in the relationship between human social behavior and health remains unexplored.

Potential interventions relying on intranasal oxytocin sprays have also been identified (Bartz and Hollander, 2006). As an example, clinical trials have begun investigating the potential therapeutic use of oxytocin to enhance social cognition such as in the treatment for autism. Hollander and colleagues (Hollander et al., 2003; Hollander et al., 2007) demonstrated that peripherally infused synthetic oxytocin (pitocin) given to patients with autistic and Asperger's disorders resulted in fewer behavioral symptoms. These researchers suggest that pitocin might alter the permeability of the blood-brain barrier or cross the blood-brain barrier in fragments to affect the central nervous system (given the assumption that oxytocin and vasopressin were not able to cross the blood-brain barrier).

Already, synthetic oxytocin (for example, pitocin) has a strong biomedical hold in the birth process and, to a lesser degree, as an agent facilitating the initiation of lactation. However, it remains unclear whether the administration of high doses of pitocin to facilitate uterine contractions has permanent behavioral effects on a neonate; oxytocin (and pitocin) can cross the placenta, but it is uncertain whether it reaches and has effects on the neonatal brain (Bales, Pfeifer, and Carter, 2004). Clinically, a vasopressin analogue (desmopressin) is commonly used to treat patients with diabetes insipidus. However, it remains unclear whether the use of vasopressin sprays has measurable behavioral effects. Research on these topics might better show both risks and benefits to treatments relying on interventions to oxytocin and vasopressin systems. Such research may also ground the potentially exaggerated marketing campaigns of commercially produced oxytocin and vasopressin analogues. "Liquid Trust," a recently marketed product, claims to increase the trustworthiness of the user, and vasopressin has been marketed as a "smart" drug because of its memory-enhancing effects.

The behavioral effects of oxytocin and vasopressin have clear implications for human sociality. Both peptides appear to serve similar functions, including those linked to social behavior, in humans as in various nonhuman animals. There appear to be clinical and mental health implications associated with oxytocin and vasopressin. Still, the breadth of unanswered research questions is remarkable, with improved methodological techniques and convergence in evolutionary, ecological, and endocrine concepts likely to fuel further investigation into the roles of these hormones in human social behavior.

CHAPTER FIFTEEN

Androgens and Diversity in Adult Human Partnering

Sari M. van Anders

Introduction

In this chapter, I survey testosterone (T) and diversity in adult human partnering, with attention to considerations about diversity and how various empirical findings bear on evolutionary understandings of partnering, sexual orientation, and life strategies. I begin by asking and attempting to answer the question, Why diversity? As diversity in human partnering is tied to sexual orientation in many ways, I next review research that examines whether and how prenatal and circulating hormones may be associated with sexual orientation in women and men. I then present a brief review of research on androgens and partnering in men (see Gray and Campbell, this volume, for a more comprehensive review) and move into discussions of theoretical frameworks for conceptualizing T-partnering associations. I subsequently explore contributions to our understanding of partnering and hormones from research that incorporates diverse populations, including research that addresses gender, sexual orientation, and relationship type. I close with a summary that emphasizes the value of diversity to behavioral neuroendocrine understandings of partnering.

Research into social relationships and hormones is grounded in evolutionary theory and can be conducted using a variety of levels of analysis. In humans, this generally has translated into behavioral and endocrine analyses, with a focus on ultimate and proximate mechanisms. At the proximate level, we can ask: Do hormones affect partnering? Does partnering affect hormones? At the ultimate level, we can ask: Why would hormones affect partnering, or partnering hormones, and how might this be adaptive? While theorizing at the ultimate level can guide our empirical research at the proximate level in terms of hypothesis generation, understanding the proximate mechanisms is really a prerequisite to proposing ultimate explanations. While this may seem tautological to some, it is no less a derivative of scientific method than in any other field, with its reliance on theory-derived hypotheses that are tested empirically, leading to evidence for or against a theoretical position and resulting in revised or strengthened theory.

Why Diversity?

I use the term *diversity* to refer to the broad spectrum of partnering styles and behaviors that are present in humans. Diversity is often used euphemistically when referring to people and is most often understood to imply sexual minorities, women, ethnic minorities or any minority, underrepresented, or "othered" groups. In contrast, diversity is a foundational aspect of behavioral neuroendocrine research, and the term is employed to refer to the naturally occurring broad spectrum of behaviors, phenotypes, and strategies apparent throughout the animal kingdom. Quick perusals of behavioral neuroendocrine textbooks (for example, Becker et al., 2002; Nelson, 2005) attest to the striking variety in gender morphs, sexual behaviors, and sexual differentiation under study. Without this naturally occurring diversity, behavioral neuroendocrinology would arguably lose much of its content and unquestionably lose a great deal of its most fascinating subject matter.

The study of partnering and hormones has generally focused on the diversity of pair-bonding apparent within and between animal species as well as heterosexual men in both North American and international populations. Additionally, researchers have long focused on clinical populations with conditions affecting their endocrine function or circulating hormones to address endocrine questions that cannot be answered in humans exposed to typical endogenous hormones. There is an additional approach to addressing those endocrine questions, and that is to look to the naturally occurring diversity in human relationship styles. Just as zoologists and ecologists have learned from the stunning variety of behavioral phenotypes, a creative and resourceful approach to human diversity should lead to some answers and more questions about partnering and hormones. Why not, occasionally, use "human models" for human research?

Research has largely focused on men, possibly because major theoretical perspectives focus on male-male competition for mates and male paternal/mate investment. Another possible reason is the large number of women who are either pregnant, postmenopausal, lactating, or using hormonal contraceptives and who therefore have altered endogenous endocrine profiles that can confound research and limit the potential pool of women participants. A third possible reason is that women show menstrual variation in hormones in addition to the seasonal (Dabbs, 1990b; Wisniewski and Nelson, 2000; van Anders, Hampson, and Watson, 2006) and diurnal (Rose et al., 1972) patterns evidenced by both women and men. A fourth reason might be limited access to women in international research. A fifth possible reason is that the field is relatively new, and testing men has just come before testing women. Additionally, researchers interested in androgens often focus on males (but see Ketterson, Nolan, and Sandell, 2005). However, as articulated elsewhere (van Anders and Watson, 2006b), including women is really not that difficult and is largely warranted by theoretical considerations.

If women have not been included, it should not be surprising that the focus has not been merely on men but on heterosexual men or at least men in opposite-sex pairings. Again, there are several possible reasons. There is a relatively small number of nonheterosexual individuals in society. Also, implicit theoretical positions that focus on male-female pairbonds because of their reproductive potential might limit the scope. Related might be the continuing view of same-sex sexual orientations as evolutionarily paradoxical, with the unresolved (though arguably constructed) paradox being that same-sex sexual orientation appears to be somewhat "biological" in origin for at least some people but is theorized by some to lead to decreased reproduction (and fitness). Other possibilities include discomfort with including a minority group for reasons of personal bias that could be of a prejudicial origin or of a wish to "leave alone" a group that has not always benefited from scientific attention. Inclusion of nonheterosexual groups is, however, crucial to understanding human partnering and hormones for a variety of reasons that I detail later and briefly allude to here. There are a consistent and significant number of people in same-sex relationships, and so to understand human partnering and hormones in entirety, we should include the entirety of the human experience (though, maybe, not all in one study). Including more than one mode of partnering is likely to shed light not only on same- *and* opposite-sex partnering and hormones but on broad understandings of partnering and hormones, providing insights that could not be gleaned otherwise.

There are a variety of diverse relationship styles that could be included in research in adult partnering and hormones, and these could be deemed human models, though it might be apparent how little any person or group would appreciate being so labeled or perceived. Humans have sometimes been described as serially monogamous, but some people stay with only one person. Some people have multiple partners simultaneously—or would if they could find willing individuals or accepting social structures. Some people live with their partners; some live apart. Some partner for love, some partner for loneliness avoidance, and some partner for family concerns. There is variety and diversity in how and why adults partner that can be related to choice, accident, or opportunity, affording us myriad relationship statuses, styles, and desires to study. Someone who longs for a long-term, committed relationship with one person is qualitatively different from someone who has little interest in commitment, at least in terms of commitment desire. Do they differ in other ways? Do they differ in endocrine parameters?

Including groups in research necessitates looking for them, and looking for them necessitates knowing that they exist and postulating that they are relevant to the topic at hand. However, some linguistic conventions may obscure forms of partnering that are less visible than heterosexual pair-bonds but important for understanding how partnering and hormones are associated. For example, *mating relationship* may seem like an appropriate descriptor stemming from nonhuman research, but not all relationship types are covered by this terminology. Though *mate* can refer to a spouse, counterpart, or one of a pair, I think it connotes a reproductive partner when used in biologically oriented research. Just as empirical studies show that *man* brings images of men to mind, and not humans more generally or women (Crawford, 2001), I think that *mate* brings reproductive partners to mind, even when users have less specific intentions, and therefore obscures forms of relationships that we do not intentionally mean to exclude.

These may seem like semantic quibbles, but semantics can predispose us to focus on certain types of relationships, and this can prevent us from including populations that are likely to be helpful in understanding how hormones and partnering are associated. Thus semantics (or not attending to them) can impede scientific progress. How can this be addressed? One possibility is to use *mate* expansively, since English is an infamously flexible language. Another option is to use more inclusive terminology. In this chapter, I use *partnering*, which has a similar meaning to *mating* but does not exclusively denote reproductive relationships.

Conducting scientific research with groups who are less visible brings its own challenges and considerations. It is impossible for experimenters to consider the inclusion of a group of which they have no awareness.

Potentially most useful are open-ended questions on questionnaires in ongoing studies. Participants can then write in their own responses, which can alert the researcher to linguistic conventions that indicate unknown (to the scientist) groupings. Similarly, taking participants' comments as helpful and potentially "expert" input (for example, "I don't fit on your questionnaire, and I think you should include this other category") can be instructive. Allowing for verbal or written feedback can also be helpful. Finally, at this early stage, sensitivity in terminology and communications regarding the study and group should go far toward reducing unintended and undesired negative repercussions.

I will focus on literature that reflects diversity in associations between androgens and partnering, including relationship status and relationship "orientation." Most people are more familiar with sexual orientation as a topic of scientific inquiry than with relationship status or orientation, as media reports of biological bases of sexual orientation are widely dispersed and attended to. Diversity in relation to androgens and partnering likely brings sexual orientation to mind, and this more established line of inquiry is an appropriate starting point for a discussion of hormones and partnering in humans.

Androgens and Sexual Orientation

One additional feature of attraction that is an important determinant of relationships is sexual orientation. *Sexual orientation* has been defined by some as a "dispositional sexual attraction towards persons of the opposite sex or same sex" (Rahman, 2005b, 1057) but is understood by others to reflect a less tangible and more contextualized concept (for example, L. Diamond, 2003). Certainly, scientists and laypeople alike use the term to refer to the direction of a person's attraction to same- or opposite-sexed individuals in conjunction with their desire to partner with these individuals and identify as someone who partners with these individuals. *Sexual orientation* has largely come to replace *sexual preference,* reflecting a shift in perception that the construct is a trait with at least some biological/innate causation.

While *sexual orientation* has come to connote a somewhat fixed and biological predisposition to same- or opposite-sex individuals (not to reify sex boundaries), researchers have studied what links exist between biology and sexual orientation (for example, Rahman, 2005b). Hormones are attended to, stemming from the idea that a gay/lesbian sexual orientation represents a sex-atypical orientation and thus sex-atypical sex hormone exposure (see Gorman, 1994, for an appropriately critical discussion of this approach). This line of theorizing is based on animal literature, where exposure to "cross-sex" hormones can affect sexual behaviors and interests toward same-sexed animals (for example, Baum, 2006).

Researchers generally divide endocrine effects into those that are organizational and those that are activational. In humans, organizational effects are generally prenatal and permanent and have a "hard-wiring" effect (for more on organizational verses activational effects, see Wallen and Hassett, this volume; McIntyre and Hooven, this volume). The prenatal androgen hypothesis has thus been proposed as an explanation for the development of same-sex sexual orientations. The hypothesis holds that higher-than-typical prenatal androgens predispose women to be sexually oriented toward women, and lower-than-typical prenatal androgens predispose men to be sexually oriented toward men.

Prenatal Hormones and Sexual Orientation

MEASURING PRENATAL HORMONES

Understanding the contributions of prenatal hormones in humans to any aspect of behavior is challenging because we cannot reliably measure prenatal hormones in random samples of people. Amniocentesis is one way to measure prenatal hormones, but the procedure leads to an increased risk of spontaneous abortion, limiting its use to clinical populations of women for whom the benefits (for example, potentially discovering a serious medical condition) outweigh the risks. Because amniocentesis is performed at varying points during pregnancy, gestational age and associated variation in T are difficult to control, leading to extreme variability and often difficulties of interpretation.

Maternal hormones have also been used as potential measures of fetal endocrine exposure and could theoretically represent maternal-plus-infant circulating hormones, but the broad changes and fluctuations in maternal hormones over gestation limit this approach. However, there is something of a consensus that this approach is not well supported (for example, Cohen-Bendahan et al., 2005). Some researchers have assayed hormones from umbilical cords, and this could be useful, though there are obvious caveats. Umbilical cord measures likely represent circulating hormones at time of parturition (when hormones are askew already). And umbilical cords link the fetus and mother, so cord measurements thus reflect both maternal and fetal hormonal contributions (for a review of these methods, see Cohen-Bendahan, van de Beek, and Berenbaum, 2005).

An additional approach is to include people who have clinical conditions affecting their hormone release, either prenatally (that is, organizationally) or postnatally (that is, activationally). Several conditions have been identified, with those affecting prenatal hormones receiving more attention. In congenital adrenal hyperplasia (CAH), cortisol cannot be produced because of a missing or faulty enzyme that converts precursors to cortisol. As a result, the fetal pituitary releases increasing amounts of adrenocorticotropic hormone (ACTH) in an attempt to increase cortisol production in the adrenal cortex. Though cortisol cannot be made or released, other adrenal hormones—like androgens—can. As a result, fetuses with CAH are exposed to higher-than-typical androgens, and females can be born with genitals showing various degrees of virilization. Androgen insensitivity syndrome (AIS) has also received attention; in this genetic condition, XY individuals have androgen receptors that are nonfunctional (complete AIS) or only partially functional (partial AIS). Androgen levels are high in people with AIS but exert no or partial effect because of the nonfunctional or faulty androgen receptors. As a result, fetuses with complete AIS are not exposed to androgens (despite high circulating levels) and are born with female-typical genitals, and fetuses with partial AIS are exposed to inconsistent amounts of androgens and are born with genitals that show various degrees of virilization. In XY individuals, another condition called 5-alpha-reductase deficiency can be present, and since the 5-alpha-reductase enzyme converts T to the more potent dihydroT (which mediates genital virilization), individuals are born with female-appearing genitals. Individuals with other clinical conditions have also received attention (for a review, see Gooren, 2006), as have people exposed to exogenous hormones prenatally. For example, diethylstilbestrol (DES) is a potent synthetic estrogen that has masculinizing effects and was administered to pregnant women.

Research on sexual orientation and hormones via clinical conditions has been valuable to scientists (though not overly helpful or welcome to the individuals themselves), but there is the possibility that findings are not generalizable beyond clinical populations to adults exposed to typical hormones. It is unclear whether findings from clinical populations exposed to higher- or lower-than-typical hormones can be generalized to healthy populations of adults exposed to hormones that are high or low but still in the typical range. As well, populations with relevant endocrine conditions are not large, limiting sample size.

Psychologists have thus also turned to potential markers of fetal endocrine exposure, and generally any variable showing sex differences has been hypothesized to be associated with prenatal androgens, for example, otoacoustic emissions (OAEs). OAEs are quiet clicks made by the inner ear. They can occur naturally or in response to external sounds and are more frequent in females, with evidence supporting prenatal androgen masculinization effects (for example, McFadden, 1993; McFadden, Loehlin, and Pasanen, 1996). Another measure is digit ratio, which is a ratio of the length of the second to fourth digit (that is, finger). Digit ratios show sexual dimorphism, with male ratios lower than females' (Manning et al., 1998), and there is some evidence that digit ratios are associated with prenatal androgens (Brown, Hines, et al., 2002; Lutchmaya et al., 2004; van Anders, Wilbur, and Vernon, 2006), though this putative association is still controversial. The theoretical grounding for such an association is that fingers develop under the same genetic control as external genitals (*HoxD* and *HoxA:* Kondo et al., 1997) with coordinated expression (Peichel, Prabhakaran, and Vogt, 1997). They thus develop at the same time, when prenatal androgen levels are high to promote sexual differentiation and masculinization of bipotential gonads. Whether the digits respond to androgens has not been examined (for example, whether they are rich in androgen receptors).

PRENATAL HORMONES AND SEXUAL ORIENTATION IN WOMEN

Studies of possible effects of prenatal hormones on sexual orientation in women have often focused on women with CAH. There is some evidence that women with CAH are more likely to report same-sex sexual fantasies and behavior (Dittmann et al., 1992) or less opposite-sex sexual fantasy and behavior (Zucker et al., 1996; Hines, Brook, and Conway, 2004) than unaffected controls. Similarly, women exposed to exogenous hormones like DES administered during their mother's pregnancies show slight increases in same-sex sexual orientation in addition to affected genital development (Ehrhardt et al., 1985; Meyer-Bahlburg et al., 1985). As well, women with complete androgen insensitivity syndrome (CAIS) who are exposed to no circulating androgens show no difference in sexual orientation compared to unaffected control women (Wisniewski et al., 2000; Hines, Ahmed, and Hughes, 2003).

Using putative markers of prenatal androgens, women with same-sex sexual orientations tend to have lower (that is, more male-typical) digit ratios than women with opposite-sex sexual orientations (Williams et al., 2000; Rahman and Wilson, 2003; Rahman, 2005a; compare Lippa, 2003; van Anders and Hampson, 2005). And there is some evidence that gendered variation within women might be associated with prenatal androgens. For example, Brown, Finn, et al. (2002) found that digit ratios were masculinized in self-identified butch lesbians compared to femme lesbians. In addition, research with OAEs show that they are more female typical in heterosexual women compared to bisexual women and lesbians (McFadden and Pasanen, 1998, 1999).

Thus research does appear to support some role for prenatal androgens in women's sexual orientation, though there are conflicting and null findings. It seems highly unlikely that higher prenatal androgens are necessary for same-sex or bisexual sexual orientation to develop in women, and so it is unclear whether prenatal androgens contribute to some women's sexual orientation but not others and whether the strength of contribution might differ between women.

PRENATAL HORMONES AND SEXUAL ORIENTATION IN MEN

Studies on men's sexual orientation in response to clinical prenatal conditions or exposure are less common. However, men exposed to DES prenatally do not appear to differ in sexual orientation compared to controls (Kester et al., 1980), suggesting that exposure to masculinizing prenatal hormones in addition to typical male levels may not affect sexual orientation development. The 5-alpha-reductase deficiency has been studied internationally and is relatively common in some contexts, including localized areas within the Dominican Republic (Imperato-McGinley et al., 1986). These XY individuals are generally reared as girls from birth, but at puberty T is sufficiently high to induce virilization, and most individuals, depending on cultural considerations, take on male roles (which generally includes partnering with women) in Papua New Guinea and the Dominican Republic (Imperato-McGinley et al., 1979; Imperato-McGinley et al., 1991). Because of the prevalence of the disorder, however, families may rear their children accordingly.

While digit ratios seem to show some evidence of prenatal masculinization in women with same-sex sexual orientations, evidence in men is mixed. Some studies point to prenatal masculinization in men (in opposite direction to the prenatal androgen hypothesis) such that men with same-sex sexual orientations exhibit lower (that is, more male-typical) digit ratios than men with opposite-sex sexual orientations (Robinson and Manning, 2000; Rahman and Wilson, 2003; Rahman, 2005a). Others have reported the converse, with higher digit ratios in men with same-sex sexual orientations (McFadden and Shubel, 2002; Lippa, 2003). In contrast to women, there is no evidence that more masculine gay men exhibit masculinized digit ratios relative to more feminine gay men (Rahman and Wilson, 2003). There has been recent discussion about what this evidence suggests—that is, hyper- or hypomasculinization—and ethnic differences between populations have been suggested as one source of the differences between results. Even if (or especially if) these differences can be explained by ethnic differences, evidence that sexual orientation in men is associated with prenatal androgens is not supported by research using digit ratios as a putative marker of prenatal androgens. OAEs also have not supported the prenatal androgen hypothesis in men, though there has been slight suggestion of hypermasculinization when evoked OAEs are measured (McFadden and Pasanen, 1998, 1999).

Adult Circulating Hormones and Sexual Orientation

A common lay assumption holds that circulating androgens are associated with sexual orientation, with same-sex sexual orientations associated with higher T in women and lower T in men. Measurement of circulating androgens in adulthood is much more straightforward than direct or estimated measurement of prenatal hormones, but sampling issues are still important. If only certain subgroups of gay men and lesbians are willing and/or able to openly identify as gay, lesbian, queer, and/or same-sex oriented, and these groups do not represent all gay men and lesbians, then one could potentially be comparing a nonrepresentative group of gay men and lesbians to a potentially representative group of heterosexual counterparts.

There has been little evidence that circulating T is higher in women with a same-sex sexual orientation compared to women with an opposite-sex sexual orientation. Instead, there is some evidence that self-identified butch lesbians have higher T than heterosexual women (Singh et al., 1999) or self-identified femme lesbians (Pearcey, Docherty, and Dabbs, 1996; Singh et al., 1999). This could be seen as suggestive that it is not sexual orientation per se that is associated with circulating T but instead gender/sex. And unlike prenatal T, causality is generally unclear. Engaging in "male-typical" behaviors or behaviors intended to lead to a more "masculine" appearance over a long term could lead to higher androgens (for example, weight lifting: Linnamo et al., 2005). As with women, there has been little evidence that circulating T is higher in heterosexual men than in gay men. Unlike in women, there has been no evidence that T is higher in more masculine gay men compared to more feminine gay men.

Are Hormones Associated with Sexual Orientation
in Women and Men?

Findings provide little evidence associating prenatal T with sexual orientation development in men. The evidence that does exist is generally contradictory. In women, research with clinical populations suggests that higher prenatal T might be associated with increased same-sex sexual orientations. As women are typically exposed to very low prenatal T, it may be that variations are large enough to affect neural development, whereas men's higher typical exposure to prenatal T (Nagamani et al.,

1979) precludes effects on neural development unless exposure is varied dramatically. Evidence from women who are XY with CAIS who show typical female development and sexual orientation supports this, as they are exposed to considerably different levels of prenatal T (that is, none) compared to unaffected XY individuals. Still, the overall lack of consistent evidence should not be interpreted as suggesting that there are no prenatal or biological influences on sexual orientation.

Research does not support an association between adult circulating T and sexual orientation but is suggestive that adult T may be associated with gender/sex roles in women. It would be helpful to know whether more masculine heterosexual women show higher T than more feminine-identified counterparts. If so, that would confirm that it is gender sub-groupings in women that are associated with circulating T more than sexual orientation per se. In contrast, T does show consistent associations with partnering, as I detail in this next section.

Androgens and Partnering in Men

Contrasting Single and Partnered Heterosexual Men's T

Research on partnering and hormones with North American humans has reliably shown that single heterosexual men have higher T than heterosexual men who are married (Booth and Dabbs, 1993; Mazur and Michalek, 1998; Gray et al., 2002) or in long-term relationships (Burnham et al., 2003; Gray, Campbell, et al., 2004; Gray, Chapman, et al., 2004). This research (see Gray and Campbell, this volume) represents an interdisciplinary endeavor, with anthropologists, sociologists, and psychologists involved, resulting in data from North American societies and beyond.

This interdisciplinary diversity is also matched by diversity of populations, as anthropologists have begun to conduct important research with international populations. These findings are less consistent, which is not surprising given that patterns of partnering can be culturally specific (see Gray and Campbell, this volume). In research conducted in Beijing, China, Gray, Yang, and Pope (2006) have found, for example, that married fathers have lower T than married nonfathers, suggesting that fatherhood may decrease T or men with lower T may be more likely to be fathers. International approaches provide for research that attends to the diversity of human experiences and patterns of affiliation. This literature is based on theoretical perspectives that focus on trade-offs in male mating effort. This work is also directly or loosely theoretically associated with the challenge hypothesis (Wingfield et al., 1990; Wingfield, this volume).

The challenge hypothesis (Wingfield et al., 1990) posits that androgens should be high around times of social challenge, with a focus on seasonal rhythms (for example, high T during the breeding season). Though originally proposed in relation to avian endocrinology, and receiving the majority of its empirical attention from bird researchers, the challenge hypothesis has been used fruitfully by researchers focusing on many other species (including humans: see Archer, 2006, for a review), as its authors intended.

Testosterone Trade-off Framework

Based on a synthesis of these studies, theories, and related bodies of literature, a framework for trade-offs associated with T and social behaviors has been posited (van Anders and Watson, 2006b). The testosterone trade-off framework hypothesizes a trade-off between high T and competitive behaviors/states, on one hand, and low T and bond maintenance behaviors/states, on the other hand. *Competitive* is associated with resource acquisition, including defense of a resource in response to a real or imagined threat. *Bond maintenance* is associated with developing intimate and/or caring social bonds with others, including partners, infants, friends, or family. This framework is conceptually related to the challenge hypothesis (Wingfield et al., 1990; Wingfield, this volume). One key difference between the testosterone trade-off framework and others is that the testosterone trade-off framework focuses on behavioral "intentions" as the key differentiation between competitive and bond maintenance behaviors as opposed to behavioral targets. For example, I hypothesize that infant defense is competitive behavior (defending a resource, that is, offspring) associated with higher T based on the testosterone trade-off framework, instead of positioning it as representative of low parental care as per the challenge hypothesis. Another difference is that the testosterone trade-off framework does not focus exclusively on males or on breeding and reproductive behaviors. Like other frameworks, the testosterone trade-off framework is useful for psychological studies because it queries and allows for state/trait effects. Trait effects need not be conceptualized as innately predetermined, and trait levels of T may not be inborn or innate. It is possible that some events along a developmental trajectory have led to both stable characteristic behaviors and T levels. Evidence does support a link between T and social behaviors in humans, with higher T associated with competitive behaviors/states and lower T associated with bond maintenance behaviors/states (for reviews, see Archer, 2006; van Anders and Watson, 2006b).

Contributions from Diversity to Partnering and Hormones

When findings are reported in one group (for example, men), it is difficult to resist the idea that they occur only in that group (that is, men) and
not in others (that is, women), turning no findings into null findings. The
converse is the difficulty of resisting the urge to extend findings that are
reported in one group (for example, men) to all groups (for example,
women and men). Both could lead to hypothesis generation of an ultimate or proximate nature based on an incomplete foundation. As noted,
research with international populations has allowed us to see how generalizable findings on partnering and hormones from North American heterosexual men are (see Gray and Campbell, this volume) and shows that
the association is complex and sensitive to cultural idiosyncrasies. In this
section, I detail the contributions from research with diverse North
American populations. Research including diverse relationship styles or
types in North America can help to expand the foundation for theorizing
hormone-partnering associations, as I hope to show.

Considering Partnering and T with Sexual Orientation and Gender/Sex

Previous studies (for example, Booth and Dabbs, 1993; Mazur and
Michalek, 1998; Gray, Campbell, et al., 2004; Gray, Chapman, et al.,
2004) have compared single heterosexual men with heterosexual men
in long-term relationships. Based on testosterone trade-offs, theoretical considerations suggested comparing two groups using a competi-
tive/bond maintenance distinction. In the context of partnering,
competitive might refer to trying to find or attract partners, and *bond
maintenance* might refer to trying to develop and maintain close intimate
bonds with partners. I was thus interested in contrasting partnered with
unpartnered people. *Partnered* included people in long-term, committed
relationships with one person (for example, marriage, cohabitation,
common-law, long-term relationships). *Unpartnered* included single
people, people who were dating, and people in multiple relationships.
This was because dating, by definition, denotes lower commitment to a
partner and the possibility of having additional dating partners. As well,
I included people in multiple relationships because they had the possibility of having additional partners. I hypothesized that unpartnered people
would have higher T than partnered people, because being unpartnered
is akin to a competitive state, while being partnered is akin to a bond
maintenance state.

A crucial issue relates to state/trait effects and whether T decreases on entering a relationship (suggesting state effects) or lower T predicted the likelihood of entering a relationship (suggesting trait effects). A better understanding of causality in hormone-partnering associations would lead to more directed theorizing of mechanisms and functions. To address this, I included a longitudinal aspect whereby I could follow people's relationship status and T. If entering a relationship decreases T, researchers might look toward aspects of being partnered that could affect T, including commitment, physical partner presence, or lifestyle, and look to possible adaptive functions of decreased T. If lower T predicts entering relationships, researchers might look to preferences for partners with lower T or individual preferences for long-term relationships with lower T individuals.

Including nonheterosexual men and women and heterosexual women in addition to heterosexual men was valuable. Understanding the populations in which relationship status and T are associated could help direct us more effectively to hypotheses of mechanisms and function. For example, if the effect is only seen in heterosexual men, that would be suggestive that it occurs only in men with opposite-sex partners. This might direct us to question something about male-specific attributes of nervous systems and female partnering cues.

Our study (van Anders and Watson, 2006a) gathered a sample from the community, nearby universities, and the local Pride Parade. Participants were divided into heterosexual and nonheterosexual based on "Kinsey's questions" (Kinsey et al., 1948) of directed sexual fantasy and behavior. Heterosexual individuals scored exclusively or nearly exclusively on opposite-sex sexual fantasy and behavior. Nonheterosexual individuals scored exclusively or moderately on same-sex sexual fantasy and behavior. Thus, nonheterosexual is a better and more apt qualifier than gay or lesbian, since our participants were not exclusively oriented toward same-sex sexuality.

Our results replicated findings of higher T in single heterosexual men compared to heterosexual men in long-term relationships (for example, Gray, Campbell, et al., 2004; Gray, Chapman, et al., 2004), by comparing unpartnered versus partnered people (see Figure 15.1). That is, heterosexual unpartnered men had significantly higher T than heterosexual partnered men. In contrast, nonheterosexual men's T did not differ as a function of partnered status, suggesting that the effect is not generalizable to all men. There was also no significant difference in T between partnered and unpartnered heterosexual women, suggesting that the effect may not be generalizable to all heterosexual individuals. Unpartnered

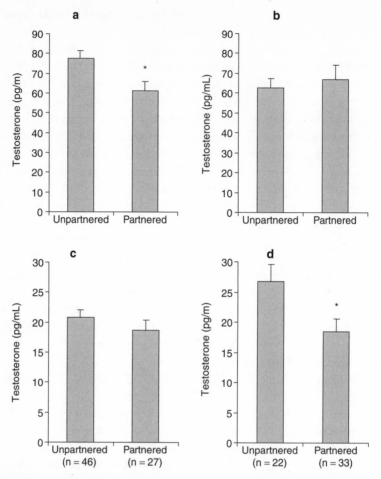

Figure 15.1   Mean baseline testosterone with standard error bars in partnered and unpartnered individuals in *(a)* heterosexual men; *(b)* nonheterosexual men; *(c)* heterosexual women; and *(d)* nonheterosexual women. * (asterisk) indicates a significant difference at $\alpha < 0.05$. From van Anders and Watson, 2006a.

nonheterosexual women had significantly higher T than partnered non-heterosexual women, duplicating heterosexual men. This suggests that the effect may occur in people who partner (or could partner) with women.

The longitudinal sample was small, and all participants showed changes in T that were due to seasonality, with higher T in the autumn (van Anders, Hampson, and Watson, 2006). Still, there was no evidence that T changes on entering a relationship, as individuals who entered a relationship showed variation in T over time that was similar to individuals who were in a relationship or unpartnered the entire time (van Anders and Watson, 2006a). Follow-up was 6 to 12 months after baseline testing, so it remains possible that longer time periods would result in a different pattern of findings.

Though there was no evidence of state effects in partnering and hormones, there was evidence that T predicted entering a long-term relationship. At baseline, unpartnered individuals who would go on to be partnered had significantly lower T than unpartnered individuals who would remain unpartnered. This is suggestive that individuals with higher T are less likely to enter long-term relationships and that the effect may be a trait one. This has led me to question whether we are examining associations between hormones and relationship *status* (that is, effects of current relationship status on T) or hormones and relationship *orientation* (effects of T on later relationships) and whether there is such a thing as relationship orientation. There is evidence linking higher T levels with less need for long-term commitment (Cashdan, 1995), more frequent extramarital sex (Booth and Dabbs, 1993), more sexual partners (Bogaert and Fisher, 1995; Cashdan, 1995), and more interest in extramarital sexual partners (McIntyre et al., 2006), suggesting possible trait effects with relationship orientations associated with T.

The findings from van Anders and Watson (2006a) can be compared with a retrospective study with very large sample sizes; Booth and Dabbs (1993) found that military men with higher T were less likely to have been married in the past and were more likely to have divorced. It could not be ascertained from their data whether these unmarried men had higher T because they were currently unmarried or because they were less likely to marry—that is, directionality of effect was unclear. In a longitudinal study with very large sample sizes, Mazur and Michalek (1998) compared T levels between consistently wed, unwed, and divorced military men. Consistently wed men had lower T than consistently unwed men, matching our findings. Divorced men exhibited similar T levels to the unmarried men. In addition, higher T was associated with the likelihood of divorcing, and T was transiently high around divorce. But if changes in relationship status did change T, then any increase in T occurring around divorce should still have been apparent later (and was not). Still, the data do demonstrate state effects from divorce on T. One difference between our results and Mazur and Michalek's (in addition to their much larger sample and time between initial and follow-up points) is that our study included a younger population who were not divorcing or remarrying but instead were finding relationship partners. Likely, further longitudinal research will clarify our understanding of possible state effects.

The results in van Anders and Watson (2006a) suggest that relationship orientation may be associated with T, as lower T individuals appear more likely to enter committed relationships. I can speculate that this might be associated with co-parenting, as long-term partnering often (though not always) is associated with childbearing and child rearing. If lower T is associated with better bond maintenance behaviors, it would be advantageous for women (and possibly men) to pick low T (and high bond maintenance) partners for committed, long-term relationships. In support, lower T has been associated with better parental responsiveness in men (Storey et al., 2000; Fleming et al., 2002), better father-child relationships (Julian and McKenry, 1989), and time spent with spouses (Gray et al., 2002). As well, evidence suggests that fathers have lower T than nonfathers (see Gray and Campbell, this volume). It may be that low T individuals are more likely to be selected for long-term, committed relationships, and Roney et al. (2006) have found that low T men are viewed as more attractive for long-term relationships and more interested in infants. It may also be that high T individuals are more likely to select into short-term relationships, and evidence supports this as well (for example, Cashdan, 1995).

Considering Partner Presence: Long-Distance Relationships

Examining individuals who are single, in long-distance relationships, or in same-city relationships allows us to explore diversity in relationship types and further facilitates addressing the issue of relationship orientation versus status (van Anders and Watson, 2007). I hypothesized that if relationship orientation is associated with T, then partnered individuals should have lower T than single individuals, regardless of the physical presence of their partners. If relationship status (that is, current affiliation) is associated with T, then individuals in same-city relationships might differ from individuals in long-distance relationships, because physical partner presence (a salient cue to partner status) would differ. To test this, we recruited men and women from our university and the community.

Based on expectations from van Anders and Watson (2006a) where significant effects in women had only been seen with nonheterosexual women, it was surprising to find that women in same-city relationships had lower T than women who were single, even when sexual orientation was covaried. Women in long-distance relationships had T levels that were intermediate to single women and women in same-city relationships (see Figure 15.2). There was no evidence that other variables (for example, age, body mass index [BMI], sleep-wake cycles) accounted for the findings. Also, same-city and long-distance partnered women did not differ in any measured relationship variables (excepting physical contact with partner) including commitment, sexual attraction, sexual contact with nonpartners, relationship length, and plans to be with their partner forever. In van Anders and Watson (2006a), participants were not asked whether their relationships were long distance, and it is possible that some of them were, potentially obscuring any association between relationships and T in women. The findings described here suggest that relationship status and T are associated in women and that partner presence mediates this effect. Another possibility is that women in same-city relationships are more likely to behave in female-stereotyped ways or behaviors because their partners are present, and this gendered behavior affects T. In possible support, self-identified femme partners have lower T than self-identified butch partners (Pearcey, Docherty, and Dabbs, 1996; Singh et al., 1999).

A different pattern of associations is apparent in men, such that men in same-city and long-distance relationships exhibit lower T than single men. This suggests that partnered men have lower T regardless of partner presence. These data are consistent with a relationship orientation explanation, since men in same-city and long-distance relationships would be subject to different partner cues. However, the similarity between commitment levels between men in same-city and long-distance relationships does not preclude the possibility that the *state* of commitment may lead to lower T. I do think, though, that commitment within a relationship may represent a separate variable from partnered status and may be more relevant to relationship orientation in some contexts. Interestingly, the results show that current partnered sexual activity could not be related to differences between partnered and single men's T, since partnered men have lower T even when they differ in frequency of partnered sexual contact as partner presence necessitates. Investigations with long-distance and same-city partnered individuals have thus provided important insights, as has research with multiple partners, as I next discuss.

Multiple Partners and Hormone-Partnering Associations

According to testosterone trade-offs, people in multiple relationships should have higher T than people in coupled relationships, in part

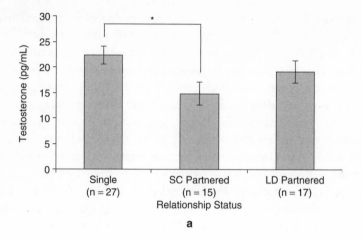

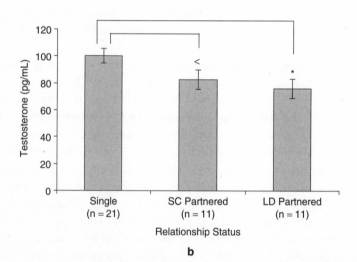

Figure 15.2   Mean testosterone levels by relationship status with standard error bars for *(a)* women (adjusted for age) and *(b)* men (adjusted for age and time of waking). *SC Partnered,* same city partnered; *LD Partnered,* long-distance partnered; * (asterisk) indicates a significant difference at $p < 0.05$; < indicates a trend with $p < 0.10$. From van Anders and Watson, 2007.

because multiple relationships should afford the possibility and probability of still further relationships (a competitive-type state). To empirically address this, I studied polyamorous, monoamorous, and single people (van Anders, Hamilton, and Watson, 2007). *Polyamory* refers to "many loves" and characterizes multiple committed relationships or multipartner relating, generally in a context of openness. Monoamorous individuals could be said to engage in monogamy (when they do), and polyamorous individuals could be said to engage in "polyfidelity," according to polyamory sources.

Despite the bonded nature of polyamorous relationships, I hypothesized that polyamorous individuals would exhibit higher T than monoamorous individuals, because polyamory should be associated with the possibility of new partners. In addition, I hypothesized that polyamorous individuals would have higher T than single individuals because polyamory is also associated with the probability of multiple and new partners in a way that being single is not.

I recruited participants from the community and polyamory groups. Polyamorous participants could be subdivided into people who currently had multiple partners (polyamorous) and people who were single or in monoamorous relationships but part of the polyamorous lifestyle (poly lifestyle). I also recruited people who were single or in monoamorous relationships. Previous research had shown that sexual orientation is relevant (van Anders and Watson, 2006a), so I covaried for sexual orientation, as there was not a large enough subsample of nonheterosexual people for analyses similar to the previous study.

Findings (see Figure 15.3) showed that monoamorously partnered men exhibit lower T than single, polyamorous, or poly lifestyle men. Interestingly, research with Swahili Kenyan men (Gray, 2003) supports this to some extent, as polygynously married men had higher T than monogamously married men. However, monogamously married men did not have lower T than men who were single, perhaps reflecting cultural considerations. In van Anders, Hamilton, and Watson (2007), single men did have higher T than monoamorously partnered men. Also contrasting with Gray, we found that polyamorous men had higher T than single men, confirming all of our hypotheses. Thus, men with multiple partners or who are more likely to be looking for potential partners appear to be more likely to have higher T, suggesting further a possible association between relationship orientation and T in men. Studies that are contextualized within varying cultures are likely to inform us further.

Findings in women also confirmed our hypotheses. Polyamorous women exhibited higher T than both monoamorously partnered and single women. As with the men, this remained true when we controlled for possibly relevant variables including sexual orientation. In contrast, however, single women did not exhibit higher T than monoamorously partnered women. The findings with women thus may point to an association between relationship status and T, such that only women currently with multiple partners exhibit higher T.

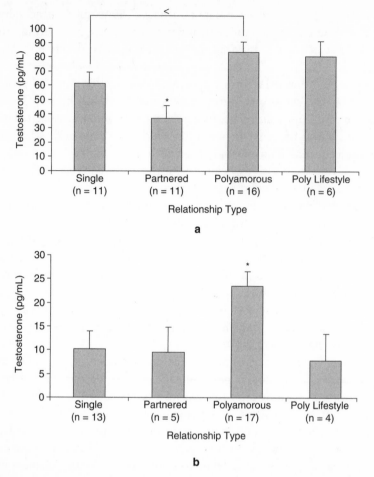

Figure 15.3   Mean testosterone levels and standard errors by relationship type, adjusted for age, sampling month, and sampling time for *(a)* men and *(b)* women. *Poly lifestyle* indicates participants not currently with multiple partners but identifying as having a poly approach to relationships, * (asterisk) indicates a significant difference from all other means at $p < 0.05$; < indicates a trend toward a significant difference from other means at $p < 0.10$. From van Anders, Hamilton, and Watson, 2007.

Summary and Conclusion

Findings on partnering and hormones show that monoamorously partnered men have lower T than men who are in polygynous relationships (Gray, 2003), polyamorous relationships (van Anders, Hamilton, and Watson, 2007), or no relationships (van Anders and Watson, 2006a). This does not depend on physical partner cues, as men in long-distance relationships have lower T than single men and similar T to men in same-city relationships (van Anders and Watson, 2007). Men with lower T are also more likely to enter relationships (van Anders and Watson, 2006a). These findings appear to apply only to heterosexual men (van Anders and Watson, 2006a), suggesting that partnering with women is somehow involved. There is evidence that women are more likely to pick lower T men for long-term partners (for example, Roney et al., 2006; see also Roney, this volume), so this may be an effect of female choice. However, men with higher T also appear less likely in a variety of ways to self-select into long-term monoamorous relationships (for example, Cashdan, 1995).

The findings in heterosexual women are much less clear-cut and therefore less amenable to theorizing about ultimate explanations. So far, we know that partner presence is associated with lower T in women (van Anders and Watson, 2007); multiple partners are associated with higher T (van Anders, Hamilton, and Watson, 2007); single nonheterosexual women have higher T than partnered nonheterosexual women (van Anders and Watson, 2006a); and women with lower T may be more likely to enter committed relationships (van Anders and Watson, 2006a). These data conflict at times, and future studies should clarify what mediates and moderates associations between T and partnering in women and whether these associations are state, trait, or both.

Behavioral neuroendocrinology relies on the diversity of the natural world to deepen and develop understandings of hormone-behavior associations. It is not surprising, then, that this approach has been helpful in elucidating how partnering and androgens may be associated in humans. Attending to gender/sex, sexual orientation, and time has pointed my research in somewhat unexpected and certainly interesting directions. Though these may moderate the association between partnering and hormones, it appears that target partner's gender may also be implicated (van Anders and Watson, 2006a) and that T may influence partnering. Multitudes of questions remain. Are higher T individuals less interested in long-term relationships? Are women who are less interested in long-term partnering less likely to attend to cues of low T in their potential partners? Do female partners with lower T show higher parenting responsivity?

Findings are suggestive that relationship *orientation* more than *status* is associated with T in men, but perhaps the reverse in women. In a way, this contrasts with sexual orientation, as there appears to be little consistent evidence associating prenatal or circulating androgens with sexual orientation in men but some support for an association in women. Some evidence for relationship orientation interpretations in men include that physical partner cues do not appear to be necessary for partnered men to exhibit lower T (van Anders and Watson, 2007); and lower T appears to predict entering committed relationships (van Anders and Watson, 2006a), staying wed (Mazur and Michalek, 1998), fewer sexual partners (Bogaert and Fisher, 1995), lower probability of extramarital sex (Booth and Dabbs, 1993), and more need for long-term commitment (Cashdan, 1995). Still, state effects are apparent in transient changes in T around divorce (Mazur and Michalek, 1998), as well as early stage love (Marazziti and Canale, 2004) and flirting (Roney, Mahler, and Maestripieri, 2003). Women in same-city relationships do show T levels that are lower than single women and women in long-distance relationships (van Anders and Watson, 2007), suggesting that T may be associated with relationship status and orientation in women and that partner-related cues may affect women's T in a statelike way. What are these cues? Do they stem from the partner, some physiological by-product in the women themselves, or from different lifestyles? Additional longitudinal studies and other studies that can address the state/trait issue are certainly warranted and would be helpful. The slash between state/trait does not indicate dichotomous thinking, and evidence will likely support various intermingled effects.

Research examining diverse relationship styles is likely to be informative. International research shows that having multiple female partners (for example, polygynous marriage) is associated with higher T in men (Gray, 2003), expanding our thinking about how and why partnering and T are associated. North American research also shows that multiple partners are associated with higher T in women and men (van Anders, Hamilton, and Watson, 2007) and that this extends to men who have a polyamorous approach to relationships but are not currently multipartnered. Would men who are oriented more toward multiple sexual contacts (for example, in swinging) as opposed to multiple committed sexual/romantic contacts, as in polyamory, show even higher T? Would women who are currently swingers have even higher T?

Exciting insights are likely to be gleaned from approaches that examine differences within broader groupings. Not all single individuals are cast from the same mold, and testable hypotheses abound for expectations of differences in T between subgroups of single, partnered, and dating individuals that may differ in temperament, experience, and interest. Diversity should be understood in its most inclusive form, including diversity within major categorical divisions.

Researchers use various theoretical approaches to study partnering and hormones in humans; though this may be seen as detrimental, it is advantageous. Informative research is conducted under the aegis of the challenge hypothesis (Wingfield et al., 1990), in which high T in males is associated with more challenge behaviors (for example, aggression) and fewer behaviors indicating parental investment (for example, mate defense). Another perspective is the testosterone trade-off framework (van Anders and Watson, 2006b) in which high T in women and men is associated with more competitive behaviors (for example, infant defense, searching for partners) and fewer bond maintenance behaviors (for example, caring for partners or offspring). Additionally, researchers from various disciplines bring different viewpoints to similar questions, including state effects (how partnering affects hormones), trait effects (how hormones affect partnering), and reciprocal effects. The variety of perspectives allows for questioning and complementary insights.

Research that includes diverse human populations can benefit our understandings of the associations between androgens and partnering even for those not interested in diversity per se. Including these groups requires recognizing the value of diversity and inclusive research practices. Since individuals in these groups are often minority holders in power structures, it is crucial that research does not contribute to their marginalization or disempowerment. One reason is admittedly selfish: groups that are distrustful of science and scientists are unlikely to volunteer to participate in scientific studies. How could research continue in this case? Others are less so: the information gathered will hopefully further our understanding of human behavior and wonder at the complexity of human nature. Behavioral neuroendocrinologists who study geographically restricted species have come to appreciate conservation efforts both for environmental reasons and for the protection of future research and demonstrate convincingly that concerns about populations need not be divorced from science.

Attending to diversity has already provided us with further understandings of how androgens and partnering are associated in humans. Excitingly, including diversity has led to the generation of even more questions and testable hypotheses. By casting the net widely, researchers are increasingly likely to make sense of this hormone-partnering puzzle and its complicated pattern. It may be that the most improbable-seeming corners, the most seldom-viewed recesses, or the most overlooked spots hold the needed pieces.

CHAPTER SIXTEEN

Early Life Influences on the Ontogeny of the Neuroendocrine Stress Response in the Human Child

Pablo Nepomnaschy and Mark Flinn

Introduction

Living organisms are flexible; they can respond to changing conditions through a variety of morphological, physiological, and behavioral mechanisms. The processes that organisms use to change and respond to environmental challenges are posited to be evolved adaptations (West-Eberhard, 2003). Flexibility involves both immediate, temporary responses and longer-term developmental changes. The ability to generate a variety of phenotypes from a single genotype to adapt to environmental variations is called *phenotypic plasticity*.

Humans exhibit a most complex form of phenotypic plasticity. Our brain has unique information-processing capacities that we use to master the fast-paced dynamics of social networks and culture (Adolphs, 2003; Roth and Dicke, 2005). The stress axis appears to play an important role in modulating the aforementioned plasticity. Humans present an extraordinary sensitivity to stress that appears to arise at the earliest stages of development. Stress response may at first sight seem paradoxical because release of cortisol and other stress hormones may have attendant somatic costs and important consequences for human health (Flinn, 2007, 2008). Maternal depression and anxiety during pregnancy, for example, are associated with low birth weight, elevated stress reactivity, and subsequent health risks for the offspring (Barker, 1998; Weinstock, 2005; Gluckman and Hanson, 2006).

In this chapter, we discuss this apparent paradox. We also evaluate possible mechanisms and developmental trajectories that link early life events to physiological stress response, psychological development, and

health outcomes from the very moment of conception. We start by analyzing stress as a modulator of women's reproductive function and the
conflicts that can arise between mother and fetus regarding the ultimate
fate of the pregnancy during stressful times. Next, we discuss the effects
that stress has on fetal development once pregnancy is firmly established
(after the placentation process has taken place). The short- and long-
term developmental consequences of stress appear to vary according to
the time in which the stress challenge takes place during ontogeny. We
review the role of the environment as a factor in "developmental programming" during ontogenetic processes. Then we focus on the specific
effects that social challenges appear to have on the postnatal ontogeny
of neuroendocrine stress response and subsequent health outcomes. We
illustrate our ideas with brief overviews of our respective work in
Guatemala and Dominica. We conclude by proposing the "depatholo-
gization" of the study of the developmental consequences of stress.

Stress and the Timing of Pregnancy

The quality of the mother's environment during pregnancy and soon after parturition can critically affect her offspring's chances of survival
and overall quality. Consequently, a pregnancy conceived or maintained
under stressful conditions may affect the mothers' lifetime reproductive
success and that of her offspring (Penn and Smith, 2007). Being able to
optimize the timing of new reproductive ventures would, therefore, be
a valuable adaptation (Nepomnaschy et al., 2006). To test whether
stress plays any role in regulating reproductive function in humans,
Nepomnaschy and colleagues conducted research in a Mayan community in the highlands of southwest Guatemala. People in this community live under stringent conditions, enduring intervals of restricted
food supply, associated threats of infectious diseases, and other seasonal, psychological, and environmental stressors. Longitudinal analyses of interview data and urine specimens collected continuously for a
full year from this population uncovered several interesting relationships. First, these women's most important expressed concerns had to
do with health issues affecting them or their immediate family and interpersonal problems (Nepomnaschy et al., in prep). Considering that
in this population common health problems such as infant diarrhea,
respiratory problems, and obstetric complications are not uncommon
causes of death, and that in this society individuals are highly interde-
pendent, the loss of a member of a woman's support network due to
illness or conflict can carry with it serious practical, social, and economic consequences.

Importantly, women's self-reports of concerns were associated with elevated cortisol levels in their urine samples (Nepomnaschy et al., 2007). Cortisol is a key mediator in the body's response to a variety of psychosocial, energetic, and health challenges and is, therefore, frequently used as a marker of stress (Sapolsky, Romero, and Munck, 2000; Tilbrook, Tunner; and Clarke, 2000; Roberti, 2003). In turn, women's increases in cortisol levels were linked with changes in the profiles of the participants' reproductive hormones. Specifically, raised cortisol levels were associated with untimely increases in gonadotrophin and progestin levels during the follicular phase of the menstrual cycle. Increased cortisol levels were also associated with significantly lower progestin levels during the middle of the luteal phase (Nepomnaschy et al., 2004). These are important results because all of these hormonal changes have been previously found to negatively affect a female's chances to conceive (Baird et al., 1999; Ferin, 1999). Furthermore, during the first three weeks of gestation (the period in which the placenta begins to develop and becomes functional), pregnancy loss was almost three times higher in those women with increased cortisol levels (see Figure 16.1) (Nepomnaschy et al., 2006; compare Ellison et al., 2007).

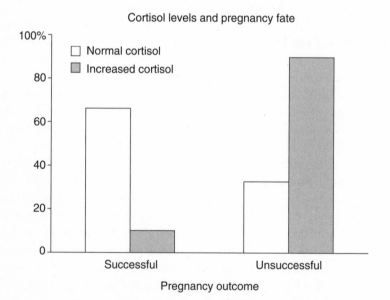

Figure 16.1   Cortisol levels and pregnancy outcome. Pregnancy outcomes stratified by average cortisol values between estimated time of ovulation and miscarriage or three weeks after ovulation, whichever came first. Pregnancies exposed to "high cortisol" were 2.7 times more likely to result in miscarriage *(unsuccessful)* than those exposed to "normal cortisol." Rao-Thomas $F(1,16)=3.4$; $p=0.03$.

In sum, what Nepomnaschy and colleagues uncovered was a connection between a woman expressing concerns, stress axis activation, and reproductive suppression. They argue that this connection could have an adaptive value: in unfavorable circumstances, avoiding or interrupting reproduction allows females to focus scarce resources on survival, improvement in overall condition, and investment in existing offspring (Nepomnaschy et al., 2004; Nepomnaschy et. al., 2006).

While the mechanisms described above may reduce a woman's chances to conceive or increase her chances of early spontaneous abortion, their effectiveness may decrease as her fetus begins to develop. After conception a fetus becomes an actor in determining the fate of its own gestation; yet mother and fetus may have different interests. Each fetus is a genetically unique entity, with only one opportunity to be born. Being alive is a prerequisite to achieving any other goals, and thus being born should be a fetus's first priority. Here is where conflicts of interest with the mother may arise. There will be conditions under which the mother would benefit from the interruption of gestation, and the fetus would benefit from promoting its continuation. These conflicts of interest might help explain why, within just hours after fertilization, embryos begin secreting a battery of metabolites that reduce the risk of miscarriage. In line with this argument Nepomnaschy and colleagues (2006) recently suggested that maternally derived abortive mechanisms may lose efficiency as the fetus progressively gains physiologic control of its own gestation. In other words, the "older" a fetus gets, the more capable it should be of surviving periods of maternal stress.

The Cost of Living: Consequences of Prenatal Stress

Whether the abortifacient effects of stress diminish as the fetus develops remains to be tested. A number of studies, however, suggest that pregnancies can continue despite acute stressors deriving from challenges experienced after placentation has taken place (for recent reviews, see de Kloet et al., 2005; Kaiser and Sachser, 2005; Murphy et al., 2006). Surviving stressful maternal conditions, though, can have serious physiologic and behavioral consequences for the fetus and lead to important variations in the resulting postnatal phenotype.

Alternative Phenotypes

Waddington (1959) termed the mysteries of the translation of information from genetic materials during development to the phenotype as the "great gap in biology." Ontogeny is an astonishingly complex process. All living organisms have significant portions of their genomes and epigenomes that perform developmental "regulatory" functions, turning some genes off and others on during development. These regulatory "switches" are attuned to environmental conditions in ways that can result in phenotypic modifications.

The putative objective of developmental plasticity is to modify the phenotype so as to better meet present and future challenges. In terms of the future, the key problem is predictability. How reliable are the cues that are used to assess future contingencies? The difficulty of prediction increases with distances in time and space. Early preparation, however, allows for more specialization and economy of development. The sooner one can adjust development to fit future environments, the better; in this way resources need not be wasted covering other options. The balance between predictability and specialization during development influences the ontogenetic trajectories of phenotypic plasticity. "Critical periods" for environmental input involve this inherent temporal trade-off between the reliability of cues and the advantages of earlier specialization. In terms of development of the stress axis and all its associated phenotypic traits, gestation appears to be one of those critical periods.

Stress and Fetal Development

Experiments conducted in nonhuman mammals, mainly rodents and primates, suggest that prenatal stress can result in low birth weight and delayed physical growth postnatally and can affect motor and cognitive development. Significantly, one of the important consequences of prenatal stress appears to be a "reprogramming" of the hypothalamic-pituitary-adrenal (HPA) axis basal function and regulation (de Kloet et al., 2005). Animals experimentally exposed to prenatal stress show a shift in their circadian cortisol secretion profile (Koehl et al., 1999) and increased basal cortisol and adrenocorticotropic hormone (ACTH) levels (for example, Schneider, 1992; Weinstock et al., 1992; Clarke et al., 1994; Weinstock, 1997; Welberg, Seckl, and Holmes, 2000; Buitelaar et al., 2003). When prenataly stressed rats are postnatally challenged, they present a faster and stronger physiologic response to stressors than controls (Fride et al., 1986; McCormick, 1995). Furthermore, these rats also take longer to recover and to habituate to stressors (Takahashi, Turner, and Kalin, 1992). In line with these observations, prenatally stressed animals tend to have higher plasma glucose levels than controls (Vallée et al., 1997). Prenatally stressed animals also present fewer hippocampal glucocorticoid and mineralocorticoid receptors (Szuran et al., 2000; Weinstock, 2005).

Neurophysiologic changes observed in prenatally stressed animals have clear behavioral correlates. Animals experimentally subjected to stress during pregnancy display increased anxiety, shorter attention spans, and depressive-like symptoms such as learned helplessness and anhedonia. They also differ from controls in their behavioral strategies when facing novel situations such as the inclusion of new individuals in their social group, physical changes to their environment, or when exposed to an open field (for example, Frye and Wawrzycki, 2003; Morley-Fletcher et al., 2003).

Although human studies have been logically more limited in scope and design, available reports are consistent with what has been observed in animal experiments. Prenatal stress in humans has been linked with premature delivery; low birth weight for gestational age (Hedegaard et al., 1996; Wadhwa et al., 1998; Ruiz et al., 2002; Rondo et al., 2003); small head circumferences (Lou et al., 1994; Buitelaar et al., 2003); and variations in the rates of physical, mental, and motor development (Gluckman, Hanson, and Beedle, 2007). Importantly, in humans prenatal stress appears to also affect HPA axis regulation (example, Andrews and Matthews, 2004; Glover et al., 2004; Halligan et al., 2004). Gitau and colleagues (2001), for example, report that piercing of the fetal trunk around the time of pituitary maturation (eighteenth gestational week) leads to detectable increases of fetal beta-endorphin levels and that the same challenge two weeks later, when the adrenal glands mature, leads to increases in fetal cortisol. The effects of prenatal stress on the HPA axis can also be observed postnatally. Field and colleagues (2004) compared depressed and nondepressed pregnant women and their children as they develop postnatally. They report a strong association between the cortisol, catecholamines, and serotonin levels in their mother-children dyads. Similarly, Gutteling, de Weerth, and Buitelaar (2004) report that maternal cortisol during gestation and pregnancy-related fears were positively associated with their children's cortisol levels before and after being vaccinated at ages four to six and during the first days of the school year (Gutteling, de Weerth, and Buitelaar, 2005). In that study, prenataly stressed children also presented steeper circadian slopes.

Even while still in utero, stress exposure may affect the fetus's neurological development and its ability to habituate to challenges. Maternal stress levels during the third trimester have been associated with changes in fetal heart rate after vibroacoustic stimulation (Sandman et al., 1999). The behavioral effects of prenatal stress in the postnatal phenotype appear to be broad. Prenatally stressed children have been reported to present an increased tendency to depression, attention deficits (Buitelaar

et al., 2003), difficult temperament (Van den Bergh, 1992), emotional problems, hyperactivity (Linnet et al., 2003; O'Connor, Heron, et al., 2003),
and difficulties in adapting to novel situations (Van den Bergh, 1992). Furthermore, prenatal stress has been linked to schizophrenia (Van Os and Selten, 1998; Hultman et al., 1999; Imamura et al., 1999; Watson et al., 1999;
Koenig, Kirkpatrick, and Lee, 2002), criminal behavior, autism, and social
withdrawal (Huttunen and Niskanen, 1978; Meijer, 1986; McIntosh,
Mulkins, and Dean, 1995; Schneider, Moore, and Kraemer, 2004).

There are also studies that do not find the purported associations between prenatal stress and changes in HPA functioning, motor and cognitive development, and postnatal behavior. For example, while
individuals prenatally exposed to the Dutch famine (1944–1945) show
higher salivary cortisol baselines than controls who were not exposed,
response to psychological stress appears not to differ between the two
(de Rooij et al., 2006). Furthermore, there are studies that report positive
effects of moderate gestational stress on development. After studying a
group of healthy women, for example, DiPietro and colleagues (2006)
found that mild maternal anxiety and depression were associated with
enhanced motor maturation in two-year-old children.

These inconsistencies may be related to the broad range of differences
in design between studies and to the multiple limitations affecting human
research on this subject. Furthermore, these inconsistencies confirm that
the relationship between prenatal stress and fetal programming is complex and highlight how little we still know about this phenomenon.

Mechanisms Mediating Maternal Effects on Fetal Development

The pathways through which prenatal stress may exert the effects described above are also poorly understood (Huizink, Mulder, and Buitelaar,
2004; de Weerth and Buitelaar, 2005). Glucocorticoids are logically suspected to be involved. Maternal cortisol levels are known to gradually increase during pregnancy. A placental physiologic barrier, however, keeps
fetal cortisol levels 5 to 10 times lower than in the mother (Predline et al.,
1979; Gitau et al., 1998). The placental enzyme 11β-hydroxysteroid dehydrogenase type 2 (11β-HSD2) inactivates maternal cortisol by metabolizing it into cortisone before it crosses the placenta (Brown et al.,
1993). Nonetheless, some maternal cortisol (about 10%) is believed to
cross the placenta as cortisol (Gitau et al., 1998). Given the difference in
basal levels between mother and fetus, even small amounts of maternal
cortisol can cause significant changes in fetal physiology (de Weerth and
Buitelaar, 2005).

Direct exposure to cortisol increases have been reported to affect fetal growth (Banks et al., 1999), birth weight (Bloom et al., 2001), head circumference (French et al., 1999), HPA axis functioning (Gitau et al., 1998), coronary function (Rotmensch et al., 1999; Subtil et al., 2003), and gastric function (Chin, Brodsky, and Bhandari, 2003). Stress metabolites may have additional effects on fetal development through less direct pathways. Placental production of corticotropin-releasing hormone (CRH) has been reported to increase with maternal stress (de Weerth and Buitelaar, 2005). CRH enters fetal circulation and can influence the development, distribution, and abundance of glucocorticoid receptors in the fetus's central nervous system (Meaney et al., 1996; Majzoub and Karalis, 1999). Increases in glucocorticoids and catecholamines can reduce uterine blood flow, which, in turn, may restrict fetal growth and contribute to fetal hypoxia. This mechanism has been proposed as a potential explanation for the link between prenatal stress, low birth weight, prematurity, and poor neurologic outcomes (Teixeira, Fisk, and Glover, 1999; Schneider, Moore, and Kraemer, 2004). Furthermore, prematurity and low birth weight have been linked with increased HPA reactivity in the adult (Phillips et al., 2000; Reynolds et al., 2001; Kajantie et al., 2002; Kajantie et al., 2003).

Another important factor to consider is interindividual variability in stress responses between mothers. There is evidence suggesting that pregnant women differ in their physiologic responses to stress, and that may differentially affect their fetuses' development (de Weerth and Buitelaar, 2005; Murphy et al., 2006). Furthermore, how much cortisol reaches the fetus is not only a function of the mothers' circulating levels of this glucocorticoid; it is also a function of the activity levels of placental 11β-HSD2. Placental 11β-HSD2 activity also varies between women and within each individual according to gestational stage, protein content of diet, oxygen levels, and hormonal levels related to placental function such as estradiol, progesterone, prostaglandins, and catecholamines (Welberg, Seckl, and Holmes, 2000; Bertram et al., 2001). Variations in placental 11β-HSD2 activity levels have been linked to changes in fetal HPA development; fetal growth; survivorship; birth weight; and adult coronary, renal, and hepatic functions (Murphy et al., 2006). Animal experiments suggest that alterations in placental function involving changes in the levels of prostaglandins, pro-opiomelanocortin (POMC), progesterone, and 11β-HSD2 activity may mediate some of the effects stress has on fetal development (Bloomfield et al., 2004).

Prenatal stress may exert its effects through a variety of other genetic, epigenetic, and ontogenetic pathways. Fetal development depends on the synchronized occurrence of a large number of complex processes. Stress-related alterations of any of those processes or their tempos are likely to have an effect on the final outcome. At the neurological level, for example, disruptions of cell differentiation, migration, positioning, and connection appear to be linked to outcomes such as dyslexia, schizophrenia, and mental retardation (Watson et al., 1999). Finally, fetal factors are also likely to play an important role in any pathway explaining the relationship between gestational stress and fetal development (Murphy et al., 2006).

Understanding the multiple mechanisms through which prenatal stress affects fetal development and the resulting postnatal phenotype is a complex task that will require high-quality longitudinal research and a lot of patience. Future studies will need to be thorough in assessing the various factors that might be mediating the observed relationships. It would be important, for example, to (1) explore the genetic heritability of HPA function from mothers to children, (2) separate the influence of the maternal HPA functioning from stress-related behaviors (such as alcohol or caffeine consumption), and (3) assess the differential effects of prenatal stress on HPA functioning at the different stages of fetal development.

Postnatal Effects of Stress on Development

Adjustments to developmental trajectories begin with the fetus's adaptation to the maternal uterine environment and continue throughout infancy and childhood. There are, however, important differences in developmental responses to the pre- and postnatal environments. After birth, children have much more information available for phenotypic adjustments. They can see, hear, feel, taste, and smell the environment to a much greater extent. They become active participants in the social network, with new opportunities to influence the actions of others via communication. Compared with other social species, human infants are mentally precocial and experience a lengthy childhood and adolescence (Bogin, 1999; Bjorklund and Pellegrini, 2002). The evolutionary point of this costly extension of the juvenile period may be related to the development of a social brain that will have to master complex dynamic tasks such as learning the personalities and social biases of peers and adults and developing appropriate emotional responses to them (R. Alexander, 1990; Flinn, 2004, 2006a; Flinn and Coe, 2007).

Parents and other kin may be especially important for the child's mental development of social and cultural maps because they can be relied on as landmarks that provide relatively honest information. Human mothers appear to have especially important roles in the development of their offsprings' sociocognitive development (Deater-Deckard, Atzaba-Poria, and Pike, 2004). Learning, practice, and experience are imperative for social success. The link between physiologic stress and responses to it may guide both the acute and long-term neurological plasticity necessary for adapting to the dynamic aspects of human sociality.

Psychosocial Stress and the Development of the Human Child

Natural selection appears to have favored the development of links between the neuropsychological mechanisms involved with assessment of the social environment and the neuroendocrine mechanisms that regulate stress response. Social challenges can trigger the activation of the HPA axis (Kirschbaum and Hellhammer, 1994; Gunnar, Bruce, and Donzella, 2000; Dickerson and Kemeny, 2004; Flinn, 2006b). The link between emotional domains and the stress axis may help manage the direction of mental processes to solving specific problems. For example, when dealing with the threat of an approaching bully, a child needs to allocate her cognitive efforts to the task at hand: prepare for immediate contingencies by recalling salient information, enhancing relevant sensory input, and activating circuits for appropriate actions.

ONTOGENY: EARLY POST NATAL TRAUMA → HPA
DYSFUNCTION HYPOTHESIS

> [T]he development of individual differences in behavioral and neuroendocrine responses to stress can be influenced by events occurring at multiple stages in development. (Francis, Diorio, et al., 2002, 7843)

As discussed in previous sections, early experiences can have profound and permanent effects on HPA regulation and stress response (Suomi, 1997; Meaney, 2001; Maccari et al., 2003; Cameron et al., 2005; cf. Levine, 2005). Research on the developmental pathways has targeted the homeostatic mechanisms of the HPA system, which appear sensitive to exposure to high levels of glucocorticoids and CRH during ontogeny. Glucocorticoid receptors (GRs) and neurons in the hippocampus, which are part of the negative feedback loop regulating release of CRH and ACTH, can be damaged by the neurotoxic levels of cortisol or CRH associated with traumatic events (Sapolsky, 1990a, 2005). Hence early trauma is posited to result in permanent HPA dysregulation and hypercortisolemia, with consequent deleterious effects on the hippocampus, thymus, and other key neural, metabolic, and immune system components (Mirescu, Peters, and

Gould, 2004; Zhang et al., 2004). In primates these effects have additional consequences resulting from a high density of GRs in the prefrontal cortex (de Kloet, Oitzl, and Joels, 1999; Patel et al., 2000; Sanchez et al., 2000).

Finer-grained analysis of the epigenetic mechanisms involved with maternal effects on glucocorticoid negative feedback on CRH release indicates that DNA methylation affects hippocampal GR exon 17 promoter activity (Weaver, Diorio, et al., 2004). The permanence of DNA methylation, set during a sensitive period in the first week after birth in the rat, is a mechanism connecting diminished maternal care (licking, grooming, and arched-back nursing) with long-term elevations of HPA stress response. While the relationship between early trauma and variation in HPA development in humans has not been as well documented as in animal studies, similar effects and intervening mechanisms appear plausible (for example, Heim et al., 2000; Essex et al., 2002; Heim et al., 2002; O'Connor, Heron, et al., 2003; Teicher et al., 2003; Lupien et al., 2005).

Adaptive Phenotypic Plasticity: Programming the Limbic System and Neocortex

Neuroendocrine stress response may guide adaptive neural reorganization, such as enhancing predator detection and avoidance mechanisms (LeDoux, 2000; Meaney, 2001; Dal Zatto, Marti, and Armario, 2003; Wiedenmayer, 2004; Buwalda et al., 2005; Rodríguez Manzanares et al., 2005). Exposure to cats can have long-term effects on the central amygdala (right side) in mice, resulting in increased fear sensitization (Ademec, Blundell, and Burton, 2005). The potential evolutionary advantages of this neural phenotypic plasticity are apparent (Rodriguez Manzanares et al., 2005). Prey benefit from adjusting alertness to match the level of risk from predators in their environments. Post-traumatic stress disorder (PTSD) appears analogous to these fear-conditioning models and involves similar effects of noradrenergic (Pitman et al., 2002) and glucocorticoid systems (Roozendaal, Quirarte, and McGaugh, 2002) on associative long-term potentiation (LTP) of the amygdala and other neurological structures that underlie the emotional state of fear.

Social defeat also affects the amygdala and hippocampus, but in different locations (Koolhaas et al., 1997; Bartolomucci et al., 2005), suggesting that neural remodeling and LTP are targeted and domain specific (for example, Rumpel, et al., 2005). Glucocorticoids, perhaps in combination with peptide hormones and catecholamines, appear to facilitate the targeting of domain-specific remodeling and LTP. The potentiating effects of cortisol on emotional memories and other socially salient information may be of special significance in humans (Pitman, 1989; Fenker et al., 2005; Lupien et al., 2005; Jackson et al., 2006; Roelofs, et al., 2007). The neurological effects of stress response may underlie adaptation to short-term contingencies and guide long-term ontogenetic adjustments of emotional regulation and associated behavioral strategies.

If physiological stress response promotes adaptive modification of neural circuits in the limbic and higher associative centers that function to solve psychosocial problems (Huether et al., 1999), then the apparent paradox of psychosocial stress would be partly resolved. Temporary elevations of cortisol in response to social challenges could have advantageous developmental effects involving synaptogenesis and neural reorganization (Huether, 1996, 1998; Buchanan and Lovallo, 2001). Such changes may be useful and necessary for coping with the demands of an unpredictable and dynamic social environment. Elevating stress hormones in response to social challenges makes evolutionary sense if it enhances specific acute mental functions and helps guide cortical remodeling, including the neurological structures involved with emotional regulation.

Chronic destabilization of neuronal networks in the hippocampus or cerebral cortex, combined with enhanced fear circuits in the amygdala (for example, Bauer, Ledoux, and Nader, 2001; Phan et al., 2006), however, could result in apparently pathological conditions such as PTSD (Yehuda, 2002) and some types of depression (Preussner et al., 2005). Even normal (but rather novel) everyday stressors in modern societies, such as social discordance between what we desire and what we have (Dressler and Bindon, 2000), might generate some maladaptive HPA responses. Individual differences in perception, emotional control, rumination, reappraisal, self-esteem, and social support networks seem likely cofactors (Chisholm et al., 2005; Ellis et al., 2006).

Hypotheses Involving Ontogenetic Processes

Testing ideas about relations among physiological stress response, neural remodeling, and adaptation to the social environment is not a simple or easy task (for example, Pine et al., 2001). Cortisol can affect cognitive functioning and emotional states; and cognitive processing and emotional regulation can affect cortisol response, all in an ongoing ontogenetic dance. Teasing out the causes and effects in ontogenetic sequence requires sequential data on physiological response profiles, environmental context, and perception. Extensive research on hormonal stress response has been conducted in clinical, experimental, school, and work settings (for reviews, see Weiner, 1992; Stansbury and Gunnar, 1994;

Panter-Brick and Pollard, 1999; Dickerson and Kemeny, 2004). We, however, know relatively little about stress neuroendocrinology among children in normal, everyday ("naturalistic") environments, particularly in nonindustrial societies. Investigation of childhood stress and its effects on development has been hampered in the past by the lack of noninvasive techniques to measure stress hormones.

The development of immunoassay techniques for saliva samples presents new opportunities to research stress response in everyday life. Saliva is relatively easy to collect and store, especially under the adverse field conditions typical of naturalistic research settings (Ellison, 1988). Longitudinal monitoring of a child's daily activities, stress hormones, and psychological conditions provides a powerful research design for investigating naturally occurring stressors. Analyzing hormone levels from saliva can be a useful tool for examining the child's imperfect world and its developmental consequences, especially when accompanied by detailed ethnographic, medical, and psychological information. Unfortunately, we do not yet have field techniques for assessment of corresponding ontogenetic changes in the relevant neurological mechanisms.

Assessment of relations among psychosocial stressors, emotional states, hormonal stress response, and health during child development is complex, requiring (1) longitudinal monitoring of social environment, emotional expressions, hormone levels, immune measures, and health; (2) control of extraneous effects from physical activity, circadian rhythms, and food consumption; (3) knowledge of individual differences in temperament, experience, and perception; and (4) awareness of specific social and cultural contexts. Multidisciplinary research that integrates human biology, psychology, and ethnography is necessary for meeting these demands. Physiological and medical assessment in concert with ethnography and co-residence with children and their families can provide intimate, prospective, longitudinal information that is not feasible to collect in clinical studies.

The Dominica Child Stress Project

For the past 20 years (1988–present), Flinn and colleagues have conducted research on childhood stress and health in the community of Bwa Mawego, located on the east coast of Dominica. In this study, Flinn and colleagues use sequential longitudinal monitoring to assess children's physiological stress response to everyday events, including social challenges. Their analyses indicate that social challenges are important stressors, with an emphasis on the family environment as both a primary source and a mediator of stressful stimuli (Flinn and England, 1995, 2003).

High-stress events (cortisol increases from 100% to 2,000%) most commonly involved trauma from family conflict or change (Flinn et al., 1996). Punishment, quarreling, and residence change substantially increased cortisol levels, whereas calm, affectionate contact was associated with diminished (−10% to −50%) cortisol levels. Of all cortisol values that were more than two standard deviations above mean levels (that is, indicative of substantial stress), 19.2% were temporally associated with traumatic family events (residence change of child or parent/caretaker, punishment, "shame," serious quarreling, and/or fighting) within a 24-hour period. In addition, 42.1% of traumatic family events were temporally associated with substantially elevated cortisol (that is, at least one of the saliva samples collected within 24 hours was greater than two standard deviations above mean levels).

There was considerable variability among children in cortisol response to family disturbances. Not all individuals had detectable changes in cortisol levels associated with family trauma. Some children had significantly elevated cortisol levels during some episodes of family trauma but not during others. Cortisol response is not a simple or uniform phenomenon. Numerous factors, including preceding events, habituation, specific individual histories, context, and temperament, might affect how children respond to particular situations. Nonetheless, traumatic family events were associated with elevated cortisol levels for all ages of children more than any other factor that we examined. These results suggest that family interactions were a critical psychosocial stressor in most children's lives, although the sample collection during periods of relatively intense family interaction (early morning and late afternoon) may have exaggerated this association.

Chronic elevations of cortisol levels are also most often associated with family difficulties. Children usually became habituated to stressful events, but absence of a parent often resulted in abnormal patterns of elevated and/or subnormal cortisol levels. Children living in families with high levels of marital conflict (observed and reported serious quarreling, fighting, residence absence) were more likely to have abnormal cortisol profiles than children living in more amiable families (Flinn and England, 2003). Long-term stress, however, may result in diminished cortisol response. In some cases, chronically stressed children had blunted response to physical activities that normally evoked cortisol elevation. Comparison of cortisol levels during "non-stressful" periods (no reported or observed crying, punishment, anxiety, residence change, family conflict, or health problem during 24-hour period before saliva collection) indicates a striking reduction and, in many cases, reversal of the family environment–stress association (Flinn and England, 2003). Chronically stressed children sometimes had subnormal cortisol levels when they were not in stressful situations. For example, cortisol levels immediately after school (walking home from school) and during noncompetitive play were lower among some chronically stressed children. Some chronically stressed children appeared socially "tough" or withdrawn and exhibited little or no arousal to the novelty of the first few days of the saliva collection procedure. These subnormal profiles may be similar in some respects to those of individuals with PTSD (for example, Yehuda et al., 2005). The relation between cortisol and level of arousal or interest is also apparent in the high reactivity of both shy (introverted) and surgent (extroverted) children to some types of social challenges.

Although elevated cortisol levels in children are usually associated with negative affect such as fear, anxiety, and anger, events that involve excitement and positive affect also stimulate stress response. For example, cortisol levels on the day before Christmas were more than one standard deviation above normal, with some of the children from two-parent households and those with the most positive expectations having the highest cortisol. Cortisol response appears sensitive to social challenges with different affective states. Other studies further suggest that the cognitive effects of cortisol may vary with affective states, such as perceived social support (Ahnert et al., 2004; Quas, Baver, and Boyce, 2004). There are, also, some age and sex differences in cortisol profiles, but it is difficult to assess the extent to which this is a consequence of neurological differences (for example, Butler et al., 2005), physical maturation processes, or the different social environments experienced, for example, during adolescence as compared with early childhood (Flinn et al., 1996).

The emerging picture of HPA stress response in the naturalistic context from the Dominica study is a combination of physical exertion and metabolic demands, on the one hand, and sensitivity to social challenges, on the other, consistent with clinical and experimental studies. The results further suggest that family environments and their developmental sequelae of affiliation, attachment, and security are an especially important source and mediator of stressful social challenges for children, consistent with other sources (for example, Garmezy, 1983; Gottman and Katz, 1989; Hetherington, 2003a, 2003b; Dunn, 2004).

Children in the Bwa Mawego study who were exposed to the stress of hurricanes and political upheavals during infancy or in utero do not have any apparent differences in cortisol profiles in comparison with children who were not exposed to such stressors. Children exposed to the stress of parental divorce, death, or abuse (hereafter *early family trauma,* or EFT), however, have significantly higher cortisol levels at age 10 than other children (Figure 16.2). Based on analogy with the nonhuman research discussed previously, two key factors could be involved: (1) diminished hippocampal GR functioning, resulting in less effective negative feedback regulation of cortisol levels; and (2) enhanced sensitivity to perceived social threats, mediated in part by emotional regulation. Cortisol increases in response to common activities such as eating meals, active play, and hard work (for example, carrying loads of wood to bay oil stills) among healthy children but within an hour or two returns to normal levels. If EFT has affected the negative feedback loop, then recovery to normal cortisol levels would be slower. Resumption of normal cortisol levels after physical stressors, however, is similar regardless of early experience of family trauma. In contrast, cortisol profiles following social stressors indicate that EFT children sustain elevated cortisol levels longer than non-EFT children (Figure 16.3). Hence, the enhanced HPA

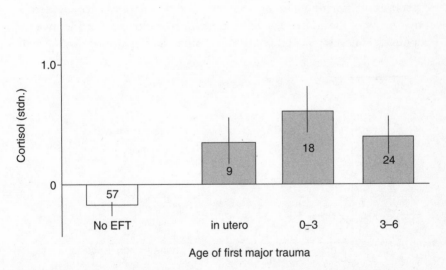

Figure 16.2 Early family trauma and cortisol at age 10 and older. Children exposed to early family trauma in utero or postnatally have higher average (means for each child) cortisol levels at ages 10 and above than children who were not exposed to early trauma (no EFT). Sample sizes (number of children) are in *bars. Vertical lines* represent 95% confidence intervals. Figure adapted from Flinn, 2006b.

stress response of children in this community who were exposed to EFT appears primarily focused on social challenges, suggesting that the ontogenetic effects of early trauma on stress response may be domain specific and even context specific. These results are consistent with studies of the effects of social defeat with nonhuman models (for example, Kaiser and Sachser, 2005).

Discussion and Concluding Remarks

Understanding the effects that early stress has on human ontogeny may have significant consequences for public health (Marmot, 2004; Flinn, 2007) because it could provide new insights into associations among stress response, social disparities, and perinatal programming, among other outcomes (Barker, 1998; Heim and Nemeroff, 2001; Maccari et al., 2003; Gluckman, Hanson, and Beedle, 2007). The analyses of the relationship between developmental stress and its postnatal consequences should, however, be guided by a more comprehensive theoretical framework than the one currently offered by the traditional medical sciences.

Many of the developmental consequences of stress are currently labeled *pathologies*. Pathologies, however, are usually understood as the result of malfunctions. In this chapter, we evaluated evidence suggesting that many of the consequences of stress may be the result of adaptive responses to developmental constraints rather than malfunctions. Several

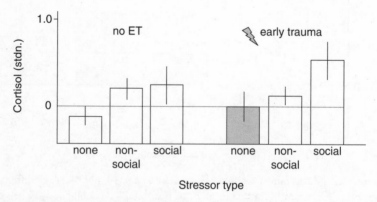

Figure 16.3   EFT and domain-specific stress response. Children exposed to EFT have higher cortisol levels in response to social stressors, but not nonsocial stressors, than no-EFT children. *Vertical lines* represent 95% confidence intervals. Figure adapted from Flinn, 2006b.

of these adaptations appear to have costly consequences later in life. True. But surviving is a prerequisite for everything else. Thus, while a depressive adult phenotype may not necessarily sound like a desirable outcome, from an evolutionary perspective a depressive adult is better than no adult at all. Furthermore, depression does not preclude reproduction and, within certain individual contexts, may actually enhance it.

The environmental constraints experienced by the developing fetus may contain important information regarding the world outside the uterus. Traits developed to face postnatal challenges may carry some costs but may be necessary for survival. The presence of these mechanisms in other species leads us to pose an obvious question: Do any of the so-called negative outcomes associated with prenatal stress in humans increase the carriers' chances of survival and reproduction? Appropriately answering this question in humans will require complex analyses of the true biologic costs and benefits of those "negative outcomes" across the entire life span of the individual, beginning from the moment of conception.

One of several challenges of such a project would be to control for both the postnatal environment of development and the environment in which individuals finally attempt reproduction. The postnatal environment of development continues to modify ontogeny as well as dictate some of the costs and benefits of the phenotypic changes triggered by prenatal stress. If individuals survive to reproductive age, the environments they face at that stage will dictate the final costs and benefits for their reproductive fitness of each particular trait affected by prenatal stress. Furthermore, as a species in which parental investment continues beyond the reproductive years (Hawkes et al., 1998), any stress-led phenotypic modifications that help the individual survive and reproduce but shorten postreproductive life may have a negative impact on inclusive fitness. One of several challenges of such a project would be to control for both the postnatal environment of development and the environment in which individuals finally attempt reproduction.

Nine months of gestation represents quite a short period compared to the length of the human life span, but information obtained during the prenatal period can still be critically relevant in the short and long terms. As discussed by Ellison (2005), certain aspects of the human environment and their related stressors are quite stable; others are not. Yet temporal domains of adaptive adjustment are not mutually exclusive. A mother's HPA axis could provide useful information to the fetus regarding both secular trends and abrupt changes. The mother's HPA baseline functioning, its reactivity, length of its refractory period, and other parameters may provide "integrated" information about the stress landscape that the mother has faced over her lifetime. For example, despite the common perception to the contrary, socioeconomic level is highly inheritable (Duncan, Kalil, and Mayer, 2005). In turn, the socioeconomic status (SES) in which an individual is born is likely to affect the type, frequency, and length of many of the postnatal challenges an individual will face throughout the life span. Thus, in preparation for his or her future postnatal environment, it may be advantageous for the fetus to adjust the baseline functioning of its own HPA axis to that of its mother. Following a similar logic, sudden alterations to the mother's HPA baseline functioning (acute stress) could also indicate relevant changes in the conditions to be faced postnatally, and the fetus should benefit from adjusting its stress response to those changes as well. A simple example of this type of scenario could concern the loss of a contributing partner/father taking place during gestation. Such an event could trigger modifications in the mother's HPA functioning and also affect the prenatal and postnatal environments of development for the fetus. Again, neurophysiological changes that help the developing fetus to survive a fatherless gestation, first, and a fatherless childhood, second, should be positively selected. Whether the resulting adult phenotype enjoys his or her life or whether peers deem the individual a carrier of one pathology or another is irrelevant to the process of natural selection.

Some of the undesirable outcomes associated with stress may represent the unavoidable costs of adaptations that allowed the individual to survive exogenous challenges at some point during development or to be better adapted for the current environment. Labeling all stress outcomes as pathologies ignores the adaptive role of stress function, reduces our ability to achieve a complete understanding of the role stress plays in the unusual life history and ontogeny of the human fetus and child, and fails to help us curtail environments that lead to those undesirable outcomes.

REFERENCES / CONTRIBUTORS / INDEX

References

Abbott, D. H., 1984. Behavioral and physiological suppression of fertility in subordinate marmoset monkeys. *Am. J. Primatol.* 6, 169–186.

———, 1993. Social conflict and reproductive suppression in marmoset and tamarin monkeys. In: Mason, W. A., Mendoza, S., (Eds.), *Primate Social Conflict*. State University of New York Press, Albany, pp. 331–372.

Abitbol, J., Abitbol, P., Abitbol, B., 1999. Sex hormones and the female voice. *J. Voice* 13, 424–446.

Achenbach, G. G., Snowdon, C. T., 2002. Costs of caregiving: Weight loss in captive adult male cotton-top tamarins *(Saguinus oedipus)* following the birth of infants. *Int. J. Primatol.* 23, 179–189.

Ademec, R. E., Blundell, J., Burton, P., 2005. Neural circuit changes mediating lasting brain and behavioral response to predator stress. *Neurosci. Biobehav. Rev.* 29, 1225–1241.

Adkins-Regan, E., 2005. *Hormones and Animal Social Behavior*. Princeton University Press, Princeton, NJ.

Adolphs, R., 2003. Cognitive neuroscience of human social behavior. *Nat. Rev. Neurosci.* 4, 165–178.

Ahnert, L., Gunnar, M. R., Lamb, M. E., Barthel, M., 2004. Transition to child care: Associations with infant-mother attachment, negative emotion, and cortisol elevations. *Child Dev.* 75, 639–650.

Aikey, J. L., Nyby, J. G., Anmuth, D. M., James, P. J., 2002. Testosterone rapidly reduces anxiety in male house mice *(Mus musculus)*. *Horm. Behav.* 42, 448–460.

Ainsworth, M., 1979. Attachment as related to mother-infant interaction. In: Rosenblatt, J. S., Hinde, R. A., Been, C., Busnel, M., (Eds.), Academic Press, New York, pp. 1–51. *Advances in the Study of Behavior.* Vol. 9.

Albers, H. E., Karom, M., Smith, D., 2002. Serotonin and vasopressin interact in the hypothalamus to control communicative behavior. *Neuroreport* 13, 931–933.

Alberts, S. C., Altmann, J., 1995. Balancing costs and opportunities-dispersal in male baboons. *Am. Nat.* 145, 279–306.

Alberts, S. C., Sapolsky, R. M., Altmann J., 1992. Behavioral, endocrine, and immunological correlates of immigration by an aggressive male into a natural primate group. *Horm. Behav.* 26, 167–178.

Alexander, G. M., 2003. An evolutionary perspective of sex-typed toy preferences: Pink, blue, and the brain. *Arch. Sex. Behav.* 321, 7–14.

———, 2006. Associations among gender-linked toy preferences, spatial ability, and digit ratio: Evidence from eye-tracking analysis. *Arch. Sex. Behav.* 35, 699–709.

Alexander, G. M., Hines, M. 1994. Gender labels and play styles: Their relative contribution to children's selection of playmates. *Child Dev.* 65. 869–879.

———, 2002. Sex differences in response to children's toys in nonhuman primates *(Cercopithecus aethiops sabaeus). Evol. Hum. Behav.* 23, 467–479.

Alexander, G. M., Swerdloff, R. S., Wang, C., Davidson, T., McDonald, V., Steiner, B., Hines, M., 1997. Androgen-behavior correlations in hypogonadal men and eugonadal men. 1. Mood and response to auditory sexual stimuli. *Horm. Behav.* 31, 110–119.

Alexander, R. D., 1987. *The Biology of Moral Systems.* Aldine de Gruyter, Hawthorne, NY.

———, 1990. Epigenetic rules and Darwinian algorithms: The adaptive study of learning and development. *Ethol. Sociobiol.* 11, 1–63.

Allen, N. E., Appleby, P. N., Davey, G. K., Key, T. J., 2002. Lifestyle and nutritional determinants of bioavailable androgens and related hormones in British men. *Cancer Causes Control* 13, 353–363.

Almli, C. R., Ball, R. H., Wheeler, M. E., 2001. Human fetal and neonatal movement patterns: Gender differences and fetal-to-neonatal continuity. *Dev. Psychobiol.* 38, 252–273.

Almond, R. E. A., Ziegler, T. E., Snowdon, C. T., 2008. Changes in prolactin and glucocorticoid levels in cotton-top tamarin fathers during their mate's pregnancy: The effect of infants and paternal experience. *Am. J. Primatol.* 70, 560–565.

Altemus, M., Deuster, P. A., Galliven, E., Carter, C. S., Gold, P. W., 1995. Suppression of hypothalamic-pituitary-adrenal axis responses to stress in lactating women. *J. Clin. Endocrinol. Metab.* 80, 2954–2959.

Altmann, J., 1980. *Baboon Mothers and Infants.* Harvard University Press, Cambridge, MA.

Altmann, J., Alberts, S. C., Haines, S. A., Dubach, J., Muruthi, P., Coote, T., Geffen, E., Cheesman, D. J., Mututua, R. S., Saiyalel, S. N., Wayne, R. K., Lacy, R. C., Bruford, M. W., 1996. Behavior predicts genetic structure in a wild primate group. *Proc. Natl. Acad. Sci. U. S. A.* 9, 5797–5801.

Altmann, J., Sapolsky, R., Licht, P., 1995. Baboon fertility and social status. *Nature* 377, 688–689.

Aluja, A., Torrubia, R., 2004. Hostility-aggressiveness, sensation seeking, and sex hormones in men: Re-exploring the relationships. *Neuropsychobiology* 50, 102–107.

Amodio, D. M., Frith, C. D., 2006. Meeting of minds: The medial frontal cortex and social cognition. *Nat. Rev. Neurosci.* 7, 268–277.

Amos, W., Pomeroy, P. P., Twiss, S. D., Anderson, S. S., 1995. Evidence for mate fidelity in the grey seal. *Science* 268, 1897–1899.

Amos, W., Schlotterer, C., Tautz, D., 1993. Social structuring of pilot whales revealed by analytical DNA profiling. *Science* 260, 670–672.

Amstislavskaya, T. G., Popova, N. K., 2004. Female-induced sexual arousal in male mice and rats: Behavioral and testosterone response. *Horm. Behav.* 46, 544–550.

Anand, S., Losee-Olson, S., Turek, F. W., Horton, T. H., 2002. Differential regulation of luteinizing hormone and follicle-stimulating hormone in male Siberian hamsters by exposure to females and photoperiod. *Endocrinology* 143, 2178–2188.

Andelman, S. A., 1986. Ecological and social determinants of cercopithecine mating patterns. In: Rubenstein, D. I., Wrangham, R. W., (Eds.), *Ecological Aspects of Social Evolution*. Princeton University Press, Princeton, NJ, pp. 201–216.

Anderson, K. G., 2000. The life histories of American stepfathers in evolutionary perspective. *Hum. Nat.* 11, 307–333.

Anderson, K. G., Kaplan, H., Lancaster, J., 1999. Paternal care by genetic fathers and stepfathers I: Reports from Albuquerque men. *Evol. Hum. Behav.* 20, 405–432.

———, 2001. Men's financial expenditures on genetic children and stepchildren from current and former relationships. Population Studies Center Research Report No. 01-484, University of Michigan.

———, 2007. Confidence of paternity, divorce, and investment in children by Albuquerque men. *Evol. Hum. Behav.* 28, 1–10.

Anderson, M., Dixson, A., 2002. Motility and the midpiece. *Nature* 418, 496.

Anderson, R. A., Bancroft, J., Wu, F. C. W., 1992. The effects of exogenous testosterone on sexuality and mood of normal men. *J. Clin. Endocrinol. Metab.* 75, 1503–1507.

Andersson, A.-M., Toppari, J., Haavisto, A.-M., Petersen, J. H., Simell, T., Simell, O., Skakkbæk, N. E., 1998. Longitudinal reproductive hormone profiles in infants: Peak of inhibin B levels in infant boys exceeds levels in adult men. *J. Clin. Endocrinol. Metab.* 83, 675–681.

Andersson, M., 1994. *Sexual Selection*. Princeton University Press, Princeton, NJ.

Andrews, M. H., Matthews, S. G., 2004. Programming of the hypothalamo-pituitary adrenal axis: Serotonergic involvement. *Stress* 7, 15–27.

Anestis, S. F., 2006. Testosterone in juvenile and adolescent male chimpanzees *(Pan troglodytes)*: Effects of dominance rank, aggression, and behavioral style. *Am. J. Phys. Anthropol.* 130, 536–545.

Anonymous, 1970. Effects of sexual activity on beard growth in man. *Nature* 226, 869–870.

Apter, D., Raisanen, I., Ylostalo, P., Vihko, R., 1987. Follicular growth in relation to serum hormonal patterns in adolescent compared with adult menstrual cycles. *Fertil. Steril.* 47, 82–88.

Aragona, B. J., Liu, Y., Curtis, J. T., Stephan, F. K., Wang, Z. X., 2003. A critical role for nucleus accumbens dopamine in partner-preference formation in male prairie voles. *J. Neurosci.* 23, 3483–3490.

Aragona, B. J., Liu, Y., Yu, Y. J., Curtis, J. T., Detwiler, J. M., Insel, T. R., Wang, Z. X., 2006. Nucleus accumbens dopamine differentially mediates the formation and maintenance of monogamous pair bonds. *Nat. Neurosci.* 9, 133–139.

Arcese, P., 1989. Territory acquisition and loss in male song sparrows. *Anim. Behav.* 37, 45–55.

Archer, J., 2006. Testosterone and human aggression: An evaluation of the challenge hypothesis. *Neurosci. Biobehav. Rev.* 30, 319–345.

Archer, J., Birring, S. S., Wu, F. C. W., 1998. The association between testosterone and aggression among young men: Empirical findings and a meta-analysis. *Aggrew. Behav.* 24, 411–420.

Aron, A., Fisher, H., Mashek, D. J., Strong, G., Li, H., Brown, L. L., 2005. Reward, motivation, and emotion systems associated with early stage intense romantic love. *J. Neurophysiol.* 94, 327–337.

Atmoko, S. U., van Hooff, J. A. R. A. M., 2004. Alternative male reproductive strategies: Male bimaturism in orangutans. In: Kappeler, P. M., van Schaik, C. P., (Eds.), *Sexual Selection in Primates.* Cambridge University Press, Cambridge, pp. 196–207.

Aubin-Horth, N., Landry, C. R., Letcher, B. H., Hofmann, H. A., 2005. Alternative life histories shape brain gene expression profiles in males of the same population. *Proc. R. Soc. Lond. B Biol. Sci.* 272, 1655–1662.

Aujard, F., Heistermann, M., Thierry, B., Hodges, J. K., 1998. Functional significance of behavioral, morphological, and endocrine correlates across the ovarian cycle in semifree ranging female Tonkean macaques. *Am. J. Primatol.* 46, 285–309.

Austad, S., 1997. *Why We Age.* Wiley, New York.

Austin, E. J., Manning, J. T., McInroy, K., Mathews, E., 2002. A preliminary investigation of the association between personality, cognitive ability and digit ratio. *Pers. Ind. Diff.* 33, 1115–1124.

Axelesson, J., Inge, M., Akerstedt, T., Holmback, U., 2005. Effects of acutely displaced sleep on testosterone. *J. Clin. Endocrinol. Metab.* 90, 4530–4535.

Bachner-Melman, R., Dina, C., Zohar, A. H., Constantini, N., Lerer, E., Hoch, S., Sella, S., Nemanov, L., Gritsenko, I., Lichtenberg, P., Granot, R., Ebstein, R. P., 2005. AVPR1a and SLC6A4 gene polymorphisms are associated with creative dance performance. *PLoS Genet.* 1, e42.

Bachner-Melman, R., Zohar, A. H., Elizur, Y., Nemanov, L., Gritensko, I., Konis, D., Ebstein, R. P., 2004. Association between a vasopressin receptor AVPR1A promoter region microsatellite and eating behavior measured by a self-report questionnaire (eating attitudes test) in a family-based study of a nonclinical population. *Int. J. Eat. Disord.* 36, 451–460.

Bagatell, C. J., Heiman, J. R., Matsumoto, A. M., Rivier, J. E., Bremner, W. J., 1994. Metabolic and behavioral effects of high-dose exogenous testosterone in healthy men. *J. Clin. Endocrinol. Metab.* 79, 561–566.

Bagatell, C. J., Heiman, J. R., Rivier, J. E., Bremner, W. J., 1994. Effects of endogenous testosterone and estradiol on sexual behavior in normal young men. *J. Clin. Endocrinol. Metab.* 78, 711–716.

Baghai, T. C., Schule, C., Zwanzger, P., Minov, C., Holme, C., Padberg, F., Bidlingmaier, M., Strasburger, C. J., Rupprecht, R., 2002. Evaluation of a salivary based combined dexamethasone/CRH test in patients with major depression. *Psychoneuroendocrinology* 27, 385–399.

Bailey, A. A., Hurd, P. L., 2005. Depression in men is associated with more feminine finger length ratios. *Pers. Ind. Diff.* 39, 829–836.

Baird, D. D., Weinberg, C. R., Zhou, H., Kamel, F., McConnaughey, D. R., Kesner, J. S., Wilcox, A. J., 1999. Preimplantation urinary hormone profiles and the probability of conception in healthy women. *Fertil. Steril.* 71, 40–49.

Baird, R. W., 2000. The killer whale: Foraging specializations and group hunting. In: Mann, J., Connor, R. C., Tyack, P. L., Whitehead, H., (Eds.), *Cetacean Societies.* University of Chicago Press, Chicago, pp. 127–153.

Bakker, J., Baum, M. J., 2000. Neuroendocrine regulation of GnRH release in induced ovulators. *Front. Neuroendocrinol.* 21, 220–262.

Bale, T. L., Dorsa, D. M., 1997. Cloning, novel promoter sequence, and estrogen regulation of a rat oxytocin receptor gene. *Endocrinology* 138, 1151–1158.

Bales, K. L., Abdelnabi, M., Cushing, B. S., Ottinger, M. A., Carter, C. S., 2004. Effects of neonatal oxytocin manipulations on male reproductive potential in prairie voles. *Physiol. Behav.* 81, 519–526.

Bales, K. L., Carter, C. S., 2003a. Sex differences and developmental effects of oxytocin on aggression and social behavior in prairie voles *(Microtus ochrogaster).* *Horm. Behav.* 44, 178–184.

———, 2003b. Developmental exposure to oxytocin facilitates partner preferences in male prairie voles *(Microtus ochrogaster).* *Behav. Neurosci.* 117, 854–859.

———, n.d. Unpublished data.

Bales, K. L., Kim, A. J., Lewis-Reese, A. D., Carter, C. S., 2004. Both oxytocin and vasopressin may influence alloparental behavior in male prairie voles. *Horm. Behav.* 45, 354–361.

Bales, K. L., Kramer, K. M., Lewis-Reese, A. D., Carter, C. S., 2006. Effects of stress on parental care are sexually dimorphic in prairie voles. *Physiol. Behav.* 87, 424–429.

Bales, K. L., Lewis-Reese, A. D., Pfeifer, L. A., Kramer, K. M., Carter, C. S., 2007. Early experience affects the traits of monogamy in a sexually dimorphic manner. *Dev. Psychobiol.* 49, 335–342.

Bales, K. L., Pfeifer, L. A., Carter, C. S., 2004. Sex differences and developmental effects of manipulations of oxytocin on alloparenting and anxiety in prairie voles. *Dev. Psychobiol.* 44, 123–131.

Bales, K. L., Plotsky, P. M., Young, L. J., Lim, M. M., Grotte, N. D., Carter, C. S., 2004. Neonatal manipulations of oxytocin have sexually dimorphic effects on vasopressin receptor binding in monogamous prairie voles. *Soc. Neurosci. Abst.* 758, 13.

Bales, K. L., Plotsky, P. M., Young, L. J., Lim, M. M., Grotte, N. D., Ferrer, E., Carter, C. S., 2007. Neonatal oxytocin manipulations have long-lasting, sexually dimorphic effects on vasopressin receptors. *Neuroscience* 144, 38–45.

Ball, G., 1999. Neuroendocrine basis of seasonal changes in vocal behavior among songbirds. In: Hauser, M., Konishi, M., (Eds.), *The Design of Communication*. MIT Press, Cambridge, MA, pp. 213–254.

Balthazart, J., 1983. Hormonal correlates of behavior. In: Farner, D. S., King, J. R., Parkes, K. S., (Eds.), *Avian Biology*. Vol. 7. Academic Press, New York, pp. 221–365.

Balthazart, J., Foidart, A., Baillien, M., Silverin, B., 1999. Brain aromatase in laboratory and free-living songbirds: Relationships with reproductive behaviour. In: Adams, N. J., Slotow, R. H., (Eds.), *Proceedings of the 22nd International Ornithological Congress*. Birdlife, Durban, South Africa, pp. 1257–1289.

Bamshad, M., Novak, M., De Vries, G., 1994. Cohabitation alters vasopressin innervation and paternal behavior in prairie voles *(Microtus orchogaster)*. *Physiol. Behav.* 56, 751–758.

Bancroft, J., 2005. The endocrinology of sexual arousal. *J. Endocrinol.* 186, 411–427.

Banks, B. A., Cnaan, A., Morgan, M. A., Parer, J. T., Merrill, J. D., Ballard, P. L., Ballard, R. A., 1999. Multiple courses of antenatal corticosteroids and outcome of premature neonates. North American Thyrotropin-Releasing Hormone Study Group. *Am. J. Obstet. Gynecol.* 181, 709–717.

Barelli, C., Heistermann, M., In press. Monitoring female reproductive status using fecal hormone analysis and patterns of genital skin swellings in relation to female cycle stage in white-handed gibbon *(Hylobates lar)*. In: Lappan, S. M., Whittaker, D., Geissmann, T., (Eds.), *The Gibbons: New Perspectives on Small Ape Socioecology and Population Biology*. Springer, Berlin.

Barelli, C., Heistermann, M., Boesch, C., Reichard, U., 2007. Sexual swellings in wild white-handed gibbon females *(Hylobates lar)* indicate the probability of ovulation. *Horm. Behav.* 51, 221–230.

Barker, D. J., 1998. In utero programming of chronic disease. *Clin. Sci. (Lond.)* 95, 115–128.

Baron-Cohen, S., 1995. *Mindblindness: An Essay on Autism and Theory of Mind*. MIT/Bradford Books, Boston, MA.

———, 2002. The extreme male brain theory of autism. *Trends Cogn. Sci.* 6, 248–254.

Baron-Cohen, S., Knickmeyer, R. C., Belmonte, M. K., 2005. Sex differences in the brain: Implications for explaining autism. *Science* 310, 819–823.

Barr, R. G., Elias, M. F., 1988. Nursing interval and maternal responsivity: Effect on early infant crying. *Pediatrics* 81, 529–536.

Barrett, G. M., Shimizu, K., Bardi, M., Asaba, S., Mori, A., 2002. Endocrine correlates of rank, reproduction, and female-directed aggression in male Japanese macaques *(Macaca fuscata)*. *Horm. Behav.* 42, 85–96.

Bartels, A., Zeki, S., 2000. The neural basis of romantic love. *Neuroreport.* 11, 3829–3834.

————, 2004. The neural correlates of maternal and romantic love. *Neuroimage* 21, 1155–1166.

Bartolomucci, A., Palanza, P., Sacerdote, P., Panerai, A. E., Sgoifo, A., Dantzer, R., Parmigiani, S., 2005. Social factors and individual vulnerability to chronic stress exposure. *Neurosci. Biobehav. Rev.* 29, 67–81.

Barton, R., 1999. The evolutionary ecology of the primate brain. In: Lee, P. C., (Ed.), *Comparative Primate Socioecology.* Cambridge University Press, Cambridge, pp. 167–194.

Bartz, J. A., Hollander E., 2006. The neuroscience of affiliation: Forging links between basic and clinical research on neuropeptides and social behavior. *Horm. Behav.* 50, 518–528.

Bass, A. H., Grober, M. S., 2001. Social and neural modulation of sexual plasticity in teleost fish. *Brain Behav. Evol.* 57, 293–300.

Bateman, A. J., 1948. Intrasexual selection in Drosophila. *Heredity* 2, 349–368.

Batista, M. C., Cartledge, T. P., Zellmer, A. W., Nieman, L. K., Merriam, G. R., Loriaux, D. L., 1992. Evidence for a critical role of progesterone in the regulation of the midcycle gonadotropin surge and ovulation. *J. Clin. Endocrinol. Metab.* 74, 565–570.

Batty, J., 1978. Acute changes in plasma testosterone levels and their relation to measures of sexual behavior in the male house mouse *(Mus musculus).* *Anim. Behav.* 26, 349–357.

Bauer, J. D., Ledoux, J. E., Nader, K., 2001. Fear conditioning and LTP in the lateral amygdala are sensitive to the same stimulus contingencies. *Nat. Neurosci.* 4, 687–688.

Baulieu, E., 1998. Neurosteroids: A novel function of the brain. *Psychoneuroendocrinology* 23, 963–987.

Baum, M. J., 2006. Mammalian animal models of psychosexual differentiation: When is "translation" to the human situation possible? *Horm. Behav.* 50, 579–588.

Baumeister, R. F., Campbell, J. D., Krueger, J. I., Vohs, K. D., 2003. Does high self-esteem cause better performance, interpersonal success, happiness, or healthier lifestyles? *Psychol. Sci.* 4, 1–44.

Beach, F. A., 1975. Behavioral endocrinology: An emerging discipline. *Am. Sci.* 63, 178–187.

————, 1976. Sexual attractivity, proceptivity, and receptivity in female mammals. *Horm. Behav.* 7, 105–138.

Beach, F. A., Sturn, B., Carmichael, M., Ranson, E., 1976. Comparisons of sexual receptivity and proceptivity in female hamsters. *Behav. Biol.* 18, 473–487.

Beagley, K., Gockel, C. M., 2003. Regulation of innate and adaptive immunity by the female sex hormones oestradiol and progesterone. *FEMS Immunol. Med. Microbiol.* 38, 13–22.

Beall, C., Worthman, C., Stallings, J., Strohl, K., Brittenham, G., Barragan, M., 1992. Salivary testosterone concentration of Aymara men native to 3600 m. *Ann. Hum. Biol.* 19, 67–78.

Beck, C. T., 1998. The effects of postpartum depression on child development: A meta-analysis. *Arch. Psychiatr. Nurs.* 12, 12–20.

———, 2006. Postpartum depression: It isn't just the blues. *Am. J. Nurs.* 106, 40–50.

Becker, J. B., Arnold, A. P., Berkley, K. J., Blaustein, J. D., Eckel, L. A., Hampson, E., Herman, J. P., Marts, S., Sadee, W., Steiner, M., Taylor, J., Young, E., 2005. Strategies and methods for research on sex differences in brain and behavior. *Endocrinology* 146, 1650–1673.

Becker, J. B., Breedlove, S. M., Crews, D., 1992. *Behavioral Endocrinology.* MIT Press, Cambridge, MA.

Becker, J. B., Breedlove, S. M., Crews, D., McCarthy, M. M., 2002. *Behavioral Endocrinology.* 2nd ed. MIT Press, Cambridge, MA.

Beckerman, S., Valentine, P., 2002. *Cultures of Multiple Fathers: The Theory and Practice of Partible Paternity in Lowland South America.* University Press of Florida, Gainesville.

Beckwith, B., Petros, T., Bergloff, P., Staebler, R., 1987. Vasopressin analog (DDAVP) facilitates recall of narrative prose. *Behav. Neurosci.* 101, 429–432.

Beckwith, B., Till, R., Schneider, V., 1984. Vasopressin analog (DDAVP) improves memory in human males. *Peptides* 5, 819–822.

Beehner, J. C., Bergman, T. J., Cheney, D. L., Seyfarth, R. M., Whitten, P. L., 2006. Testosterone predicts future dominance rank and mating activity among chacma baboons. *Behav. Ecol. Sociobiol.* 59, 469–479.

Beehner, J. C., Phillips-Conroy, J. E., Whitten, P. L., 2005. Female testosterone, dominance rank, and aggression in an Ethiopian population of hybrid baboons. *Am. J. Primatol.* 67, 101–119.

Bell, C. J., Nicholson, H., Mulder, R. T., Luty, S. E., Joyce, P. R., 2006. Plasma oxytocin levels in depression and their correlation with the temperament dimension of reward dependence. *J. Psychopharmacol.* 20, 656–660.

Bell, S., Ainsworth, M., 1972. Infant crying and maternal responsiveness. *Child Dev.* 43, 1171–1190.

Bellieni, C. V., Sistos, R., Cordelli, D. M., Buonocore, G., 2004. Cry features reflect pain intensity in term newborns: An alarm threshold. *Pediatr. Res.* 55, 142–146.

Belsky, J., Steinberg, L., Draper, P., 1991. Childhood experience, interpersonal development, and reproductive strategy: An evolutionary theory of socialization. *Child Dev.* 62, 647–670.

Bentley, G. R., Harrigan, B., Campbell, B. C., 1993. Seasonal effects on salivary testosterone levels among Lese males of the Ituri forest, Zaire. *Am. J. Hum. Biol.* 5, 711–717.

Bentley, G. R., Harrigan, A. M., Ellison, P. T., 1998. Dietary composition and ovarian function among Lese horticulturist women of the Ituri forest, Democratic Republic of Congo. *Eur. J. Clin. Nutr.* 52, 261–270.

Berard, J., 1994. Alternative reproductive tactics and reproductive success in male rhesus macaques. *Behaviour* 129, 177–201.

Bercovitch, F. B., 1993. Dominance rank and reproductive maturation in male rhesus macaques *(Macaca mulatta). J. Reprod. Fertil.* 99, 113–120.

Berg, S. J., Wynne-Edwards, K. E., 2001. Changes in testosterone, cortisol, and estradiol levels in men becoming fathers. *Mayo Clin. Proc.* 76, 582–592.

Bergadá, I., Milani, C., Bedecarrás, P., Andreone, L., Ropelato, M. G., Gottlieb, S., Bergadá, C., Campo, S., Rey, R. A., 2006. Time course of the serum gonadotropin surge, inhibins, and anti-Müllerian hormone in normal newborn males during the first month of life. *J. Clin. Endocrinol. Metab.* 91, 4092–4098.

Bergman, T. J., Beehner, J. C., Cheney, D. L., Seyfarth, R. M., Whitten, P. L., 2006. Interactions in male baboons: The importance of both males' testosterone. *Behav. Ecol. Sociobiol.* 59, 480–489.

Bernstein, I. S., Rose, R. M., Gordon, T. P., Grady, C. L., 1979. Agonistic rank, aggression, social context, and testosterone in male pigtail monkeys. *Aggress Behav.* 5, 329–339.

Bernstein, I. S., Ruehlmann, T. E., Judge, P. G., Lindquist, T., Weed, J. L., 1991. Testosterone changes during the period of adolescence in male rhesus monkeys *(Macaca mulatta)*. *Am. J. Primatol.* 24, 29–38.

Berthold, P., Helbig, A. J., Mohr, G., Querner, U., 1992. Rapid microevolution of migratory behavior in a wild bird species. *Nature* 360, 668–670.

Bertram, C., Trowern, A. R., Copin, N., Jackson, A. A., Whorwood, C. B., 2001. The maternal diet during pregnancy programs altered expression of the glucocorticoid receptor and type 2 11β-hydroxysteroid dehydrogenase: Potential molecular mechanisms underlying the programming of hypertension in utero. *Endocrinology* 142, 2841–2853.

Bester-Meredith, J. K., Young, L. J., Marler, C. A., 1999. Species differences in paternal behavior and aggression in *Peromyscus* and their associations with vasopressin immunoreactivity and receptors. *Horm. Behav.* 36, 25–38.

Betzig, L., 1986. *Despotism and Differential Reproduction: A Darwinian View of History.* Aldine de Gruyter, Hawthorne, NY.

———, 1989. Causes of conjugal dissolution: A cross-cultural study. *Curr. Anthropol.* 30, 654–676.

———, 1992. Roman monogamy. *Ethol. Sociobiol.* 13, 351–383.

———, 1993. Sex, succession and stratification in the first six civilizations: How powerful men reproduced, passed power on to their sons, and used their power to defend their wealth, women and children. In: Ellis, L., (Ed.), *Social Stratification and Socioeconomic Inequality.* Praeger, New York, pp. 37–74.

———, 1995. Medieval monogamy. *J. Fam. Hist.* 20, 181–216.

Betzig, L., Weber, S., 1993. Polygyny in American politics. *Politics Life Sci.* 12, 1–8.

Bhasin, S., Woodhouse, L., Casaburi, R., Singh, A. B., Bhasin, D., Berman, N., Chen, X., Yarasheski, K. E., Magliano, L., Dzekov, C., Dzekov, J., Bross, R., Phillips, J., Sinha-Hikim, I., Shen, R., Storer, T. W., 2001. Testosterone dose-response relationships in healthy young men. *Am. J. Physiol. Endocrinol. Metab.* 281, E11172–E11181.

Bielert, C., Anderson, C. M., 1985. Baboon sexual swellings and male response: A possible operational mammalian supernormal stimulus and response interaction. *Int. J. Primatol.* 6, 377–393.

Bielsky, I. F., Hu, S.-B., Ren, X., Terwilliger, E. F., Young, L. J., 2005. The V1a vasopressin receptor is necessary and sufficient for normal social recognition: A gene replacement study. *Neuron* 47, 503–513.

Bielsky, I. F., Hu, S.-B., Young, L. J. 2005. Sexual dimorphism in the vasopressin system: Lack of an altered behavioral phenotype in female V1a receptor knockout mice. *Behav. Brain Res.* 164, 132–136.

Bielsky, I. F., Young, L. J., 2004. Oxytocin, vasopressin, and social recognition in mammals. *Peptides* 25, 1565–1574.

Birkhead, T. R., Parker, G. A., 2005. Sperm competition and mating systems. In: Krebs, J., Davies, N., (Eds.), *Behavioural Ecology*. 4th ed. Blackwell, Oxford, pp. 121–148.

Bjorklund, D. F., Pellegrini, A. D., 2002. *The Origins of Human Nature: Evolutionary Developmental Psychology*. American Psychological Association Press, Washington, DC.

Bleichfeld, B., Moely, B. E., 1984. Psychophysiological responses to an infant cry: Comparison of groups of women in different phases of the maternal cycle. *Dev. Psychol.* 20, 1082–1091.

Bloom, S. L., Sheffield, J. S., McIntire, D. D., Leveno, K. J., 2001. Antenatal dexamethasone and decreased birth weight. *Obstet. Gynecol.* 97, 485–490.

Bloomfield, F. H., Oliver, M. H., Hawkins, P., Holloway, A. C., Campbell, M., Gluckman, P. D., Harding, J. E., Challis, J. R., 2004. Periconceptional undernutrition in sheep accelerates maturation of the fetal hypothalamic-pituitary-adrenal axis in late gestation. *Endocrinology* 145, 4278–4285.

Blurton Jones, N., Marlowe, F., Hawkes, K., Ober, C., Connell, J., 2000. Hunter-gatherer divorce rates and the paternal investment theory of human pair-bonding. In: Cronk, L., Chagnon, N., Irons, W., (Eds.), *Human Behavior and Adaptation: An Anthropological Perspective*. Aldine de Gruyter, Hawthorne, NY.

Bock, J., Johnson, S. E., 2004. Subsistence ecology and play among the Okavango Delta peoples of Botswana. *Hum. Nat.* 15, 63–81.

Boehm, C., 1999. *Hierarchy in the Forest: The Evolution of Egalitarian Behavior*. Harvard University Press, Cambridge, MA.

Boer, G. J., Quak, J., Devries, M. C., Heinsbroek, R. P. W., 1994. Mild sustained effects of neonatal vasopressin and oxytocin treatment on brain growth and behavior of the rat. *Peptides* 15, 229–236.

Boesch, C., Boesch-Achermann, H., 2000. *The Chimpanzees of the Taï Forest: Behavioural Ecology and Evolution*. Oxford University Press, Oxford.

Bogaert, A. F., Fisher, W. A., 1995. Predictors of university men's number of sexual partners. *J. Sex Res.* 32, 119–130.

Bogin, B., 1999. *Patterns of Human Growth*. 2nd ed. Cambridge University Press, Cambridge.

Bonilla-Jaime, H., Vazquez-Palacios, M., Artega-Silva, M., Retana-Marquez, S., 2006. Hormonal responses to different sexually related conditions in male rats. *Horm. Behav.* 49, 376–382.

Book, A. S., Quinsey, V. L., 2005. Re-examining the issues: A response to Archer et al. *Aggress. Violent Behav.* 10, 637–646.

Book, A. S., Starzyk, K. B., Quinsey, V. L., 2001. The relationship between testosterone and aggression: A meta-analysis. *Aggress. Violent Behav.* 6, 579–599.

Boone, E., Sanzenbacher, L. L., Carter, C. S., Bales, K. L., 2006. Sexually-dimorphic effects of early experience on alloparental care and adult social behaviors in voles. *Soc. Neurosci. Abst.* 578, 6.

Boonstra, R., 2005. Equipped for life: The adaptive role of the stress axis in male mammals. *J. Mammal.* 86, 236–247.

Boonstra, R., Hik, D., Singleton, G. R., Tinnikov, A., 1998. The impact of predator-induced stress on the snowshoe hare cycle. *Ecol. Monog.* 79, 371–394.

Booth, A., Dabbs, J. M., 1993. Testosterone and men's marriages. *Soc. Forces* 72, 463–477.

Booth, A., Granger, D. A., Mazur, A., Kivlighan, K. T., 2006. Testosterone and social behavior. *Soc. Forces* 85, 167–191.

Booth, A., Johnson, D. R., Granger, D., 2005. Testosterone, marital quality, and role overload. *J. Marriage Fam.* 67, 483–498.

Booth, A., Shelley, G., Mazur, A., Tharp, G., Kittok, R., 1989. Testosterone, and winning and losing in human competition. *Horm. Behav.* 23, 556–571.

Borgerhoff-Mulder, M., Milton, M., 1985. Factors affecting infant care in the Kipsigis. *J. Anthropol. Res.* 41, 231–262.

Born, J., Lange, T., Kern, W., McGregor, G., Bickel, U., Fehm, H., 2002. Sniffing neuropeptides: A transnasal approach to the human brain. *Nat. Neurosci.* 5, 514–516.

Born, J., Pietrowsky, R., Fehm, H. L., 1998. Neuropsychological effects of vasopressin in healthy humans. *Prog. Brain Res.* 119, 619–643.

Bornstein, M. H., 2002. *Handbook of Parenting.* 2nd ed. Vols. 1–5. Erlbaum, Mahwah, NJ.

Bornstein, M. H., Cote, L. R., 2004. Mothers' parenting cognitions in cultures of origin, acculturating cultures, and cultures of destination. *Child Dev.* 75, 221–235.

Botchin, M. B., Kaplan, J. R., Manuck, S. B., Mann, J. J., 1993. Low versus high prolactin responders to fenfluramine challenge: Marker of behavioral differences in adult male cynomolgus macaques. *Neuropsychopharmacology* 9, 93–99.

Both, C., Visser, M. E., 2001. Adjustment of climate change is constrained by arrival date in a long-distance migrant bird. *Nature* 411, 296–298.

Bouman, A., Heineman, M. J., Faas, M. M., 2005. Sex hormones and immune responses in humans. *Hum. Reprod. Update* 11, 411–423.

Bowlby, J., 1969. *Attachment and Loss. Vol. 1, Attachment.* Hogarth Press, London.

Bradley, B. J., Doran-Sheehy, D. M., Lukas, D., Boesch, C., Vigilant, L., 2004. Dispersed male networks in western gorillas. *Curr. Biol.* 14, 510–513.

Bradley, B. J., Robbins, M. M., Williamson, E. A., Steklis, H. D., Steklis, N. G., Eckhardt, N., Boesch, C., Vigilant, L., 2005. Mountain gorilla tug-of-war: Silverbacks have limited control over reproduction in multimale groups. *Proc. Natl. Acad. Sci. U. S. A.* 102, 9418–9423.

Brazelton, T. B., 1977. Implications of infant development among Mayan Indians of Mexico. In: Leiderman, P. H., Tulkin, S. R., Rosenfeld, A., (Eds.),

Culture and Infancy: Variations in Human Experience. Academic Press, New York, pp. 151–188.

Bribiescas, R. G., 1996. Testosterone levels among Ache hunter-gatherer men. *Hum. Nat.* 7, 163–188.

———, 2001a. Reproductive ecology and life history of the human male. *Am. J. Phys. Anthropol.* 44, 148–176.

———, 2001b. Serum leptin levels and anthropometric correlates in Ache Amerindians of eastern Paraguay. *Am. J. Phys. Anthropol.* 115, 297–303.

———, 2005. Serum leptin levels in Ache Amerindian females with normal adiposity are not significantly different from American anorexia nervosa patients. *Am. J. Hum. Biol.* 17, 207–210.

———, 2006. *Men: Evolutionary and Life History.* Harvard University Press, Cambridge, MA.

Bribiescas, R. G., Hickey, M. S., 2006. Population variation and differences in serum leptin independent of adiposity: A comparison of Ache Amerindian men of Paraguay and lean American male distance runners. *Nutr. Metab. (Lond.)* 3, 34.

Bridges, R. S., 1996. Biochemical basis of parental behavior in the rat. In: Rosenblatt, J. S., Snowdon, C. T., (Eds.), *Parental Care: Evolution, Mechanisms, and Adaptive Significance.* Academic Press, San Diego, pp. 215–242.

Brockman, D. K., Whitten, P. L., Richard, A. F., Schneider, A., 1998. Reproduction in free-ranging male *Propithecus verreauxi:* The hormonal correlates of mating and aggression. *Am. J. Phys. Anthropol.* 105, 137–151.

Brown, D., Perkowski, S., 1998. Oxytocin content of the cerebrospinal fluid of dogs and its relationship to pain induced by spinal cord compression. *Vet. Surg.* 27, 607–611.

Brown, J. K., 1970. A note on the division of labor by sex. *Am. Anthropol.* 72, 1073–1078.

Brown, P. J., 1991. Culture and the evolution of obesity. *Hum. Nat.* 2, 31–57.

Brown, R. E., 1993. Hormonal and experiential factors influencing parental behaviour in male rodents: An integrative approach. *Behav. Proc.* 30, 1–28.

Brown, R. P., Zeigler-Hill, V., 2004. Narcissism and the non-equivalence of self-esteem measures: A matter of dominance? *J. Res. Pers.* 38, 585–592.

Brown, R. W., Chapman, K. E., Edwards, C. R., Seckl, J. R., 1993. Human placental 11β-hydroxysteroid dehydrogenase: Evidence for and partial purification of a distinct NAD-dependent isoform. *Endocrinology* 132, 2614–2621.

Brown, W. M., Finn, C. J., Cooke, B. M., Breedlove, S. M., 2002. Differences in finger length ratios between self-identified "butch" and "femme" lesbians. *Arch. Sex. Behav.* 31, 123–127.

Brown, W. M., Hines, M., Fanes, B. A., Breedlove, S. M., 2002. Masculinized finger length patterns in human males and females with congenital adrenal hyperplasia. *Horm. Behav.* 42, 380–386.

Brownstein, M. J., Russell, J. T., Gainer, H., 1980. Synthesis, transport, and release of posterior pituitary hormones. *Science* 207, 373–378.

Bruce, J., Davis, E. P., Gunnar, M. R., 2002. Individual differences in children's cortisol response to the beginning of a new school year. *Psychoneuroendocrinology* 27, 635–650.

Bruins, J., Hijman, R., Van Ree, J. M., 1992. Effect of a single dose of des-glycinamide-[Arg8] vasopressin or oxytocin on cognitive processes in young healthy subjects. *Peptides* 13, 461–468.

Brunner, R., Schaefer, D., Hess, K., Parzer, P., Resch, F., Schwab, S., 2006. Effect of high-dose cortisol on memory functions. *Ann. N. Y. Acad. Sci.* 1071, 434–437.

Buchan, J., Alberts, S. C., Silk, J. B., Altmann, J., 2003. True paternal care in a multi-male primate society. *Nature* 425, 179–181.

Buchanan, T. W. Lovallo, W. R., 2001. Enhanced memory for emotional material following stress-level cortisol treatment in humans. *Psychoneuroendocrinology*. 26, 307–317.

Bucher, T., Ryan, M., Bartholomew, G., 1982. Oxygen consumption during resting, calling, and nest building in the frog, *Physalaemus pustulosus*. *Physiol. Zool.* 55, 10–22.

Buitelaar, J. K., Huizink, A. C., Mulder, E. J., Robles De Medina, P. G., Visser, G. H., 2003. Pre-natal stress and cognitive development and temperament in infants. *Neurobiol. Aging Suppl.* 1, 24.

Bullen, B. A., Skrinar, G. S., Beitins, I. Z., von Mering, G., Turnbull, B. A., McArthur, J. W., 1985. Induction of menstrual disorders by strenuous exercise in untrained women. *N. Engl. J. Med.* 312, 1349–1353.

Buntin, J. D., 1996. Neural and hormonal control of parental behavior in birds. In: Rosenblatt, J. S., Snowdon, C. T., (Eds.), *Parental Care: Evolution, Mechanisms, and Adaptive Significance*. Academic Press, San Diego, pp. 161–202.

Burger, A. E., Millar, R. P., 1980. Seasonal changes of sexual and territorial behavior and plasma testosterone levels in male lesser sheathbills *(Chionis minor)*. *Z. Tierpsychol.* 52, 397–406.

Burley, N., 1981. Sex ratio manipulation and selection for attractiveness. *Science* 211, 721–722.

Burnham, T. C., Chapman, J. F., Gray, P. B., McIntyre, M. H., Lipson, S. F., Ellison, P. T., 2003. Men in committed, romantic relationships have lower testosterone. *Horm. Behav.* 44, 119–122.

Burton, C. L., Chatterjee, D., Chatterjee-Chakraborty, M., Lovic, V., Grella, S. L., Steiner, M., Fleming, A. S., 2007. Prenatal restraint stress and motherless rearing disrupts expression of plasticity markers and stress-induced corticosterone release in adult female Sprague-Dawley rats. *Brain Res.* 1158, 28–38.

Burton, F. D., 1972. The integration of biology and behavior in the socialization of *Macaca sylvana* of Gibraltar. In: Poirier, F. (Ed.), *Primate Socialization*. Random House, New York, pp. 29–62.

Buss, D. M., 2003. *The Evolution of Desire: Strategies of Human Mating*. Rev. ed. Basic Books, New York.

Butler, T., Pan, H., Epstein, J., Protopopescu, X., Tuescher, O., Goldstein, M., Cloitre, M., Yang, Y., Phelps, E., Gorman, J., Ledoux, J. E., Stern, E., Silbersweig, D., 2005. Fear-related activity in subgenual anterior cingulate differs between men and women. *Neuroreport* 16, 1233–1236.

Buwalda, B., Kole, M. H. P., Veenema, A. H., Huininga, M., De Boer, S. F., Korte, S. M., Koolhas, J. M., 2005. Long term effects of social stress on brain and behavior: A focus on hippocampal functioning. *Neurosci. Biobehav. Rev.* 29, 83–97.

Cachel, S., 2006. *Primate and Human Evolution.* Cambridge University Press, Cambridge.

Calkins, S. D., Fox, N. A., Marshall, T. R., 1996. Behavioral and physiological antecedents of inhibited and uninhibited behavior. *Child Dev.* 67, 523–540.

Cameron, N. M., Champagne, F. A., Parent, C., Fish, E. W., Ozaki-Kuroda, K., Meaney, M. J., 2005. The programming of individual differences in defensive responses and reproductive strategies in the rat through variations in maternal care. *Neurosci. Biobehav. Rev.* 29, 843–865.

Campbell, A., 2006. Sex differences in direct aggression: What are the psychological mediators? *Aggress. Violent Behav.* 11, 237–264.

Campbell, A., Muncer, S., Odber, J., 1997. Aggression and testosterone: Testing a bio-social model. *Aggress. Behav.* 23, 229–238.

Campbell, B. C., 2006. Adrenarche and the evolution of human life history. *Am. J. Hum. Biol.* 18, 569–589.

Campbell, B. C., Gray, P. B., Eisenberg, D. T., Ellison, P. T., Sorenson, M., 2008. Testosterone and androgen receptor CAG repeat length predict variation in body composition among Ariaal men of northern Kenya. *Int. J. Androl.* 2008.

Campbell, B. C., Gray, P. B., Ellison, P. T., 2007. Age-related patterns of body composition and salivary testosterone among Ariaal men of northern Kenya. *Aging Clin. Exp. Res.* 18, 1–7.

Campbell, B. C., Leslie, P. W., 1995. Reproductive ecology of human males. *Yrbk. Phys. Anthropol.* 38, 1–26.

Campbell, B. C., O'Rourke, M. T., Lipson, S. F., 2003. Salivary testosterone and body composition among Ariaal males. *Am. J. Hum. Biol.* 15, 697–708.

Campbell, D. W., Eaton, W. O., 1999. Sex differences in the activity level of infants. *Infant Child Dev.* 8, 1–17.

Campbell, D. W., Eaton, W. O., McKeen, N. A., 2002. Motor activity level and behavioural control in young children. *Int. J. Behav. Dev.* 26, 289–296.

Campbell, K. L., 1994. Blood, urine, saliva and dip-sticks: Experiences in Africa, New Guinea, and Boston. *Ann. N. Y. Acad. Sci.* 709, 312–330.

Canoine, V., Fusani, L., Schlinger, B., Hau, M., 2007. Low sex steroids, high steroid receptors: Increasing the sensitivity of the non-reproductive brain. *J. Neurobiol.* 67, 57–67.

Canoine, V., Gwinner, E., 2002. Seasonal differences in the hormonal control of territorial aggression in free-living European stonechats. *Horm. Behav.* 4, 1–8.

Carani, C., Bancroft, J., Del Rio, G., Granata, A. R. M., Facchinetti, F., Marrama, P., 1990. The endocrine effects of visual erotic stimuli in normal men. *Psychoneuroendocrinology* 15, 207–216.

Carani, C., Granata, A., Fustini, M. F., Marrama, P., 1996. Prolactin and testosterone: Their role in male sexual function. *Int. J. Androl.* 19, 48–54.

Carlson, A. A., Ziegler, T. E., Snowdon, C. T., 1997. Ovarian function of pygmy marmoset daughters *(Cebuella pygmaea)* in intact and motherless families. *Am. J. Primatol.* 43, 347–355.

Carter, C. S., 1992a. Hormonal influences on human sexual behavior. In: Becker, J. B., Breedlove, S. M., Crews, D., (Eds.), *Behavioral Endocrinology*. MIT Press, Cambridge, MA, pp. 131–142.

———, 1992b. Oxytocin and sexual behavior. *Neurosci. Biobehav. Rev.* 16, 131–144.

———, 1998. Neuroendocrine perspectives on social attachment and love. *Psychoneuroendocrinology* 23, 779–818.

———, 2003. Developmental consequences of oxytocin. *Physiol. Behav.* 79, 383–397.

———, 2007. Sex differences in oxytocin and vasopressin: Implications for autism spectrum disorders? *Behav. Brain Res.* 176, 170–186.

Carter, C. S., Altemus, M., Chrousos, G. P., 2001. Neuroendocrine and emotional changes in the postpartum period. *Prog. Brain Res.* 133, 241–249.

Carter, C. S., Bales, K. L., Porges, S. W., 2005. Neuropeptides influence expression of and capacity to form social bonds. *Behav. Brain Sci.* 28, 353–354.

Carter, C. S., DeVries, A. C., Getz, L. L., 1995. Physiological substrates of mammalian monogamy: The prairie vole model. *Neurosci. Biobehav. Rev.* 19, 303–314.

Carter, C. S., DeVries, A. C., Taymans, S. E., Roberts, R. L., Williams, J. R., Chrousos, G. P., 1995. Adrenocorticoid hormones and the development and expression of mammalian monogamy. *Ann. N. Y. Acad. Sci.* 771, 82–91.

Carter, C. S., Keverne, E. B., 2002. The neurobiology of social affiliation and pair bonding. In: Pfaff, D., (Ed.), *Hormones, Brain, and Behavior*. Vol. 1, Academic Press, San Diego, pp. 299–337.

Carter, C. S., Pournajafi-Nazarloo, H., Kramer, K. M., Ziegler, T. E., White-Traut, R., Bello, D., Schwertz, D., 2007. Oxytocin: Behavioral associations and potential as a salivary biomarker. *Ann. N. Y. Acad. Sci.* 1098, 312–322.

Carter, C. S., Roberts, R. L., 1997. The psychobiological basis of cooperative breeding in rodents. In: Solomon, N. G., French, J. A., (Eds.), *Cooperative Breeding in Mammals*. Cambridge University Press, New York, pp. 231–266.

Carter, C. S., Yamamoto, Y., Kramer, K. M., Bales, K. L., Hoffman, G. E., Cushing, B. S., 2003. Early handling produces long-lasting increases in oxytocin and alters the response to separation. *Soc. Neurosci. Abst.* 191, 14.

Cashdan, E., 1995. Hormones, sex, and status in women. *Horm. Behav.* 29, 354–366.

Cavigelli, S. A., Pereira, M. E., 2000. Mating season aggression and fecal testosterone levels in male ring-tailed lemurs *(Lemur catta)*. *Horm. Behav.* 37, 246–255.

Cerda-Molina, A. L., Hernández-López, L., Chavira, R., Cárdenas, M., Paez-Ponce, D., Cervantes-De la Luz, H., Mondragón-Ceballos, R., 2006. Endocrine changes in male stumptailed macaques *(Macaca arctoides)* as a response to odor stimulation with vaginal secretions. *Horm. Behav.* 49, 81–87.

Cerqueira, J. J., Pego, J. M., Taipa, R., Bessa, J. M., Almeida, O. F., Sousa, N., 2005. Morphological correlates of corticosteroid-induced changes in prefrontal cortex-dependent behaviors. *J. Neurosci.* 25, 7792–7800.

Champagne, F. A., Diorio, J., Sharma, S., Meaney, M. J., 2001. Naturally occurring variations in maternal behavior in the rat are associated with differences in estrogen-inducible central oxytocin receptors. *Proc. Natl. Acad. Sci. U. S. A.* 98, 12736–12741.

Champagne, F. A., Meaney, M., 2001. Like mother, like daughter: Evidence for non-genomic transmission of parental behavior and stress responsivity. *Prog. Brain Res.* 133, 287–302.

Champagne, F. A., Weaver, I. C. G., Diorio, J., Sharma, S., Meaney, M. J., 2003. Natural variations in maternal care are associated with estrogen receptor alpha expression and estrogen sensitivity in the medial preoptic area. *Endocrinology* 144, 4720–4724.

Chang, C. L., Hsu, S. Y., 2004. Ancient evolution of stress-regulating peptides in vertebrates. *Peptides* 25, 1681–1688.

Charnov, E. L., 2002. Reproductive effort, offspring size and benefit-cost ratios in the classification of life histories. *Evol. Ecol. Res.* 4, 749–758.

Chen, J. Y., Simpson, J. A., Rholes, W. S., in prep. Oxytocin and social support in romantic relationships: An attachment perspective.

Cheney, D. L., Seyfarth, R. M., Fischer, J., Beehner, J., Bergman, T., Johnson, S. E., Kitchen, D. M., Palombit, R. A., Rendall, D., Silk, J. B., 2004. Factors affecting reproduction and mortality among baboons in the Okavango Delta, Botswana. *Int. J. Primatol.* 25, 401–428.

Chepko-Sade, B. D., Halpin, Z. T., 1987. *Mammalian Dispersal Patterns.* University of Chicago Press, Chicago.

Cheyne, S. M., Chivers, D. J., 2006. Sexual swellings of female gibbons. *Folia Primatol.* 77, 345–352.

Chin, S. O., Brodsky, N. L., Bhandari, V., 2003. Antenatal steroid use is associated with increased gastroesophageal reflux in neonates. *Am. J. Perinatol.* 20, 205–213.

Chisholm, J. S., 1999. *Death, Hope and Sex: Steps to an Evolutionary Ecology of Mind and Morality.* Cambridge University Press, Cambridge.

Chisholm, J. S., Burbank, V. K., Coall, D. A., Gemmiti, F., 2005. Early stress: Perspectives from developmental evolutionary ecology. In: Ellis, B. J., Bjorklund, D. F., (Eds.), *Origins of the Social Mind: Evolutionary Psychology and Child Development.* Guilford Press, New York, pp. 76–107.

Chivers, D. J., 1974. The siamang in Malaya: A field study of a primate in tropical rain forest. *Contrib. Primatol.* 4, 1–335.

Cho, M. M., DeVries, A. C., Williams, J. R., Carter, C. S., 1999. The effects of oxytocin and vasopressin on partner preferences in male and female prairie voles *(Microtus ochrogaster). Behav. Neurosci.* 113, 1071–1079.

Choi, P. Y. L., Pope, H. G., Jr., 1994. Violence toward women and illicit androgenic-anabolic steroid use. *Ann. Clin. Psychiatry* 6, 21–25.

Choong, C. S., Kemppainen, J. A., Wilson, E. M., 1998. Evolution of the primate androgen receptor: A structural basis for disease. *J. Mol. Evol.* 47, 334–342.

Christensson, K., Nilsonn, B. A., Stock, S., Matthiesen A. S., Unväs-Moberg, K., 1989. Effect of nipple stimulation on uterine activity and on plasma levels of oxytocin in full term, healthy, pregnant women. *Acta Obstet. Gynecol. Scand.* 68, 205–210.

Christiansen, K., 2002. Anthropology of human reproduction: The male factor. *Evol. Anthropol.* 11, 200–203.

Christiansen, K., Winkler, E., 1992. Hormonal, anthropometrical, and behavioral correlates of physical aggression in !Kung San men of Namibia. *Aggress. Behav.* 18, 271–280.

Clark, A. S., Kelton, M. C., Guarraci, F. A., Clyons, E. Q., 2004. Hormonal status and test condition, but not sexual experience, modulate partner preference in female rats. *Horm. Behav.* 45, 314–323.

Clarke, A. S., Kammerer, C. M., George, K. P., Kupfer, D. J., McKinney, W. T., Spence, M. A., Kraemer, G. W., 1995. Evidence for heritability of biogenic amine levels in the cerebrospinal fluid of rhesus monkeys. *Biol. Psychiatry* 38, 572–577.

Clarke, A. S., Wittwer, D. J., Abbott, D. H., Schneider, M. L., 1994. Long-term effects of pre-natal stress on HPA axis activity in juvenile rhesus monkeys. *Dev. Psychobiol.* 27, 257–269.

Clarke, M. R., 1983. Infant-killing and infant disappearance following male takeovers in a group of free-ranging howling monkeys *(Alouatta palliata)* in Costa Rica. *Am. J. Primatol.* 5, 241–247.

Clarke, M. R., Kaplan, J. R., Bumstede, P. R., Koritnik, D. R., 1986. Social dominance and serum testosterone concentration in dyads of male *Macaca fascicularis. J. Med. Primatol.* 15, 419–432.

Clarke-Stewart, K. A., Allhusen, V. D., 2002. Nonparental caregiving. In: Bornstein, M. H., (Ed.), *Handbook of Parenting.* 2nd ed., Vol. 3. Erlbaum, Mahwah, NJ, pp. 215–252.

Cleveland, H. H., Udry, J. R., Chantala, K., 2001. Environmental and genetic influences on sex-typed behaviors and attitudes of male and female adolescents. *Pers. Soc. Psychol. Bull.* 27, 1587–1598.

Clinton, J. F., 1986. Expectant fathers at risk for couvade. *Nurs. Res.* 35, 290–295.

Clutton-Brock, T. H., 1982. Sons and daughters. *Nature* 298, 11–13.

———, (Ed.), 1988. *Reproductive Success: Studies of Individual Variation in Contrasting Breeding Systems.* University of Chicago Press, Chicago.

———, 1989. Female transfer, male tenure and inbreeding avoidance in social mammals. *Nature* 337, 70–72.

———, 1991 *The Evolution of Parental Care.* Princeton University Press, Princeton, NJ.

———, 1994. The costs of sex. In: Short, R. V., Balaban, E., (Eds.), *The Difference between the Sexes.* Cambridge University Press, Cambridge, pp. 347–362.

———, 2002. Breeding together: Kin selection and mutualism in cooperative vertebrates. *Science* 296, 69–72.

———, 2004. What is sexual selection? In: Kappeler, P. M., van Schaik, C. P., (Eds.), *Sexual Selection in Primates.* Cambridge University Press, Cambridge, pp. 24–36.

Clutton-Brock, T. H., Brotherton, P., O'Riain, M., Griffin, A., Gaynor, D., Kansky, R., Sharpe, L., McIlrath, G., 2001. Contributions to co-operative rearing in meerkats. *Anim. Behav.* 61, 705–710.

Coccaro, E., Kavoussi, R., Hauger, R., Cooper, T., Ferris, C., 1998. Cerebrospinal fluid vasopressin levels: Correlates with aggression and serotonin function in personality-disordered subjects. *Arch. Gen. Psychiatry* 55, 708–714.

Cohan, C. L., Booth, A., Granger, D. A., 2003. Gender moderates the relationship between testosterone and marital interaction. *J. Fam. Psychol.* 17, 29–40.

Cohen-Bendahan, C. C. C., van de Beek, C., Berenbaum, S. A., 2005. Prenatal sex hormone effects on child and adult sex-typed behavior: Methods and findings. *Neurosci. Biobehav. Rev.* 29, 353–384.

Cohen-Bendahan, C. C. C., van Goozen, S. H., Buitelaar, J. K., Cohen-Kettenis, P. T., 2005. Maternal serum steroid levels are unrelated to fetal sex: A study in twin pregnancies. *Twin Res. Hum. Genet.* 8, 173–177.

Cohn, J. F., Campbell, S. B., Matias, R., Hopkins, J., 1990. Face-to-face interactions of postpartum depressed and nondepressed mother-infant pairs at 2 months. *Dev. Psychol.* 35, 119–123.

Coirini, H., Johnson, A. E., McEwen, B. S., 1989. Estradiol modulation of oxytocin binding in the ventromedial hypothalamic nucleus of male and female rats. *Neuroendocrinology* 50, 193–198.

Coirini, H., Johnson, A. E., Schumacher, M., McEwen, B. S., 1992. Sex differences in the regulation of oxytocin receptors by ovarian steroids in the ventromedial hypothalamus of the rat. *Neuroendocrinology* 55, 269–275.

Coirini, H., Schumacher, M., Flanagan, L. M., McEwen, B. S., 1991. Transport of estrogen-induced oxytocin receptors in the ventromedial hypothalamus. *J. Neurosci.* 11, 3317–3324.

Collins, S. A., 2000. Men's voices and women's choices. *Anim. Behav.* 60, 773–780.

Collins, S. A., Missing, C., 2003. Vocal and visual attractiveness are related in women. *Anim. Behav.* 65, 997–1004.

Coltman, D. W., 2005. Differentiation by dispersal: Evolutionary genetics. *Nature* 433, 23–24.

Coltman, D. W., Festa-Bianchet, M., Jorgenson, J. T., Strobeck, C., 2002. Age-dependent sexual selection in bighorn rams. *Proc. R. Soc. Lond. B Biol. Sci.* 269, 165–172.

Connor, R. C., Wells, R., Mann, J., Read, A., 2000. The bottlenose dolphin, *Tursiops* spp.: Social relationships in a fission-fusion society. In: Mann, J., Connor, R. C., Tyack, P. L., Whitehead, H., (Eds.), *Cetacean Societies: Field Studies of Whales and Dolphins.* University of Chicago Press, Chicago, pp. 91–126.

Consiglio, A. R., Borsoi, A., Pereira, G. A., Lucion, A. B., 2005. Effects of oxytocin microinjected into the central amygdaloid nucleus and bed nucleus of the stria terminalis on maternal aggressive behavior in rats. *Physiol. Behav.* 85, 354–362.

Constable, J. L., Ashley, M. V., Goodall, J., Pusey, A. E., 2001. Noninvasive paternity assignment in Gombe chimpanzees. *Mol. Ecol.* 10, 1279–1300.

Cooper, P. J., Murray, L., 1995. The course and recurrence of postnatal depression: Evidence for the specificity of the diagnostic concept. *Br. J. Psychiatry* 166, 191–195.

Corter, C., Fleming, A. S., 1995. Psychobiology of maternal behavior in human beings. In: Bornstein, M., (Ed.), *Handbook of Parenting: Biology and Ecology of Parenting*. Erlbaum, Mahwah, NJ, pp. 87–116.

———, 2002. Psychobiology of maternal behavior in human beings. In: Bornstein, M. H., (Ed.), *Handbook of Parenting: Biology and Ecology of Parenting*. 2nd ed. Erlbaum, Mahwah, NJ, pp. 141–182.

Cote, L., Bornstein, M., 2005. Child and mother play in cultures of origin, acculturating cultures, and cultures of destination. *Int. J. Behav. Dev.* 29, 479–488.

Coulter, C. L., Jaffe, R. B., 1998. Functional maturation of the primate fetal adrenal *in vivo*: 3. Specific zonal localization and developmental regulation of CYP21A1 (P450c21) and CYP11B1/CYP11B2 (P450c11/aldosterone synthase) lead to integrated concept of zonal and temporal steroid biosynthesis. *Endocrinology* 139, 5144–5150.

Couzinet, B., Schaison, G., 1993. The control of gonadotrophin secretion by ovarian steroids. *Hum. Reprod.* 8, 97–101.

Crawford, M., 2001. Gender and language. In: Unger, R. K., (Ed.), *Handbook of the Psychology of Women and Gender*. Wiley, New York, pp. 228–244.

Creel, S., 2001. Social stress and dominance. *Trends Ecol. Evol.* 16, 491–497.

———, 2005. Dominance, aggression, and glucocorticoid levels in social carnivores. *J. Mammal.* 86, 255–264.

Creel, S., Creel, N. M., Monfort, S. L., 1996. Social stress and dominance. *Nature* 379, 212.

Creutz, L. M., Kritzer, M. F., 2004. Mesostrial and mesolimbic projections of midbrain neurons immunoreactive for estrogen receptor beta or androgen receptors in rats. *J. Comp. Neurol.* 476, 348–362.

Crews, D., 1987. Diversity and evolution of behavioral controlling mechanisms. In: Crews, D., (Ed.), *Psychobiology of Reproductive Behavior: An Evolutionary Perspective*. Prentice Hall, Englewood Cliffs, NJ, pp. 88–119.

Cristóbal-Azkarate, J., Chavira, R., Boeck, L., Rodríguez-Luna, E., Veà, J. J., 2006. Testosterone levels of free-ranging resident mantled howler monkey males in relation to the number and density of solitary males: A test of the challenge hypothesis. *Horm. Behav.* 49, 261–267.

Crockett, C. M., Pope, T. R., 1993. Consequences of sex differences in dispersal for juvenile red howler monkeys. In: Pereira, M. E., Fairbanks, L. A., (Eds.), *Juvenile Primates*. Oxford University Press, New York, pp. 104–118.

Crofoot, M. C., Knott, C. D., in press. What we do and do not know about orangutan male dimorphism. In: Galdikas, B. M. F., Briggs, N., Sheeran, L. K., Shapiro, G. L., (Eds.), *Great and Small Apes of the World*. Springer, New York.

Crowley, W., Rodriguez-Sierra, J., Komisaruk, B., 1977. Analgesia induced by vaginal stimulation in rats is apparently independent of a morphine-sensitive process. *Psychopharmacology* 54, 223–225.

Cunningham, M. R., 1986. Measuring the physical in physical attractiveness: Quasiexperiments on the sociobiology of female facial beauty. *J. Pers. Soc. Psychol.* 50, 925–935.

Curley, J. P., Keverne, E. B., 2005. Genes, brains and mammalian social bonds. *Trends Ecol. Evol.* 20, 561–567.

Cushing, B. S., Kramer, K. M., 2005. Mechanisms underlying epigenetic effects of early social experience: The role of neuropeptides and steroids. *Neurosci. Biobehav. Rev.* 29, 1089–1105.

Cutler, W. B., Preti, G., Krieger, A., Huggins, G. R., Garcia, C. R., Lawley, H. J., 1986. Human axillary secretions influence women's menstrual cycles: The role of donor extract from men. *Horm. Behav.* 20, 463–473.

Czekala, N. M., Benirschke, K., McClure, H. M., Lasley, B. L., 1983. Urinary estrogen excretion during pregnancy in the gorilla *(Gorilla gorilla),* orangutan *(Pongo pygmaeus),* and the human *(Homo sapiens). Biol. Reprod.* 28, 289–294.

Czekala, N. M., Shideler, S. E., Lasley, B. L., 1988. Comparisons of female reproductive hormone patterns in the hominoids. In: Schwartz, J. H., (Ed.), *Orang-utan Biology.* Oxford University Press, New York, pp. 117–122.

da Silva Mota, M. T., Franci, C. R., de Sousa, M. B., 2006. Hormonal changes related to paternal and alloparental care in common marmosets *(Callithrix jacchus). Horm. Behav.* 49, 293–302.

Daan, S., Tinbergen, J. M., 1997. Adaptation of life histories. In: Krebs, J. R., Davies, N. B., (Eds.), *Behavioural Ecology.* 4th ed. Blackwell, Oxford, pp. 311–333.

Dabbs, J. M., Jr., 1990a. Age and seasonal variation in serum testosterone concentrations among men. *Chronobiol. Int.* 7, 245–249.

———, 1990b. Salivary testosterone measurements: Reliability across hours, days, weeks. *Physiol. Behav.* 48, 83–86.

———, 1992. Testosterone and occupational achievement. *Soc. Forces* 70, 813–824.

———, 1997. Testosterone, smiling, and facial appearance. *J. Nonverbal Behav.* 21, 45–55.

Dabbs, J. M., Jr., Carr, T. S., Prady, R. L., Riad, J. K., 1995. Testosterone, crime, and misbehavior among 692 male prison inmates. *Pers. Ind. Diff.* 18, 627–633.

Dabbs, J. M., Dabbs, M. G., 2000. *Heroes, Rogues, and Lovers: Testosterone and Behavior.* McGraw-Hill, New York.

Dabbs, J. M., Hopper, C. H., Jurkovic, G. J., 1991. Testosterone and personality among college students and military veterans. *Pers. Ind. Diff.* 11, 1263–1269.

Dabbs, J. M., Mallinger, A., 1999. High testosterone levels predict low voice pitch among men. *Pers. Ind. Diff.* 27, 801–804.

Dahl, J. F., Gould, K. G., Nadler, R. D., 1993. Testicle size of orang-utans in relation to body size. *Am. J. Phys. Anthropol.* 90, 229–236.

Dahl, J. F., Nadler, R. D., 1992. Genital swelling in females of the monogamous gibbon, *Hylobates (H.) lar. Am. J. Phys. Anthropol.* 89, 101–108.

Dahl, J. F., Nadler, R. D., Collins, D. C., 1991. Monitoring the ovarian cycles of *Pan troglodytes* and *P. paniscus*: A comparative approach. *Am. J. Primatol.* 24, 195–209.

Daitzman, R. J., Zuckerman, M., 1980. Disinhibitory sensation seeking, personality, and gonadal hormones. *Pers. Ind. Diff.* 1, 103–110.

Daitzman, R. J., Zuckerman, M., Sammelwitz, P., Ganjam, V., 1978. Sensation seeking and gonadal hormones. *J. Biosoc. Sci.* 10, 401–408.

Dal Zatto, S., Marti, O., Armario, A., 2003. Glucocorticoids are involved in the long-term effects of a single immobilization stress on the hypothalamic-pituitary-adrenal axis. *Psychoneuroendocrinology,* 28, 992–1009.

Daly, M., Wilson, M., 1983. *Sex, Evolution, and Behavior.* Wadsworth, Belmont, CA.

———, 1988. *Homicide.* Aldine de Gruyter, Hawthorne, NY.

———, 1999. Darwinism and the roots of machismo. *Sci. Am. Pres.* 10, 8–14.

Dantzer, R., 2000. Cytokine-induced sickness behavior: Where do we stand? *Brain Behav. Immun.* 10, 1–18.

Darwin, C., 1871. *The Descent of Man, and Selection in Relation to Sex.* John Murray, London. Facsimile Princeton University Press ed. Princeton University Press, Princeton, NJ, 1981.

Davies, N. B., 1989. Sexual conflict and the polygyny threshold. *Anim. Behav.* 38, 226–234.

Dawson, A., King, V. M., Bentley, G. E., Ball, G. F., 2001. Photoperiodic control of seasonality in birds. *J. Biol. Rhythms* 16, 365–380.

de Bree, F. M., Burbach, J. P., 1998. Structure-function relationships of the vasopressin prohormone domains. *Cell Mol. Neurobiol.* 18, 173–191.

de Bruin, E. I., Verheij, F., Wiegman, T., Ferdinand, R. F., 2006. Differences in finger length ratio between males with autism, pervasive developmental disorder–not otherwise specified, ADHD, and anxiety disorders. *Dev. Med. Child Neurol.* 48, 962–965.

De Clercq, B., De Fruyt, F., Van Leeuwen, K., Mervielde, I., 2006. The structure of maladaptive personality traits in childhood: A step toward an integrative developmental perspective for DSM-V. *J. Abnorm. Psychol.* 115, 639–657.

de Kloet, E. R., Oitzl, M. S., Joels, M., 1999. Stress and cognition: Are corticosteroids good or bad guys? *Trends Neurosci.* 22, 422–426.

de Kloet, E. R., Sibug, R. M., Helmerhorst, F. M., Schmidt, M., 2005. Stress, genes, and the mechanism for programming the brain for later life. *Neurosci. Biobehav. Rev.* 29, 271–281.

de Kloet, E. R., Voorhuis, D. A., Boschma, Y., Elands, J., 1986. Estradiol modulates density of putative "oxytocin receptors" in discrete rat brain regions. *Neuroendocrinology* 44, 415–421.

de Lathouwers, M., van Elsacker, L., 2005. Reproductive parameters of female *Pan paniscus* and *P. troglodytes:* Quality versus quantity. *Int. J. Primatol.* 26, 55–71.

de Rooij, S. R., Painter, R. C., Phillips, D. I., Osmond, C., Michels, R. P., Bossuyt, P. M., Bleker, O. P., Roseboom, T. J., 2006. Hypothalamic-pituitary-adrenal axis activity in adults who were pre-natally exposed to the Dutch famine. *Eur. J. Endocrinol.* 55, 153–160.

de Ruiter, J. R., van Hooff, J. A. R. A. M., 1993. Male dominance rank and reproductive success in primate groups. *Primates* 34, 513–523.

de Vos, G. J., 1983. Social behaviour of black grouse: An observational and experimental field study. *Ardea* 71, 1–103.

De Vries, G. J., Miller, M. A., 1998. Anatomy and function of extrahypothalamic vasopressin systems in the brain. *Prog. Brain Res.* 119, 3–20.

De Vries, G. J., Simerly, R. B., 2002. Anatomy, development, and function of sexually dimorphic neural circuits in the mammalian brain. In: Pfaff, D. W., (Ed.), *Hormones, Brain, and Behavior.* Vol. 4. Academic Press, San Diego, pp. 137–192.

de Weerth, C., Buitelaar, J. K., 2005. Physiological stress reactivity in human pregnancy—a review. *Neurosci. Biobehav. Rev.* 29, 295–312.

Deater-Deckard, K., Atzaba-Poria, N., Pike, A., 2004. Mother- and father-child mutuality in Anglo and Indian British families: A link with lower externalizing behaviors. *J. Abnorm. Child Psychol.* 32, 609–620.

DeBold, C. R., Sheldon, W. R., DeCherney, G. S., Jackson, R. V., Alexander, A. N., Vale, W., Rivier, J., Orth, D. N., 1984. Arginine vasopressin potentiates adrenocorticotropin release induced by ovine corticotropin-releasing factor. *J. Clin. Invest.* 73, 533–538.

Delbende, C., Delarue, C., Lefebvre, H., Bunel, D. T., Szafarczyk, A., Mocaer, E., Kamoun, A., Jegou, S., Vaudry, H., 1992. Glucocorticoids, transmitters and stress. *Br. J. Psychiatry Suppl.* 15, 24–35.

Delgado, R., 2006. Sexual selection in the loud calls of male primates: Signal content and function. *Int. J. Primatol.* 27, 5–25.

Delville, Y., Mansour, K., Ferris, C., 1996. Testosterone facilitates aggression by modulating vasopressin receptors in the hypothalamus. *Physiol. Behav.* 60, 25–29.

Dempsey, E. W., Hertz, R., Young, W. C., 1936. The experimental induction of oestrus (sexual receptivity) in the normal and ovariectomized guinea pig. *Am. J. Physiol.* 116, 201–209.

Depue, R. A., Collins, P. F., 1999. Neurobiology of the structure of personality, dopamine, facilitation of incentive motivation, and extraversion. *Behav. Brain Sci.* 22, 491–517.

Depue, R. A., Morrone-Strupinsky, J. V., 2005. A neurobehavioral model of affiliative bonding: Implications for conceptualizing a human trait of affiliation. *Behav. Brain Sci.* 28, 313–350.

Deschner, T., Heistermann, M., Hodges, K., Boesch, C., 2003. Timing and probability of ovulation in relation to sex skin swelling in wild West African chimpanzees, *Pan troglodytes verus. Anim. Behav.* 66, 551–560.

———, 2004. Female sexual swelling size, timing of ovulation, and male behavior in wild West African chimpanzees. *Horm. Behav.* 46, 204–215.

Detillion, C., Craft, T., Glasper, E., Pendergast, B., DeVries, A., 2004. Social facilitation of wound healing. *Psychoneuroendocrinology* 29, 1004–1011.

Deutsch, F. M., Ruble, D. N., Fleming, A. S., Brooks-Gunn, J., Stangor, C., 1988. Information-seeking and maternal self-definition during the transition to motherhood. *J. Pers. Soc. Psychol.* 55, 420–431.

Deutsch, J. C., 1994. Lekking by default: Female habitat preferences and male strategies in Uganda kob. *J. Anim. Ecol.* 63, 101–115.

DeVries, A. C., Carter, C. S., 1999. Sex differences in temporal parameters of partner preference in prairie voles *(Microtus ochrogaster)*. *Can. J. Zool.* 77, 885–889.

DeVries, A. C., DeVries, M. B., Taymans, S., Carter, C. S., 1995. The modulation of pair bonding by corticosterone in female prairie voles *(Microtus ochrogaster)*. *Proc. Natl. Acad. Sci. U. S. A.* 92, 7744–7748.

———, 1996. The effects of stress on social preferences are sexually dimorphic in prairie voles. *Proc. Natl. Acad. Sci. U. S. A.* 93, 11980–11984.

DeVries, A. C., Guptaa, T., Cardillo, S., Cho, M., Carter, C. S., 2002. Corticotropin-releasing factor induces social preferences in male prairie voles. *Psychoneuroendocrinology* 27, 705–714.

Dewsbury, D. A., 1982. Ejaculate cost and mate choice. *Am. Nat.* 119, 601–610.

Diamond, A., 1985. Development of the ability to use recall to guide action, as indicated by infants' performance on AB. *Child Dev.* 56, 868–883.

———, 1988. Abilities and neural mechanisms underlying AB performance. *Child Dev.* 59, 523–527.

Diamond, A., Gilbert, J., 1989. Development as progressive inhibitory control of action-retrieval of a contiguous object. *Cogn. Dev.* 4, 223–249.

Diamond, L. M., 2003. What does sexual orientation orient? A biobehavioral model distinguishing romantic love and sexual desire. *Psychol. Rev.* 110, 173–192.

Dickerson, S. S., Kemeny, M. E., 2004. Acute stressors and cortisol responses: A theoretical integration and synthesis of laboratory research. *Psychol. Bull.* 130, 355–391.

Diego, M. A., Field, T., Jones, N. A., Hernandez-Reif, M., 2006. Withdrawn and intrusive maternal interaction style and infant frontal EEG asymmetry shifts in infants of depressed and non-depressed mothers. *Infant Behav. Dev.* 29, 220–229.

DiFiore, A., Rendall, D., 1994. Evolution of social organization: A reappraisal for primates by using phylogenetic methods. *Proc. Natl. Acad. Sci. U. S. A.* 91, 9941–9945.

DiPietro, J. A., Bornstein, M. H., Costigan, K. A., Pressman, E. K., Hahn, C. S., Painter, K., Smith, B. A., Yi, L. J., 2002. What does fetal movement predict about behavior during the first two years of life? *Dev. Psychobiol.* 40, 358–371.

DiPietro, J. A., Novak, M. F., Costigan, K. A., Atella, L. D., Reusing, S. P., 2006. Maternal psychological distress during pregnancy in relation to child development at age two. *Child. Dev.* 77, 573–587.

Dittmann, R. W., Kappes, M. E., Kappes, M. H., 1992. Sexual behaviour in adolescent and adult females with congenital adrenal hyperplasia. *Psychoneuroendocrinology* 17, 153–170.

Diver, M. J., Imtiaz, K. E., Ahmad, A. M., Vora, J. P., Fraser, W. D., 2003. Diurnal rhythms of serum total, free and bioavailable testosterone and of SHBG in middle-aged men compared to those in young men. *Clin. Endocrinol.* 58, 710–717.

Dixson, A. F., 1987. Observations on the evolution of the genitalia and copulatory behaviour in male primates. *J. Zool. (Lond.)* 213, 423–443.

———, 1998. *Primate Sexuality: Comparative Studies of the Prosimians, Monkeys, Apes, and Human Beings.* Oxford University Press, Oxford.

Dixson, A. F., Everitt, G. J., Herbert, J., Rugman, S. M., Scruton, D. M., 1973. Hormonal and other determinants of sexual attractiveness and receptivity in rhesus and talapoin monkeys. In: Phoenix, C. H., (Ed.), *Symposia of the IVth International Congress of Primatology.* Vol. 2, *Primate Reproductive Behavior.* Karger, Basel, pp. 36–63.

Dixson, A. F., George, L., 1982. Prolactin and parental behaviour in a male New World primate. *Nature* 299, 551–553.

Dixson, A. F., Halliwell, G., East, R., Wignarajah, P., Anderson, M. J., 2003. Masculine somatotype and hirsuteness as determinants of sexual attractiveness to women. *Arch. Sex. Behav.* 32, 29–39.

Dixson, A. F., Nevison, C. M., 1997. The socioendocrinology of adolescent development in male rhesus monkeys *(Macaca mulatta). Horm. Behav.* 31, 126–135.

Domb, L. G., Pagel, M., 2001. Sexual swellings advertise female quality in wild baboons. *Nature* 410, 204–206.

DonCarlos, L. L., Sarkey, S., Lorenz, B., Azcoitia, I., Garcia-Ovejero, D., Huppenbauer, C., Garcia-Segura, L. M., 2006. Novel cellular phenotypes and subcellular sites for androgen action in the forebrain. *Neuroscience* 138, 801–807.

Double, M. C., Peakall, R., Beck, L. N. R., Cockburn, A., 2005. Dispersal, philopatry, and infidelity: Dissecting local genetic structure in superb fairywrens *(Malurus cyaneus). Evolution* 59, 625–635.

Draper, P., 1992. Room to maneuver: !Kung women cope with men. In: Counts, D., Brown, J. K., Campbell, J., (Eds.), *Sanctions and Sanctuary: Cultural Perspectives on the Beating of Wives.* Westview Press, Boulder, CO, pp. 43–63.

Drent, R. H., Daan, S., 1981. The prudent parent: Energetic adjustments in avian breeding. *Ardea* 68, 225–252.

Dressler, W. W., Bindon, J. R., 2000. The health consequences of cultural dissonance: Cultural dimensions of lifestyle, social support, and arterial blood pressure in an African American community. *Am. Anthropol.* 102, 244–260.

Dunbar, R. I. M., 1984. *Reproductive Decisions.* Princeton University Press, Princeton, NJ.

———, 1995. The mating system of callitrichid primates: I. Conditions for the co-evolution of paired bonding and twinning. *Anim. Behav.* 50, 1057–1070.

———, 1998. The social brain hypothesis. *Evol. Anthropol.* 6, 178–190.

Duncan, G., Kalil, A., Mayer, S., Tepper, R., Payne, M. R., 2005. The apple does not fall from the Tree. In: Bowles, S., Gintis, H., Osborne Groves, M., (Eds.), *Unequal Chances: Family Background and Economic Success.* Princeton University Press, pp. 23–79.

Dunn, J., 2004. Understanding children's family worlds: Family transitions and children's outcome. *Merrill-Palmer Q.* 50, 224–235.

Eaton, G. G., Resko, J. A., 1974. Plasma testosterone and male dominance in a Japanese macaque *(Macaca fuscata)* troop compared with repeated measures of testosterone in laboratory males. *Horm. Behav.* 5, 251–259.

Eaton, S. B., Pike, M. C., Short, R. V., Lee, N. C., Trussell, J., Hatcher, R. A., Wood, J. W., Worthman, C. M., Blurton Jones, Konner, M. J., Hill, K. R., Bailey, R., Hurtado, A. M., 1994. Women's reproductive cancers in evolutionary context. *Q. Rev. Biol.* 69, 353–367.

Eaton, W. O., Saudino, K. J., 1992. Prenatal activity level as a temperament dimension? Individual differences and developmental functions in fetal movement. *Infant Behav. Dev.* 15, 57–70.

Eaton, W. O., Yu, A. P., 1989. Are sex-differences in child motor-activity level a function of sex-differences in maturational status? *Child Dev.* 60, 1005–1011.

Ebner, K., Bosch, O. J., Kromer, S. A., Singewald, N., Neumann, I. D., 2005. Release of oxytocin in the rat central amygdala modulates stress-coping behavior and the release of excitatory amino acids. *Neuropsychopharmacology* 30, 223–230.

Ebner, K., Wotjak, C. T., Landgraf, R., Engelmann, M., 2005. Neuroendocrine and behavioral response to social confrontation: Residents versus intruders, active versus passive coping styles. *Horm. Behav.* 47, 14–21.

Edwards, D. A., Pfeifle, J. K., 1983. Hormonal control of receptivity, proceptivity, and sexual motivation. *Physiol. Behav.* 30, 437–443.

Edwards, D. A., Wetzel, K., Wyner, D. R., 2006. Intercollegiate soccer: Saliva cortisol and testosterone are elevated during competition, and testosterone is related to status and social connectedness with team mates. *Physiol. Behav.* 87, 135–143.

Ehrhardt, A. A., Meyer-Bahlburg, H. F., Rosen, L. R., Feldman, J. F., Veridiano, M. P., Zimmerman, M. J., McEwen, B. S., 1985. Sexual orientation after prenatal exposure to exogenous estrogen. *Arch. Sex. Behav.* 14, 57–77.

Elias, M. F., 1981. Serum cortisol, testosterone, and testosterone-binding globulin responses to competitive fighting in human males. *Aggress Behav.* 7, 215–224.

Elias, M. F., Teas, J., Johnston, J., Bora, C., 1986. Nursing practices and lactation amenorrhoea. *J. Biosoc. Sci.* 18, 1–10.

Ellefson, J. O., 1974. A natural history of white-handed gibbons in the Malayan Peninsula. In: Rumbaugh, D. M., (Ed.), *Gibbon and Siamang*. Vol. 3, *Natural History, Social Behavior, Reproduction, Vocalizations, Prehension*. Karger, Basel, pp. 1–136.

Ellis, B. J., Essex, M. J., Boyce, W. T., 2005. Biological sensitivity to context: II. Empirical explorations of an evolutionary-developmental theory. *Dev. Psychopathol.* 17, 303–328.

Ellis, B. J., Jackson, J. J., Boyce, W. T., 2006. The stress response systems: Universality and adaptive individual differences. *Devel. Rev.* 26. 175–212.

Ellison, P. T., 1988. Human salivary steroids: Methodological considerations and applications in physical anthropology. *Yrbk. Phys. Anthropol.* 31, 115–142.

———, 1994. Salivary steroids and natural variation in human ovarian function. *Ann. N. Y. Acad. Sci.* 709, 287–298.

———, 1995. Breastfeeding, fertility, and maternal condition. In: Dettwyler, K. A., Stuart-Macadam, P., (Eds.), *Breastfeeding: Biocultural Perspectives.* Aldine de Gruyter, Hawthorne, NY, pp. 305–345.

———, 2001. *On Fertile Ground: A Natural History of Reproduction.* Harvard University Press, Cambridge, MA.

———, 2003. Energetics and reproductive effort. *Am. J. Hum. Biol.* 15, 342–351.

———, 2005. Evolutionary perspectives on the fetal origins hypothesis. *Am. J. Hum. Biol.* 17, 113–118.

Ellison, P. T., Bribiescas, R. G., Bentley, G. R., Campbell, B. C., Lipson, S. F., Panter-Brick, C., Hill, K., 2002. Population variation in age-related decline in male salivary testosterone. *Hum. Reprod.* 17, 3251–3253.

Ellison, P. T., Lager, C., 1985. Exercise-induced menstrual disorders. *N. Engl. J. Med.* 313, 825–826.

Ellison, P. T., Lipson, S. F., Jasianska, G., Ellison, P. L., 2007. Moderate anxiety, whether acute or chronic, is not associated with ovarian suppression in healthy, well-nourished, Western women. *Am. J. Phys. Anthropol.* 134, 513–519.

Ellison, P. T., Panter-Brick, C., Lipson, S. F., O'Rourke, M. T., 1993. The ecological context of human ovarian function. *Hum. Reprod.* 8, 2248–2258.

Ellison, P. T., Peacock, N. R., Lager, C., 1986. Salivary progesterone and luteal function in two low-fertility populations of northeast Zaire. *Hum. Biol.* 58, 473–483.

———, 1989. Ecology and ovarian function among Lese females of the Ituri Forest, Zaire. *Am. J. Phys. Anthrop.* 78. 519–526.

Ellison, P. T., Valeggia, C. R., 2003. C-peptide levels and the duration of lactational amenorrhea. *Fertil. Steril.* 80, 1279–1280.

Else-Quest, N. M., Hyde, J. S., Goldsmith, H. H., Van Hulle, C. A., 2006. Gender differences in temperament: A meta-analysis. *Psychol. Bull.* 132, 33–72.

Emerson, S. B., 2001. Male advertisement calls: Behavioral variation and physiological processes. In: Ryan, M. J., (Ed.), *Anuran Communication.* Smithsonian Institution Press, Washington, DC, pp. 36–44.

Emerson, S. B., Hess, D. L., 2001. Glucocorticoids, androgens, testis mass, and the energetics of vocalization in breeding male frogs. *Horm. Behav.* 39, 59–69.

Emery, M. A., Whitten, P. L., 2003. Size of sexual swellings reflects ovarian function in chimpanzees *(Pan troglodytes). Behav. Ecol. Sociobiol.* 54, 340–351.

———, n.d. Unpublished data.

Emery Thompson, M., 2005. Reproductive endocrinology of wild female chimpanzees *(Pan troglodytes schweinfurthii):* Methodological considerations and the role of hormones in sex and conception. *Am. J. Primatol.* 67, 137–158.

Emery Thompson, M., Wrangham, R. W., 2008. Diet and reproduction in wild chimpanzees *(Pan troglodytes schweinfurthii)* in Kibale National Park, Uganda. *Am. J. Phys. Anthropol.* 135, 171–181.

————, in press. Male mating interest varies with female fecundity in chimpanzees of Kanyawara, Kibale National Park. *Int. J. Primatol.*

Emlen, S. T., 1982. The evolution of helping: An ecological constraints model. *Am. Nat.* 119, 29–39.

————, 1997. Predicting family dynamics in social vertebrates. In: Krebs, J. R., Davies, N. B., (Eds.), *Behavioural Ecology.* 4th ed. Blackwell, Oxford, pp. 228–253.

Emlen, S. T., Oring, L. W., 1977. Ecology, sexual selection, and the evolution of mating systems. Science 197, 215–223.

Engelmann, M., Wotjak, C. T., Neumann, I., Ludwig, M., Landgraf, R., 1996. Behavioral consequences of intracerebral vasopressin and oxytocin: Focus on learning and memory. *Neurosci. Biobehav. Rev.* 20, 341–358.

Epple, G., 1986. Communication by chemical signals. In: Mitchell, G., Erwin, J., (Eds.), *Comparative Primate Biology.* Vol. 2, *Part A: Behavior, Conservation and Ecology.* Alan R. Liss, New York, pp. 531–580.

Erickson, K., Drevets, W., Schulkin, J., 2003. Glucocorticoid regulation of diverse cognitive functions in normal and pathological emotional states. *Neurosci. Biobehav. Rev.* 27, 233–246.

Essex, M. J., Klein, M. H., Cho, E., Kalin, N. H., 2002. Maternal stress beginning in infancy may sensitize children to later stress exposure: Effects on cortisol and behavior. *Biol. Psychiatry* 52, 776–784.

Evans, M. R., Goldsmith, A. R., Norris, S. R. A., 2000. The effects of testosterone on antibody production and plumage coloration in male house sparrows *(Passer domesticus). Behav. Ecol. Sociobiol.* 47, 156–163.

Fairbanks, L. A., 2001. Individual differences in response to a stranger: Social impulsivity as a dimension of temperament in vervet monkeys. *J. Comp. Psychol.* 115, 22–28.

————, n.d. Unpublished data.

Fairbanks, L. A., Fontenot, M. B., Phillips-Conroy, J. E., Jolly, C. J., Kaplan, J. R., Mann, J. J., 1999. CSF monoamines, age and impulsivity in wild grivet monkeys *(Cercopithecus aethiops aethiops). Brain Behav. Evol.* 53, 305–312.

Fairbanks, L. A., Jorgensen, M. J., Huff, A., Blau, K., Hung, Y.-Y., Mann, J. J., 2004. Adolescent impulsivity predicts adult male dominance attainment in male vervet monkeys. *Am. J. Primatol.* 64, 1–17.

Fairbanks, L. A., McGuire, M. T., 1995. Maternal condition and the quality of maternal care in vervet monkeys. *Behaviour* 132, 733–754.

Fairbanks, L. A., Melega, W. P., Jorgensen, M. J., Kaplan, J. R., McGuire, M. T., 2001. Social impulsivity inversely associated with CSF 5-HIAA and fluoxetine exposure in vervet monkeys. *Neuropsychopharmacology* 24, 370–378.

Fairbanks, L. A., Newman, T. K., Bailey, J. N., Jorgensen, M. J., Breidenthal, S. E., Ophoff, R. A., Comuzzie, A. G., Martin, L. J., Rogers, J., 2004. Genetic contributions to social impulsivity and aggressiveness in vervet monkeys. *Biol. Psychiatry* 55, 642–647.

Faulkes, O. G., Bennett, N. C., 2001. Family values: Group dynamics and social control of reproduction in African mole-rats. *Trends Ecol. Evol.* 16, 184–190.

Faulkner, K., Valeggia, C., Ellison, P. T., 2000. Infant growth status in a Toba community of Formosa, Argentina. *Soc. Biol. Hum. Affairs* 65, 8–19.

Febo, M., Numan, M., Ferris, C. F., 2005. Functional magnetic resonance imaging shows oxytocin activates brain regions associated with mother-pup bonding during suckling. *J. Neurosc.* 25, 11637–11644.

Fedigan, L. M., Zohar, S., 1997. Sex differences in mortality of Japanese macaques: Twenty-one years of data from the Arashiyama West population. *Am. J. Phys. Anthropol.* 102, 161–175.

Feek, C. M., Tuzi, N. L., Edwards, C. R., 1989. The adrenal gland and progesterone stimulates testicular steroidogenesis in the rat in vivo. *J. Steroid Biochem. Mol. Biol.* 32, 573–579.

Feek, C. M., Tuzi, N. L., Williams, B. C., Burt, D., Wu, F. C., Wallace, A. M., Gray, C. E., Edwards, C. R., 1985. The adrenal secretion of progesterone stimulates testicular steroidogenesis in the rat in vitro. *J. Steroid Biochem. Mol. Biol.* 23, 617–623.

Fehm-Wolfsdorf, G., Born, J., 1991. Behavioral effects of neurohypophyseal peptides in healthy volunteers: 10 years of research. *Peptides* 12, 1399–1406.

Fehm-Wolfsdorf, G., Born, J., Voigt, K. H, Fehm, H. L., 1984. Behavioral effects of vasopressin. *Neuropsychobiology* 11, 49–53.

Feinberg, D. R., Jones, B. C., DeBruine, L. M., Moore, F. R., Law-Smith, M. J., Cornwell, E., Tiddeman, B. P., Boothroyd, L. G., Perrett, D. I., 2005. The voice and face of woman: One ornament that signals quality? *Evol. Hum. Behav.* 26, 398–408.

Feinberg, D. R., Jones, B. C., Law-Smith, M. J., Moore, F. R., DeBruine, L. M., Cornwell, R. E., Hillier, S. G., Perrett, D. I., 2006. Menstrual cycle, trait estrogen level, and masculinity in the human voice. *Horm. Behav.* 49, 215–222.

Fenker, D. T., Schott, B. J., Richardson-Klavehn, A., Heinze, H-J., Düzel, E., 2005. Recapitulating emotional context: Activity of amygdala, hippocampus and fusiform cortex during recollection and familiarity. *Eur. J. Neurosci.* 21, 1–7.

Ferguson, J. N., Young, L. J., Insel, T. R., 2002. The neuroendocrine basis of social recognition. *Front. Neuroendocrinol.* 23, 200–224.

Ferin, M., 1999. Stress and the reproductive cycle. *J. Clin. Endocrinol. Metab.* 84, 1768–1774.

Ferkin, M. H., Gorman, M. R., 1992. Photoperiod and gonadal hormones influence odor preferences of the male meadow vole, *Microtus pennsylvanicus.* *Physiol. Behav.* 51, 1087–1091.

Ferkin, M. H., Sorokin, E. S., Renfroe, M. W., Johnston, R. E., 1994. Attractiveness of male odors to females varies directly with plasma testosterone concentration in meadow voles. *Physiol. Behav.* 55, 347–353.

Ferkin, M. H., Zucker, I., 1991. Seasonal control of odour preferences of meadow voles *(Microtus pennsylvanicus)* by photoperiod and ovarian hormones. *J. Reprod. Fertil.* 92, 433–441.

Ferris, C. F., 2000. Adolescent stress and neural plasticity in hamsters: A vasopressin-serotonin model of inappropriate aggressive behaviour. *Exp. Physiol.* 85, 85s–90s.

Ferris, C. F., Snowdon, C. T., King, J. A., Dung, T. Q., Ziegler, T. E., Ugurbil, K., Ludwig, R., Schultz-Darken, N. J., Wu, Z., Olson, D. P., Sullivan, J. M., Jr., Tannenbaum, P. L., Vaughan, J. T. 2001. Functional imaging of brain activity in conscious monkeys responding to sexually arousing cues. *Neuroreport* 12, 2231–2236.

Ferris, C. F., Snowdon, C. T., King, J. A., Sullivan, J. M. Jr., Ziegler, T. E., Olson, D. P., Schultz-Darken, N. J., Tannenbaum, P. L., Ludwig, R., Wu, Z., Einspanier, A., Vaughan, J. T., Duong, T. Q., 2004. Imaging neural pathways associated with stimuli for sexual arousal in non-human primates. *J. Magn. Reson. Imaging* 19, 168–175.

Festa-Bianchet, M., 1989. Individual differences, parasites and the costs of reproduction for bighorn ewes *(Ovis candadensis)*. *J. Anim. Ecol.* 58, 585–795.

Field, T., Diego, M., Hernanadez-Reif, M., 2006. Prenatal depression effects on the fetus and newborn: A review. *Inf. Beh. Dev.* 29, 445–455.

Field, T., Diego, M., Hernandez-Reif, M., Vera, Y., Gil, K., Schanberg, S., Kuhn, C., Gonzalez-Garcia, A., 2004. Prenatal maternal biochemistry predicts neonatal biochemistry. *Int. J. Neurosci.* 114, 933–945.

Field, T., Healy, B., Goldstein, S., Guthertz, M., 1990. Behavior-state matching and synchrony in mother-infant interactions of non-depressed versus depressed dyads. *Dev. Psychol.* 26, 7–14.

Field, T., Healy, B., Goldstein, S., Perry, S., Bendell, D., Schanberg, S., Zimmerman, Kuhn, C., 1988. Infants of depressed mothers show "depressed" behavior even with nondepressed adults. *Child Dev.* 59, 1569–1579.

Fietz, J., Dausmann, K. H., 2003. Costs and potential benefits of parental care in the nocturnal fat-tailed dwarf lemur *(Cheirogaleus medius)*. *Folia Primatol.* 74, 246–258.

Filippi, S., Vignozzi, L., Vannelli, G. B., Ledda, F., Forti, G., Maggi, M., 2003. Role of oxytocin in the ejaculatory process. *J. Endocrinol. Invest.* 26s, 82–86.

Fink, B., Manning, J. T., Neave, N., 2004. Second to fourth digit ratio and the "Big Five" personality factors. *Pers. Ind. Diff.* 37, 495–503.

Fink, B., Manning, J. T., Williams, J. H. G., Podmore-Nappin, C., 2007. The 2nd to 4th digit ratio and developmental psychopathology in school-aged children. *Pers. Ind. Diff.* 42, 369–379.

Fink, B., Neave, N., Laughton, K., Manning, J. T., 2006. Second to fourth digit ratio and sensation seeking. *Pers. Ind. Diff.* 41, 1253–1262.

Fink, S., Excoffier, L., Heckel, G., 2006. Mammalian monogamy is not controlled by a single gene. *Proc. Natl. Acad. Sci. U. S. A.* 103, 10956–10960.

Fisher, H., Aron, A., Brown, L. L., 2005. Romantic love: An fMRI study of a neural mechanism for mate choice. *J. Comp. Neurol.* 493, 58–62.

Fisher, R. A., 1930. *The Genetical Theory of Evolution.* Dover, New York.

Fleming, A. S., 2005. Plasticity of innate behavior: Experiences throughout life affect maternal behavior and its neurobiology. In: Carter, C. S., Ahnert, L., Grossmann, K. E., Hrdy, S. B., Lamb, M. E., Porges, S. W., Sachser, N., (Eds.), *Attachment and Bonding: A New Synthesis.* MIT Press, Cambridge, MA, pp. 135–168.

Fleming, A. S., Corter, C., Franks, P., Surbey, M., Schneider, B., Steiner, M., 1993. Postpartum factors related to mother's attraction to newborn infant odors. *Dev. Psychobiol.* 26, 115–132.

Fleming, A. S., Corter, C., Stallings, J., Steiner, M., 2002. Testosterone and prolactin are associated with emotional responses to infant cries in new fathers. *Horm. Behav.* 42, 399–413.

Fleming, A. S., Corter, C., Surbey, M., Franks, P., Steiner, M., 1995. Postpartum factors related to mother's recognition of newborn infant odours. *J. Reprod. Infant Psychol.* 13, 197–210.

Fleming, A. S., Klein, E., Corter, C., 1992. The effects of a social support group on depression, maternal attitudes and behavior in new mothers. *J. Child Psychol. Psychiatry* 33, 685–698.

Fleming, A. S., Morgan, H. D., Walsh, C., 1996. Experiential factors in postpartum regulation of maternal care. In: Rosenblatt, J. S., Snowdon, C. T., (Eds.), *Parental Care: Evolution, Mechanisms, and Adaptive Significance.* Academic Press, San Diego, pp. 295–332.

Fleming, A. S., Ruble, D. N., Flett, G. L., Shaul, D., 1988. Postpartum adjustment in first-time mothers: Relations between mood, maternal attitudes and mother-infant interactions. *Dev. Psychol.* 24, 71–81.

Fleming, A. S., Ruble, D., Krieger, H., Wong, P. Y., 1997. Hormonal and experiential correlates of maternal responsiveness during pregnancy and the puerperium in human mothers. *Horm. Behav.* 31, 145–158.

Fleming, A. S., Steiner, M., Anderson, V., 1987. Hormonal and attitudinal correlates of maternal behaviour during the early postpartum period. *J. Reprod. Infant Psychol.* 5, 193–205.

Fleming, A. S., Steiner, M., Corter, C., 1997. Cortisol, hedonics, and maternal responsiveness in human mothers. *Horm. Behav.* 32, 85–98.

Fletcher, G. J. O., Simpson, J. A., Thomas, G., 2000. The measurement of perceived relationship quality components: A confirmatory factor analytic approach. *Pers. Soc. Psychol. Bull.* 26, 340–354.

Flinn, M. V., 2004. Culture and developmental plasticity: Evolution of the social brain. In: Burgess, R. L., MacDonald, K., (Eds.), *Evolutionary Perspectives on Child Development.* Sage, Thousand Oaks, CA, pp. 73–98.

———, 2006a. Cross-cultural universals and variations: The evolutionary paradox of informational novelty. *Psychol. Inq.* 17, 118–123.

———, 2006b. Evolution and ontogeny of stress response to social challenge in the human child. *Dev. Rev.* 26, 138–174.

———, 2007. Evolution of stress response to social-evaluative threat. In: Dunbar, R., Barrett, L., (Eds.), *Oxford Handbook of Evolutionary Psychology.* Oxford University Press, Oxford, pp. 272–296.

———, 2008. Why words can hurt us: Social relationships, stress, and health. In: Trevathan, W., Smith, E. O., McKenna, J., (Eds.), *Evolutionary Medicine and Health.* Oxford University Press, Oxford, pp. 247–258.

Flinn, M. V., Alexander, R. D., 2007. Runaway social selection. In: Gangestad, S. W., Simpson, J. A., (Eds.), *The Evolution of Mind.* Guilford Press, New York, pp. 249–255.

Flinn, M. V., Baewald, C., Decker, S. A., England, B. G., 1998. Evolutionary functions of neuroendocrine response to social environment. *Behav. Brain Sci.* 21, 372–374.

Flinn, M. V., Coe, K., 2007. The linked red queens of human cognition, reciprocity, and culture. In: Gangestad, S. W., Simpson, J. A., (Eds.), *The Evolution of Mind.* Guilford Press, New York, pp. 339–347.

Flinn, M. V. England, B. G., 1995. Childhood stress and family environment. *Cur. Anthropol.* 36. 854–866.

———, 2003. Childhood stress: Endocrine and immune responses to psychosocial events. In: Wilce, J. M., (Ed.), *Social and Cultural Lives of Immune Systems.* Routledge, London, pp. 107–147.

Flinn, M. V., Geary, D. C., Ward, C. V., 2005. Ecological dominance, social competition, and coalitionary arms races: Why humans evolved extraordinary intelligence. *Evol. Hum. Behav.* 26, 10–46.

Flinn, M. V., Quinlan, R. J., Turner, M. T., Decker, S. D., England, B. G., 1996. Male-female differences in effects of parental absence on glucocorticoid stress response. *Hum. Nat.* 7, 125–162.

Flinn, M. V., Quinlan, R. J., Ward, C. V., Coe, K., 2007. Evolution of the human family: Cooperative males, long social childhoods, smart mothers, and extended kin networks. In: Salmon, C., Shackelford, T., (Eds.), *Evolutionary Family Psychology.* Oxford University Press, Oxford, pp. 16–38.

Foley, R. A., Lee, P. C., 1989. Finite social space, evolutionary pathways and reconstructing hominid behaviour. *Science* 243, 901–906.

Follett, B. K., 1984. Birds. In: Lamming, G. E., (Ed.), *Marshall's Physiology of Reproduction I. Reproductive Cycles of Vertebrates.* Churchill Livingstone, Edinburgh, pp. 283–350.

Folstad, I., Karter, A. J., 1992. Parasites, bright males, and the immunocompetence handicap. *Am. Nat.* 139, 603–622.

Forcada-Guex, M., Pierrehumbert, B., Borghini, A., Moessinger, A., Muller-Nix, C., 2006. Early dyadic patterns of mother-infant interactions and outcomes of prematurity at 18 months. *Pediatrics* 118, 107–114.

Forest, M. G., 1990. Pituitary gonadotropin and sex steroid secretion during the first two years of life. In: Grumbach, M. M., Sizonenko, P. C., Aubert, M. L., (Eds.), *Control of the Onset of Puberty.* Williams & Wilkins, Baltimore, pp. 451–477.

Formby, D., 1967. Maternal recognition of infant's cry. *Dev. Med. Child Neurol.* 9, 293–298.

Forsling, M. L., 2000. Diurnal rhythms in neurohypophysial function. *Exp. Physiol.* 85S, 179s–186s.

Fox, E. A., 1998. The function of female mate choice in the Sumatran orangutan *(Pongo pygmaeus abelii).* Ph.D. diss., Duke University.

———, 2002. Female tactics to reduce sexual harassment in the Sumatran orangutan *(Pongo pygmaeus abelii).* Behav. Ecol. Sociobiol. 52, 93–101.

Francis, D., Diorio J., Liu, D., Meaney, M. J., 1999. Nongenomic transmission across generations of maternal behavior and stress responses in the rat. *Science* 286, 1155–1158.

Francis, D. D., Diorio, J., Plotsky, P. M., Meaney, M. J., 2002. Environmental enrichment reverses the effects of maternal separation on stress reactivity. *J. Neurosci.* 22, 7840–7843.

Francis, D. D., Young, L. J., Meaney, M. J., Insel, T. R., 2002. Naturally occurring differences in maternal care are associated with the expression of oxytocin and vasopressin (V1a) receptors: Gender differences. *J. Neuroendocrinol.* 14, 349–353.

Franks, S., 1998. Growth hormone and ovarian function. *Baillières Clin. Endocrinol. Metab.* 12, 331–340.

Franks, S., Gilling-Smith, C., Watson, H., Willis, D., 1999. Insulin action in the normal and polycystic ovary. *Endocrinol. Metab. Clin. North Am.* 28, 361–378.

Freimer, N. B., Service, S. K., Ophoff, R. A., Jasinska, A. J., McKee, K., Villeneuve, A., Belisle, A., Bailey, J. N., Breidenthal, S. E., Jorgensen, M. J., Mann, J. J., Cantor, R. M., Dewar, K., Fairbanks, L. A., 2007. Quantitative trait locus for variation in dopamine metabolism mapped in a primate model using reference sequences from related species. *Proc. Natl. Acad. Sci. U. S. A.* 104, 15811–15816.

French, N. P., Hagan, R., Evans, S. F., Godfrey, M., Newnham, J. P., 1999. Repeated antenatal corticosteroids: Size at birth and subsequent development. *Am. J. Obstet. Gynecol.* 180, 114–121.

Fride, E., Dan, Y., Feldon, J., Halevy, G., Weinstock, M., 1986. Effects of prenatal stress on vulnerability to stress in prepubertal and adult rats. *Physiol. Behav.* 37, 681–687.

Fries, A. B. W., Ziegler, T. E., Kurian, J. R., Jacoris, S., Pollak, S. D., 2005. Early experience in humans is associated with changes in neuropeptides critical for regulating social behavior. *Proc. Natl. Acad. Sci. U. S. A.* 102, 17237–17240.

Frodi, A. M. L., 1980. Child abusers' responses to infant smiles and cries. *Child Dev.* 51, 238–241.

Frodi, A. M. L., Lamb, M. E., 1978a. Fathers' and mothers' response to the faces and cries of normal and premature infants. *Dev. Psychol.* 14, 490–498.

———, 1978b. Sex differences in responsiveness to infants: A developmental study of psychophysiological and behavioral responses. *Child Dev.* 49, 1182–1188.

Frye, C. A., Wawrzycki, J., 2003. Effect of pre-natal stress and gonadal hormone condition on depressive behaviors of female and male rats. *Horm. Behav.* 44, 319–326.

Fuentes, A., 2000. Hylobatid communities: Changing views on pair bonding and social organization in hominoids. *Yrbk. Phys. Anthropol.* 43, 33–60.

———, 2002. Patterns and trends in primate pair bonds. *Int. J. Primatol.* 23, 953–978.

Furedy, J., Fleming, A. S., Ruble, D., Scher, H., Daly, J., Day, D., Loewen, R., 1989. Sex differences in small-magnitude heart-rate respones to sexual and infant-related stimuli: A psychophysiological approach. *Physiol. Behav.* 46, 903–905.

Furuichi, T., 1987. Sexual swelling, receptivity and grouping of wild pygmy chimpanzee females at Wamba, Zaire. *Primates* 28, 309–318.

———, 1997. Agonistic interactions and matrifocal dominance rank of wild bonobos *(Pan paniscus)* at Wamba. *Int. J. Primatol.* 18, 855–875.

Furuichi, T., Hashimoto, C., 2004. Sex differences in copulation attempts in wild bonobos at Wamba. *Primates* 45, 59–62.

Fusani, L., Van't Hof, T., Hutchison, J. B., Gahr, M., 2000. Seasonal expression of androgen receptors, estrogen receptors, and aromatase in the canary brain in relation to circulating androgens and estrogens. *J. Neurobiol.* 43, 254–268.

Gagneux, P., Gonder, M. K., Goldberg, T., Morin, P., 2001. Gene flow in wild chimpanzee populations: What genetic data tells us about chimpanzee movement over space and time. *Trans. R. Soc. Lond. B Biol. Sci.* 356, 889–897.

Gahr, M., Metzdorf, R., 1997. Distribution and dynamics in the expression of androgen and estrogen receptors in vocal control systems of songbirds. *Brain Res. Bull.* 44, 509–517.

Gaillard, J. M., Pontier, D., Allaine, D., Loison, A., Herve, J. C., Heizmann, A., 1997. Variation in growth form and precocity at birth in eutherian mammals. *Proc. Biol. Sci.* 264, 859–868.

Galdikas, B. M. F., 1981. Orangutan reproduction in the wild. In: Graham, C. E., (Ed.), *Reproductive Biology of the Great Apes: Comparative and Biomedical Perspectives.* Academic Press, New York, pp. 281–300.

———, 1985. Subadult male orangutan sociality and reproductive behavior at Tanjung Puting. *Am. J. Primatol.* 8, 87–99.

Galdikas, B. M. F., Wood, J. W., 1990. Birth spacing patterns in humans and apes. *Am. J. Phys. Anthropol.* 83, 185–192.

Gallese, V., Keysers, C., Rizzolatti, G. 2004. A unifying view of the basis of social cognition. *Trends Cogn. Sci.* 8, 396–403.

Gandelman, R., 1983. Gonadal hormones and sensory function. *Neurosci. Biobehav. Rev.* 7, 1–17.

Gangestad, S. W., 2000. Human sexual selection, good genes, and special design. *Ann. N. Y. Acad. Sci.* 907, 50–61.

———, 2006. Evidence for adaptations for female extra-pair mating in humans: Thoughts on current status and future directions. In: Platek, S. M., Shackelford, T. K., (Eds.), *Female Infidelity and Paternal Uncertainty: Evolutionary Perspectives on Male Anti-Cuckoldry Tactics.* Cambridge University Press, Cambridge, pp. 37–57.

Gangestad, S. W., Simpson, J. A., Cousins, A. J., Garver-Apgar, C. E., Christensen, N. P., 2004. Women's preferences for male behavioral displays change across the menstrual cycle. *Psychol. Sci.* 15, 203–207.

Gangestad, S. W., Thornhill, R., 1998. Menstrual cycle variation in women's preferences for the scent of symmetrical men. *Proc. R. Soc. Lond. B Biol. Sci.* 265, 927–933.

———, 2004. Female multiple mating and genetic benefits in humans: Investigations of design. In: Kappeler, P. M., van Schaik, C. P., (Eds.), *Sexual Selection in Primates: New and Comparative Perspectives.* Cambridge University Press, Cambridge, pp. 90–116.

Gangestad, S. W., Thornhill, R., Flinn, M. V., Dane, L. K., Flacon, R. G., Garver-Apgar, C. E., Franklin, M., England, B. C., 2005. Men's testosterone and life history in a Caribbean rural village. Paper presented at the Human Behavior and Evolution Society meeting, Austin, TX.

Gangestad, S. W., Thornhill, R., Garver, C. E., 2002. Changes in women's sexual interests and their partners' mate-retention tactics across the menstrual cycle: Evidence for shifting conflicts of interest. *Proc. R. Soc. Lond. B Biol. Sci.* 269, 975–982.

Gangestad, S. W., Thornhill, R., Garver-Apgar, C. E., 2006. Adaptations to ovulation. *Curr. Dir. Psychol. Sci.* 14, 312–316.

Ganzhorn, J. U., Klaus, S., Ortmann, S., Schmid, J., 2003. Adaptations to seasonality: Some primate and nonprimate examples. In: Kappeler, P. M, Pereira, M. E., (Eds.), *Primate Life Histories and Socioecology.* University of Chicago Press, Chicago, pp. 132–144.

Garamszegi, L. Z., Eens, M., Hurtrez-Boussés, S., Møller, A. P., 2005. Testosterone, testes size, and mating success in birds: A comparative study. *Horm. Behav.* 47, 389–409.

Garas, A., Trypsianis, G., Kallitsaris, A., Milingos, S., Messinis, I. E., 2006. Oestradiol stimulates prolactin secretion in women through oestrogen receptors. *Clin. Endocrinol.* 65, 638–642.

Garber, P. A., Leigh, S. R., 1997. Ontogenetic variation in small-bodied New World primates: Implications for patterns of reproduction and infant care. *Folia Primatol.* 68, 1–22.

Garmezy, N., 1983. Stressors of childhood. In: Garmezy, N., Rutter, M., (Eds.), *Stress, Coping, and Development in Children.* McGraw-Hill, New York, pp. 43–83.

Garver-Apgar, C. E., Gangestad, S. W., Thornhill, R., Miller, R. D., Olp, J. J., 2006. Major histocompatibility complex alleles, sexual responsivity, and unfaithfulness in romantic couples. *Psychol. Sci.* 17, 830–835.

Geary, D. C., 1998. *Male, Female: The Evolution of Human Sex Differences.* American Psychological Association Press, Washington, DC.

Genuth, S. M., 1993. The endocrine system. In: Berne, R. M., Levy, M. N., (Eds.), *Physiology.* Mosby Year Book, St. Louis, MO, pp. 813–1024.

Gerald, M. S., Higley, S., Lussier, I. D., Westergaard, G. C., Suomi, S. J., Higley, J. D., 2002. Variation in reproductive outcomes for captive male rhesus macaques *(Macaca mulatta)* differing in CSF 5-hydroxyindoleacetic acid concentrations. *Brain Behav. Evol.* 60, 117–124.

Gerloff, U., Hartung, B., Fruth, B., Hohman, G., Tautz, D., 1999. Intracommunity relationships, dispersal pattern and paternity success in a wild living community of bonobos *(Pan paniscus)* determined from DNA analysis of faecal samples. *Proc. Natl. Acad. Sci. U. S. A.* 266, 1189–1195.

Gerra, G., Avanzini, P., Zaimovic, A., Sartori, R., Bocchi, C., Timpano, M., Zambelli, U., Delsignore, R., Gardini, F., Talarico, E., Brambilla, F., 1999. Neurotransmitters, neuroendocrine correlates of sensation-seeking temperament in normal humans. *Neuropsychobiology* 39, 207–213.

Gesquiere, L. R., Altmann, J., Khan, M. Z., Couret, J., Yu, J. C., Endres, E. S., Lynch, J. W., Ogola, P., Fox, E. A., Alberts, S., Wango, E. O., 2005. Coming of age: Steroid hormones of wild immature baboons *(Papio cynocephalus)*. *Am. J. Primatol.* 67, 83–100.

Giardino, J., Gonzalez, A., Steiner, M., Fleming, A. S., 2008. Effects of motherhood on physiological and subjective responses to infant cries in teenage mothers: A comparison with non-mothers and adult mothers. *Horm. Behav.* 53, 149–158.

Giedd, J. N., Clasen, L. S., Lenroot, R., Greenstein, D., Wallace, G. L., Ordaz, S., Molloy, E. A., Blumenthal, J. D., Tossell, J. W., Stayer, C., Samango-Sprouse, C. A., Shen, D. G., Davatzikos, C., Merke, D., Chrousos, G. P., 2006. Puberty-related influences on brain development. *Mol. Cell. Endocrinol.* 254, 154–162.

Gilbert, P., 2001. Evolutionary approaches to psychopathology: The role of natural defences. *Aust. N. Z. J. Psychiatry* 35, 17–27.

———, 2005. Social mentalities: A biopsychosocial and evolutionary approach to social relationships. In: Baldwin, M. W., (Ed.), *Interpersonal Cognition*. Guilford Press, New York, pp. 299–333.

Gimpl, G., Fahrenholz, F., 2001. The oxytocin receptor system: Structure, function and regulation. *Physiol. Rev.* 81, 629–683.

Ginther, A. J., Carlson, A. A., Ziegler, T. E., Snowdon, C. T., 2002. Neonatal and pubertal development in males of a cooperatively breeding primate, the cotton-top tamarin *(Saguinus oedipus oedipus)*. *Biol. Reprod.* 66, 282–290.

Gitau, R., Cameron, A., Fisk, N. M., Glover, V., 1998. Fetal exposure to maternal cortisol. *Lancet* 352, 707–708.

Gitau, R., Fisk, N. M., Teixeira, J. M. A., Cameron, A., Glover, V., 2001. Fetal hypothalamic-pituitary-adrenal stress responses to invasive procedures are independent of maternal responses. *J. Clin. Endocrinol. Metab.* 86, 104–109.

Gittleman, J. L., 1994. Female brain size and parental care in carnivores. *Proc. Natl. Acad. Sci. U. S. A.* 91, 5495–5497.

Glover, V., O'Connor, T. G., Heron, J., Golding, J., ALSPAC Study Team, 2004. Antenatal maternal anxiety is linked with atypical handedness in the child. *Early Hum. Dev.* 79, 107–118.

Gluckman, P. D., Hanson, M. A., 2006. Adult disease: Echoes of the past. *Eur. J. Endocrinol.* 155, s47–s50.

Gluckman, P. D., Hanson, M. A., Beedle, A. S., 2007. Early life events and their consequences for later disease: A life history and evolutionary perspective. *Am. J. Hum. Biol.* 19, 1–19.

Goldbort, J., 2006. Transcultural analysis of postpartum depression. *MCN AM. J. Matern. Child Nurs.* 31, 121–126.

Goldsmith, H. H., Lemery, K. S., Buss, K. A., Campos, J. J., 1999. Genetic analyses of focal aspects of infant temperament. *Dev. Psychol.* 35, 972–985.

Gomendio, M., Harcourt, A. H., Roldan, E. R. S., 1998. Sperm competition in mammals. In: Birkhead, T. R., Møller, A. P., (Eds.), *Sperm Competition and Sexual Selection*. Academic Press, San Diego, pp. 667–756.

Gompper, M. E., Gittleman, J. L., Wayne, R. K., 1998. Dispersal, philopatry, and genetic relatedness in a social carnivore: Comparing males and females. *Mol. Ecol.* 7, 157–163.

Gonzaga, G. C., Turner, R. A., Keltner, D., Campos, B., Altemus, M., 2006. Romantic love and sexual desire in close relationships. *Emotion* 6, 163–179.

Gonzalez, A., Lovic, V., Ward, G., Wainwright, P., Fleming, A., 2001. Intergenerational effects of complete maternal deprivation and replacement stimulation on maternal behavior and emotionality in female rats. *Dev. Psychobiol.* 38, 11–32.

Gonzalez, A., Steiner, M., Fleming, A. S., in pre-pa. Depressed mothers show altered physiological responsiveness to infant cries.

————, in prep-b. Neurobiology and neuropsychology as mediators between early maltreatment and parenting behaviours.

Gonzalez-Mariscal, G., Rosenblatt, J. S., 1996. Maternal behavior in rabbits: A historical and multidisciplinary perspective. In: Rosenblatt, J. S., Snowdon, C. T., (Eds.), *Parental Care: Evolution, Mechanisms, and Adaptive Significance*. Academic Press, San Diego, pp. 333–360.

Goodall, J., 1986. *The Chimpanzees of Gombe: Patterns of Behavior*. Belknap Press, Cambridge.

Goodson, J. L., Bass, A. H., 2001. Social behavior functions and related anatomical characteristics of vasotocin/vasopressin systems in vertebrates. *Brain Res. Rev.* 35, 246–265.

Goodspeed, R. B., Gent, J. F., Catalanotto, F. I., Cain, W. S., Zagraniski, R. T., 1986. Corticosteroids in olfactory dysfunction. In: Meiselman, H. L., Rivlin, R. S., (Eds.), *Clinical Measurement of Taste and Smell*. Macmillan, New York, pp. 514–518.

Gooren, L., 2006. The biology of human psychosexual differentiation. *Horm. Behav.* 50, 589–601.

Gopnik, A., Meltzoff, A. N., Kuhl, P. K., 1999. *The Scientist in the Crib: Minds, Brains, and How Children Learn*. William Morrow, New York.

Gordon, T. P., Bernstein, I. S., Rose, R. M., 1978. Social and seasonal influences on testosterone secretion in the male rhesus monkey. *Physiol. Behav.* 21, 623–627.

Gordon, T. P., Rose, R. M., Bernstein, I. S., 1976. Seasonal rhythm in plasma testosterone levels in the rhesus monkey *(Macaca mulatta):* A three year study. *Horm. Behav.* 7, 229–243.

Gordon, T. P., Rose, R. M., Grady, C. L., Bernstein, I. S., 1979. Effects of increased testosterone secretion on the behavior of adult male rhesus living in a social group. *Folia Primatol.* 32, 149–160.

Gorman, M. R., 1994. Male homosexual desire: Neurological investigations and scientific bias. *Perspect Biol Med* 38(1), 61–81.

Gottman, J. M., Katz, L. F., 1989. Effects of marital discord on young children's peer interaction and health. *Dev. Psychol.* 25, 373–381.

Gould, L., Ziegler, T. E., 2007. Variation in fecal testosterone levels, inter-male aggression, dominance rank and age during mating and post-mating periods in wild adult male ring-tailed lemurs *(Lemur catta). Am. J. Primatol.* 69, 1–15.

Gowaty, P. A., 2004. Sex roles, contests for the control of reproduction, and sexual selection. In: Kappeler, P., van Schaik, C., (Eds.), *Sexual Selection in Primates*. Cambridge University Press, Cambridge, pp. 37–54.

Goymann, W., Moore, I. T., Scheuerlein, A., Hirschenhauser, K., Grafen, A., Wingfield, J. C., 2004. Testosterone in tropical birds: Effects of environmental and social factors. *Am. Nat.* 164, 327–333.

Goymann, W., Wingfield, J. C., 2004. Allostatic load, social status, and stress hormones—the costs of social status matter. *Anim. Behav.* 67, 591–602.

Graham, C. E., 1988. Reproductive physiology. In: Schwartz, J. H., (Ed.), *Orang-utan Biology.* Oxford University Press, New York, pp. 91–103.

Graham, C. E., Warner, H., Misener, J., Collins, D. C., Preedy, J. R. K., 1977. The association between basal body temperature, sexual swelling and urinary gonadal hormone levels in the menstrual cycle of the chimpanzee. *J. Reprod. Fertil.* 50, 23–28.

Graham, M. D., Rees, S. L., Steiner, M., Fleming, A. S., 2006. The effects of adrenalectomy and corticosterone replacement on maternal memory in postpartum rats. *Horm. Behav.* 49, 353–361.

Granger, D. A., Schwartz, E. B., Booth, A., Arentz, M., 1999. Salivary testosterone determination in studies of child health and development. *Horm. Behav.* 35, 18–27.

Granger, D. A., Shirtcliff, E. A., Zahn-Waxler, C., Usher, B., Klimes-Dougan, B., Hastings, P., 2003. Salivary testosterone diurnal variation and psychopathology in adolescent males and females: Individual differences and developmental effects. *Dev. Psychopathol.* 15, 431–449.

Gray, P. B., 2003. Marriage, parenting and testosterone variation among Kenyan Swahili men. *Am. J. Phys. Anthropol.* 122, 279–286.

Gray, P. B., Campbell, B. C., Marlowe, F. W., Lipson, S. F., Ellison, P. T., 2004. Social variables predict between-subject but not day-to-day variation in the testosterone of US men. *Psychoneuroendocrinology* 29, 1153–1162.

Gray, P. B., Chapman, J. F., Burnham, T. C., McIntyre, M. H., Lipson, S. F., Ellison, P. T., 2004. Human male pair bonding and testosterone. *Hum. Nat.* 15, 119–131.

Gray, P. B., Ellison, P. T., Campbell, B. C., 2007. Testosterone and marriage among Ariaal men of northern Kenya. *Curr. Anthropol.* 48, 750–755.

Gray, P. B., Kahlenberg, S. M., Barrett, E. S., Lipson, S. F., Ellison, P. T., 2002. Marriage and fatherhood are associated with lower testosterone in human males. *Evol. Hum. Behav.* 23, 193–201.

Gray, P. B., Kruger, A., Huisman, H. W., Wissing, M. P., Vorster, H. H., 2006. Predictors of South African male testosterone levels: The THUSA study. *Am. J. Hum. Biol.* 18, 123–132.

Gray, P. B., Parkin, J. C., Samms-Vaughan, M. E. 2007. Hormonal correlates of human paternal interactions: A hospital-based investigation in urban Jamaica. *Horm. Behav.* 52, 499–507.

Gray, P. B., Singh, A. B., Woodhouse, L. J., Storer, T. W., Casaburi, R., Dzekov, J., Dzekov, C., Sinha-Hikim, I., Bhasin, S., 2005. Dose-dependent effects of testosterone on sexual function, mood, and visuospatial cognition in older men. *J. Clin. Endocrinol. Metab.* 90, 3838–3846.

Gray, P. B., Yang, C. J., Pope, H. G., Jr., 2006. Fathers have lower salivary testosterone levels than unmarried men and married non-fathers in Beijing, China. *Proc. R. Soc. Lond. B Biol. Sci.* 273, 333–339.

Gray, R. H., Campbell, O. M., Apelo, R., Eslami, S. S., Zacur, H., Ramos, R. M., Gehret, J. C., Labbok, M. H., 1990. Risk of ovulation during lactation. *Lancet* 335, 25–29.

Green, R., 1985. Gender identity in childhood and later sexual orientation: Follow-up of 78 males. *Am. J. Psychiatry* 142, 339–341.

Green, R., Roberts, C. W., Williams, K., Goodman, M., Mixon, A., 1987. Specific cross-gender behaviour in boyhood and later homosexual orientation. *Br. J. Psychiatry* 151, 84–88.

Greenspan, F. S., Gardner, D. G., 2001. *Basic and Clinical Endocrinology*. McGraw-Hill, New York.

Greenwood, P. J., 1980. Mating systems, philopatry and dispersal in birds and mammals. *Anim. Behav.* 28, 1140–1162.

Grewen, K., Girdler, S., Amico, J., Light, K., 2005. Effects of partner support on resting oxytocin, cortisol, norepinephrine, and blood pressure before and after warm partner contact. *Psychosom. Med.* 67, 531–538.

Grippo, A. J., Lamb, D. G., Carter, C. S., Porges, S. W., 2007. Social isolation disrupts autonomic regulation of the heart and influences negative affective behaviors. *Biol. Psychiatry.* 62, 1162–1170.

Grippo, A. J., Cushing, B. S., Carter, C. S., 2007. Depression-like behavior and stressor-induced neuroendocrine activation in female prairie voles exposed to chronic social isolation. *Psychosom. Med.* 69, 149–157.

Groome, L. J., Swiber, M. J., Holland, S. B., Bentz, L. S., Atterbury, J. L., Trimm, R. F., 1999. Spontaneous motor activity in the perinatal infant before and after birth: Stability in individual differences. *Dev. Psychobiol.* 35, 15–24.

Gubernick, D. J., Nelson, R. J., 1989. Prolactin and paternal behavior in the biparental California mouse, *Peromyscus californicus. Horm. Behav.* 23, 203–210.

Guillemin, R., 1980. Hypothalamic hormones: Releasing and inhibiting factors. In: Krieger, D. T., Hughes, J. C., (Eds.), *Neuroendocrinology*. Sinaeur, Sunderland, MA, pp. 23–32.

Gunnar, M. R., Bruce, J., Donzella, B., 2000. Stress physiology, health, and behavioral development. In: Thornton, A., (Ed.), *The Well-Being of Children and Families: Research and Data Needs*. University of Michigan Press, Ann Arbor, pp. 188–212.

Gunnar, M. R., Donzella, B., 2002. Social regulation of the cortisol levels in early human development. *Psychoneuroendocrinology* 27, 199–220.

Gurven, M., Kaplan, H., 2006. Determinants of time allocation to production across the lifespan among the Machiguenga and Piro Indians of Peru. *Hum. Nat.* 17, 1–49.

Gurven, M., Kaplan, H., Gutierrez, M., 2006. How long does it take to become a proficient hunter? Implications for the evolution of extended development and long life span. *J. Hum. Evol.* 51, 454–470.

Gurven, M., Walker, R., 2006. Energetic demand of multiple dependents and the evolution of slow human growth. *Proc. R. Soc. Lond. B Biol. Sci.* 273, 835–841.

Gutteling, B. M., de Weerth, C., Buitelaar, J. K., 2004. Maternal prenatal stress and 4–6 year old children's salivary cortisol concentrations pre- and post-vaccination. *Stress* 7, 257–260.

———, 2005. Pre-natal stress and children's cortisol reaction to the first day of school. *Psychoneuroendocrinology* 30, 541–549.

Gwinner, E., 1986. *Circannual Rhythms*. Springer, Berlin.

———, 1996. Circannual clocks in avian reproduction and migration. *Ibis* 138, 47–63.

Gwinner, E., Rödl, T., Schwabl, H., 1994. Pair territoriality of wintering stonechats: Behaviour, function and hormones. *Behav. Ecol. Sociobiol.* 34, 321–327.

Halbreich, U., Karkun, S., 2006. Cross-cultural and social diversity of prevalence of postpartum depression and depressive symptoms. *J. Affect. Disord.* 91, 97–111.

Halem, H. A., Cherry, J. A., Baum, M. J., 1999. Vomeronasal neuroepithelium and forebrain fos responses to male pheromones in male and female mice. *J. Neurobiol.* 39, 249–263.

Halliday, T. R., 1994. Sex and evolution. In: Slater, P. J. B., Halliday, T. R., (Eds.), *Behaviour and Evolution*. Cambridge University Press, Cambridge, pp. 150–192.

Halligan, S. L, Herbert, J., Goodyer, I. M., Murray, L., 2004. Exposure to postnatal depression predicts elevated cortisol in adolescent offspring. *Biol. Psychiatry* 55, 376–381.

Halpern, C. T., Udry, J. R., Campbell, B., Suchindran, C., Mason, G. A., 1994. Testosterone and religiosity as predictors of sexual attitudes and activity among adolescent males—a biosocial model. *J. Biosoc. Sci.* 26, 217–234.

Halpern, C. T., Udry, J. R., Suchindran, C., 1998. Monthly measures of salivary testosterone predict sexual activity in adolescent males. *Arch. Sex. Behav.* 27, 445–465.

Hamilton, J. B., Mestler, G. E., 1969. Mortality and survival: Comparison of eunuchs with intact men and women in a mentally retarded population. *J. Gerontol.* 24, 395–411.

Hamilton, W. D., Zuk, M., 1982. Heritable true fitness and bright birds: A role for parasites? *Science* 218, 384–387.

Hamilton, W. J., 1984. Significance of paternal investment by primates to the evolution of adult male-female associations. In: Taub, D. M., (Ed.), *Primate Paternalism*. Van Nostrand Reinhold, New York, pp. 309–335.

Hammock, E. A., Young, L. J., 2002. Variation in the vasopressin V1a receptor promoter and expression: Implications for inter- and intraspecific variation in social behaviour. *Eur. J. Neurosci.* 16, 399–402.

———, 2005. Microsatellite instability generates diversity in brain and sociobehavioral traits. *Science* 308, 1630–1634.

Handley, S. L., Dunn, T. L., Baker, S. M., Cockshott, C., Goulds, S., 1977. Mood changes in the puerperium and plasma tryptophan and cortisol. *BMJ* 2, 18–22.

Handley, S. L., Dunn, T. L., Waldron, G., Baker, J. M., 1980. Tryptophan, cortisol and puerperal mood. *Br. J. Psychiatry* 136, 498–508.

Harcourt, A. H., Fossey, D., Stewart, K. J., Watts, D. P., 1980. Reproduction in wild gorillas and some comparisons with the chimpanzee. *J. Reprod. Fertil. Suppl.* 28, 59–70.

Harcourt, A. H., Harvey, P. H., Larson, S. G., Short, R. V., 1981. Testis weight, body weight and breeding system in primates. *Nature* 293, 55–57.

Harcourt, A. H., Stewart, K. J., 1978. Sexual behavior of wild mountain gorillas. In: Chivers, D. J., Herberts, J., (Eds.), *Recent Advances in Primatology*. Academic Press, London, pp. 611–612.

Harcourt, A. H., Stewart, K. J., Fossey, D., 1981. Gorilla reproduction in the wild. In: Graham, C. E., (Ed.), *Reproductive Biology of the Great Apes: Comparative and Biomedical Perspectives*. Academic Press, New York, pp. 265–280.

Harding, A., Halliday, G., Ng, J., Harper, C., Kril, J., 1996. Loss of vasopressin-immunoreactive neurons in alcoholics is dose-related and time-dependent. *Neuroscience* 72, 699–708.

Harding, C. F., 1983. Hormonal influences on avian aggressive behavior. In: Svare, B., (Ed.), *Hormones and Aggressive Behavior*. Plenum Press, New York, pp. 435–467.

Harkness, S., Super, C. M., 2002. Culture and parenting. In: Bornstein, M. H., (Ed.), *Handbook of Parenting*. Vol. 2, *Biology and Ecology of Parenting*. 2nd ed. Erlbaum, Mahwah, NJ, pp. 3–43.

Harris, B., Lovett, L., Newcombe, R. G., Read, G. F., Walker, R., Riad-Fahmy, D., 1994a. Cardiff puerperal mood and hormone study III. Postnatal depression at 5 to 6 weeks pastpartum, and its hormonal correlates across the peripartum period. *Br. J. Psychiatry* 168, 739–744.

———, 1994b. Maternity blues and major endocrine changes: Cardiff puerperal mood and hormone study II. *Br. Med. J.* 308, 949–953.

Hart, B. L., 1988. Biological basis of the behavior of sick animals. *Neurosci. Biobehav. Rev.* 12, 123–137.

Hart, J., Gunnar, M., Cicchetti, D., 1995. Salivary cortisol in maltreated children: Evidence of relations between neuroendocrine activity and social competence. *Dev. Psychopathol.* 7, 11–26.

Haselton, M. G., Gangestad, S. W., 2006. Conditional expression of women's desires and men's mate guarding across the ovulatory cycle. *Horm. Behav.* 49, 509–518.

Hashimoto, C., Furuichi, T., 2006. Comparison of behavioral sequence of copulation between chimpanzees and bonobos. *Primates* 47, 51–55.

Hau, M., 2007. Regulation of male traits by testosterone: Implications for the evolution of vertebrate life histories. *Bioessays* 29, 133–144.

Hau, M., Wikelski, M., Soma, K., Wingfield, J., 2000. Testosterone and year-round territorial aggression in a tropical bird. *Gen. Comp. Endocrinol.* 117, 20–33.

Hauner, H., 2005. Secretory factors from human adipose tissue and their functional role. *Proc. Nutr. Soc.* 64, 163–169.

Havlíček, J., Dvořáková, R., Bartoš, L., Flegr, J., 2006. Non-advertised does not mean concealed: Body odour changes across the human menstrual cycle. *Ethology* 112, 81–90.

Hawkes, K., Bliege Bird, R., 2002. Showing off, handicap signaling, and the evolution of men's work. *Evol. Anthropol.* 11, 58–67.

Hawkes, K., O'Connell, J. F., Jones, N. G. B., Alvarez, H., Charnov, E. L., 1998. Grandmothering, menopause, and the evolution of human life histories. *Proc. Natl. Acad. Sci. U. S. A.* 95, 1336–1339.

Hawley, P. H., 2003. Prosocial and coercive configurations of resource control in early adolescence: A case for the well-adapted Machiavellian. *Merrill-Palmer Q.* 49, 279–309.

Hayes, F. J., Seminara, S. B., Decruz, S., Boepple, P. A., Crowley, W. F., Jr., 2000. Aromatase inhibition in the human male reveals a hypothalamic site of estrogen feedback. *J. Clin. Endocrinol. Metab.* 85, 3027–3035.

Hedegaard, M., Henriksen, T. B., Secher, N. J., Hatch, M. C., Sabroe, S., 1996. Do stressful life events affect duration of gestation and risk of preterm delivery? *Epidemiology* 7, 3339–3345.

Hedge, G. A., Colby, H. D., Goodman, R. L., 1987. *Clinical Endocrine Physiology*. Saunders, Philadelphia.

Hegner, R. E., Wingfield, J. C., 1987. Effects of experimental manipulation of testosterone levels on parental investment in male house sparrows. *Auckland* 104, 462–469.

Hegstrom, C. D., Breedlove, S. M., 1999. Social cues attenuate photoresponsiveness of the male reproductive system in Siberian hamsters *(Phodopus sungorus)*. *J. Biol. Rhythms* 14, 54–61.

Heim, C., Nemeroff, C. B., 2001. The role of childhood trauma in the neurobiology of mood and anxiety disorders: Preclinical and clinical studies. *Biol, Psychiatry.* 49, 1023–1039.

Heim, C., Newport, D. J., Heit, S., Graham, Y. P., Wilcox, M., Bonsall, R., Miller, A. H., Nemeroff, C. B., 2000. Pituitary-adrenal and autonomic responses to stress in women after sexual and physical abuse in childhood. *JAMA* 284, 592–597.

Heim, C., Newport, D. J., Wagner, D., Wilcox, M. M., Miller, A. H., Nemeroff, C. B., 2002. The role of early adverse experience and adulthood stress in the prediction of neuroendocrine stress reactivity in women: A multiple regression analysis. *Depress. Anxiety* 15, 117–125.

Heinrichs, M., Baumgartner, T., Kirschbaum, C., Ehlert, U., 2003. Social support and oxytocin interact to suppress cortisol and subjective responses to psychosocial stress. *Biol. Psychiatry* 54, 1389–1398.

Heinrichs, M., Meinlschmidt, G., Wippich, W., Ehlert, U., Hellhammer, D., 2004. Selective amnesic effects of oxytocin on human memory. *Physiol. Behav.* 83, 31–38.

Heistermann, M., Möle, U., Vervaecke, H., van Elsacker, L., Hodges, J. K., 1996. Application of urinary and fecal steroid measurements for monitoring ovarian function and pregnancy in the bonobo *(Pan paniscus)* and evaluation of perineal swelling patterns in relation to endocrine events. *Biol. Reprod.* 55, 844–853.

Heistermann, M., Ziegler, T., van Schaik, C. P., Launhardt, K., Winkler, P., Hodges, J. K., 2001. Loss of oestrus, concealed ovulation and paternity confusion in free-ranging Hanuman langurs. *Proc. R. Soc. Lond. B Biol. Sci.* 268, 2445–2451.

Hellhammer, D. H., Hubert, W., Schurmeyer, T., 1985. Changes in saliva testosterone after psychological stimulation in men. *Psychoneuroendocrinology* 10, 77–81.

Henry, J. P., 1992. Biological basis of the stress response. *Integr. Physiol. Behav. Sci.* 27, 66–83.

Henry, J. P., Stephens, P. M., 1977. *Stress, Health, and the Social Environment.* Springer, Berlin.

Henzi, S. P., Lucas, J. W., 1980. Observations on the inter-troop movement of adult vervet monkeys *(Cercopithecus aethiops). Folia Primatol.* 33, 220–235.

Herman, L. M., 2002. Exploring the cognitive world of the bottlenose dolphin. In: Bekoff, M., Allen, C., Burghardt, G. M., (Eds.), *The Cognitive Animal: Empirical and Theoretical Perspectives.* MIT Press, Cambridge, MA, pp. 275–283.

Hermans, E. J., Putman, P., Baas, J. M., Koppeschaar, H. P., van Honk, J., 2006. A single administration of testosterone reduces fear-potentiated startle in humans. *Biol. Psychiatry* 59, 872–874.

Hermans, E. J., Putman, P., van Honk, J., 2006. Testosterone administration reduces empathetic behavior: A facial mimicry study. *Psychoneuroendocrinology* 31, 859–866.

Het, S., Ramlow, G., Wolf, O. T., 2005. A meta-analytic review of the effects of acute cortisol administration on human memory. *Psychoneuroendocrinology* 30, 771–784.

Hetherington, E. M., 2003a. Intimate pathways: Changing patterns in close personal relationships across time. *Fam. Relations* 52, 318–331.

——, 2003b. Social support and the adjustment of children in divorced and remarried families. *Childhood* 10, 217–236.

Hews, D. K., Thompson, C. W., Moore, I. T., Moore, M. C., 1997. Population frequencies of alternative male phenotypes in tree lizards: Geographic variation and common-garden rearing studies. *Behav. Ecol. Sociobiol.* 41, 371–380.

Higley, J. D., King, S. T., Hasert, M. F., Champoux, M., Suomi, S. J., Linnoila, M., 1996. Stability of interindividual differences in serotonin function and its relationship to severe aggression and competent social behavior in rhesus macaque females. *Neuropsychopharmacology* 14, 67–76.

Higley, J. D., Linnoila, M., 1997. Low central nervous system serotonergic activity is traitlike and correlates with impulsive behavior. *Ann. N. Y. Acad. Sci.* 836, 39–56.

Higley, J. D., Mehlman, P. T., Poland, R. E., Taub, D. M., Vickers, J., Suomi, S. J., Linnoila, M., 1996. CSF testosterone and 5-HIAA correlate with different types of aggressive behavior. *Biol. Psychiatry* 40, 1067–1082.

Higley, J. D., Suomi, S. J., Linnoila, M., 1992. A longitudinal assessment of CSF monoamine metabolite and plasma cortisol concentrations in young rhesus monkeys. *Biol. Psychiatry* 32, 127–145.

Higley, J. D., Thompson, W. W., Champoux, M., Goldman, D., Hasert, M. F., Kraemer, G. W., Scanlon, J. M., Suomi, S. J., Linnoila, M., 1993. Paternal and maternal genetic and environmental contributions to cerebrospinal fluid monoamine metabolites in rhesus monkeys *(Macaca mulatta). Arch. Gen. Psychiatry* 50, 615–623.

Hill, K., 1988. Macronutrient modifications of optimal foraging theory: An approach using indifference curves applied to some modern foragers. *Hum. Ecol.* 16, 157–197.

Hill, K., Hurtado, A. M., 1996. *Ache life history: The ecology and demography of a foraging people.* Aldine de Gruyten, Hawthorne, NY.

Hill, K., Kaplan, H., Hawkes, K., Hurtado, A. M., n.d. Unpublished data.

Hill, R. A., Lee, P. C., 1998. Predation risk as an influence on group size and composition in primates. *J. Zool. (Lond.)* 245, 447–456.

Hinde, R. A., 1965. Interactions of internal and external factors in integration of canary reproduction. In: Beach, F., (Ed.), *Sex and Behaviour.* Wiley, New York, pp. 381–415.

———, 1983. *Primate Social Relationships: An Integrated Approach.* Sinauer, Sunderland, MA.

Hines, M., 2003. Sex steroids and human behavior: Prenatal androgen exposure and sex-typical play behavior in children. *Ann. N. Y. Acad. Sci.* 1007, 272–282.

Hines, M., Ahmed, S. F., Hughes, I. A., 2003. Psychological outcomes and gender-related development in complete androgen insensitivity syndrome. *Arch. Sex. Behav.* 32, 93–101.

Hines, M., Brook, C., Conway, G. S., 2004. Androgen and psychosexual development: Core gender identity, sexual orientation and recalled childhood gender role behavior in women and men with congenital adrenal hyperplasia (CAH). *J. Sex Res.* 41, 75–81.

Hirschenhauser, K., Oliveira, R., 2006. Social modulation of androgens in male vertebrates: Meta-analyses of the challenge hypothesis. *Anim. Behav.* 71, 265–277.

Hirschenhauser, K., Winkler, H., Oliveira, R. F., 2003. Comparative analysis of male androgen responsiveness to social environment in birds: The effects of mating system and paternal incubation. *Horm. Behav.* 43, 508–519.

Hoff, J. D., Quigley, M. E., Yen, S. S., 1983. Hormonal dynamics at midcycle: A reevaluation. *J. Clin. Endocrinol. Metab.* 57, 792–796.

Hoffmann, M. L., Powlishta, K. K., 2001. Gender segregation in childhood: A test of the interaction style theory. *J. Gen. Psychol.* 162, 298–313.

Hohmann, G., Fruth, B., 2003. Intra- and inter-sexual aggression by bonobos in the context of mating. *Behaviour* 140, 1389–1413.

Hohmann, G., Gerloff, U., Tautz, D., Fruth, B., 1999. Social bonds and genetic ties: Kinship, association and affiliation in a community of bonobos *(Pan paniscus). Behaviour* 136, 1219–1235.

Hollander, E., Bartz, J., Chaplin, W., Phillips, A., Sumner, J., Soorya, L., Anagnostou, E., Wasserman, S., 2007. Oxytocin increases retention of social cognition in autism. *Biol. Psychiatry* 61, 498–503.

Hollander, E., Novotny, S., Hanratty, M., Yaffe, R., DeCaria, C., Aronowitz, B., Mosovich, S., 2003. Oxytocin infusion reduces repetitive behaviors in adults with autistic and Asperger's disorders. *Neuropsychopharmacology* 28, 193–198.

Holloway, C. C., Clayton, D. F., 2001. Estrogen synthesis in the male brain triggers development of the avian song control pathway in vitro. *Nat. Neurosci.* 4, 170–175.

Holman, D. J., Wood, J. W., 2001. Pregnancy loss and fecundability in women. In: Ellison, P. T., (Ed.), *Reproductive Ecology and Human Evolution.* Aldine de Gruyter, Hawthorne, NY, pp. 15–39.

Holman, S., Goy, R., 1995. Experimental and hormonal correlates of care-giving in rhesus macaques. In: Pryce, C., Skuse, D., (Eds.), *Motherhood in Human and Nonhuman Primates.* Karger, Basel, pp. 87–93.

Holmgren, S., Jensen, J., 2001. Evolution of vertebrate neuropeptides. *Brain Res. Bull.* 55, 723–735.

Hooven, C. K., Chabris, C. F., Ellison, P. T., Kosslyn, S. M., 2004. The relationship of male testosterone to components of mental rotation. *Neuropsychologia* 42, 782–790.

Horvat-Gordon, M., Granger, D. A., Schwartz, E. B., Nelson, V. J., Kivlighan, K. T., 2005. Oxytocin is not a valid biomarker when measured in saliva by immunoassay. *Physiol. Behav.* 16, 445–448.

House, J. S., Landis, K. R., Umberson, D., 1988. Social relationships and health. *Science* 241, 540–545.

Howell, N., 1979. *Demography of the Dobe !Kung.* Academic Press, New York.

Howell, S., Westergaard, G., Hoos, B., Chavanne, T. J., Shoaf, S. E., Cleveland, A., Snoy, P. J., Suomi, S. J., Higley, J. D., 2007. Serotonergic influences on life-history outcomes in free-ranging male rhesus macaques. *Am. J. Primatol.* 69, 1–15.

Hrdy, S. B., 1981. *The Woman That Never Evolved.* Harvard University Press, Cambridge, MA.

———, 2000. *Mother Nature: Maternal Instincts and How They Shape the Human Species.* Ballantine Books, New York.

———, 1999. *Mother Nature.* Pantheon Books, New York.

Hu, Y., Goldman, N., 1990. Mortality differentials by marital status: An international comparison. *Demography* 27, 233–250.

Huck, M., Lotter, P., Bohle, U. R., Heymann, E. W., 2005. Paternity and kinship patterns in polyandrous mustached tamarins *(Saguinus mystax). Am. J. Phys. Anthropol.* 127, 449–464.

Huether, G., 1996. The central adaptation syndrome: Psychosocial stress as a trigger for adaptive modifications of brain structure and brain function. *Prog. Neurobiol.* 48, 568–612.

———, 1998. Stress and the adaptive self organization of neuronal connectivity during early childhood. *Int. J. Dev. Neurosci.* 16, 297–306.

Huether, G., Doering, S., Ruger, U., Ruther, E., Schussler, G., 1999. The stress-reaction process and the adaptive modification and reorganization of neuronal networks. *Psychiatry Res.* 87, 83–95.

Huffman, S. L., Chowdhury, A., Allen, H., Nahar, L., 1987. Suckling patterns and post-partum amenorrhea in Bangladesh. *J. Biosoc. Sci.* 19, 171–179.

Huizink, A. C., Mulder, E. J., Buitelaar, J. K., 2004. Prenatal stress and risk for psychopathology: Specific effects or induction of general susceptibility? *Psychol. Bull.* 130, 115–142.

Hull, K. L., Harvey, S., 2000. Growth hormone: A reproductive endocrine-paracrine regulator? *Rev. Reprod.* 5, 175–182.

Hultman, C. M., Sparen, P., Takei, N., Murray, R. M., Cnattingius, S., 1999. Pre-natal and perinatal risk factors for schizophrenia, affective psychosis, and reactive psychosis of early onset: Case-control study. *Br. Med. J.* 318, 421–426.

Hunt, K. E., Hahn, T. P., Wingfield, J. C., 1997. Testosterone implants increase song but not aggression in male Lapland longspurs. *Anim. Behav.* 54, 1177–1192.

———, 1999. Endocrine influences on parental care during a short breeding season: Testosterone and male parental care in Lapland longspurs *(Calcarius lapponicus). Behav. Ecol. Sociobiol.* 45, 360–369.

Hurtado, A. M., Hawkes, K., Hill, K., Kaplan, H., 1985. Female subsistence strategies among Ache hunter-gatherers of eastern Paraguay. *Hum. Ecol.* 13, 29–47.

Hurtado, A. M., Hill, K., 1990. Seasonality in a foraging society: Variation in diet, work effort, fertility, and the sexual division of labor among the Hiwi of Venezuela. *J. Anthropol. Res.* 46, 293–345.

Hurtado, A. M., Hill, K., Kaplan, H., Huntado, I., 1992. Trade-offs between female food acquisition and child care among Hiwi and Ache foragers. *Hum. Nat.* 3, 185–216.

Hutchison, J., Steimer, T., Jaggard, D., 1986. Effects of photoperiod on formation of oestradiol-17b in the dove brain. *J. Endocrinol.* 109, 371–377.

Huttunen, M. O., Niskanen, P., 1978. Pre-natal loss of father and psychiatric disorders. *Arch. Gen. Psychiatry* 35, 429–431.

Ihobe, H., 1992. Male-male relationships among wild bonobos *(Pan paniscus)* at Wamba, Republic of Zaire. *Primates* 33, 163–179.

Imamura, Y., Nakane, Y., Ohta, Y., Kondo, H., 1999. Lifetime prevalence of schizophrenia among individuals pre-natally exposed to atomic bomb radiation in Nagasaki City. *Acta Psychiatr. Scand.* 100, 344–349.

Imperato-McGinley, J., Gautier, T., Peterson, R. E., Shackleton, C., 1986. The prevalence of 5 alpha-reductase deficiency in children with ambiguous genitalia in the Dominican Republic. *J. Urol.* 136, 867–873.

Imperato-McGinley, J., Miller, M., Wilson, J., Peterson, R. E., Shackleton, C., Gajdusek, D. C., 1991. A cluster of male pseudohermaphrodites with 5 alpha-reductase deficiency in Papua New Guinea. *Clin. Endocrinol.* 34, 293–298.

Imperato-McGinley, J., Peterson, R. E., Gautier, T., Sturla, E., 1979. Androgens and the evolution of male-gender identity among male pseudohermaphrodites with 5 alpha-reductase deficiency. *N. Engl. J. Med.* 300, 1233–1237.

Insel, T. R., 1992. Oxytocin—a neuropeptide for affiliation: Evidence from behavioral, receptor autoradiographic, and comparative studies. *Psychoneuroendocrinology* 17, 3–35.

Insel, T. R., Fernald, R. D., 2004. How the brain processes social information: Searching for the social brain. *Annu. Rev. Neurosci.* 27, 687–722.

Insel, T. R., Preston, S., Winslow, J. T., 1995. Mating in the monogamous vole: Behavioral consequences. *Physiol. Behav.* 57, 615–627.

Insel, T. R., Shapiro, L. E., 1992. Oxytocin receptor distribution reflects social organization in monogamous and polygamous voles. *Proc. Natl. Acad. Sci. U. S. A.* 89, 5981–5985.

Insel, T. R., Young, L. J., 2001. The neurobiology of attachment. *Nat. Rev. Neurosci.* 2, 129–136.

Isabella, R. A., Belsky, J., 1991. Interactional synchrony and the origins of infant-mother attachment: A replication study. *Child Dev.* 62, 373–384.

Isbell, L. A., Young, T. P., 2002. Ecological models of female social relationships in primates: Similarities, disparities, and some directions for future clarity. *Behaviour* 139, 177–202.

Işeri, S., Ener, G., Saglam, B., Gedik, N., Ercan, F., Yegen, B., 2005. Oxytocin protects against sepsis-induced multiple organ damage: Role of neutrophils. *J. Surg. Res.* 126, 73–81.

Ivell, R., Schmale, H., Richter, D., 1983. Vasopressin and oxytocin precursors as model preprohormones. *Neuroendocrinology* 37, 235–240.

Jacklin, C. N., DiPietro, J. A., Maccoby, E. E., 1984. Sex-typing behavior and sex-typing pressure in child/parent interaction. *Arch. Sex. Behav.* 13, 413–425.

Jacklin, C. N., Maccoby, E. E., Doering, C. H., 1983. Neonatal sex-steriod hormones and timidity in 6–18-month-old boys and girls. *Dev. Psychobiol.* 16, 163–168.

Jackson, E. D., Payne, J. D., Nadel, L., Jacobs, W. J., 2006. Stress differentially modulates fear conditioning in healthy men and women. *Biol. Psychiatry* 59, 516–522.

Jakobwitz, S., Egan, V., 2006. The dark triad and normal personality traits. *Pers. Ind. Diff.* 40, 331–339.

James, P. J., Nyby, J. G., 2002. Testosterone rapidly affects the expression of copulatory behavior in house mice *(Mus musculus)*. *Physiol. Behav.* 75, 287–294.

James, P. J., Nyby, J. G., Saviolakis, G. A., 2006. Sexually stimulated testosterone release in male mice *(Mus musculus):* Roles of genotype and sexual arousal. *Horm. Behav.* 50, 424–431.

Jankowiak, W., 1995. *Romantic Passion: The Universal Experience?* Columbia University Press, New York.

Jarman, P., 1974. The social organisation of antelope in relation to their ecology. *Behaviour* 58, 167–215.

Jasienska, G., Ellison, P. T., 1998. Physical work causes suppression of ovarian function in women. *Proc. R. Soc. Lond. B Biol. Sci.* 265, 1847–1851.

Jasienska, G., Thune, I., Ellison, P., 2006. Fatness at birth predicts adult susceptibility to ovarian suppression: An empirical test of the "Predictive Adaptive Response" hypothesis. *Proc. Natl. Acad. of Sci. U. S. A.* 103, 12759–12762.

Jasienska, G., Ziomkiewicz, A., Ellison, P. T., Lipson, S. F., Thune, I., 2004. Large breasts and narrow waists indicate high reproductive potential in women. *Proc. R. Soc. Lond. B Biol. Sci.* 271, 1213–1217.

Jasienska, G., Ziomkiewicz, A., Lipson, S. F., Thune, I., Ellison, P. T., 2006. High ponderal index at birth predicts high estradiol levels in adult women. *Am. J. Hum. Biol.* 18, 133–140.

Jeffery, R. W., Forster, J. L., Folsom, A. R., Luepker, R. V., Jacobs, D. R., Jr., Blackburn, H., 1989. The relationship between social status and body mass index in the Minnesota Heart Health Program. *Int. J. Obes.* 13, 59–67.

Jennings, D. H., Moore, M. C., Knapp, R., Matthews, L., Orchinik, M., 2000. Plasma steroid-binding globulin mediation of differences in stress reactivity in alternative male phenotypes in tree lizards, *Urosaurus ornatus. Gen. Comp. Endocrinol.* 120, 289–299.

Jennions, M. D., Macdonald, D. W., 1994. Cooperative breeding in mammals. *Trends Ecol. Evol.* 9, 89–93.

Jennions, M. D., Petrie, M., 2000. Why do females mate multiply? A review of the genetic benefits. *Biol. Rev. Camb. Philos. Soc.* 75, 21–64.

Johnson, A. E., 1992. The regulation of oxytocin receptor binding in the ventromedial hypothalamic nucleus by gonadal steroids. *Ann. N. Y. Acad. Sci.* 652, 357–373.

Johnson, A. M., Wadsworth, J., Wellings, K., Field, J., 1994. *Sexual Attitudes and Lifestyles.* Blackwell, London.

Johnson, D. D. P., McDermott, R., Barrett, E. S., Cowden, J., Wrangham, R., McIntyre, M. H., Rosen, S. P., 2006. Overconfidence in wargames: Experimental evidence on expectations, aggression, gender and testosterone. *Proc. R. Soc. Lond. B Biol. Sci.* 273, 2513–2520.

Johnson, M. H., Everitt, B. J., 1988. *Essential Reproduction.* Blackwell, Oxford.

Johnson, N. J., Backland, E., Sorlie, P. D., Loveless, C. A., 2000. Marital status and mortality: The national mortality study. *Ann. Epidemiol.* 10, 224–238.

Johnson, R. W., Curtis, S. E., Dantzer, R., Bahr, J. M., Kelley, K. W., 1993. Sickness behavior in birds caused by peripheral or central injection of endotoxin. *Physiol. Behav.* 53, 343–348.

Johnson, R. W., Curtis, S. E., Dantzer, R., Kelley, K. W., 1993. Central and peripheral prostaglandins are involved in sickness behavior in birds. *Physiol. Behav.* 53, 127–131.

Johnston, R. E., 1979. Olfactory preference, scent marking, and "proceptivity" in female hamsters. *Horm. Behav.* 13, 21–39.

Johnston, V. S., Hagel, R., Franklin, M., Fink, B., Grammer, K., 2001. Male facial attractiveness: Evidence for hormone-mediated adaptive design. *Evol. Hum. Behav.* 22, 251–267.

Jolley, S. N., Elmore, S., Barnard, K. E., Carr, D. B., 2007. Dysregulation of the hypothalamic-pituitary-adrenal axis in postpartum depression. *Biol. Res. Nurs.* 8, 210–222.

Jolly, A., 1998. Pair-bonding, female aggression and the evolution of lemur societies. *Folia Primatol.* 69, 1–13.

Jones, A., Godfrey, K. M., Wood, P., Osmond, C., Goulden, P., Phillips, D. I. W., 2006. Fetal growth and the adrenocortical response to psychological stress. *J. Clin. Endocrinol. Metab.* 91, 1868–1871.

Jones, D., 1995. Sexual selection, physical attractiveness, and facial neoteny. *Curr. Anthropol.* 36, 723–748.

Josephs, R. A., Newman, M. L., Brown, R. P., Beer, J. M., 2003. Status, testosterone, and human intellectual performance: Stereotype threat as status concern. *Psychol. Sci.* 14, 158–163.

Jost, A., 1970. Hormonal factors in the sex differentiation of the mammalian foetus. *Philos. Trans. R. Soc. Lond. B Biol. Sci.* 259, 119–130.

Julian, T., McKenry, P., 1989. Relationship of testosterone to men's family functioning at mid-life. *Aggress. Behav.* 15, 281–289.

Jurke, M. H., Hagey, L. R., Czekala, N. M., Harvey, N. C., 2001. Metabolites of ovarian hormones and behavioral correlates in captive female bonobos *(Pan paniscus)*. In: Galdikas, B. M. F., Briggs, N. E., Sheeran, L. K., Shapiro, G. L., Goodall, J. (Eds.), *All Apes Great and Small.* Vol. 1, *African Apes.* Springer, New York, pp. 217–229.

Kaiser, S., Sachser, N., 2005. The effects of pre-natal social stress on behaviour: Mechanisms and function. *Neurosci. Biobehav. Rev.* 29, 283–294.

Kaitz, M., Chriki, M., Bear-Scharf, L., Nir, T., Eidelman, A. I., 2000. Effectiveness of primiparae and multiparae at soothing their newborn infants. *J. Gen. Psychol.* 161, 203–215.

Kaitz, M., Good, A., Rokem, A. M., Eidelman, A. I., 1988. Mothers' and fathers' recognition of their newborns' photographs during the postpartum period. *J. Dev. Behav. Pediatr.* 9, 223–226.

Kaitz, M., Lapidot, P., Bronner, R., Eidelman, A., 1992. Parturient women can recognize their infants by touch. *Dev. Psychol.* 28, 35–39.

Kaitz, M., Rokem, A. M., Eidelman, A. I., 1988. Infants' face-recognition by primiparous and multiparous women. *Percept. Mot. Skills* 67, 495–502.

Kajantie, E., Eriksson, J., Barker, D. J., Forsen, T., Osmond, C., Wood, P. J., Andersson, S., Dunkel, L., Phillips, D. I., 2003. Birth size, gestational age and adrenal function in adult life: Studies of dexamethasone suppression and ACTH1–24 stimulation. *Eur. J. Endocrinol.* 149, 569–575.

Kajantie, E., Phillips, D. I., Andersson, S., Barker, D. J., Dunkel, L., Forsen, T., Osmond, C., Tuominen, J., Wood, P. J., Eriksson, J., 2002. Size at birth, gestational age and cortisol secretion in adult life: Foetal programming of both hyper- and hypocortisolism? *Clin. Endocrinol.* 57, 635–641.

Kamel, F., Frankel, A. I., 1978. The effect of medial preoptic area lesions on sexually stimulated hormone release in the male rat. *Horm. Behav.* 10, 10–21.

Kamel, F., Mock, E. J., Wright, W. W., Frankel, A. I., 1975. Alterations in plasma concentrations of testosterone, LH, and prolactin associated with mating in the male rat. *Horm. Behav.* 6, 277–288.

Kanda, N., Tsuchida, T., Tamaki, K., 1996. Testosterone inhibits immunoglobulin production by human peripheral blood mononuclear cells. *Clin. Exp. Immunol.* 106, 410–415.

Kano, T., 1989. The sexual behavior of pygmy chimpanzees. In: Heltne, P. G., Marquardt, L. A., (Eds.), *Understanding Chimpanzees.* Harvard University Press, Cambridge, MA, pp. 176–183.

———, 1992. *The Last Ape: Pygmy Chimpanzee Behavior and Ecology.* Stanford University Press, Stanford, CA.

————, 1996. Male rank order and copulation rate in a unit-group of bonobos at Wamba, Zaire. In: McGrew, W. C., Marchant, L. F., Nishida, T., (Eds.), *Great Ape Societies*. Cambridge University Press, Cambridge, pp. 135–145.

Kaplan, H., 1997. The evolution of the human life course. In: Wachter, K., Finch, C., (Eds.), *Between Zeus and Salmon: The Biodemography of Aging*. National Academy of Sciences, Washington, DC, pp. 175–211.

Kaplan, H., Gurven, M., 2005. The natural history of human food sharing and cooperation: A review and a new multi-individual approach to the negotiation of norms. In: H. Gintis, H., Bowles, S., Boyd, R., Fehr, E., (Eds.), *Moral Sentiments and Material Interests: On the Foundations of Cooperation in Economic Life*. MIT Press, Cambridge, MA, pp. 75–113.

Kaplan, H., Gurven, M., Lancaster, J. B., 2007. Brain evolution and the human adaptive complex: An ecological and social theory. In: Gangestad, S. W., Simpson, J. A., (Eds.), *The Evolution of Mind: Fundamental Questions and Controversies*. Guilford Press, New York, pp. 269–279.

Kaplan, H., Hill, K., Hurtado, A. M., Lancaster, J., 2001. The embodied capital theory of human evolution. In: Ellison, P. T., (Ed.), *Reproductive Ecology and Human Evolution*. Aldine de Gruyter, Hawthorne, NY, pp. 293–318.

Kaplan, H., Hill, K., Lancaster, J. B., Hurtado, A. M., 2000. A theory of human life history evolution: Diet, intelligence, and longevity. *Evol. Anthropol.* 9, 1–30.

Kaplan, H., Lancaster, J. B., 2000. The evolutionary economics and psychology of the demographic transition to low fertility. In: Cronk, L., Irons, W., Chagnon, N., (Eds.), *Human Behavior and Adaptation: An Anthropological Perspective*. Aldine de Gruyter, Hawthorne, NY, pp. 238–322.

————, 2003. An evolutionary and ecological analysis of human fertility, mating patterns, and parental investment. In: Wachter, K. W., Bulatao, R. A., (Eds.), *Offspring: Human Fertility Behavior in Biodemographic Perspective*. National Academies Press, Washington, DC, pp. 170–223.

Kaplan, H., Lancaster, J. B., Anderson, K. G., 1998. Human parental investment and fertility: The life histories of men in Albuquerque. In: Booth, A., Crouter, A., (Eds.), *Men in Families*. Erlbaum, Mahwah, NJ, pp. 55–109.

Kaplan, H., Mueller, T., Gangestad, S., Lancaster, J., 2003. Neural capital and lifespan evolution among primates and humans. In: Finch, C. E., Robine, J. M., Christen, Y., (Eds.), *Brain and Longevity*. Springer, New York, pp. 69–98.

Kaplan, H., Robson, A., 2002. The emergence of humans: The coevolution of intelligence and longevity with intergenerational transfers. *Proc. Natl. Acad. Sci. U. S. A.* 99, 10221–10226.

Kaplan, J. R., Fontenot, M. B., Berard, J., Manuck, S. B., Mann, J. J., 1995. Delayed dispersal and elevated monoaminergic activity in free-ranging rhesus monkeys. *Am. J. Primatol.* 35, 229–234.

Kaplan, J. R., Manuck, S. B., Fontenot, M. B., Mann, J. J., 2002. Central nervous system monoamine correlates of social dominance in cynomolgus monkeys *(Macaca fascicularis)*. *Neuropsychopharmacology* 26, 431–443.

Kaplan, J. R., Phillips-Conroy, J., Fontenot, M. B., Jolly, C. J., Fairbanks, L. A., Mann, J. J., 1999. Cerebrospinal fluid monoaminergic metabolites differ in wild anubis and hybrid *(Anubis-hamadryas)* baboons: Possible relationships to life history and behavior. *Neuropsychopharmacology* 20, 517–524.

Kappeler, P., 2000. *Primate Males.* Cambridge University Press, Cambridge.

Kappeler, P., van Schaik, C., 2004. *Sexual Selection in Primates.* Cambridge University Press, Cambridge.

Kendrick, K. M., Dixson, A. F., 1983. The effect of the ovarian cycle on the sexual behaviour of the common marmoset *(Callithrix jacchus). Physiol. Behav.* 30, 735–742.

Kendrick, K. M., Keverne, E. B., 1991. Importance of progesterone and estrogen priming for the induction of maternal behavior by vaginocervical stimulation in sheep: Effects of maternal experience. *Physiol. Behav.* 49, 745–750.

Kendrick, K. M., Keverne, E. B., Baldwin, B. A., 1987. Intracerebroventricular oxytocin stimulates maternal behavior in the sheep. *Neuroendocrinology* 46, 56–61.

Kennedy, K. I., Visness, C. M., 1992. Contraceptive efficacy of lactational amenorrhoea. *Lancet* 339, 227–230.

Kent, S., Bluthé, R.-M., Kelley, K. W., Dantzer, R., 1992. Sickness behavior as a new target for drug development. *Trends Pharmacol. Sci.* 13, 24–28.

Kester, P., Green, R., Finch, S., Williams, K., 1980. Prenatal "female hormone" administration and psychosexual development in human males. *Psychoneuroendocrinology* 5, 269.

Ketterson, E. D., Nolan, V., Jr., Cawthorn, M. J., Parker, P. G., Ziegenfus, C., 1996. Phenotypic engineering: Using hormones to explore the mechanistic and functional bases of phenotypic variation in nature. *Ibis* 138, 70–86.

Ketterson, E. D., Nolan, V., Jr., Sandell, M., 2005. Testosterone in females: Mediator of adaptive traits, constraint on sexual dimorphism, or both? *Am. Nat.* 166, s85–s98.

Ketterson, E. D., Nolan, V., Jr., Wolf, L., Ziegenfus, C., 1992. Testosterone and avian life histories: Effects of experimentally elevated testosterone on behavior and correlates of fitness in the dark-eyed junco *(Junco hyemalis). Am. Nat.* 140, 980–999.

Keverne, E. B., 1979. Sexual and aggressive behaviour in social groups of talapoin monkeys. *Ciba Found. Symp.* 62, 271–297.

———, 2005. Neurobiological and molecular approaches to attachment and bonding. In: Carter, C. S., Ahnert, L., Grossman, K. E., Hrdy, S. B., Lamb, M. E., Porges, S. W., Sachser, N., (Eds.), *Attachment and Bonding: A New Synthesis.* MIT Press, Cambridge, MA, pp. 101–117.

Keverne, E. B., Martel, F. L., Nevison, C. M., 1996. Primate brain evolution: Genetic and functional considerations. *Proc. R. Soc. Lond. B Biol. Sci.* 263, 689–696.

Keverne, E. B., Nevison, C. M., Martel, F. L., 1997. Early learning and the social bond. Ann. *N. Y. Acad. Sci.* 807, 329–339.

Key, C. A., Aiello, L. C., 2000. A prisoner's dilemma model of the evolution of paternal care. *Folia Primatol.* 71, 77–92.

Keyes, R. C., Botchin, M. B., Kaplan, J. R., Manuck, S. B., Mann, J. J., 1995. Aggression and brain serotonergic responsivity: Response to slides in male macaques. *Physiol. Behav.* 57, 205–208.

Kiecolt-Glaser, J. K., Glaser, R., Cacioppo, J. T., MacCallum, R. C., Synder-smith, M., Kim, C., Malarkey, W. B. 1997. Marital conflict in older adults: Endocrinological and immunological correlates. *Psychosom. Med.* 59, 339–349.

Kim, S., Young, L., Gonen, D., Veenstra-VanderWeele, J., Courchesne, R., Courchesne, E., Lord, C., Leventhal, B. L., Cook, E. H. Jr., Insel, T. R., 2002. Transmission disequilibrium testing of arginine vasopressin receptor 1A (AVPR1A) polymorphisms in autism. *Mol. Psychiatry* 7, 503–507.

Kimura, T., Saji, F., Nishimori, K., Ogita, K., Nakamura, H., Koyama, M., Murata, Y., 2003. Molecular regulation of the oxytocin receptor in peripheral organs. *J. Mol. Endocrinol.* 30, 109–115.

King, J. A., Rosal, M. C., Ma, Y., Reed, G. W., 2005. Association of stress, hostility and plasma testosterone levels. *Neuro. Endocrinol. Lett.* 26, 355–360.

Kingsley, S., 1982. Causes of non-breeding and the development of secondary sexual characteristics in the male orang-utan: A hormonal study. In: de Boer, L. E. M., (Ed.), *The Orang-utan: Its Biology and Conservation.* W. Junk, The Hague, pp. 215–229.

———, 1988. Physiological development of male orang-utans and gorillas. In: Schwartz, J. H., (Ed.), *Orang-utan Biology.* Oxford University Press, New York, pp. 123–131.

Kinsey, A. C., Pomeroy, W. B., Martin, C. E., Gebhard, P. H., 1948. *Sexual Behavior in the Human Male.* Saunders, Philadelphia.

Kirkwood, T. B. L., Rose, M. R., 1991. Evolution of senescence: Late survival sacrificed for reproduction. *Trans. R. Soc. Lond. B Biol. Sci.* 332, 15–24.

Kirsch, P., Esslinger, C., Chen, Q., Mier, D., Lis, S., Siddhanti, S., Gruppe, H., Mattay, V. S., Gallhofer, B., Meyer-Lindenberg, A., 2005. Oxytocin modulates neural circuitry for social cognition and fear in humans. *J. Neurosci.* 25, 11489–11493.

Kirschbaum, C., Hellhammer, D. H., 1994. Salivary cortisol in psychoneuroendocrine research: Recent developments and applications. *Psychoneuroendocrinology* 19, 313–333.

Kitaysky, A. S., Kitaiskaia, E. V., Piatt, J. F., Wingfield, J. C., 2003. Benefits and costs of increased levels of corticosterone in seabird chicks. *Horm. Behav.* 43, 140–149.

Kitaysky, A. S., Wingfield, J. C., Piatt, J. F., 2001. Corticosterone facilitates begging and affects resource allocation in the black-legged kittiwake. *Behav. Ecol.* 12, 619–625.

Klaus, M. H., Kennell, J., 1997. The doula: An essential ingredient of childbirth rediscovered. *Acta Paediatr.* 86, 1034–1036.

Kleiman, D. G., 1977. Monogamy in mammals. *Q. Rev. Biol.* 52, 39–69.

Klein, H. 1991. Couvade syndrome: Male counterpart to pregnancy. *Int. J. Psychiatry Med.* 21, 57–69.

Klinkova, E., Heistermann, M., Hodger, J. K., 2004. Social parameters and urinary testosterone level in male chimpanzees *(Pan troglodytes).* *Horm. Behav.* 46, 474–481.

Kluger, M. J., 1979. Fever in ectotherms: Evolutionary implications. *Am. Zool.* 19, 295–304.

Knapp, R., Moore, M. C., 1996. Male morphs in tree lizards, *Urosaurus ornatus*, have different delayed responses to aggressive encounters. *Anim. Behav.* 52, 1045–1055.

Knickmeyer, R., Baron-Cohen, S., Fane, B. A., Wheelwright, S., Mathews, G. A., Conway, G. S., Brook, C. G. D., Hines, M., 2006. Androgens and autistic traits: A study of individuals with congenital adrenal hyperplasia. *Horm. Behav.* 50, 148–153.

Knickmeyer, R., Baron-Cohen, S., Raggatt, P., Taylor, K., Hackett, G., 2006. Fetal testosterone and empathy. *Horm. Behav.* 49, 282–292.

Knickmeyer, R., Wheelwright, S., Taylor, K., Raggatt, P., Hackett, G., and Baron-Cohen, S., 2005. Gender-typed play and amniotic testosterone. *Dev. Psychol.* 41, 517–528.

Knott, C. D., 1997. Field collection and preservation of urine in orangutans and chimpanzees. *Trop. Biodivers.* 4, 95–102.

———, 1998. Changes in orangutan caloric intake, energy balance, and ketones in response to fluctuating fruit availability. *Int. J. Primatol.* 19, 1061–1079.

———, 1999. Reproductive physiological and behavioral responses of orangutans in Borneo to fluctuations in food availability. PhD. Diss., Harvard University.

———, 2001. Female reproductive ecology of the apes. In: Ellison, P. T., (Ed.), *Reproductive Ecology and Human Evolution*. Aldine de Gruyter, Hawthorne, NY, pp. 429–464.

———, in press. Orangutans: Direct sexual coercion in a solitary species. In: Muller, M. N., Wrangham, R. W., (Eds.), *Sexual Coercion in Primates: An Evolutionary Perspective on Male Aggression against Females*. Harvard University Press, Cambridge, MA.

Knott, C. D., Emery Thompson, M., n.d. Unpublished data.

Knott, C. D., Emery Thompson, M. Stumpf, R. M., 2007. Sexual coercion and Mating strategies of wild Bornean orangutans. *Am. J. Phys. Anthrop. Suppl.* 44, 145.

Knott, C. D., Kahlenberg, S. M., 2006. Orangutans in perspective: Forced copulations and female mating resistance. In: Campbell, C. J., Fuentes, A., Mackinnon, K. C, Panger, M. A., Bearder, S. K., (Eds.), *Primates in Perspective*. Oxford University Press, New York, pp. 290–304.

Knox, S., Uvnäs-Moberg, K., 1998. Social isolation and cardiovascular disease: An atherosclerotic pathway? *Psychoneuroendocrinology* 23, 877–890.

Knussmann, R., Christiansen, K., Couwenbergs, C., 1986. Relations between sex hormone levels and sexual behavior in men. *Arch. Sex. Behav.* 15, 429–445.

Koehl, M., Damaudéry M., Dulluc, J., Van Reeth, O., Le Moal, M., Maccari, S., 1999. Prenatal stress alters circadian activity of hypothalamo-pituitary-adrenal axis and hippocampal corticosterold receptors in adult rats of both genders. *J. Neurobiol.* 40, 302–315.

Koenig, J. I., Kirkpatrick, B., Lee, P., 2002. Glucocorticoid hormones and early brain development in schizophrenia. *Neuropsychopharmacology* 27, 309–318.

Kolb, B., Pellis, S., Robinson, T. E., 2004. Plasticity and functions of the orbital frontal cortex. *Brain Cogn.* 55, 104–115.

Kondo, T., Zakany, J., Innis, J. W., Duboule, D., 1997. Of fingers, toes and penises. *Nature* 390, 29.

Konner, M., 1977. Infancy among Kalahari Desert San. In: Leiderman, P. H., Tulkin, S. R., Rosenfeld, A., (Eds.), *Culture and I infancy: Variations in Human Experience*. Academic Press, New York, pp. 287–328.

Kanner, M., Worthman, C., 1980. Nursing frequency, gonadal function, and birth spacing among! Kung huntes-gatherers. *Science* 207, 788–791.

Koolhaas, J. M., de Boer, S. F., Buwalda, B., Van der Vegt, B. J., Carere, C., Groothuis, A. G. G., 2001. How and why coping systems vary among individuals. In: Broom, D. M., (Ed.), *Coping with Challenge: Welfare in Animals Including Humans*. Dahlem Workshop Report #87. Dahlem UP, Dahlem, pp. 197–209.

Koolhaas, J. M., de Boer, S. F., de Ruiter, A. J. H., Meerlo, P., Sgoifo, A., 1997. Social stress in rats and mice. *Acta Physiol. Scand.* 161, 69–72.

Koolhaas, J. M., Korte, S. M., Boer, S. F., Van der Vegt, B. J., Van Renen, C. G., Hopster, H., De Jong, I. C., Ruis, M. A. W., Blokhuis, H. J., 1999. Coping styles in animals: Current status in behavior and stress-physiology. *Neurosci. Biobehav. Rev.* 23, 925–935.

Korte, S. M., Koolhaas, J. M., Wingfield, J. C., McEwen, B. S., 2005. The Darwinian concept of stress: Benefits of allostasis and costs of allostatic load and the trade-offs in health and disease. *Neurosci. Biobehav. Rev.* 29, 3–38.

Kosfeld, M., Heinrichs, M., Zak, P. J., Fischbacher, U., Fehr, E., 2005. Oxytocin increases trust in humans. *Nature* 435, 673–676.

Kostoglou-Athanassiou, I., Treacher, D., Wheeler, T., Forsling, M., 1998. Bright light exposure and pituitary hormone secretion. *Clin. Endocrinol.* 48, 73–79.

Kozorovitskiy, Y., Hughes, M., Lee, K., Gould, E., 2006. Fatherhood affects dendritic spines and vasopressin V1a receptors in the primate prefrontal cortex. *Nat. Neurosci.* 9, 1094–1095.

Kraemer, B., Noll, T., Delsignore, A., Milos, G., Schnyder, U., Hepp, U., 2006. Finger length ratio (2D:4D) and dimensions of sexual orientation. *Neuropsychobiology* 53, 210–214.

Kramer, K. L., 2006. Children's help and the pace of reproduction: Cooperative breeding in humans. *Evol. Anthropol.* 14, 224–237.

Kramer, K. M., Choe, C., Carter, C. S., Cushing, B. S., 2006. Developmental effects of oxytocin on neural activation and neuropeptide release in response to social stimuli. *Horm. Behav.* 49, 206–214.

Kraus, C., Heistermann, M., Kappeler, P. M., 1999. Physiological suppression of sexual function of subordinate males: A subtle form of competition. *Physiol. Behav.* 66, 855–861.

Krebs, C. J., Boonstra, R., Boutin, S., Sinclair, A. R. E., 2001. What drives the 10-year cycle of snowshoe hares? *BioScience* 51, 25–35.

Kritzer, M. F., 1997. Selective colocalization of immunoreactivity for intracellular gonadal hormone receptors and tyrosine hydroxylase in the ventral tegmental area, substantia nigra, and retrorubral fields in the rat. *J. Comp. Neurol.* 379, 247–260.

Krpan, K. M., Coombs, R., Zinga, D., Steiner, M., Fleming, A. S., 2005. Experiential and hormonal correlates of maternal behavior in teen and adult mothers. *Horm. Behav.* 47, 112–122.

Kruger, T., Exton, M. S., Pawlak, C., von zur Muhlen, A., Hartmann, U., Schedlowski, M., 1998. Neuroendocrine and cardiovascular response to sexual arousal and orgasm in men. *Psychoneuroendocrinology* 23, 401–411.

Kruger, T., Haake, P., Chereath, D., Knapp, W., Janssen, O., Exton, M., Schedlowski, M., Hartmann, U., 2003. Specificity of the neuroendocrine response to orgasm during sexual arousal in men. *J. Endocrinol.* 177, 57–64.

Kruijt, J. P., de Vos, G. J., 1988. Individual variation in reproductive success in male black grouse *Tetrao tetrix* L. In: Clutton-Brock, T. H., (Ed.), *Reproductive Success*. Princeton University Press, Princeton, NJ, pp. 279–304.

Kudryavtseva, N. N., Amstislavskaya, T. G., Kucheryavy, S., 2004. Effects of repeated aggressive encounters on approach to a female and plasma testosterone in male mice. *Horm. Behav.* 45, 103–107.

Kulhmann, S., Wolf, O. T., 2006. A non-arousing test situation abolishes the impairing effects of cortisol on delayed memory retrieval in healthy women. *Neurosci. Lett.* 399, 268–272.

Kuukasjarvi, S., Eriksson, P., Koskela, E., Mappes, T., Nissinen, K., Rantala, M. J., 2004. Attractiveness of women's body odors over the menstrual cycle: The role of oral contraceptives and receiver sex. *Behav. Ecol.* 15, 579–584.

Kyes, R. C., Botchin, M. B., Kaplan, J. R., Manuck, S. B., Mann, J. J., 1995. Aggression and brain serotonergic responsivity: Response to slides in male macaques. *Physiol. Behav.* 57, 205–208.

La Spada, A. R., Wilson, E. M., Lubahn, D. B., Harding, A. E., Fischbeck, K. H., 1991. Androgen receptor gene mutations in X-linked spinal and bulbar muscular atrophy. *Nature* 352, 77–79.

Labrie, F., Bélanger, A., Simard, J., Luu-The, V., Labrie, C., 1995. DHEA and peripheral androgen and estrogen formation: Intracrinology. *Ann. N. Y. Acad. Sci.* 774, 16–28.

Lack, D., 1968. *Ecological Adaptations for Breeding in Birds*. Methuen, London.

LaGasse, L. L., Neal, A. R., Lester, B. M., 2005. Assessment of infant cry: Acoustic cry analysis and parental perception. *Ment. Retard. Dev. Disabil. Res. Rev.* 11, 83–93.

Lager, C., Ellison, P. T., 1990. Effect of moderate weight loss on ovarian function assessed by salivary progesterone measurements. *Am. J. Hum. Biol.* 2, 303–312.

Lancaster, J. B., 1986. Human adolescence and reproduction: An evolutionary perspective. In: Lancaster, J. B., Hamburg, B. A., (Eds.), *School-Age Pregnancy and Parenthood*. Aldine de Gruyter, Hawthorne, NY, pp. 17–39.

———, 1991. A feminist and evolutionary biologist looks at women. *Yrbk. Phys. Anthropol.* 34, 1–11.

———, 1997. The evolutionary history of human parental investment in relation to population growth and social stratification. In: Gowaty, P. A., (Ed.), *Feminism and Evolutionary Biology*. Chapman and Hall, New York, pp. 466–489.

Lancaster, J. B., Kaplan, H., 1992. Human mating and family formation strate-gies: The effects of variability among males in quality and the allocation of mating effort and parental investment. In: Nishida, T., McGrew, W. C., Marler, P., Pickford, M., de Waal, F., (Eds.), *Topics in Primatology: Human Origins*. University of Tokyo Press, Tokyo, pp. 21–33.

Lancaster, J. B., Kaplan, H., Hill, K., Hurtado, A. M., 2000. The evolution of life history, intelligence and diet among chimpanzees and human foragers. In: Tonneau, F., Thompson, N. S., (Eds.), *Perspectives in Ethology: Evolution, Culture and Behavior*. Vol. 13. Plenum, New York, pp. 47–72.

Landgraf, R., 1995. Mortyn Jones Memorial Lecture. Intracerebrally released vasopressin and oxytocin: Measurement, mechanisms and behavioural con-sequences. *J. Neuroendocrinol.* 7, 243–253.

Landgraf, R., Holsboer, F., 2005. The involvement of neuropeptides in evolu-tion, signaling, behavioral regulation and psychopathology: Focus on vaso-pressin. *Drug Dev. Res.* 65, 185–190.

Landgraf, R., Neumann, I., 2004. Vasopressin and oxytocin release within the brain: A dynamic concept of multiple and variable modes of neuropeptide communication. *Front. Neuroendocrinol.* 25, 150–176.

Landgraf, R., Wotjak, C. T., Neumann, I. D., Engelmann, M., 1998. Release of vasopressin within the brain contributes to neuroendocrine and behavioural regulation. *Prog. Brain Res.* 119, 201–220.

Lassek, W. D., Gaulin, S. J. C., 2006. Changes in body fat distribution in relation to parity in American women: A covert form of maternal depletion. *Am. J. Phys. Anthropol.* 131, 295–302.

Laumann, E. O., Gagnon, J. H., Michael, R. T., Michaels, S., 1994. *The Social Organization of Sexuality: Sexual Practices in the United States*. University of Chicago Press, Chicago.

Law Smith, M. J., Perrett, D. I., Jones, B. C., Cornwell, R. E., Moore, F. R., Feinberg, D. R., Boothroyd, L. G., Durrani, S. J., Stirrat, M. R., Whiten, S., Pitman, R. M., Hillier, S. G., 2006. Facial appearance is a cue to oestrogen levels in women. *Proc. R. Soc. Lond. B Biol. Sci.* 273, 135–140.

Lazarus, J., Inglis, I. R., 1986. Shared and unshared parental investment, parent-offspring conflict and brood size. *Anim. Behav.* 34, 1791–1804.

Le Boeuf, B. J., Reiter, J., 1988. Lifetime reproductive success in northern ele-phant seals. In: Clutton-Brock, T. H., (Ed.), *Reproductive Success*. Univer-sity of Chicago Press, Chicago, pp. 344–362.

Leavitt, L., Donovan, W., 1979. Perceived infant temperament, focus of control, and maternal physiological response to infant gaze. *J. Res. Pers.* 13, 267–278.

LeDoux, J. E., 2000. Emotion circuits in the brain. *Annu. Rev. Neurosci.* 23, 155–184.

Lee, A., Clancy, S., Fleming, A. S., 2000. Mother rats bar-press for pups: Effects of lesions of the MPOA and limbic sites on maternal behavior and operant responding for pup-reinforcement. *Behav. Brain Res.* 108, 215–231.

Lee, A., Li, M., Watchus, J., Fleming, A. S., 1999. Neuroanatomical basis of ma-ternal memory in postpartum rats: Selective role for the nucleus accumbens. *Behav. Neurosci.* 113, 523–538.

Lee, K. B., Ashton, M. C., 2005. Psychopathy, Machiavellianism, and narcissism in the five-factor model and the HEXACO model of personality structure. *Pers. Ind. Diff.* 38, 1571–1582.

Lee, P. C., 1987. Allomothering among African elephants. *Anim. Behav.* 35, 278–291.

———, 1989. Family structure, communal care and female reproductive effort. In: Standen, V., Foley, R., (Eds.), *Comparative Socioecology*. Blackwell, Oxford, pp. 323–340.

———, 1994. Social structure and evolution. In: Slater, P., Halliday, T., (Eds.), *Behaviour and Evolution*. Cambridge University Press, Cambridge, pp. 266–303.

———, 1999. Comparative ecology of post-natal growth and weaning among haplorhine primates. In: Lee, P. C., (Ed.), *Comparative Primate Socioecology*. Cambridge University Press, Cambridge, pp. 111–139.

Lee, P. C., Kappeler, P. M., 2003. Socio-ecological correlates of phenotypic plasticity in primate life history. In: Kappeler, P. M., Pereira, M., (Eds.), *Primate Life Histories*. University of Chicago Press, Chicago, pp. 41–65.

Lee, P. C., Majluf, P., Gordon, I. J., 1991. Growth, weaning and maternal investment from a comparative perspective. *J. Zool. (Lond.)* 225, 99–114.

Lee, P. R., Brady, D. L., Shapiro, R. A., Dorsa, D. M., Koenig, J. I.,2005. Social interaction deficits caused by chronic phencyclidine administration are reversed by oxytocin. *Neuropsychopharmacology* 30, 1883–1894.

Lehrman, D. S., 1965. Interaction between internal and external environments in the regulation of the reproductive cycle of the ring dove. In: Beach, F., (Ed.), *Sex and Behavior*. Wiley, New York, pp. 355–380.

Leibenluft, E., Gobbini, M. I., Harrison, T., Haxby, J. V., 2004. Mothers' neural activation in response to pictures of their children and other children. *Biol. Psychiatry* 56, 225–232.

Leiderman, P., Leiderman, M., 1977. Economic change and infant care in an East African agricultural community. In: Leiderman, P. H., Tulkin, S. R., Rosenfeld, A., (Eds.), *Culture and Infancy: Variations in Human Experience*. Academic Press, New York, pp. 405–438.

Leonard, W. R., 2003. Measuring human energy expenditure: What have we learned from the flex-heart rate method? *Am. J. Hum. Biol.* 15, 479–489.

Levine, S., 2001. Primary social relationships influence the development of the hypothalamic-pituitary-adrenal axis in the rat. *Physiol. Behav.* 73, 255–260.

———, 2005. Developmental determinants of sensitivity and resistance to stress. *Psychoneuroendocrinology* 30, 939–946.

Lewin, R., 1988. Life history patterns emerge in primate study. *Science* 242, 1636–1637.

Lewis, P. R., Brown, J. B., Renfree, M. B., Short, R. V., 1991. The resumption of ovulation and menstruation in a well-nourished population of women breastfeeding for an extended period of time. *Fertil. Steril.* 55, 529–536.

Li, M., Fleming, A. S., 2003. The nucleus accumbens shell is critical for normal expression of pup-retrieval in postpartum female rats. *Behav. Brain Res.* 145, 99–111.

Liberzon, I., Young, E. A., 1997. Effects of stress and glucocorticoids on CNS oxytocin receptor binding. *Psychoneuroendocrinology* 22, 411–422.

Lier, L., 1988. Mother-infant relationship in the first year of life. *Acta Paediatr. Scand. Suppl.* 344, 31–42.

Light, K. C., Grewen, K. M., Amico, J. A., 2005. More frequent hugs and higher oxytocin levels are linked to lower blood pressure and heart rate in pre-menopausal women. *Biol. Psychol.* 69, 5–21.

Light, K. C., Grewen, K., Amico, J., Boccia, M., Brownley, K., Johns, J., 2004. Deficits in plasma oxytocin responses and increased negative affect, stress, and blood pressure in mothers with cocaine exposure during pregnancy. *Addict. Behav.* 29, 1541–1564.

Light, K., Smith, T., Johns, J., Brownley, K., Hofheimer, J., 2000. Oxytocin responsivity in mothers of infants: A preliminary study of relationships with blood pressure during laboratory stress and normal ambulatory activity. *Health Psychol.* 19, 560–567.

Lim, M. M., Hammock, E. A., Young, L. J., 2004. The role of vasopressin in the genetic and neural regulation of monogamy. *J. Neuroendocrinol.* 16, 325–332.

Lim, M. M., Wang, Z., Olazabal, D. E., Ren, X., Terwilliger, E. F., Young, L. J., 2004. Enhanced partner preference in a promiscuous species by manipulating the expression of a single gene. *Nature* 429, 754–757.

Lim, M. M., Young, L. J., 2004. Vasopressin-dependent neural circuits underlying pair bond formation in the monogamous prairie vole. *Neuroscience* 125, 35–45.

———, 2006. Neuropeptidergic regulation of affiliative behavior and social bonding in animals. *Horm. Behav.* 50, 506–517.

Lindburg, D. G., 1987. Seasonality of reproduction in primates. In: Mitchell, G., Erwin, J. M., (Eds.), *Comparative Primate Biology.* Vol. 2B, *Behavior, Cognition, and Motivation.* Alan R. Liss, New York, pp. 167–218.

Linnamo, V., Pakarinen, A., Komi, P. V., Kraemer, W. J., Hakkinen, K., 2005. Acute hormonal responses to submaximal and maximal heavy resistance and explosive exercises in men and women. *J. Strength Cond. Res.* 19, 566–571.

Linnet, K. M., Dalsgaard, S., Obel, C., Wisborg, K., Henriksen, T. B., Rodriguez, A., Kotimaa, A., Moilanen, I., Thomsen, P. H., Olsen, J., Jarvelin, M. R., 2003. Maternal lifestyle factors in pregnancy risk of attention deficit hyperactivity disorder and associated behaviors: Review of the current evidence. *Am. J. Psychiatry* 160, 1028–1040.

Lipowicz, A., Gronkiewicz, S., Malina, R. M., 2002. Body mass index, overweight and obesity in married and never married men and women in Poland. *Am. J. Hum. Biol.* 14, 468–475.

Lippa, R. A., 2003. Are 2D:4D finger-length ratios related to sexual orientation? Yes for men, no for women. *J. Pers. Soc. Psychol.* 85, 179–188.

———, 2006. Finger lengths, 2D:4D ratios, and their relation to gender-related personality traits and the Big Five. *Biol. Psychol.* 71, 116–121.

Lipson, S. F., Ellison, P. T., 1996. Comparison of salivary steroid profiles in naturally occurring conception and non-conception cycles. *Hum. Reprod.* 11, 2090–2096.

Litvinova, E. A., Kudaeva, O. T., Mershieva, L. V., Moshkin, M. P., 2005. High circulating testosterone abolishes decline in scent attractiveness in antigen-treated male mice. *Anim. Behav.* 69, 511–517.

Lloyd, S. A., Dixson, A. F., 1988. Effects of hypothalamic lesions upon the sexual and social behaviour of the male common marmoset *(Callithrix jacchus).* *Brain Res.* 829, 55–68.

Logan, C. A., Wingfield, J. C., 1990. Autumnal territorial aggression is independent of plasma testosterone in mockingbirds. *Horm. Behav.* 24, 568–581.

Lonstein, J. S., De Vries, G. J., 1999a. Comparison of the parental behavior of pair-bonded female and male prairie voles *(Microtus ochrogaster).* *Physiol. Behav.* 66, 33–40.

———, 1999b. Sex differences in the parental behaviour of adult virgin prairie voles: Independence from gonadal hormones and vasopressin. *J. Neuroendocrinol.* 11, 441–449.

Lorberbaum, J., Newman, J., Horwitz, A., Dubno, J., Lydiard, R., Hamner, M., Bohning, D., George, M., 2002. A potential role for thalamocingulate circuitry in human maternal behavior. *Biol. Psychiatry* 51, 431–445.

Lou, H. C., Hansen, D., Nordentoft, M., Pryds, O., Jensen, F., Nim, J., Hemmingsen, R., 1994. Pre-natal stressors of human life affect fetal brain development. *Dev. Med. Child Neurol.* 36, 826–832.

Loup, F., Tribollet, E., Dubois-Dauphin, M., Dreifuss, J. J., 1991. Localization of high-affinity binding sites for oxytocin and vasopressin in the human brain. An autoradiographic study. *Brain Res.* 555, 220–232.

Lovejoy, J., Wallen, K., 1990. Adrenal suppression and sexual initiation in group-living female rhesus monkeys. *Horm. Behav.* 24, 256–269.

Lovic, V., Fleming, A. S., 2004. Artificially-reared female rats show reduced pre-pulse inhibition and deficits in the attentional set shifting task—reversal of effects with maternal-like licking stimulation. *Behav. Brain Res.* 148, 209–219.

Low, B. S., 2000. *Why Sex Matters.* Princeton University Press, Princeton, NJ.

Lu, P. H., Masterman, D. A., Mulnard, R., Cotman, C., Miller, B., Yaffe, K., Reback, E., Porter, V., Swerdloff, R., Cummings, J. L., 2006. Effects of testosterone on cognition and mood in male patients with mild Alzheimer disease and healthy elderly men. *Arch. Neurol.* 63, 177–185.

Luboshitzky, R., Aviv, A., Hefetz, A., Herer, P., Shen-Orr, Z., Lavie, L., Lavie, P., 2002. Decreased pituitary-gonadal secretion in men with obstructive sleep apnea. *J. Clin. Endocrinol. Metab.* 87, 3394–3398.

Ludington-Hoe, S. M., Cong, X., Hashemi, F., 2002. Infant crying: Nature, physiologic consequences, and select interventions. *Neonatal Netw.* 21, 29–36.

Ludwig, M., 1995. Functional role of intrahypothalamic release of oxytocin and vasopressin: Consequences and controversies. *Am. J. Physiol. Endocrinol. Metab.* 31, E537–E545.

Lukas, K. E., Barkauskas, R. T., Maher, S. A., Jacobs, B. A., Bauman, J. E., Henderson, A. J., Calcagno, J. M., 2002. Longitudinal study of delayed reproductive success in a pair of white-cheeked gibbons *(Hylobates leucogenys).* *Zoo Biol.* 21, 413–434.

Lukas, W. D., Campbell, B. C., Ellison, P. T., 2004. Testosterone, aging, and body composition in men from Harare, Zimbabwe. *Am. J. Hum. Biol.* 16, 704–712.

Lundy, B. L., Jones, N. A., Field, T., Nearing, G., Davalos, M., Pietro, P. A., Schanberg, S., Kuhn, C., 1999. Prenatal depression effects on neonates. *Infant Behav. Dev.* 22, 119–129.

Lunn, P. G., Austin, S., Prentice, A. M., Whitehead, R. G., 1984. The effect of improved nutrition on plasma prolactin concentrations and postpartum infertility in lactating Gambian women. *Am. J. Clin. Nutr.* 39, 227–235.

Lunn, P. G., Watkinson, M., Prentice, A. M., Morrell, P., Austin, S., Whitehead, R. G., 1981. Maternal nutrition and lactational amenorrhoea. *Lancet* 1, 1428–1429.

Lunn, S. F., McNeilly, A. S., 1982. Failure of lactation to have a consistent effect on interbirth interval in the common marmoset, *Callithrix jacchus jacchus*. *Folia Primatol.* 37, 99–105.

Lupien, S. J., Fiocco, A., Wan, N., Maheu, F., Lord, C., Schramek, T. Tu, M., 2005. Stress hormones and human memory function across the lifespan. *Psychoneuroendocrinology* 30, 225–242.

Lupien, S. J., Gillin, C. J., Hauger, R. L., 1999. Working memory is more sensitive than declarative memory to the acute effects of corticosteroids: A dose-response study in humans. *Behav. Neurosci.* 113, 420–430.

Lutchmaya, S., Baron-Cohen, S., 2002. Human sex differences in social and non-social looking preferences, at 12 months of age. *Infant Behav. Dev.* 25, 319–325.

Lutchmaya, S., Baron-Cohen, S., Raggatt, P., 2002. Foetal testosterone and eye contact in 12-month-old human infants. *Infant Behav. Dev.* 25, 327–335.

Lutchmaya, S., Baron-Cohen, S., Raggatt, P., Knickmeyer, R., Manning, J. T., 2004. 2nd to 4th digit ratios, fetal testosterone and estradiol. *Early Hum. Dev.* 77, 23–28.

Luxen, M. F., Buunk, B. P., 2005. Second-to-fourth digit ratio related to verbal and numerical intelligence and the Big Five. *Pers. Ind. Diff.* 39, 959–966.

Lycett, J. E., Henzi, S. P., Barrett, L., 1998. Maternal investment in mountain baboons and the hypothesis of reduced care. *Behav. Ecol. Sociobiol.* 42, 49–56.

Lynch, J. W., Ziegler, T. E., Strier, K. B., 2002. Individual and seasonal variation in fecal testosterone and cortisol of wild male tufted capuchin monkeys, *Cebus apella nigritus*. *Horm. Behav.* 41, 275–287.

Lynn, S. E., Hayward, L. S., Benowitz-Fredericks, Z. M., Wingfield, J. C., 2002. Behavioral insensitivity to supplementary testosterone during the parental phase in the chestnut-collared longspur *(Calcarius ornatus)*. *Anim. Behav.* 63, 795–803.

Lynn, S. E., Walker, B. G., Wingfield, J. C., 2005. A phylogenetically controlled test of hypotheses for behavioral insensitivity to testosterone in birds. *Horm. Behav.* 47, 170–177.

Lynn, S. E., Wingfield, J. C., 2003. Male chestnut-collared longspurs are essential for nestling survival: A removal study. *Condor* 105, 154–158.

Lyons, D. M., Mendoza, S. P., Mason, W. A., 1994. Psychosocial and hormonal aspects of hierarchy formation in groups of male squirrel monkeys. *Am. J. Primatol.* 32, 109–122.

Maccari, S., Darnaudery, M., Morley-Fletcher, S., Zuena, A. R., Cinque, C., Van Reeth, O., 2003. Pre-natal stress and long-term consequences: Implications of glucocorticoid hormones. *Neurosci. Biobehav. Rev.* 27, 119–227.

Maccoby, E. E., 1998. *The Two Sexes: Growing Up Apart, Coming Together.* Harvard University Press, Cambridge, MA.

——, 2000. Perspectives on gender development. *Int. J. Behav. Dev.* 24, 398–406.

——, 2002. Gender and group process: A developmental perspective. *Curr. Dir. Psychol. Sci.* 11, 54–58.

Maccoby, E. E., Snow, M. E., Jacklin, C. N., 1984. Children's dispositions and mother-child interaction at 12 and 18 months: A short-term longitudinal study. *Dev. Psychol.* 20, 459–472.

MacKinnon, J. R., 1979. Reproductive behavior in wild orangutan populations. In: Hamburg, D. A., McCown, E. R., (Eds.), *The Great Apes.* Benjamin Cummings, Menlo Park, CA, pp. 257–273.

Macrides, F., Bartke, A., Dalterio, S., 1975. Strange females increase plasma testosterone levels in male mice. *Science* 189, 1104–1105.

Macrides, F., Bartke, A., Fernandez, F., D'Angelo, W., 1974. Effects of exposure to vaginal odor and receptive females on plasma testosterone in the male hamster. *Neuroendocrinology* 15, 355–364.

Madigan, S., Bakermans-Kranenburg, M. J., Van Ijzendoorn, M. H., Moran, G., Pederson, D. R., Benoit, D., 2006. Unresolved states of mind, anomalous parental behavior, and disorganized attachment: A review and meta-analysis of a transmission gap. *Attach. Hum. Dev.* 8, 89–111.

Maestripieri, D., 2001. Female-biased maternal investment in rhesus macaques. *Folia Primatol.* 72, 44–47.

Maestripieri, D., Megna, N. L., 2000. Hormones and behavior in rhesus macaque abusive and nonabusive mothers. 2. Mother-infant interactions. *Physiol. Behav.* 71, 43–49.

Maggioncalda, A. N., Czekala, N. M., Sapolsky, R. M., 2000. Growth hormone and thyroid stimulating hormone concentrations in captive male orangutans: Implications for understanding developmental arrest. *Am. J. Primatol.* 50, 67–76.

——, 2002. Male orangutan subadulthood: A new twist on the relationship between chronic stress and developmental arrest. *Am. J. Phys. Anthropol.* 118, 25–32.

Maggioncalda, A. N., Sapolsky, R. M., Czekala, N. M., 1999. Reproductive hormone profiles in captive male orangutans: Implications for understanding developmental arrest. *Am. J. Phys. Anthropol.* 109, 19–32.

Magid, K. W., Chatterton, R. T., Uddin Ahamed, F., Bentley, G. R., 2006. No effect of marriage or fatherhood on salivary testosterone levels in Bangladeshi men. Poster presented at the Human Behavior and Evolution Society meeting, Philadelphia, PA.

Maisey, D. S., Vale, E. L. E., Cornelissen, P. L., Tovee, M. J., 1999. Characteristics of male attractiveness to women. *Lancet* 353, 1500.

Majdandzic, M., van den Boom, D. C., 2007. Multimethod longitudinal assessment of temperament in early childhood. *J. Pers.* 75, 121–167.

Majzoub, J. A., Karalis, K. P., 1999. Placental corticotropin-releasing hormone: Function and regulation. *Am. J. Obstet. Gynecol.* 180, s242–s246.

Makara, G., Sutton, S., Otto, S., Plotsky, P., 1995. Marked changes of arginine vasopressin, oxytocin, and corticotropin-releasing hormone in hypophysial portal plasma after pituitary stalk damage in the rat. *Endocrinology* 136, 1864–1868.

Manning, J. T., Baron-Cohen, S., Wheelwright, S., Sanders, G., 2001. The 2nd to 4th digit ratio and autism. *Dev. Med. Child Neurol.* 43, 160–164.

Manning, J. T., Churchill, A. J. G., Peters, M., 2007. The effects of sex, ethnicity, and sexual orientation on self-measured digit ratio (2D:4D). *Arch. Sex. Behav.* 36, 223–233.

Manning, J. T., Scutt, D., Wilson, J., Lewis-Jones, D. I., 1998. The ratio of 2nd to 4th digit length: A predictor of sperm numbers and concentrations of testosterone, luteinizing hormone and oestrogen. *Hum. Reprod.* 13, 3000–3004.

Mantzoros, C. S., Georgiadis, E. I., Trichopoulos, D., 1995. Contribution of dehydrotestosterone to male sexual behaviour. *Br. Med. J.* 310, 1289–1291.

Manuck, S. B., Kaplan, J. R., Rymeski, B. A., Fairbanks, L. A., Wilson, M. E., 2003. Approach to a stranger is associated with low central nervous system serotonergic responsivity in female cynomolgus monkeys *(Macaca fascicularis)*. *Am. J. Primatol.* 61, 187–194.

Marazziti, D., Canale, D., 2004. Hormonal changes when falling in love. *Psychoneuroendocrinology* 29, 931–936.

Marazziti, D., Osso, B. D., Baroni, S., Mugai, F., Catena, M., Rucci, P., Albanese, F., Giannaccini, G., Betti, L., Fabbrini, L., Italiani, P., Debbio, A. D., Lucacchini, A., Osso, L. D., 2006. A relationship between oxytocin and anxiety of romantic attachment. *Clin. Pract. Epidemiol. Ment. Health* 2, 28.

Marchlewska-Koj, A., Zacharczuk-Kakietek, M., 1990. Acute increase in plasma corticosterone level in female mice evoked by pheromones. *Physiol. Behav.* 48, 577–580.

Marlowe, F., 1999. Male care and mating effort among Hadza foragers. *Behav. Ecol. Sociobiol.* 46, 57–64.

———, 2003. The mating system of foragers in the standard cross-cultural sample. *Cross-cult. Res.* 37, 282–306.

Marlowe, F., Westman, A., 2001. Preferred waist-to-hip ratio and ecology. *Pers. Ind. Diff.* 30, 481–489.

Marmot, M. G., 2004. *The Status Syndrome: How Social Standing Affects Our Health and Longevity.* Times Books/Henry Holt, New York.

Marshall, A. J., Hohmann, G., 2005. Urinary testosterone levels of wild male bonobos *(Pan paniscus)* in the Lomako Forest, Democratic Republic of Congo. *Am. J. Primatol.* 65, 87–92.

Martin, R. P., Wisenbaker, J., Baker, J., Huttunen, M. O., 1997. Gender differences in temperament at six months and five years. *Infant Behav. Dev.* 20, 339–347.

Martin, S., Manning, J. T., Dowrick, C. D., 1999. Fluctuating asymmetry, relative digit length and depression in men. *Evol. Hum. Behav.* 20, 203–214.

Mason, J. W., 1968. Review of psychoendocrine research on pituitary adrenocortical system. *Psychosom. Med.* 30, 576–607.

Masters, A., Markham, R., 1991. Assessing reproductive status in orangutans by using urinary estrone. *Zoo Biol.* 10, 197–208.

Matochik, J. A., Sipos, M. L., Nyby, J. G., Barfield, R. J., 1994. Intracranial androgenic activation of male-typical behaviors in house mice: Motivation versus performance. *Behav. Brain Res.* 60, 141–149.

Matsumoto-Oda, A., 1999. Female choice in the opportunistic mating of wild chimpanzees *(Pan troglodytes schweinfurthii)*. *Behav. Ecol. Sociobiol.* 46, 258–266.

Matsumoto-Oda, A., Oda, R., Hayashi, Y., Murakami, H., Maeda, N., Kumazaki, K., Shimizu, K., Matsuzawa, T., 2002. Vaginal fatty acids produced by chimpanzees during menstrual cycles. *Folia Primatol.* 74, 75–79.

Matthiesen, A. S., Ransjö-Arvidson, A. B., Nissen, E., Uvnäs-Moberg, K., 2001. Postpartum maternal oxytocin release by newborns: Effects of infant hand massage and sucking. *Birth* 28, 13–19.

Maynard Smith, J., 1977. Parental investment—a prospective analysis. *Anim. Behav.* 25, 1–9.

———, 1978. *The Evolution of Sex.* Cambridge University Press, Cambridge.

Mazur, A., Booth, A., 1998. Testosterone and dominance in men. *Behav. Brain Sci.* 21, 353–363; discussion 363–397.

Mazur, A., Booth, A., Dabbs, J. M., 1992. Testosterone and chess competition. *Soc. Psychol. Q.* 55, 70–77.

Mazur, A., Michalek, J., 1998. Marriage, divorce, and male testosterone. *Soc. Forces* 77, 315–330.

McAuliffe, K., 2004. Interspecific variation in allomaternal behaviour among mammals. MPhil. diss., University of Cambridge.

McCarthy, M. M., Altemus, M., 1997. Central nervous system actions of oxytocin and modulation of behavior in humans. *Mol. Med. Today* 3, 269–275.

McClintock, M. K., 1971. Menstrual synchrony and suppression. *Nature* 229, 244–245.

McComb, K., Moss, C., Sayialel, S., Baker, L., 2000. Unusually extensive networks of vocal recognition in African elephants. *Anim. Behav.* 59, 1103–1109.

McCormick, C. M., Smythe, J. W., Sharma, S., Meaney, M. J., 1995. Sex-specific effects of prenatal stress on hypothalamic-pituitary-adrenal responses to stress and brain glucocorticoid receptor density in adult rats. *Brain Res. Dev. Brain Res.* 14:84, 55–61.

McEwen, B. S., 2000: Allostasis, allortatic load, and the aging nervous system: Role of excitatory amino acids and excitotoxicity. *Neurochem. Res.* 25, 1219–1231.

McEwen, B. S., Wingfield, J. C., 2003. The concept of allostasis in biology and biomedicine. *Horm. Behav.* 43, 2–15.

McFadden, D., 1993. A masculinizing effect on the auditory systems of human females having male co-twins. *Proc. Natl. Acad. Sci. U. S. A.* 90, 11900–11904.

McFadden, D., Loehlin, J. C., Pasanen, E. G., 1996. Additional findings on heritability and prenatal masculinization of cochlear mechanisms: Click-evoked otoacoustic emissions. *Hear. Res.* 97, 102–119.

McFadden, D., Pasanen, E. G., 1998. Comparison of the auditory systems of heterosexuals and homosexuals: Click-evoked otoacoustic emissions. *Proc. Natl. Acad. Sci. U. S. A.* 95, 2709–2713.

———, 1999. Spontaneous otoacoustic emissions in heterosexuals, homosexuals, and bisexuals. *J. Acoust. Soc. Am.* 105, 2403–2413.

McFadden, D., Shubel, E., 2002. Relative lengths of fingers and toes in human males and females. *Horm. Behav.* 42, 492–500.

McGinnis, M. Y., Kahn, D. F., 1997. Inhibition of male sexual behavior by intracranial implants of the protein synthesis inhibitor anisomycin into the medial preoptic area of the rat. *Horm. Behav.* 31, 15–23.

McIntosh, D. E., Mulkins, R. S., Dean, R. S., 1995. Utilization of maternal perinatal risk indicators in the differential diagnosis of ADHD and UADD children. *Int. J. Neurosci.* 81, 35–46.

McIntyre, M. H., 2003. Digit ratios, childhood gender role behavior, and erotic role preferences of gay men. *Arch. Sex. Behav.* 32, 495–497.

———, 2006. The use of digit ratios as markers for perinatal androgen action. *Reprod. Biol. Endocrinol.* 2, 10.

McIntyre, M. H., Barrett, E. S., McDermott, R., Johnson, D. D. P., Cowden, J., Rosen, S. P., 2007. Finger length ratio (2D:4D) and sex differences in aggression during a simulated war game. *Pers. Ind. Diff.* 42, 755–764.

McIntyre, M. H., Gangestad, S. W., Gray, P. B., Chapman, J. F., Burnham, T. C., O'Rourke, M. T., Thornhill, R., 2006. Romantic involvement often reduces men's testosterone levels but not always: The moderating role of extrapair sexual interest. *J. Pers. Soc. Psychol.* 91, 642–651.

McNeilly, A. S., 2001. Lactational control of reproduction. *Reprod. Fertil. Dev.* 13, 583–590.

McNeilly, A. S., Robinson, I. C., Houston, M. J., Howie, P. W., 1983. *BMJ* 286, 257–259.

Meaney, M. J., 2001. Maternal care, gene expression, and the transmission of individual differences in stress reactivity across generations. *Annu. Rev. Neurosci.* 24, 1161–1192.

Meaney, M. J., Diorio, J., Francis, D., Widdowson, J., LaPlante, P., Caldji, C., 1996. Early environmental regulation of forebrain glucocorticoid receptor gene expression: Implications for adrenocortical responses to stress. *Dev. Neurosci.* 18, 49–72.

Meaney, M. J., McEwen, B. S., 1986. Testosterone implants into the amygdala during the neonatal period masculinize the social play of juvenile female rats. *Brain Res.* 398, 324–328.

Meaney, M. J., Szyf, M., 2005. Environmental programming of stress responses through DNA methylation: Life at the interface between a dynamic environment and a fixed genome. *Dialogues Clin. Neurosci.* 7, 103–123.

Mehlman, P. T., Higley, J. D., Faucher, I., Lilly, A. A., Taub, D. M., Vickers, J., Suomi, S. J., Linnoila, M., 1994. Low CSF 5-HIAA concentrations and severe aggression and impaired impulse control in nonhuman primates. *Am. J. Psychiatry* 151, 1485–1491.

———, 1995. Correlation of CSF 5-HIAA concentration with sociality and the timing of emigration in free-ranging primates. *Am. J. Psychiatry* 152, 907–913.

Mehlman, P. T., Higley, J. D., Fernald, B. J., Sallee, F. R., Suomi, S. J., Linnoila, M., 1997. CSF 5-HIAA, testosterone, and sociosexual behaviors in free-ranging male rhesus macaques in the mating season. *Psychiatry Res.* 72, 89–102.

Meijer, A., 1986. Child psychiatric sequelae of maternal war stress. *Acta Psychiatr. Scand.* 72, 505–511.

Meisenberg, G., Simmons, W. H., 1983a. Centrally mediated effects of neurohypophyseal hormones. *Neurosci. Biobehav. Rev.* 7, 263–280.

———, 1983b. Minireview. Peptides and the blood-brain barrier. *Life Sci.* 32, 2611–2623.

Mendoza, S. P., Mason, W. A., 1986. Parental division of labour and differentiation of attachments in a monogamous primate *(Callicebus moloch)*. *Anim. Behav.* 34, 1336–1347.

Mennella, J., Pepino, M. Y., Teff, K., 2005. Acute alcohol consumption disrupts the hormonal milieu of lactating women. *J. Clin. Endocrinol. Metab.* 90, 1979–1985.

Meyer-Bahlburg, H. F. L., 1977. Sex hormones and male homosexuality in comparative perspective. *Arch. Sex. Behav.* 6, 297–325.

Meyer-Bahlburg, H. F. L., Dolezal, C., Baker, S. W., Carlson, A. D., Obeid, J. S., New, M. I., 2004. Prenatal androgenization affects gender-related behavior but not gender identity in 5–12-year-old girls with congenital adrenal hyperplasia. *Arch. Sex. Behav.* 33, 97–104.

Meyer-Bahlburg, H. L., Ehrhardt, A. A., Feldman, J. F., Rosen, L. R., Veridiano, N. P., Zimmerman, I., 1985. Sexual activity level and sexual functioning in women prenatally exposed to diethylstilbestrol. *Psychosom. Med.* 47, 497–511.

Micevych, P. E., Chaban, V., Ogi, J., Dewing, P., Lu, J. K., Sinchak, K., 2007. Estradiol stimulates progesterone synthesis in hypothalamic astrocyte cultures. *Endocrinology* 148, 782–789.

Michael, R. P., Zumpe, D., 1993. A review of hormonal factors influencing the sexual and aggressive behavior of macaques. *Am. J. Primatol.* 30, 213–241.

Michelsson, K., Eklund, K., Leppanen, P., Lyytinen, H., 2002. Cry characteristics of 172 healthy 1- to 7-day-old infants. *Folia Phoniatr. Logop.* 54, 190–200.

Millam, J. R., Zhang, B., el Halawani, M. E., 1996. Egg production of cockatiels *(Nymphicus hollandicus)* is influenced by number of eggs in nest after incubation begins. *Gen. Comp. Endocrinol.* 101, 205–210.

Miller, G. E., Cohen, S., Ritchey, A. K., 2002. Chronic psychological stress and the regulation of pro-inflammatory cytokines: A glucocorticoid-resistance model. *Health Psychol.* 21, 531–541.

Milton, K., Demment, M., 1988. Digestive and passage kinetics of chimpanzees fed high and low fiber diets and comparison with human data. *J. Nutr.* 118, 107.

Mireault, G. C., Bond, L. A., 1992. Parental death in childhood: Perceived vulnerability, and adult depression and anxiety. *Am. J. Orthopsychiatry* 62, 517–524.

Mirescu, C., Peters, J. D., Gould, E., 2004. Early life experience alters response of adult neurogenesis to stress. *Nat. Neurosci.* 7, 841–846.

Missmer, S. A., Spiegelman, D., Bertone-Johnson, E. R., Barbieri, R. L., Pollak, M. N., Hankinson, S. E., 2006. Reproducibility of plasma steroid hormones, prolactin, and insulin-like growth factor levels among premenopausal women over a 2- to 3-year period. *Cancer Epidemiol. Biomarkers Prev.* 15, 972–978.

Mitani, J. C., 1985. Mating behaviour of male orangutans in the Kutai Game Reserve, Indonesia. *Anim. Behav.* 33, 392–402.

Mitani, J. C., Gros-Louis, J., Richard, A. F., 1996. Sexual dimorphism, the operational sex-ratio, and the intensity of male competition in polygynous primates. *Am. Nat.* 147, 966–980.

Mitchell, W. R., Lindburg, D. G., Shideler, S. E., Presley, S., Lasley, B. L., 1985. Sexual behavior and urinary ovarian hormone concentrations during the lowland gorilla menstrual cycle. *Int. J. Primatol.* 6, 161–172.

Mizoguchi, K., Ishige, A., Takeda, S., Aburada, M., Tabira, T., 2004. Endogenous glucocorticoids are essential for maintaining prefrontal cortical cognitive function. *Neuroscience* 24, 5492–5499.

Mizuno, K., Mizuno, N., Shinohara, T., Noda, M., 2004. Mother-infant skin-to-skin contact after delivery results in early recognition of own mother's milk odour. *Acta Paediatr.* 93, 1640–1645.

Mohr, B. A., Bhasin, S., Link, C. L., O'Donnell, A. B., McKinlay, J. B., 2006. The effect of changes in adiposity on testosterone levels in older men: Longitudinal results from the Massachusetts Male Aging Study. *Eur. J. Endocrinol.* 155, 443–452.

Møller, A. P., 1990. Parasites and sexual selection: Current status of the Hamilton-Zuk hypothesis. *J. Evol. Biol.* 3, 319–328.

Montanes-Rada, F., Ramirez, J. M., Taracena, M. T. D., 2006. Violence in mental disorders and community sample: An evolutionary model related with dominance in social relationships. *Med. Hypotheses* 67, 930–940.

Moore, I. T., Jessop, T. S., 2003. Stress, reproduction, and adrenocortical modulation in amphibians and reptiles. *Horm. Behav.* 43, 39–47.

Moore, I. T., Walker, B. G., Wingfield, J. C., 2004. The effects of combined aromatase inhibitor and anti-androgen on male territorial aggression in a tropical population of rufous-collared sparrows, *Zonotrichia capensis. Gen. Comp. Endocrinol.* 135, 223–229.

Moore, M. R., Brooks-Gunn, J., 2002. Adolescent parenthood. In: Bornstein, M., (Ed.), *Handbook of Parenting. Vol. 3, Being and Becoming a Parent.* 2nd ed. Erlbaum, Mahwah, NJ, pp. 173–214.

Mooring, M. S., Patton, M. L., Lance, V. A., Hall, B. M., Schaad, E. W., Fortin, S. S., Jella, J. E., McPeak, K. M., 2004. Fecal androgens of bison bulls during the rut. *Horm. Behav.* 46, 392–398.

Morgan, D., Grant, K. A., Gage, H. D., Mach, R. H., Kaplan, J. R., Prioleau, O., Nader, S. H., Buchheimer, N., Ehrenkaufer, R. L., Nader, M. A., 2002. Social dominance in monkeys, dopamine D2 receptors and cocaine self-administration. *Nat. Neurosci.* 5, 169–174.

Morgan, M. A., Schulkin, J., Pfaff, D., 2004. Estrogens and non-reproductive behaviors related to anxiety and fear. *Neurosci. Biobehav. Rev.* 28, 55–63.

Morgan, P., 1996. Characteristic features of modern American fertility. *Popul. Dev. Rev. Suppl.* 22, 19–63.

Morin, P. A., Moore, J. J., Chakraborty, R., Jin, L., Goodall, J., Woodruff, D. S., 1994. Kin selection, social structure, gene flow, and the evolution of chimpanzees. *Science* 265, 1193–1201.

Morley-Fletcher, S., Darnaudery, M., Koehl, M., Casolin, P., Van Reeth, O., Maccari, S., 2003. Pre-natal stress in rats predicts immobility behavior in the forced swim test: Effects of a chronic treatment with tianeptine. *Brain Res.* 989, 246–251.

Moss, C. J., 1983. Oestrous behaviour and female choice in the African elephant. *Behaviour* 86, 167–196.

Mota, M. T., Sousa, M. B. C., 2000. Prolactin levels of fathers and helpers related to alloparental care in common marmosets, *Callithrix jacchus. Folia Primatol.* 71, 22–26.

Muehlenbein, M. P., Bribiescas, R. G., 2005. Testosterone-mediated immune functions and male life histories. *Am. J. Hum. Biol.* 17, 527–558.

Muehlenbein, M. P., Watts, D. P., Whitten, P. L., 2004. Dominance rank and fecal testosterone levels in adult male chimpanzees *(Pan troglodytes schweinfurthii)* at Ngogo, Kibale National Park, Uganda. *Am. J. Primatol.* 64, 71–82.

Müller, A. E., Thalmann, U., 2000. Origin and evolution of primate social organisation: A reconstruction. *Biol. Rev.* 75, 405–435.

Muller, M. N., 2002. Agonistic relations among Kanyawara Chimpanzees In: Boesch, C., Hohmann, G. Marchant, L. F., (Eds.), *Behavioral Diversity in Chimpanzees and Borobos.* Cambridge University Press, pp. 112–124.

Muller, M. N., Emery Thompson, M., Wrangham, R. W., 2006. Male chimpanzees prefer mating with old females. *Curr. Biol.* 16, 2234–2238.

Muller, M. N., Kahlenberg, S. M., Emery Thompson, M., Wrangham, R. W., 2007. Male coercion and the costs of promiscuous mating for female chimpanzees. *Proc. R. Soc. Lond. B Biol. Sci.* 274, 1009–1014.

Muller, M. N., Marlowe, F. W., Bugumba, R., Ellison, P. T., 2007. Fatherhood and testosterone in Hadza hunter-gatherers and neighboring Datoga pastoralists. Paper presented at the Human Behavior and Evolution Society meeting, Williamsburg, VA.

Muller, M. N., Wrangham, R. W., 2001. The reproductive ecology of male hominoids. In: Ellison, P. T., (Ed.), *Reproductive Ecology and Human Evolution.* Aldinede, Gruyter Hawthorne, NY, pp. 397–428.

———, 2002. Sexual mimicry in hyenas. *Q. Rev. Biol.* 77, 3–16.

———, 2004a. Dominance, aggression and testosterone in wild chimpanzees: A test of the "challenge hypothesis." *Anim. Behav.* 67, 113–123.

———, 2004b. Dominance, cortisol and stress in wild chimpanzees *(Pan troglodytes schweinfurthii). Behav. Ecol. Sociobiol.* 55, 332–340.

———, 2005. Testosterone and energetics in wild chimpanzees *(Pan troglodytes schweinfurthii). Am. J. Primatol.* 66, 119–130.

Munroe, R. L., Romney, A. K., 2006. Gender and age differences in same-sex aggregation and social behavior—a four-culture study. *J. Cross Cult. Psychol.* 37, 3–19.

Muroyama, Y., Shimizu, K., Sugiura, H., 2007. Seasonal variation in fecal testosterone levels in free-ranging male Japanese macaques. *Am. J. Primatol.* 69, 1–8.

Murphy, D., Wells, S., 2003. In vivo gene transfer studies on the regulation and function of the vasopressin and oxytocin genes. *J. Neuroendocrinol.* 15, 109–125.

Murphy, M. R., Seckl, J. R., Burton, S., Checkley S. A., Lightman, S. L., 1987. Changes in oxytocin and vasopressin secretion during sexual activity in men. *J. Clin. Endocrinol. Metab.* 64, 27–31.

Murphy, V. E., Gibson, P. G., Giles, W. B., Zakar, T., Smith, R., Bisits, A. M., Kessell, L. G., Clifton, V. L., 2003. Maternal asthma is associated with reduced female fetal growth. *Am. J. Respir. Crit. Care Med.* 168, 1317–1323.

Murphy, V. E., Smith, R., Giles, W. B., Clifton, V. L., 2006. Endocrine regulation of human fetal growth. The rule of the mother, placenta, and fetus. *Endocr. Rev.* 27, 141–169.

Murray, L., Fiori-Cowley, A., Hooper, R., Cooper, P., 1996. The impact of postnatal depression and associated adversity on early mother-infant interactions and later infant outcome. *Child Dev.* 67, 2512–2526.

Mutlu, G., Factor, P., 2004. Role of vasopressin in the management of septic shock. *Intensive Care Med.* 30, 1276–1291.

Nadler, R. D., 1975. Cyclicity in tumescence of the perineal labia of female lowland gorillas. *Anat. Rec.* 181, 791–798.

———, 1976. Sexual behavior of captive lowland gorillas. *Arch. Sex. Behav.* 5, 487–502.

———, 1977. Sexual behavior of captive orangutans. *Arch. Sex. Behav.* 6, 457–475.

———, 1982. Laboratory research on sexual behavior and reproduction of gorillas and orang-utans. *Am. J. Primatol. Suppl.* 1, 57–66.

———, 1988. Sexual and reproductive behavior. In: Schwartz, J. H., (Ed.), *Orang-utan Biology.* Oxford University Press, Oxford, pp. 105–116.

———, 1995. Proximate and ultimate influences on the regulation of mating in the great apes. *Am. J. Primatol.* 37, 93–102.

Nadler, R. D., Collins, D. C., 1991. Copulatory frequency, urinary pregnanediol, and fertility in great apes. *Am. J. Primatol.* 24, 167–179.

Nadler, R. D., Collins, D. C., Miller, L. C., Graham, C. E., 1983. Menstrual cycle patterns of hormones and sexual behavior in gorillas. *Horm. Behav.* 17, 1–17.

Nadler, R. D., Graham, C. E., Collins, D. C., Gould, K. G., 1979. Plasma gonadotropins, prolactin, gonadal steroids and genital swelling during the menstrual cycle of lowland gorillas. *Endocrinology* 105, 290–296.

Nadler, R. D., Graham, C. E., Gosselin, R. E., Collins, D. C., 1985. Serum levels of gonadotropins and gonadal steroids, including testosterone, during the menstrual cycles of the chimpanzee *(Pan troglodytes). Am. J. Primatol.* 9, 273–284.

Nadler, R. D., Miller, L. C., 1982. Influence of male aggression on mating of gorillas in the laboratory. *Folia Primatol.* 38, 233–239.

Nagamani, M., McDonough, P. G., Ellegood, J. O., Mahesh, V. B., 1979. Maternal and amniotic fluid steroids throughout human pregnancy. *Am. J. Obstet. Gynecol.* 134, 674–680.

Nahum, R., Thong, K. J., Hillier, S. G., 1995. Metabolic regulation of androgen production by human theca cells in vitro. *Hum. Reprod.* 10, 75–81.

Neave, N., Laing, S., Fink, B., Manning, J. T., 2003. Second to fourth digit ratio, testosterone, and perceived male dominance. *Proc. R. Soc. Lond. B Biol Sci.* 270, 2167–2172.

Nelson, R. J., 2005. *An Introduction to Behavioral Endocrinology.* 3rd ed. Sinauer, Sunderland, MA.

Nepomnaschy, P. A., 2007. *On Maya Women's Fertility and Breast Feeding.* J. B. Lancaster.

Nepomnaschy, P. A., Sheiner, E., Mastorakos, G. C., Arck, P. C., 2007. Stress, immune function, and women's reproduction. *N. Y. Acad. Sci.* 1113, 350–364.

Nepomnaschy, P. A., Welch, K. B., McConnell, D. S., Low, B. S., Strassmann, B. I., England, B. G., 2006. Cortisol levels and very early pregnancy loss in humans. *Proc. Natl. Acad. Sci. U. S. A.* 103, 3938–3942.

Nepomnaschy, P. A., Welch, K. B., McConnell, D. S., Strassmann, B. I., England, B. G., 2004. Stress and female reproductive function: A study of daily variations in cortisol, gonadotrophins, and gonadal steroids in a rural Mayan population. *Am. J. Hum. Biol.* 16, 523–532.

Neumann, I. D., Ludwig, M., Engelmann, M., Pittman, Q. J., Landgraf, R., 1993. Simultaneous microdialysis in blood and brain: Oxytocin and vasopressin release in response to central and peripheral osmotic stimulation and suckling in the rat. *Neuroendocrinology* 58, 637–645.

Neumann, I. D., Toschi, N., Ohl, F., Torner, L., Kromer, S. A., 2001. Maternal defence as an emotional stressor in female rats: Correlation of neuroendocrine and behavioural parameters and involvement of brain oxytocin. *Eur. J. Neurosci.* 13, 1016–1024.

Newman, M. L., Sellers, J. G., Josephs, R. A., 2005. Testosterone, cognition, and social status. *Horm. Behav.* 47, 205–211.

Nicholls, T. J., Goldsmith, A. R., Dawson, A., 1988. Photorefractoriness in birds and comparison with mammals. *Physiol. Rev.* 68, 133–176.

Nierop, A., Bratsikas, A., Zimmermann, R., Ehlert, U., 2006. Are stress-induced cortisol changes during pregnancy associated with postpartum depressive symptoms? *Psychosom. Med.* 68, 931–937.

Nisbett, R. E., 1996. *Culture of Honor: The Psychology of Violence in the South*. Westview Press, Boulder, CO.

Nissen, E., Gustavsson, P., Widstrom, A. M., Uvnäs-Moberg, K., 1998. Oxytocin, prolactin, milk production and their relationship with personality traits in women after vaginal delivery or Cesarean section. *J. Psychosom. Obstet. Gynaecol.* 19, 49–58.

Nissen, E., Uvnäs-Moberg, K., Svensson, K., Stock, S., Widstrom, A. M., Winberg, J., 1996. Different patterns of oxytocin, prolactin but not cortisol release during breastfeeding in women delivering by caesarean section or by the vaginal route. *Early Hum. Dev.* 45, 103–118.

Nitschke, J. B., Nelson, E. E., Rusch, B. D., Fox, A. S., Oakes, R. T. R., Davidson, R. J., 2004. Orbitofrontal cortex tracks positive mood in mothers viewing pictures of their newborn infants. *Neuroimage* 21, 583–592.

Noë, R., 1992. Alliance formation among male baboons: Shopping for profitable partners. In: Harcourt, A. H., de Waal, F. M. B., (Eds.), *Coalitions and Alliances in Animals and Humans*. Oxford University Press, Oxford, pp. 285–321.

Nordby, J. C., Campbell, S. E, Beecher, M. D., 1999. Ecological correlates of song learning in song sparrows. *Behav. Ecol.* 10, 287–297.

Nøvik, T. S., Hervas, A., Ralston, S. J., Dalsgaard, S., Pereira, R. R., Lorenzo, M. J., 2006. Influence of gender on attention-deficit/hyperactivity disorder in Europe—ADORE. *Eur. Child Adolesc. Psychiatry* 15, 15–24.

Nowicki, S., Ball, G. F., 1989. Testosterone induction of song in photosensitive and photorefractory male sparrows. *Horm. Behav.* 23, 514–525.

Numan, M., 1994. Maternal behavior. In: Knobil, K., Neill, J., (Eds.), *The Physiology of Reproduction*, 2nd ed. Raven Press, New York, pp. 221–302.

Numan, M., Fleming, A. S., Levy, F., 2006. Maternal behavior. In: Neill, J. D., (Ed.), *Knobil and Neill's Physiology of Reproduction*. 3rd ed. Elsevier, San Diego, pp. 1921–1993.

Numan, M., Insel, T. R., 2003. *The Neurobiology of Parental Behavior*. Springer, New York.

Nunes, S., Fite, J. E., Patera, K. J., French, J. A., 2001. Interactions among paternal behavior, steroid hormones and parental experience in male marmosets *(Callitrix kuhlii)*. *Horm. Behav.* 39, 70–82.

Nunn, C. L., 1999. The evolution of exaggerated sexual swellings in primates and the graded-signal hypothesis. *Anim. Behav.* 58, 229–246.

Nunn, C. L., Altizer, S. M., 2004. Sexual selection, behaviour and sexually transmitted diseases. In: Kappeler, P., van Schaik, C., (Eds.), *Sexual Selection in Primates*. Cambridge University Press, Cambridge, pp. 117–130.

Nunn, C. L., van Schaik, C. P., Zinner, D., 2001. Do exaggerated sexual swellings function in female mating competition in primates? A comparative test of the reliable indicator hypothesis. *Behav. Ecol.* 12, 646–654.

Nyby, J., Whitney, G., 1980. Experience affects behavioral response to sex odors. In: Muller-Schwarz, D., Silverstein, R. M., (Eds.), *Chemical Signals in Vertebrates and Aquatic Animals*. Plenum, New York, pp. 173–192.

Nyby, J., Wysocki, C. J., Whitney, G., Dizinno, G., 1977. Pheromonal regulation of male mouse ultrasonic courtship *(Mus musculus)*. *Anim. Behav.* 25, 333–341.

Oates, M. R., Cox, J. L., Neema, S., Asten, P., Glangeaud-Freudenthal, N., Figueiredo, B., Gorman, L. L., Hacking, S., Hirst, E., Kammerer, M. H., Klier, C. M., Senevirante, G., Smith, M., Sutter-Dallay, A. L., Valoriani, V., Wickberg, B., Yoshida, K., TCS-PND Group, 2004. Postnatal depression across countries and cultures: A qualitative study. *Br. J. Psychiatry Suppl.* 46, s10–s16.

O'Connor, D. B., Archer, J., Hair, W. M., Wu, F. C. W., 2002. Exogenous testosterone, aggression, and mood in eugonadal and hypogonadal men. *Physiol. Behav.* 75, 557–566.

O'Connor, D. B., Archer, J., Wu, F. C. W., 2004. Effects of testosterone on mood, aggression, and sexual behavior in young men: A double-blind, placebo-controlled, cross-over study. *J. Clin. Endocrinol. Metab.* 89, 2837–2845.

O'Connor, K. A., Brindle, E., Holman, D. J., Klein, N. A., Soules, M. R., Campbell, K. L., Kohen, F., Munro, C. J., Shofer, J. B., Lasley, B. L., Wood, J. W., 2003. Urinary estrone conjugate and pregnanediol 3-glucuronide enzyme immunoassays for population research. *Clin. Chem.* 49, 1139–1148.

O'Connor, T. G., Heron, J., Golding, J., Glover, V., ALSPAC Study Team, 2003. Maternal antenatal anxiety and behavioral/emotional problems in children: A test of a programming hypothesis. *J. Child Psychol. Psychiatry* 44, 1025–1036.

Odendaal, J., Meintjes, R., 2003. Neurophysiological correlates of affiliative behaviour between humans and dogs. *Vet. J.* 165, 296–301.

O'Hara, M. W., Schlechte, J. A., Lewis, D. A., Wright, E. J., 1991. Prospective study of postpartum blues. Biologic and psychosocial factors. *Arch. Gen. Psychiatry* 48, 801–806.

Olausson, H., Uvnäs-Moberg, K., Sohlstrom, A., 2003. Postnatal oxytocin alleviates adverse effects in adult rat offspring caused by maternal malnutrition. *Am. J. Physiol. Endocrinol. Metab* 284, E475–E480.

Olazabal, D. E., Young, L. J., 2006. Oxytocin receptors in the nucleus accumbens facilitate "spontaneous" maternal behavior in adult female prairie voles. *Neuroscience* 141, 559–568.

Oliveira, R. F., 1998. Of fish and men: A comparative approach to androgens and social dominance. *Behav. Brain Sci.* 21, 383.

Oliveira, R. F., Lopes, M., Carneiro, L. A., Canario, A. V. M., 2001. Watching fights raises fish hormone levels. *Nature* 409, 475.

O'Reilly, K. M., Wingfield, J. C., 1995. Spring and autumn migration in Arctic shorebirds: Same distance, different strategies. *Am. Zool.* 35, 222–233.

Orians, G. H., 1969. On the evolution of mating systems in birds and mammals. *Am. Nat.* 103, 589–603.

Ostner, J. L., Kappeler, P. M., Heistermann, M., 2002. Seasonal variation and social correlates of androgen excretion in male redfronted lemurs *(Eulemur fulvus rufus)*. *Behav. Ecol. Sociobiol.* 52, 485–495.

Overdorff, D. J., Tecot, S. R., 2006. Social pair-bonding and resource defense in wild red-bellied lemurs *(Eulemur rubriventer)*. In: Gould, L., Sauther, M. L., (Eds.), *Lemurs: Ecology and Adaptation.* Springer, New York, pp. 237–256.

Owen, G. I., Zelent, A., 2000. Origins and evolutionary diversification of the nuclear receptor superfamily. *Cell. Mol. Life Sci.* 57, 809–827.

Owen-Ashley, N. T., Turner, M., Hahn, T. P., Wingfield, J. C., 2006. Hormonal, behavioral, and thermoregulatory responses to bacterial lipopolysaccharide in captive and free-living white-crowned sparrows *(Zonotrichia leucophrys gambelii)*. *Horm. Behav.* 49, 15–19.

Owen-Ashley, N. T., Wingfield, J. C., 2006. Seasonal modulation of sickness behavior in free-living northwestern song sparrows *(Melospiza melodia morphna)*. *J. Exp. Biol.* 209, 3062–3070.

Ozasa, H., Gould, K. G., 1982. Demonstration and characterization of estrogen receptor in chimpanzee sex skin: Correlation between nuclear receptor levels and degree of swelling. *Endocrinology* 111, 125–131.

Packer, C. R., Lewis, S., Pusey, A. E., 1992. A comparative analysis of non-offspring nursing. *Anim. Behav.* 43, 265–281.

Pagel, M. D., 1994. The evolution of conspicuous oestrous advertisement in Old World monkeys. *Anim. Behav.* 47, 1333–1341.

Pagel, M. D., Harvey, P. H., 1989. Taxonomic differences in the scaling of brain on body weight among mammals. *Science* 244, 1589–1593.

Pagonis, T. A., Angelopoulos, N. V., Koukoulis, G. N., Hadjichristodoulou, C. S., 2006. Psychiatric side effects induced by supraphysiological doses of combinations of anabolic steroids correlate to the severity of abuse. *Eur. Psychiatry* 21, 551–562.

Pagonis, T. A., Angelopoulos, N. V., Koukoulis, G. N., Hadjichristodoulou, C. S., Toli, P. N., 2006. Psychiatric and hostility factors related to use of anabolic steroids in monozygotic twins. *Eur. Psychiatry* 21, 563–569.

Paland, S., Lynch, M., 2006. Transitions to asexuality result in excess amino acid substitutions. *Science* 311, 990.

Palanza, P., 2001. Animal models of anxiety and depression: How are females different? *Neurosci. Biobehav. Rev.* 25, 219–233.

Palanza, P., Morellini, F., Parmigiani, S., vom Saal, F. S., 1999. Prenatal exposure to endocrine disrupting chemicals: Effects on behavioral development. *Neurosci. Biobehav. Rev.* 23, 1011–1027.

Palombit, R. A., 1994a. Dynamic pair bonds in hylobatids: Implications regarding monogamous social systems. *Behaviour* 128, 65–101.

———, 1994b. Extra-pair copulations in a monogamous ape. *Anim. Behav.* 47, 721–723.

———, 1996. Pair bonds in monogamous apes: A comparison of the siamang *(Hylobates syndactylus)* and the white-handed gibbon *(Hylobates lar)*. *Behaviour* 133, 321–356.

———, 1999. Infanticide and the evolution of pair bonds in nonhuman primates. *Evol. Anthropol.* 7, 117–129.

Panter-Brick, C., Ellison, P. T., 1994. Seasonality of workloads and ovarian function in Nepali women. *Ann. N. Y. Acad. Sci.* 709, 234–235.

Panter-Brick, C., Lotstein, D. S., Ellison, P. T., 1993. Seasonality of reproductive function and weight loss in rural Nepali women. *Hum. Reprod.* 8, 684–690.

Panter-Brick, C., Pollard, T. M., 1999. Work and hormonal variation in subsistence and industrial contexts. In: Panter-Brick, C., Worthman, C., (Eds.), *Hormones, Health, and Behavior.* Cambridge University Press, Cambridge, pp. 139–183.

Paoli, T., Palagi, E., Tacconi, G., Tarli, S., 2006. Perineal swelling, intermenstrual cycle and female sexual behavior in bonobos *(Pan paniscus). Am. J. Primatol.* 68, 333–347.

Papoušek, H., Papoušek, M., 1995. *Intuitive Parenting.* In: Bornstein, M. H., (Ed.), *Handbook of Parenting*, Vol. 2: *Biology and Ecology of Parenting*, 1st ed. Lawrence Erlbaum Associates, Mahwah, NJ, pp. 117–136.

————, M. 2002. *Intuitive Parenting.* In: Bornstein, M. H., (Ed.), *Handbook of Parenting*, Vol. 2: *Biology and Ecology of Parenting*, 2nd ed. Lawrence Erlbaum Associates, Mahwah, NJ, pp. 183–206.

Parker, G., MacNair, M. R., 1978. Models of parent offspring conflict. I. Monogamy. *Anim. Behav.* 26, 97–110.

Parker, T. H., Knapp, R., Rosenfield, J. A., 2002. Social mediation of sexually selected ornamentation and steroid hormone levels in male junglefowl. *Anim. Behav.* 64, 291–298.

Parkin, J., n.d. Unpublished data.

Partridge, L., 1988. Lifetime reproductive success in *Drosophila.* In: Clutton-Brock, T. H., (Ed.), *Reproductive Success.* Princeton University Press, Princeton, NJ, pp. 11–23.

Patel, P. D., Lopez, J. F., Lyons, D. M., Burke, S., Wallace, M., Shatzberg, A. F., 2000. Glucocorticoid and mineralocorticoid receptor mRNA expression in squirrel monkey brain. *J. Psychiatr. Res.* 34, 383–392.

Patterson, N., Richter, D. J., Gnerre, S., Lander, E. S., Reich, D., 2006. Genetic evidence for complex speciation of humans and chimpanzees. *Nature* 441, 1103–1108.

Paulhus, D. L., Williams, K. M., 2002. The dark triad of personality: Narcissism, Machiavellianism, and psychopathy. *J. Res. Pers.* 36, 556–563.

Pearcey, S. M., Docherty, K. J., Dabbs, J. M., Jr., 1996. Testosterone and sex role identification in lesbian couples. *Physiol. Behav.* 60, 1033–1035.

Pedersen, C. A., Boccia, M. L., 2002. Oxytocin links mothering received, mothering bestowed and adult stress responses. *Stress 5*, 259–267.

————, 2006. Vasopressin interactions with oxytocin in the control of female sexual behavior. *Neuroscience* 139, 843–851.

Pedersen, C. A., Prange, A. J., Jr., 1979. Induction of maternal behavior in virgin rats after intracerebroventricular administration of oxytocin. *Proc. Natl. Acad. Sci. U. S. A.* 76, 6661–6665.

Pederson, D. R., Moran, G., Sitko, C., Campbell, K., Ghesquire, K., Acton, H., 1990. Maternal sensitivity and the security of infant-mother attachment: A Q-sort study. *Child Dev.* 61, 1974–1983.

Peel, A. J., Vogelnest, L., Finnigan, M., Grossfeldt, L., O'Brien, J. K., 2005. Noninvasive fecal hormone analysis and behavioral observations for monitoring stress responses in captive western lowland gorillas *(Gorilla gorilla gorilla)*. *Zoo Biol.* 24, 431–445.

Peichel, C. L., Prabhakaran, B., Vogt, T. F., 1997. The mouse *Ulnaless* mutation deregulates posterior *HoxD* gene expression and alters appendicular patterning. *Development* 124, 3481–3492.

Pellegrini, A. D., 2004. Sexual segregation in childhood: A review of evidence for two hypotheses. *Anim. Behav.* 68, 435–443.

Pellegrini, A. D., Bartini, M., 2001. Dominance in early adolescent boys: Affiliative and aggressive dimensions and possible functions. *Merrill-Palmer Q.* 47, 142–163.

Pellegrini, A. D., Smith, P. K., 1998. Physical activity play: The nature an function of a neglected aspect of play. *Child Dev.* 69, 577–598.

Penn, D. J., Smith, K. R., 2007. Differential fitness costs of reproduction between the sexes. *Proc. Natl. Acad. Sci. U. S. A.* 104, 553–558.

Pennington, R., Harpending, H., 1988. Fitness and fertility among Kalahari !Kung. *Am. J. Phys. Anthropol.* 77, 303–319.

Penton-Voak, I. S., Chen, J. Y., 2004. High salivary testosterone is linked to masculine male facial appearance in humans. *Evol. Hum. Behav.* 25, 229–241.

Penton-Voak, I. S., Perrett, D. I., 2000. Female preference for male faces changes cyclically: Further evidence. *Evol. Hum. Behav.* 21, 39–48.

Penton-Voak, I. S., Perrett, D. I., Castles, D. L., Kobayashi, T., Burt, D. M., Murray, L. K., Minamisawa, R., 1999. Female preference for male faces changes cyclically. *Nature* 399, 741–742.

Peredery, O., Persinger, M., Blomme, C., Parker, G., 1992. Absence of maternal behavior in rats with lithium/pilocarpine seizure-induced brain damage: Support of MacLean's triune brain theory. *Physiol. Behav.* 52, 665–671.

Pereira, M., 1995. Development and social dominance among group living primates. *Am. J. Primatol.* 37, 143–175.

Perrett, D. I., Lee, K. J., Penton-Voak, I. S., Rowland, D., Yoshikawa, S., Burt, D. M., Henzi, S. P., Castles, D. L., Akamatsu, S., 1998. Effects of sexual dimorphism on facial attractiveness. *Nature* 394, 884–887.

Petrie, M., Halliday, T., Sanders, C., 1991. Peahens prefer peacocks with elaborate trains. *Anim. Behav.* 41, 323–331.

Petrie, M., Krupa, A., Burke, T., 1999. Peacocks lek with relatives even in the absence of social and environmental cues. *Nature* 401, 155–157.

Petrovich-Bartell, N., Cowan, N., Morse, P. A., 1982. Mothers' perceptions of infant distress vocalizations. *J. Speech Hear. Res.* 25, 371–376.

Pfaff, D., Frohlich, J., Morgan, M., 2002. Hormonal and genetic influences on arousal—sexual and otherwise. *Trends Neurosci.* 25, 45–50.

Pfeifer, L. A., Tishkoff, S. A., 2007. Nucleotide sequence variation of the oxytocin receptor (OTXR) in ethnically diverse human populations. Poster presentation at the American Academy of Physical Anthropology meetings, Philadelphia, PA.

Pfeiffer, C. A, Johnston, R. E., 1992. Socially stimulated androgen surges in male hamsters: The roles of vaginal secretions, behavioral interactions, and housing conditions. *Horm. Behav.* 26, 283–293.

———, 1994. Hormonal and behavioral responses of male hamsters to females and female odors: Roles of olfaction, the vomeronasal system, and sexual experience. *Physiol. Behav.* 55, 129–138.

Phan, K. L., Fitzgerald, D. A., Nathan, P. J., Tancer, M. E., 2006. Association between amygdala hyperactivity to harsh faces and severity of social anxiety in generalized social phobia. *Biol. Psychiatry* 59, 424–429.

Phelps, S., Young, L., 2003. Extraordinary diversity in vasopressin (V1a) receptor distribution among wild prairie voles *(Microtus ochrogaster):* Patterns of variation and covariation. *J. Comp. Neurol.* 466, 564–576.

Phillips, D. I., Walker, B. R., Reynolds, R. M., Flanagan, D. E., Wood, P. J., Osmond, C., Barker, D. J., Whorwood, C. B., 2000. Low birth weight predicts elevated plasma cortisol concentrations in adults from 3 populations. *Hypertension* 35, 1301–1306.

Phillips-Conroy, J. E., Jolly, C. J., 1986. Changes in the structure of the hybrid zone in the Awash National Park, Ethiopia. *Am. J. Phys. Anthropol.* 71, 337–350.

Phoenix, C. H., Goy, R. W., Gerall, A. A., Young, W. C., 1959. Organizing action of prenatally administered testosterone propionate on the tissues mediating mating behavior in the female guinea pig. *Endocrinology* 65, 369–382.

Pike, I. L., 2005. Maternal stress and fetal responses: Evolutionary perspectives on preterm delivery. *Am. J. Hum. Biol.* 17, 55–65.

Pine, D. S., Fyer, A., Grun, J., Phelps, E. A., Szeszko, P. R., Koda, V., Li, W., Ardekani, B., Maguire, E. A., Burgess, N., Bilder, R. M., 2001. Methods for developmental studies of fear conditioning circuitry. *Biol. Psychiatry* 50, 225–228.

Pinker, S., 1994. *The Language Instinct.* William Morrow, New York.

Pinxten, R., Ridder, E., Eens, M., 2003. Female presence affects male behavior and testosterone levels in the European starling *(Sturnus vulgaris). Horm. Behav.* 44, 103–109.

Pirke, K. M., Schweiger, U., Lemmel, W., Krieg, J. C., Berger, M., 1985. The influence of dieting on the menstrual cycle of healthy young women. *J. Clin. Endocrinol. Metab.* 60, 1174–1179.

Pitman, R. K., 1989. Post-traumatic stress disorder, hormones, and memory. *Biol. Psychiatry* 26, 221–223.

Pitman, R. K., Orr, S. P., Lasko, N. B., 1993. Effects of intranasal vasopressin and oxytocin on physiologic responding during personal combat imagery in Vietnam veterans with posttraumatic stress disorder. *Psychiatry Res.* 48, 107–117.

Pitman, R. K., Sanders, K. M., Zusman, R. M., Healy, A. R., Cheema, F., Lasko, N. B., Cahill, L., Orr, S. P., 2002. Pilot study of secondary prevention of posttraumatic stress disorder with propranolol. *Biol. Psychiatry* 51, 189–192.

Plavcan, J. M., van Schaik, C. P., 1992. Intrasexual competition and canine dimorphism in anthropoid primates. *Am. J. Phys. Anthropol.* 87, 461–477.

———, 1997. Intrasexual competition and body weight dimorphism in anthropoid primates. *Am. J. Phys. Anthropol.* 103, 37–68.

Poole, J. H., 1989. Mate guarding, reproductive success and female choice in African elephants. *Anim. Behav.* 37, 842–849.

Poole, J. H., Lee, P. C., Moss, C. J., in press. Longevity, competition and musth: A long-term perspective on male reproductive strategies. In: Moss, C. J., Croze, H., (Eds.), *The Amboseli Elephant: A Long-Term Perspective on a Long-Lived Species*. University of Chicago Press, Chicago.

Pope, H. G., Jr., Cohane, G. H., Kanayama, G., Siegel, A. J., Hudson, J. I., 2003. Testosterone gel supplementation for men with refractory depression: A randomized, placebo-controlled trial. *Am. J. Psychiatry* 160, 105–111.

Pope, H. G., Jr., Kouri, E., Hudson, J., 2000. Supraphysiological doses of testosterone on mood and aggression in normal men. *Arch. Gen. Psychiatry* 57, 133–140.

Pope, H. G., Jr., Phillips, K. A., Olivardia, R., 2000. *The Adonis Complex: The Secret Crisis of Male Body Obsession*. Free Press, New York.

Popeski, N., Scherling, C., Fleming, A. S., Lydon, J., Pruessner, J. C., Meaney, M., in prep. Maternal adversity alters patterns of neural activation in response to infant cues.

Popova, N. K., Amstislavskaya, T. G., 2002. Involvement of the 5-HT1a and 5-HT1b serotonergic receptor subtypes in sexual arousal in male mice. *Psychoneuroendocrinology* 27, 609–618.

Poretsky, L., Kalin, M. F., 1987. The gonadotropic function of insulin. *Endocr. Rev.* 8, 132–141.

Porges, S. W., 2001. The polyvagal theory: Phylogenetic substrates of a social nervous system. *Int. J. Psychophysiol.* 42, 123–146.

Porges, S. W., 2003a. The polyvagal theory: Phylogenetic contributions to social behavior. *Physiol. Behav.* 79, 503–513.

———, 2003b. Social engagement and attachment: A phylogenetic perspective. *Ann. N. Y. Acad. Sci.* 1008, 31–47.

Porter, R. H., 2004. The biological significance of skin-to-skin contact and maternal odours. *Acta Paediatr.* 93, 1560–1562.

Porter, R. H., Cernoch, J. M., McLaughlin, F. J., 1983. Maternal recognition of neonates through olfactory cues. *Physiol. Behav.* 30, 151–154.

Porter, R. H., Winberg, J., 1999. Unique salience of maternal breast odors for newborn infants. *Neurosci. Biobehav. Rev.* 23, 439–449.

Potts, W. K., Manning, C. J., Wakeland, E. K., 1994. The role of infectious disease, inbreeding and mating preferences in maintaining MHC genetic diversity: An experimental test. *Trans. R. Soc. Lond. B Biol. Sci.* 346, 369–378.

Powlishta, K. K., Serbin, L. A., Moller, L. C. 1993. The stability of individual differences in gender typing: Implications for understanding gender segregation. *Sex Roles* 29, 723–737.

Pratto, F., Sidanius, J., Stallworth, L. M., Malle, B. F., 1994. Social dominance orientation: A personality variable predicting social and political attitudes. *J. Pers. Soc. Psychol.* 67, 741–763.

Predine, J., Merceron, L., Barrier, G., Sureau, C., Milgrom, E., 1979. Unbound cortisol in umbilical cord plasma and maternal plasma: A reinvestigation. *Am. J. Obstet. Gynecol.* 135, 1104–1108.

Preti, G., Wysocki, C. J., Barnhart, K. T., Sondheimer, S. J., Leyden, J. J., 2003. Male axillary extracts contain pheromones that affect pulsatile secretion of luteinizing hormone and mood in women recipients. *Biol. Reprod.* 68, 2107–2113.

Preussner, J. C., Baldwin, M. W., Dedovic, K., Renwick, R., Khalili Mahani, N., Lord, C., Meaney, M., Lupien, S., 2005. Self-esteem, locus of control, hippocampal volume, and cortisol regulation in young and old adulthood. *Neuroimage* 28, 815–826.

Pryce, C. R., 1993. The regulation of maternal behaviour in marmosets and tamarins. *Behav. Processes.* 30, 201–224.

———, 1995. Determinants of Motherhood in human and nonhuman primates. A biosocial model. In: Pryce, C. R., Martin, R. D., Skuse, D., (Eds.), *Motherhood in Human and Nonhuman Primates.* Karger, Basel, pp. 1–15.

———, 1996. Socialization, hormones, and the regulation of maternal behavior in nonhuman simian primates. In: Rosenblatt, J. S., Snowdon, C. T., (Eds.), *Parental Care: Evolution, Mechanisms, and Adaptive Significance.* Academic Press, San Diego, pp. 423–476.

Pryce, C. R., Abbott, D. H., Hodges, J. K., Martin, R. D., 1988. Maternal behaviour is related to prepartum urinary estradiol levels in the red-bellied tamarin monkey. *Physiol. Behav.* 44, 717–726.

Pryce, C. R., Dobeli, M., Martin, R. D., 1993. Effects of sex steroids on maternal motivation in the common marmoset *(Callithrix jacchus):* Development and application of an operant system with maternal reinforcement. *J. Comp. Psychol.* 107, 99–115.

Pulido, F., Berthold, P., Mohr, G., Querner, U., 2001. Heritability of the timing of autumn migration in a natural bird population. *Proc. R. Soc. Lond. B Biol. Sci.* 268, 953–959.

Purvis, K., Haynes, N. B., 1974. Short-term effects of copulation, human chorionic gonadotropin injection and non-tactile association with a female on testosterone levels in the rat. *J. Endocrinol.* 60, 429–439.

Pusey, A. E., Packer, C., 1987. Dispersal and philopatry. In: Smuts, B. B., Cheney, D. L., Seyfarth, R. M., Wrangham, R. W., Struhsaker, T. T., (Eds.), *Primate Societies.* University of Chicago Press, Chicago, pp. 250–266.

Puts, D. A., 2005. Mating context and menstrual phase affect women's preferences for male voice pitch. *Evol. Hum. Behav.* 26, 388–397.

———, 2006. Cyclic variation in women's preferences for masculine traits. *Hum. Nat.* 17, 114–127.

Puts, D. A., Gaulin, S. J. C., Verdolini, K., 2006. Dominance and the evolution of sexual dimorphism in human voice pitch. *Evol. Hum. Behav.* 27, 283–296.

Quas, J. A., Bauer, A., Boyce, W. T., 2004. Physiological reactivity, social support, and memory in early childhood. *Child Dev.* 75, 797–814.

Rahman, Q., 2005a. Fluctuating asymmetry, second to fourth finger length ratios and human sexual orientation. *Psychoneuroendocrinology* 30, 382–391.

———, 2005b. The neurodevelopment of sexual orientation. *Neurosci. Biobehav. Rev.* 29, 1057–1066.

Rahman, Q., Wilson, G. D., 2003. Sexual orientation and the 2nd to 4th finger length ratio: Evidence for organising effects of sex hormones or developmental instability? *Psychoneuroendocrinology* 28, 288–303.

Rajpurohit, L. S., Sommer, V., 1993. Juvenile male emigration from natal one-male troops in Hanuman langurs. In: Pereira, M. E., Fairbanks, L. A., (Eds.), *Juvenile Primates.* Oxford University Press, New York, pp. 86–103.

Raleigh, M. J., Brammer, G. L., McGuire, M. T., Pollack, D. P., Yuwiler, A., 1992. Individual differences in basal cisternal cerebrospinal fluid 5-HIAA and HVA in monkeys. The effects of gender, age, physical characteristics, and matrilineal influences. *Neuropsychopharmacology* 7, 295–304.

Ramos-Fernández, G., Nuñez-de-la Mora, A., Wingfield, J. C., Drummond, H., 2000. Endocrine correlates of dominance in chicks of the blue-footed booby *(Sula nebouxii):* Testing the challenge hypothesis. *Ethol. Ecol. Evol.* 12, 27–34.

Ramsey, P., Ramin, K., Ramin, S., 2000. Labor induction. *Curr. Opin. Obstet. Gynecol.* 12, 463–473.

Ranote, S., Elliott, R., Abel, K. M., Mitchell, R., Deakin, J. F., Appleby, L., 2004. The neural basis of maternal responsiveness to infants: An fMRI study. *Neuroreport* 15, 1825–1829.

Rantala, M. J., Eriksson, C. J. P., Vainikka, A., Kortet, R., 2006. Male steroid hormones and female preference for male body odor. *Evol. Hum. Behav.* 27, 259–269.

Rebuffé-Scrive, M., Enk, L., Crona, N., Lönnroth, P., Abrahamsson, L., Smith, U., Björntorp, P., 1985. Fat cell metabolism in different regions in women. *J. Clin. Invest.* 75, 1973–1976.

Redoute, J., Stoleru, S., Gregoire, M., Costes, N., Cinotti, L., Lavenne, F., LeBars, D., Forest, M. G., Pujol, J., 2000. Brain processing of visual sexual stimuli in human males. *Hum. Brain Map.* 11, 162–177.

Reed, W. L., Clark, M. E., Parker, P. G., Raouf, S. A., Arguedas, N., Monk, D. S., Snajdr, E., Nolan, V., Jr., Ketterson, E. D., 2006. Physiological effects on demography: A long term experimental study of testosterone's effects on fitness. *Am. Nat.* 167, 667–683.

Rees, S., Panesar, S., Steiner, M., Fleming, A., 2004. The effects of adrenalectomy and corticosterone replacement on maternal behavior in the postpartum rat. *Horm. Behav.* 46, 411–419.

———, 2006. The effects of adrenalectomy and corticosterone replacement on induction of maternal behavior in the virgin female rat. *Horm. Behav.* 49, 337–345.

Reichard, U., 1995. Extra-pair copulations in a monogamous gibbon. *Ethology* 100, 99–112.

Reichert, K. E., Heistermann, M., Hodges, J. K., Boesch, C., Hohmann, G., 2002. What females tell males about their reproductive status: Are morphological and behavioural cues reliable signals of ovulation in bonobos *(Pan paniscus)? Ethology* 108, 583–600.

Reynolds, R. M., Walker, B. R., Syddall, H. E., Andrew, R., Wood, P. J., Whorwood, C. B., Phillips, D. I., 2001. Altered control of cortisol secretion in adult men with low birth weight and cardiovascular risk factors. *J. Clin. Endocrinol. Metab.* 86, 245–250.

Rhen, T., Crews, D., 2002. Variation in reproductive behavior within a sex: Neural systems and endocrine activation. *J. Neuroendocrinol.* 14, 517–531.

Rhodes, G., 2006. The evolutionary psychology of facial beauty. *Annu. Rev. Psychol.* 57, 199–226.

Richard, S., Zingg, H. H., 1990. The human oxytocin gene promoter is regulated by estrogens. *J. Biol. Chem.* 265, 6098–6103.

Richardson, H. N., Nelson, A. L., Ahmed, E. I., Parfitt, D. B., Romeo, R. D., Sisk, C. L., 2004. Female pheromones stimulate release of luteinizing hormone and testosterone without altering GnRH mRNA in adult male Syrian hamsters *(Mesocricetus auratus)*. *Gen. Comp. Endocrinol.* 138, 211–217.

Richter, D., 1983. Synthesis, processing, and gene structure of vasopressin and oxytocin. *Prog. Nucleic Acid Res. Mol. Biol.* 30, 245–266.

Righetti-Veltema, M., Bousquet, A., Manzano, J., 2003. Impact of postpartum depressive symptoms on mother and her 18-month-old infant. *Eur. Child Adolesc. Psychiatry* 12, 75–83.

Righetti-Veltema, M., Conne-Perreard, E., Bousquet, A., Manzano, J., 2002. Postpartum depression and mother-infant relationship at 3 months old. *J. Affect. Disord.* 70, 291–306.

Rijksen, H., 1978. *A Field Study on Sumatran Orang-utans (Pongo pygmaeus abelli, Lesson 1827): Ecology, Behaviour and Conservation.* Veenman and Zonen, Wageningen, Netherlands.

Ring, R. H., 2005. The central vasopressonergic system: Examining the opportunities for psychiatric drug development. *Curr. Pharm. Design* 11, 205–225.

Rissman, E. F., 1990. The musk shrew, *Suncus murinus,* a unique animal model for the study of female behavioral endocrinology. *J. Exp. Zool. Suppl.* 4, 207–209.

Riters, L. V., Ball, G. F., 1999. Lesions to the medial preoptic area affect singing in the male European starling *(Sturnus vulgaris)*. *Horm. Behav.* 36, 276–286.

Robbins, M. M., 1999. Male mating patterns in wild multimale mountain gorilla groups. *Anim. Behav.* 57, 1013–1020.

Robbins, M. M., Czekala, N. M., 1997. A preliminary investigation of urinary testosterone and cortisol levels in wild male mountain gorillas. *Am. J. Primatol.* 43, 51–64.

Robbins, M. M., McNeilage, A., 2003. Home range and frugivory patterns of mountain gorillas in Bwindi Impenetrable National Park, Uganda. *Int. J. Primatol.* 24, 467–491.

Robbins, T. W., Cador, M., Taylor, J. R., Everett, B. J., 1989. Limbic-striatal interactions in reward-related processes. *Neurosci. Biobehav. Rev.* 13, 155–162.

Robel, P., Baulieu, E., 1995. Dehydroepiandrosterone (DHEA) is a neuroactive neurosteroid. *Ann. N. Y. Acad. Sci.* 774, 82–110.

Roberti, J. W., 2003. Biological responses to stressors and the role of personality. *Life Sci.* 73, 2527–2531.

Roberts, C. W., Green, R., Williams, K., Goodman, M., 1987. Boyhood gender identity development: A statistical contrast of two family groups. *Dev. Psychol.* 23, 544–557.

Roberts, R. L., Williams, J. R., Wang, A. K., Carter, C. S., 1998. Cooperative breeding and monogamy in prairie voles: Influence of the sire and geographical variation. *Anim. Behav.* 55, 1131–1140.

Roberts, S. C., Havlicek, J., Flegr, J., Hruskova, M., Little, A. C., Jones, B. C., Perrett, D. I., Petrie, M., 2004. Female facial attractiveness increases during the fertile phase of the menstrual cycle. *Proc. R. Soc. Lond. B Suppl.* 271, s270–s272.

Robinson, G., Evans, J. J., 1990. Oxytocin has a role in gonadotrophin regulation in rats. *J. Endocrinol.* 125, 425–432.

Robinson, S. J., Manning, J. T., 2000. The ratio of 2nd to 4th digit length and male homosexuality. *Evol. Hum. Behav.* 21, 333–345.

Robles, T. F., Kiecolt-Glaser, J. K., 2003. The physiology of marriage. *Physiol. Behav.* 79, 409–416.

Robles, T. F., Schaffer, V. A., Malarkey, W. B., Kiecolt-Glaser, J. K., 2006. Positive behaviors during marital conflict: Influences on stress hormones. *J. Soc. Pers. Rel.* 23, 305–325.

Robson, A., Kaplan, H., 2003. The co-evolution of longevity and intelligence in hunter-gatherer economies. *Am. Econ. Rev.* 93, 150–169.

Rochira, V., Zirilli, L., Genazzani, A. D., Balestrieri, A., Aranda, C., Fabre, B., Antunez, P., Diazzi, C., Carani, C., Maffei, L., 2006. Hypothalamic-pituitary-gonadal axis in two men with aromatase deficiency: Evidence that circulating estrogens are required at the hypothalamic level for the integrity of gonadotropin negative feedback. *Eur. J. Endocrinol.* 155, 513–522.

Rodríguez Manzanares, P. A, Isoardi, N. A., Carrer, H. F., Molina, V. A., 2005. Previous stress facilitates fear memory, attenuates GABAergic inhibition, and increases synaptic plasticity in the rat basolateral amygdala. *J. Neurosci.* 25 8725–8734.

Rodseth, L., Wrangham, R. W., Harrigan, A., Smuts, B. B., 1991. The human community as a primate society. *Curr. Anthropol.* 32, 221–254.

Roelofs, K., Bakvis, P., Hermans, E. J., van Pelt, J., van Honk, J., 2007. The effects of social stress and cortisol responses on the preconscious selective attention to social threat. *Biol. Psychol.* 75, 1–7.

Rogers, J., Martin, L. J., Comuzzie, A. G., Mann, J. J., Manuck, S. B., Leland, M., Kaplan, J. R., 2004. Genetics of monoamine metabolites in baboons: Overlapping sets of genes influence levels of 5-hydroxyindoleacetic acid, 3-hydroxy-4-methoxyphenylglycol, and homovanillic acid. *Biol. Psychiatry* 55, 739–744.

Rogoff, B., 2003. *The Cultural Nature of Human Development.* Oxford University Press, Oxford.

Romero, L. M., 2002. Seasonal changes in plasma glucocorticoid concentrations in free-living vertebrates. *Gen. Comp. Endocrinol.* 128, 1–24.

Rondo, P. H., Ferreira, R. F., Nogueira, F., Ribeiro, M. C., Lobert, H., Artes, R., 2003. Maternal psychological stress and distress as predictors of low birth weight, prematurity and intrauterine growth retardation. *Eur. J. Clin. Nutr.* 57, 266–272.

Roney, J. R., 2005. An alternative explanation for menstrual phase effects on women's psychology and behavior. Paper presented at the Human Behavior and Evolution Society, Austin, TX.

Roney, J. R., Hanson, K. N., Durante, K. M., Maestripieri, D., 2006. Reading men's faces: Women's mate attractiveness judgments track men's testosterone and interest in infants. *Proc. R. Soc. Lond. B Biol. Sci.* 273, 2169–2175.

Roney, J. R., Lukaszewski, A. W., Simmons, Z. L., 2007. Rapid endocrine responses of young men to social interactions with young women. *Horm. Behav.* 52, 326–333.

Roney, J. R., Mahler, S. V., Maestripieri, D., 2003. Behavioral and hormonal responses of men to brief interactions with women. *Evol. Hum. Behav.* 24, 365–375.

Roney, J. R., Simmons, Z. L., 2008. Women's estradiol predicts preference for facial cues of men's testosterone. *Horm. Behav.* 53, 14–19.

Roopnarine, J. L., 2004. African American and African Caribbean fathers: Level, quality, and meaning of involvement. In: Lamb, M. E., (Ed.), *The Role of the Father in Child Development.* 4th ed. Wiley, Hoboken, NJ, pp. 58–97.

Roopnarine, J. L., Gielen, U., 2005. *Families in Global Perspective.* Allyn and Bacon, Boston.

Roozendaal, B., de Quervain, D. J. F., 2005. Glucocorticoid therapy and memory function: Lessons learned from basic research. *Neurology* 64, 184–185.

Roozendaal, B., Quirarte, G. L., McGaugh, J. L., 2002. Glucocorticoids interact with the basolateral amygdala beta-adrenoceptor–cAMP/cAMP/PKA system in influencing memory consolidation. *Eur. J. Neurosci.* 15, 553–560.

Rose, R. M., Bernstein, I. S., Gordon, T. P., 1975. Consequences of social conflict on plasma testosterone levels in rhesus monkeys. *Psychosom. Med.* 37, 50–61.

Rose, R. M., Bernstein, I. S., Gordon, T. P., Lindsley, J. G., 1978. Changes in testosterone and behavior during adolescence in the male rhesus monkey. *Psychosom. Med.* 40, 60–70.

Rose, R. M., Holaday, J. W., Bernstein, I. S., 1971. Plasma testosterone, dominance rank and aggressive behavior in male rhesus monkeys. *Nature* 231, 366–368.

Rose, R. M., Kreuz, L. E., Holaday, J. W., Sulak, K. J., Johnson, C. E., 1972. Diurnal variation of plasma testosterone and cortisol. *J. Endocrinol.* 54, 177–178.

Rosenblatt, J. S., 1990. Landmarks in the physiological study of maternal behavior with special reference to the rat. In: Krasnegor, N. A., Bridges, R. S., (Eds.), *Mammalian Parenting: Biochemical, Neurobiological and Behavioral Determinants.* Oxford University Press, New York, pp. 40–60.

Roth, G., Dicke, U., 2005. Evolution of the brain and intelligence. *Trends Cogn. Sci.* 9, 250–257.

Rotmensch, S., Celentano, C., Liberati, M., Sadan, O., Glezerman, M., 1999. The effect of antenatal steroid administration on the fetal response to vibroacoustic stimulation. *Acta Obstet. Gynecol. Scand.* 78, 847–851.

Rowe, R., Maughan, B., Worthman, C. M., Costello, E. J., Angold, A., 2004. Testosterone, antisocial behavior, and social dominance in boys: Pubertal development and biosocial interaction. *Biol. Psychiatry* 55, 546–552.

Rubenstein, D. I., Hack, M., 2004. Natural and sexual Selection and the evolution of multi-tiered societies: Insights from zebras with comparisons to primates. In: Kappeler, P., van Schaik, C., *Sexual Selection in Primates.* Cambridge University Press, Cambridge, pp. 266–279.

Rubinow, D., Post, R. M., Savard, R., Gold, P. W., 1984. Cortisol hypersecretion and cognitive impairment in depression. *Arch. Gen. Psychiatry* 41, 279–283.

Ruff, C. B., 2000. Body size, body shape, and long bone strength in modern humans. *J. Hum. Evol.* 38, 269–290.

———, 2006. Gracilization of the modern human skeleton. *Am. Sci.* 94, 508–514.

Ruiz, J. M., Smith, T. W., Rhodewalt, F., 2001. Distinguishing narcissism and hostility: Similarities and differences in interpersonal circumplex and five-factor correlates. *J. Pers. Assess.* 76, 537–555.

Ruiz, R. J, Fullerton, J., Brown, C. E., Dudley, D. J., 2002. Predicting risk of preterm birth: The roles of stress, clinical risk factors, and corticotropin-releasing hormone. *Biol. Res. Nurs.* 4, 54–64.

Rumpel, S., LeDoux, J. E., Zador, A., Malinow, R., 2005. Postsynaptic receptor trafficking underlying a form of associative learning. *Science* 308, 83–88.

Runefors, P., Arnbjornsson, E., Elander, G., Michelsson, K., 2000. Newborn infants' cry after heel-prick: Analysis with sound spectrograph. *Acta Paediatr.* 89, 68–72.

Ruscio, M. G., Sweeny, T. D., Hazelton, J. L., Suppatkul, P., Booth, E., Carter, C. S., 2008. Pup exposure elicits hippocampal cell proliferation in the prairie vole. *Behavioural Brain Research* 187, 9–16.

Ruscio, M. G., Sweeny, T., Hazelton, J. L., Suppatkul, P., Carter, C. S., 2007. Social environment regulates corticotropin releasing factor, corticosterone and vasopressin in juvenile prairie voles. *Horm. Behav.* 51, 54–61.

———, n.d. Unpublished data. Sadalla, E. K., Kenrick, D. T., Vershure, B., 1987. Dominance and heterosexual attraction. *J. Pers. Soc. Psychol.* 52, 730–738.

Saito, T., Ishikawa, S., Abe, K., Kamoi, K., Yamada, K., Shimizu, K., Saruta, T., Yoshida, S., 1997. Acute aquaresis by the nonpeptide arginine vasopressin (AVP) antagonist OPC-31260 improves hyponatremia in patients with syndrome of inappropriate secretion of antidiuretic hormone (SIADH). *J. Clin. Endocrinol. Metab.* 82, 1054–1057.

Sakaguchi, K., Oki, M., Hasegawa, T., Honma, S., 2006. Influence of relationship status and personality traits on salivary testosterone among Japanese men. *Pers. Ind. Diff.* 41, 1077–1087.

Sakaguchi, K. Oki, M., Honma, S., Uehara, H., Hasegawa, T., 2007. The lower salivary testosterone levels among unmarried and married sexually active men. *J. Ethol.* 25, 223–229.

Salonia, A., Nappi, R. E., Pontillo, M., Daverio, R., Smeraldi, A., Briganti, A., Fabbri, F., Zanni, G., Rigatti, P., Montorsi, F., 2005. Menstrual-cycle changes in plasma oxytocin are relevant to normal sexual function in healthy women. *Horm. Behav.* 47, 164–169.

Salvador, A., 2005. Coping with competitive situations in humans. *Neurosci. Biobehav. Rev.* 29, 195–205.

Salzmann, W., Severin, J. M., Schulz-Darken, N. J., Abbott, D. H., 1997. Behavioral and social correlates of escape from suppression of ovulation in female common marmosets housed with the natal family. *Am. J. Primatol.* 41, 1–21.

Sanchez, M. M., Ladd, C. O., Plotsky, P. M., 2001. Early adverse experience as a developmental risk factor for later psychopathology: Evidence from rodent and primate models. *Devel. Psychopathol.* 13, 419–449.

Sanchez, S., Pelaez, F., Gil-Burmann, C., Kaumanns, W., 1999. Costs of infant-carrying in the cotton-top tamarin *(Sagunius oedipus)*. *Am. J. Primatol.* 48, 99–111.

Sandman, C. A., Kastin, A. J., 1981. The influence of fragments of the LPH chains on learning, memory and attention in animals and man. *Pharmacol. Ther.* 13, 39–60.

Sandman, C. A., Wadhwa, P., Glynn, L., Glynn, L., Chicz-Demet, A., Porto, M., Garite, T. J., 1999. Corticotrophin-releasing hormone and fetal responses in human pregnancy. *Ann. N. Y. Acad. Sci.* 897, 66–75.

Sannen, A., Heistermann, M., van Elsacker, L., Mohle, U., Eens, M., 2003. Urinary testosterone metabolite levels in bonobos: A comparison with chimpanzees in relation to social system. *Behaviour* 140, 683–696.

Sannen, A., van Elsacker, L., Heistermann, M., Eens, M., 2004a. Urinary testosterone metabolite levels and aggressive behaviors in male and female bonobos *(Pan paniscus)*. *Aggress. Behav.* 30, 425–434.

———, 2004b. Urinary testosterone-metabolite levels and dominance rank in male and female bonobos *(Pan paniscus)*. *Primates* 45, 89–96.

———, 2005. Certain aspects of bonobo female sexual repertoire are related to urinary testosterone metabolite levels. *Folia Primatol.* 76, 21–32.

Sapolsky, R. M., 1983. Endocrine aspects of social instability in the olive baboon *(Papio anubis)*. *Am. J. Primatol.* 5, 365–379.

———, 1990a. Adrenocortical function, social rank, and personality among wild baboons. *Biol. Psychiatry* 28, 862–878.

———, 1990b. Stress in the wild. *Sci. Am.* 262, 116–123.

———, 1991. Testicular function, social rank and personality among wild baboons. *Psychoneuroendocrinology* 16, 281–293.

———, 1992. Neuroendocrinology of the stress response. In: Becker, J. B., Breedlove, S. M., Crews, D., (Eds.), *Behavioral Endocrinology*. MIT Press, Cambridge, MA, pp. 284–324.

———, 1993. The physiology of dominance in stable versus unstable social hierarchies. In: Mason, W. A., Mendoza, S. P., (Eds.), *Primate Social Conflict*. State University of New York Press, Albany, pp. 171–204.

———, 1997. *The Trouble with Testosterone*. Simon and Schuster, New York.

———, 2002. Endocrinology of the stress response. In: Becker, J. B., Breedlove, S. M., Crews, D., McCarthy, M. M., (Eds.), *Behavioral Endocrinology*. 2nd ed. MIT Press, Cambridge, MA, pp. 409–450.

———, 2005. The influence of social hierarchy on primate health. *Science* 308, 648–652.

Sapolsky, R. M., Ray, J. C., 1989. Styles of dominance and their endocrine correlates among wild olive baboons *(Papio anubis)*. *Am. J. Primatol.* 18, 1–13.

Sapolsky, R. M., Romero, L. M., Munck, A. U., 2000. How do glucocorticoids influence stress responses? Integrating permissive, suppressive, stimulatory, and preparative actions. *Endocr. Rev.* 21, 55–89.

Savage, A., Snowdon, C. T., Giraldo, L. H., Soto, L. H., 1996. Parental care patterns and vigilance in wild cotton-top tamarins *(Saguinus oedipus)*. In: Norconk, M. A., Rosenberger, A. L., Garber, P. A., (Eds.), *Adaptive Radiations of Neotropical Primates*. Plenum, New York.

Sawchenko, P. E., Swanson, L. W., 1985. Relationship of oxytocin pathways to the control of neuroendocrine and autonomic function. In: Amico, J. A., Robinson, A. G., (Eds.), *Oxytocin: Clinical and Laboratory Studies.* Elsevier, Amsterdam, pp. 87–103.

Schaal, B., Porter, R. H., 1991. Microsomatic humans revisited: The generation and perception of chemical signals. *Adv. Study Behav.* 20, 135–199.

Schäfer-Somi, S., 2003. Cytokines during early pregnancy of mammals: A review. *Anim. Reprod. Sci.* 75, 73–94.

Schenker, N. M., Desgouttes, A. M., Semendeferi, K., 2005. Neural connectivity and cortical substrates of cognition in hominoids. *J. Hum. Evol.* 49, 447–469.

Schiavi, R. C., White, D., Mandeli, J., 1992. Pituitary-gonadal function during sleep in healthy aging men. *Psychoneuroendocrinology* 17, 599–609.

Schlinger, B. A., 1994. Estrogens to song: Picograms to sonograms. *Horm. Behav.* 28, 191–198.

Schlinger, B. A., Callard, G. V., 1990. Aromatization mediates aggressive behavior in quail. *Gen. Comp. Endocrinol.* 79, 39–53.

Schlinger, B. A., Lane, N. I., Grisham, W., Thompson, L. 1999. Androgen synthesis in a songbird: A study of Cyp17 (17a-hydroxylase/C17,20-lyase) activity in the zebra finch. *Gen. Comp. Endocrinol.* 113, 46–58.

Schneider, H. J., Pickel, J., Stalla, G. K., 2006. Typical female 2nd-4th finger length (2D:4D) ratios in male-to-female transsexuals: Possible implications for prenatal androgen exposure. *Psychoneuroendocrinology* 31, 265–269.

Schneider, J. E., Wade, G. N., 1990. Effects of diet and body fat content on cold-induced anestrus in Syrian hamsters. *Am. J. Physiol. Regul. Integr. Comp. Physiol.* 259, 1198–1204.

Schneider, M. L., 1992. Pre-natal stress exposure alters postnatal behavioral expression under conditions of novelty challenge in rhesus monkey infants. *Dev. Psychobiol.* 25, 529–540.

Schneider, M. L., Moore, C. F., Kraemer, G. W., 2004. Moderate level alcohol during pregnancy, pre-natal stress, or both and limbic-hypothalamic-pituitary-adrenocortical axis response to stress in rhesus monkeys. *Child Dev.* 75, 96–109.

Schradin, C., Anzenberger, G., 2004. Development of prolactin levels in marmoset males: From adult son to first-time father. *Horm. Behav.* 46, 670–677.

Schradin, C., Reeder, D. M., Mendoza, S. P., Anzenberger, G., 2003. Prolactin and paternal care: Comparison of three species of monogamous New World monkeys. *J. Comp. Psychol.* 117, 166–175.

Schultheiss, O. C., Campbell, K. L., McClelland, D. C., 1999. Implicit power motivation moderates men's testosterone response to imagined and real dominance success. *Horm. Behav.* 36, 234–241.

Schürmann, C. L., 1982. Mating behaviour of wild orang-utans. In: de Boer, L. E. M., (Ed.), *The Orang-utan, Its Biology and Conservation.* W. Junk, The Hague, pp. 269–284.

Schürmann, C. L., van Hooff, J. A. R. A. M., 1986. Reproductive strategies of the orang-utan: New data and a reconsideration of existing sociosexual models. *Int. J. Primatol.* 7, 265–287.

Schwabe, J. W., Teichmann, S. A., 2004. Nuclear receptors: The evolution of diversity. *Sci. STKE* 217, 4.

Schwabl, H., Kriner, E., 1991. Territorial aggression and song of male European robins *(Erithacus rubecula)* in autumn and spring: Effects of antiandrogen treatment. *Horm. Behav.* 25, 180–194.

Schwabl, H., Mock, D. W., Gieg, J. A., 1997. A hormonal mechanism for parental favouritism. *Nature* 386, 231.

Schweiger, U., Herrmann, F., Laessle, R., Riedel, W., Schweiger, M., Pirke, K., 1988. Caloric intake, stress, and menstrual function in athletes. *Fertil. Steril.* 49, 447–450.

Seamans, J. K., Yang, C. R., 2004. The principal features and mechanisms of dopamine modulation in the prefrontal cortex. *Prog. Neurobiol.* 74, 1–58.

Seifritz, E., Esposito, F., Neuhoff, J. G., Luthi, A., Mustovic, H., Dammann, G., von Bardeleben, U., Radue, E. W., Cirillo, S., Tedeschi, G., Di Salle F., 2003. Differential sex-independent amygdala response to infant crying and laughing in parents versus nonparents. *Biol. Psychiatry* 54, 1367–1375.

Sellen, D. W., 2006. Lactation, complementary feeding, and human life history. In: Hawkes, K., Paine, R. R., (Eds.), *The Evolution of Human Life History.* School of American Research Press, Santa Fe, NM, pp. 155–196.

Sellen, D. W., Smay, D. B., 2001. Relationship between subsistence and age at weaning in "pre-industrial" societies. *Hum. Nat.* 12, 47–87.

Seltzer, L. J., Ziegler, T. E., 2007. Non-invasive measurement of small peptides in the common marmoset *(Callithrix jacchus):* A radiolabeled clearance study and endogenous excretion under varying social conditions. *Horm. Behav.* 51, 436–442.

Semendeferi, K., Armstrong, E., Schleicher, A., Zilles, K., Vas Hoesen, G. W., 1998. Limbic frontal cortex in hominoids: A comparative study of area 13. *Am. J. Phys. Anthropol.* 106, 129–155.

Serbin, L. A., Moller, L. C., Gulko, J., Powlishta, K. K., Colburne, K. A., 1994. The emergence of gender segregation in toddler playgroups. *New Dir. Child Dev.* 65, 7–17.

Serbin, L. A., Poulin-Dubois, D., Colburne, K. A., Sen, M. G., Eichstedt, J. A., 2001. Gender stereotyping in infancy: Visual preferences for and knowledge of gender-stereotyped toys in the second year. *Int. J. Behav. Dev.* 25, 7–15.

Serio, M., Forti, G., 1989. IGF-I, IGF-II and gonadal function. *J. Endocrinol. Invest.* 12, 97–99.

Servin, A., Nordenstrom, A., Larsson, A., Bohlin, G., 2003. Prenatal androgens and gender-typed behavior: A study of girls with mild and severe forms of congenital adrenal hyperplasia. *Dev. Psychol.* 39, 440–450.

Setchell, J., Lee, P. C., 2004. Development and sexual selection in primates. In: Kappeler, P. M., van Schaik, C. P., *Sexual Selection in Primates: Causes, Mechanisms and Consequences.* Cambridge University Press, Cambridge, pp. 175–195.

Shackelford, T. K., Weekes-Shackelford, V. A., Schmitt, D. P., 2005. An evolutionary perspective on why some men refuse or reduce their child support payments. *Basic Appl. Soc. Psychol.* 27, 297–306.

Shafer, A. B., 2006. Meta-analysis of the factor structures of four depression questionnaires: Beck, CES-D, Hamilton, and Zung. *J. Clin. Psychol.* 62, 123–146.

Share, L., Crofton, J. T., Ouchi, Y., 1988. Vasopressin: Sexual dimorphism in secretion, cardiovascular actions and hypertension. *Am. J. Med. Sci.* 295, 314–319.

Shea, A. K., Kamath, M. V., Fleming, A. S., Streiner, D. L., Broad, K., Steiner, M., 2007. The effect of depression and anxiety on the cortisol awakening response during pregnancy. *Psychoneuroendocrinology* 32, 1013–1020.

Sheehan, T., Numan, M., 2002. Estrogen, progesterone and pregnancy termination alter neural activity in brain regions that control maternal behavior in rats. *Neuroendocrinolgy* 75, 12–23.

Shelton, S. E., Kalin, N. H., Gluck, J. P., Keresztury, M. F., Schneider, V. A., Lewis, M. H., 1988. Effect of age on cisternal cerebrospinal fluid concentrations of monoamine metabolites in nonhuman primates. *Neurochem. Int.* 13, 353–357.

Sherry, D. S., Ellison, P. T., 2007. Potential applications of urinary C-peptide of insulin for comparative energetics research. *Am. J. Phys. Anthropol.* 133, 771–778.

Shideler, S. E., Lasley, B. L., 1982. A comparison of primate ovarian cycles. *Am. J. Primatol. Suppl.* 1, 171–180.

Shih, J. H., Eberhart, N. K., Hammen, C. L., Brennan, P. A., 2006. Differential exposure and reactivity to interpersonal stress predict sex differences in adolescent depression. *J. Clin. Child Adolesc. Psychol.* 35, 103–115.

Shimizu, K., Douke, C., Fujita, S., Matsuzawa, T., Tomonaga, M., Tanaka, M., Matsubayashi, K., Hayashi, M., 2003. Urinary steroids, FSH and CG measurements for monitoring the ovarian cycle and pregnancy in the chimpanzee. *J. Med. Primatol.* 32, 15–22.

Shingo, T., Gregg, C., Enwere, E., Fujikawa, H., Hassam, R., Geary, C., Cross, J. C., Weiss, S., 2003. Pregnancy-stimulated neurogenesis in the adult female forebrain mediated by prolactin. *Science* 299, 117–120.

Shively, C. A., Friedman, D. P., Gage, H. D., Bounds, M. C., Brown-Proctor, C., Blair, J. B., Henderson, J. A., Smith, M. A., Buchheimer, N., 2006. Behavioral depression and positron emission tomography–determined serotonin 1A receptor binding potential in cynomolgus monkeys. *Arch. Gen. Psychiatry* 63, 396–403.

Short, R. V., 1987. The biological basis for the contraceptive effect of breast feeding. *Int. J. Gynaecol. Obstet. Suppl.* 25, 207–217.

——, 1994. What the breast does for the baby, and what the baby does for the breast. *Aust. N. Z. J. Obstet. Gynaecol.* 34, 262–264.

Sicotte, P., 1993. Inter-group encounters and female transfer in mountain gorillas: Influence of group composition on male behavior. *Am. J. Primatol.* 30, 21–36.

——, 1994. Effect of male competition on male-female relationships in bi-male groups of mountain gorillas. *Ethology* 97, 47–64.

——, 2002. The function of male aggressive displays towards females in mountain gorillas. *Primates* 43, 277–289.

Sikkel, P. C., 1993. Changes in plasma androgen levels associated with changes in male reproductive behavior in a brood cycling marine fish. *Gen. Comp. Endocrinol.* 89, 229–237.

Silk, J. B., 1987. Social behavior in evolutionary perspective. In: Smuts, B. B., Cheney, D. L., Seyfarth, R. M., Wrangham, R. W., Struhsaker, T. T., (Eds.), *Primate Societies*. University of Chicago Press, Chicago, pp. 318–329.

——, 2002. Practice random acts of aggression and senseless acts of intimidation: The logic of status contests in social groups. *Evol. Anthropol.* 11, 221–225.

Sillen-Tullberg, B., Møller, A. P., 1993. The relationship between concealed ovulation and mating systems in anthropoid primates: A phylogenetic analysis. *Am. Nat.* 141, 1–25.

Silverin, B., 1980. Effects of long-acting testosterone treatment on free-living pied flycatchers, *Ficedula hypoleuca*, during the breeding period. *Anim. Behav.* 28, 906–912.

Silverin, B., Deviche, P., 1991. Biochemical characterization and seasonal changes in the concentration of testosterone-metabolizing enzymes in the European great tit *(Parus major)* brain. *Gen. Comp. Endocrinol.* 81, 146–159.

Singh, D., 1993. Adaptive significance of female physical attractiveness: Role of waist-to-hip ratio. *J. Pers. Soc. Psychol.* 65, 293–307.

Singh, D., Bronstad, P. M., 2001. Female body odour is a potential cue to ovulation. *Proc. R. Soc. Lond. B Biol. Sci.* 268, 797–801.

Singh, D., Vidaurri, M., Zambarano, R. J., Dabbs, J. M., 1999. Lesbian erotic role identification: Behavioral, morphological, and hormonal correlates. *J. Pers. Soc. Psychol.* 76, 1035–1049.

Sipos, M. L., Nyby, J. G., 1996. Concurrent androgenic stimulation of the ventral tegmental area and medial preoptic area: Synergistic effects on male-typical reproductive behaviors in house mice. *Brain Res.* 729, 29–44.

Smith, G. T., Brenowitz, E. A., Beecher, M. D., Wingfield, J. C., 1997. Seasonal changes in testosterone, neural attributes of song control nuclei, and song structure in wild songbirds. *J. Neurosci.* 17, 6001–6010.

Smith, H. J., 2005. *Parenting for Primates*. Harvard University Press, Cambridge, MA.

Smith, R., Cubis, J., Brinsmead, M., Lewin, T., Singh, B., Owens, P., Chan, E. C., Hall, C., Adler, M., Lovelock, M., Hurt, D., Rowley, M., Nolan, M., 1990. Mood changes, obstetric experience and alterations in plasma cortisol, beta-endorphin and corticotrophin releasing hormone during pregnancy and the puerperium. *J. Psychosom. Res.* 34, 53–69.

Smith, T. D., Bhatnagar, K. P., Bonar, C. J., Shimp, K. L., Mooney, M. P., Siegle, M. I., 2003. Ontogenetic characteristics of the vomeronasal organ in *Saguinus geoffroyi* and *Leontopithecus rosaila*, with comparisons to other primates. *Am. J. Phys. Anthropol.* 121, 342–353.

Smuts, B. B., 1985. *Sex and Friendship in Baboons*. Aldine de Gruyter, Hawthorne, NY.

Snyder, P. J., Peachey, H., Berlin, J. A., Hannoush, P., Haddad, G., Dlewati, A., Santanna, J., Loh, L., Lehrow, D. A., Holmes, J. H., Kapoor, S. C., Atkinson, L. E., Strom, B. L., 2000. Effects of testosterone replacement in hypogonadal men. *J. Clin. Endocrinol. Metab.* 85, 2670–2677.

Sobrinho, L. G., 2003. Prolactin, psychological stress and environment in humans: Adaptation and maladaption. *Pituitary* 6, 35–39.

Sohlstrom, A., Forsum, E., 1995. Changes in adipose tissue volume and distribution during reproduction in Swedish women as assessed by magnetic resonance imaging. *Am. J. Clin. Nutr.* 61, 287–295.

Solomon, N. G., French, J. F., 1997. The study of mammalian cooperative breeding. In: Solomon, N. G., French, J. F., (Eds.), *Cooperative Breeding in Mammals.* Cambridge University Press, Cambridge, pp. 1–10.

Soltis, J., 2004. The signal functions of early infant crying. *Behav. Brain Sci.* 27, 443–458; discussion 459–490.

Soma, K. K., 2006. Testosterone and aggression: Berthold, birds and beyond. *J. Neuroendocrinol.* 18, 543–551.

Soma, K. K., Bindra, R. K., Gee, J., Wingfield, J. C., Schlinger, B. A., 1999. Androgen metabolizing enzymes show region-specific changes across the breeding season in the brain of a wild songbird. *J. Neurobiol.* 41, 176–188.

Soma, K. K., Sullivan, K. A., Tramontin, A. D., Saldanha, C. J., Schlinger, B. A., Wingfield, J. C., 2000. Acute and chronic effects of an aromatase inhibitor on territorial aggression in breeding and non-breeding male song sparrows. *J. Comp. Physiol. A* 186, 759–769.

Soma, K. K., Sullivan, K. A., Wingfield, J. C., 1999. Combined aromatase inhibitor and antiandrogen treatment decreases territorial aggression in a wild songbird during the nonbreeding season. *Gen. Comp. Endocrinol.* 115, 442–453.

Soma, K. K., Tramontin, A. D., Wingfield, J. C., 2000. Oestrogen regulates male aggression in the non-breeding season. *Proc. R. Soc. Lond. B Biol. Sci.* 267, 1089–1096.

Soma, K. K., Wingfield, J. C., 2001. Dehydroepiandrosterone in songbird plasma: Seasonal regulation and relationship to territorial aggression. *Gen. Comp. Endocrinol.* 123, 144–155.

Soma, K. K., Wissman, A. M., Brenowitz, E. A., Wingfield, J. C., 2002. Dehydroepiandrosterone (DHEA) increases male aggression and the size of an associated brain region. *Horm. Behav.* 41, 203–212.

Sosa, R., Kennell, J., Marshall, K., Robertson, S., Urrutia, J., 1980. The effect of a supportive companion on perinatal problems, length of labor, and mother-infant interaction. *N. Engl. J. Med.* 303, 597–600.

Soubrie, P., 1986. Reconciling the role of central serotonin neurons in human and animal behavior. *Behav. Brain Sci.* 9, 319–364.

Spangler, G., Schieche, M., 1998. Emotional and adrenocortical responses of infants to the strange situation: The differential function of emotional expression. *Int. J. Behav. Dev.* 22, 681–706.

Spangler, G., Schieche, M., Ilg, U., Maier, U., Ackermann, C., 1994. Maternal sensitivity as an external organizer for biobehavioral regulation in infancy. *Dev. Psychobiol.* 27, 425–437.

Spear, L. P., 2000. The adolescent brain and age-related behavioral manifestations. *Neurosci. Biobehav. Rev.* 24, 417–463.

Spinrad, T. L., Stifter, C. A., 2006. Toddlers' empathy-related responding to distress: Predictions from negative emotionality and maternal behavior in infancy. *Infancy* 10, 97–121.

Spong, C., Stone, J., Creel, S., Björklund, M., 2002. Genetic structure of lions (*Panthera leo* L.) in the Selous Game Reserve: Implications for the evolution of sociality. *J. Evol. Biol.* 15, 945–953.

Stallings, J., Fleming, A. S., Corter, C., Worthman, C., Steiner, M., 2001. The effects of infant cries and odors on sympathy, cortisol, and autonomic responses in new mothers and nonpostpartum women. *Parenting Sci. Pract.* 1, 71–100.

Stanford, C. B., 1998. The social behavior of chimpanzees and bonobos: Empirical evidence and shifting assumptions. *Curr. Anthropol.* 39, 399–420.

Stansbury, K., Gunnar, M. R., 1994. Adrenocortical activity and emotion regulation. *Monogr. Soc. Res. Child Devel.* 59, 108–134.

Steklis, H. D., Brammer, G. L., Raleigh, M. J., McGuire, M. T., 1985. Serum testosterone, male dominance, and aggression in captive groups of vervet monkeys *(Cercopithecus aethiops sabaeus)*. *Horm. Behav.* 19, 154–163.

Stern, D., 1977. *The First Relationship: Infant and Mother.* Harvard University Press, Cambridge, MA.

Stern, J. M., 1996. Somatosensation and maternal care in Norway rats. In: Rosenblatt, J. S., Snowdon, C. T., (Eds.), *Parental Care: Evolution, Mechanisms, and Adaptive Significance.* Academic Press, San Diego, pp. 243–294.

Stevenson, J. C., Everson, P. M., Williams, D. C., Hipskind, G., Grimes, M., Mahoney, E. R., 2007. Attention deficit/hyperactivity disorder (ADHD) symptoms and digit ratios in a college sample. *Am. J. Hum. Biol.* 19, 41–50.

Stoinski, T. S., Czekala, N. M., Lukas, K. E., Maple, T. L., 2002. Urinary androgen and corticoid levels in captive, male western lowland gorillas *(Gorilla g. gorilla)*: Age- and social group–related differences. *Am. J. Primatol.* 56, 73–87.

Stoleru, S. G., Ennaji, A., Cournot, A., Spira, A., 1993. LH pulsatile secretion and testosterone blood levels are influenced by sexual arousal in human males. *Psychoneuroendocrinology* 18, 205–218.

Stoleru, S. G., Redoute, J., Costes, N., Lavenne, F., Bars, D. I., Dechaud, H., Forest, M. G., Pugeat, M., Cinotti, L., Pujol, J. F., 2003. Brain processing of visual stimuli in men with hypoactive sexual desire disorder. *Psychiatry Res.* 124, 67–86.

Storey, A. E., Walsh, C. J., Quinton, R. L., Wynne-Edwards, K. E., 2000. Hormonal correlates of paternal responsiveness in new and expectant fathers. *Evol. Hum. Behav.* 21, 79–95.

Storm, E. E., Tecott, L. H., 2005. Social circuits: Peptidergic regulation of mammalian social behavior. *Neuron* 47, 483–486.

Strasser, R., Schwabl, H., 2004. Yolk testosterone organizes behavior and male plumage coloration in house sparrows *(Passer domesticus)*. *Behav. Ecol. Sociobiol.* 56, 491–497.

Strassmann, B. I., 1997. The biology of menstruation in *Homo sapiens:* Total lifetime menses, fecundity, and nonsynchrony in a natural-fertility population. *Curr. Anthropol.* 38, 123–129.

————, 1999. Menstrual cycling and breast cancer: An evolutionary perspective. *J. Women's Health* 8, 193–202.

Stribley, J. M., Carter, C. S., 1999. Developmental exposure to vasopressin increases aggression in adult prairie voles. *Proc. Natl. Acad. Sci. U. S. A.* 96, 12601–12604.

Strier, K. B., 1992. *Faces in the Forest: The Endangered Muriquis of Brazil.* Oxford University Press, New York.

————, 2007. *Primate Behavioral Ecology.* 3rd ed. Allyn and Bacon, Boston.

Strier, K. B., Ziegler, T. E., Wittwer, D. J., 1999. Seasonal and social correlates of fecal testosterone and cortisol levels in wild male muriquis *(Brachyteles arachnoides).* *Horm. Behav.* 35, 125–134.

Stumpf, R. M., Boesch, C., 2005. Does promiscuous mating preclude female choice? Female sexual strategies in chimpanzees *(Pan troglodytes verus)* of the Taï National Park, Côte d'Ivoire. *Behav. Ecol. Sociobiol.* 57, 511–524.

————, 2006. The efficacy of female choice in chimpanzees of the Taï Forest, Côte d'Ivoire. *Behav. Ecol. Sociobiol.* 60, 749–765.

Subtil, D., Tiberghien, P., Devos, P., Therby, D., Leclerc, G., Vaast, P., Puech, F., 2003. Immediate and delayed effects of antenatal corticosteroids on fetal heart rate: A randomized trial that compares betamethasone acetate and phosphate, betamethasone phosphate, and dexamethasone. *Am. J. Obstet. Gynecol.* 188, 524–531.

Sugiyama, L. S., 2004. Is beauty in the context-sensitive adaptations of the beholder? Shiwiar use of waist-to-hip ratio in assessments of female mate value. *Evol. Hum. Behav.* 25, 51–62.

————, 2005. Physical attractiveness in adaptationist perspective. In: Buss, D. M., (Ed.), *Handbook of Evolutionary Psychology.* Wiley, Hoboken, NJ, pp. 292–343.

Suiming, W., 2004. Three "red light districts" in China. In: Micollier, E., (Ed.), *Sexual Cultures in East Asia: The Social Construction of Sexuality and Sexual Risk in a Time of AIDS.* RoutledgeCurzon, New York, pp. 23–53.

Suomi, S. J., 1997. Long-term effects of differential early experiences on social, emotional, and physiological development in nonhuman primates. In: Keshavan, M. S., Murray, R. M., (Eds.), *Neurodevelopment and Adult Psychopathology.* Cambridge University Press, Cambridge, pp. 104–116.

Sussman, R. W., Garber, P. A., 1987. A new interpretation of the social organization and mating system of the Callitrichidae. *Int. J. Primatol.* 8, 73–92.

Swain, J. E., Lorberbaum, J. P., Kose, S., Strathearn, L., 2007. Brain basis of early parent-infant interactions: Psychology, physiology, and in vivo functional neuroimaging studies. *J. Child Psychol. Psychiatry* 48, 262–287.

Szuran, T. F., Pliska, V., Pokorny, J., Welzl, H., 2000. Prenatal stress in rats: Effects on plasma corticosterone, hippocampal glucocorticoid receptors, and maze performance. *Physiol. Behav.* 71, 353–362.

Takahashi, L. K., Turner, J. G., Kalin, N. H., 1992. Pre-natal stress alters brain catecholaminergic activity and potentiates stress-induced behavior in adult rats. *Brain Res.* 574, 131–137.

Takayanagi, Y., Yoshida, M., Bielsky, I. F., Ross, H. E., Kawamata, M., Onaka, T., Yanagisawa, T., Kimura, T., Matzuk, M. M., Young, L. J., Nishimori, K., 2005. Pervasive social deficits, but normal parturition, in oxytocin receptor-deficient mice. *Proc. Natl. Acad. Sci. U. S. A.* 102, 16096–16101.

Taniguchi, K., Matsusaki, Y., Ogawa, K., Saito, T. R., 1992. Fine structure of the vomeronasal organ in the common marmoset *(Callithris jacchus)*. *Folia Primatol.* 59, 169–176.

Tay, C. C., Glasier, A. F., McNeilly, A. S., 1996. Twenty-four hour patterns of prolactin secretion during lactation and the relationship to suckling and the resumption of fertility in breast-feeding women. *Hum. Reprod.* 11, 950–955.

Taylor, S. E., 2002. *The Tending Instinct.* Holt, New York.

Taylor, S. E., Gonzaga, G. C., Klein, L. C., Hu, P., Greendale, G. A., Seeman, T. E., 2006. Relation of oxytocin to psychological stress responses and hypothalamic-pituitary-adrenocortical axis activity in older women. *Psychosom. Med.* 68, 238–245.

Taylor, S. E., Klein, L. C., Lewis, B. P., Gruenewald, T. L., Gurung, R. A. R., Updegraff, J. A., 2000. Biobehavioral responses to stress in females: Tend-and-befriend, not fight-or-flight. *Psychol. Rev.* 107, 411–429.

te Boekhorst, I. J. A., Schürmann, C. L., Sugardjito, J., 1990. Residential status and seasonal movements of wild orang-utans in the Gunung Leuser Reserve (Sumatra, Indonesia). *Anim. Behav.* 39, 1098–1109.

Teicher, M. H., Andersen, S. L., Polcari, A., Anderson, C. M., Navalta, C. P., Kim, D. M., 2003. The neurobiological consequences of early stress and childhood maltreatment. *Neurosci. Biobehav. Rev.* 27, 33–44.

Teixeira, J. M., Fisk, N. M., Glover, V., 1999. Association between maternal anxiety in pregnancy and increased uterine artery resistance index: Cohort based study. *Br. Med. J.* 318, 153–157.

Terkel, J., Bridges, R. S., Sawyer, C. H., 1979. Effects of transecting lateral neural connections of the medial preoptic area on maternal behavior in the rat: Nest building, pup retrieval and prolactin secretion. *Brain Res.* 169, 369–380.

Tessitore, C., Brunjes, P. C., 1988. A comparative study of myelination in precocial and altricial murid rodents. *Brain Res.* 471, 139–147.

Thomas, C. D., Lennon, J. J., 1999. Birds extend their ranges northwards. *Nature* 399, 213.

Thompson, C. W., Moore, M. C., 1991. Throat color reliably signals status in male tree lizards, *Urosaurus ornatus*. *Anim. Behav.* 42, 745–753.

Thompson, R., George, K., Walton, J., Orr, S., Benson, J., 2006. Sex-specific influences of vasopressin on human social communication. *Proc. Natl. Acad. Sci. U. S. A.* 103, 7889–7894.

Thompson, R., Gupta, S., Miller, K., Mills, S., Orr, S., 2004. The effects of vasopressin on human facial responses related to social communication. *Psychoneuroendocrinology* 29, 35–48.

Thorne, B., Luria, Z., 1986. Sexuality and gender in children's daily worlds. *Soc. Prob.* 33, 176–190.

Thornhill, R., Gangestad, S. W., 1999. The scent of symmetry: A human pheromone that signals fitness? *Evol. Hum. Behav.* 20, 175–201.

Thornhill, R., Gangestad, S. W., Miller, R., Scheyd, G., McCollough, J. K., Franklin, M., 2003. Major histocompatibility complex genes, symmetry, and body scent attractiveness in men and women. *Behav. Ecol.* 14, 668–678.

Thornhill, R., Grammer, K., 1999. The body and face of woman: One ornament that signals quality? *Evol. Hum. Behav.* 20, 105–120.

Tilbrook, A. J., Clarke, I. J., 2001. Negative feedback regulation of the secretion and actions of gonadotropin-releasing hormone in males. *Biol. Reprod.* 64, 735–742.

Tilbrook, A. J., Turner, A. I., Clarke, I. J., 2000. Effects of stress on reproduction in non-rodent mammals: The role of glucocorticoids and sex differences. *Rev. Reprod.* 5, 105–113.

Todd, K., Lightman, S., 1986. Oxytocin release during coitus in male and female rabbits: Effect of opiate receptor blockade with naloxone. *Psychoneuroendocrinology* 11, 367–371.

Tooby, J., Cosmides, L., 1992. The psychological foundations of culture. In: Barkow, J., Cosmides, L., Tooby, J., (Eds.), *The Adapted Mind*. Oxford University Press, New York, pp. 19–136.

Tracer, D. O., 1996. Lactation, nutrition, and postpartum amenorrhea in lowland Papua–New Guinea. *Hum. Biol.* 68, 277–292.

Trainor, B. D., Marler, C. A., 2002. Testosterone promotes paternal behaviour in a monogamous mammal via conversion to oestrogen. *Proc. R. Soc. Lond. B Biol. Sci.* 269, 823–829.

Travis, J. M. J., 2003. Climate change and habitat destruction: A deadly anthropogenic cocktail. *Proc. R. Soc. Lond. B Bio. Sci.* 270, 467–473.

Travison, T. G., Araujo, A. B., Kupelian, V., O'Donnell, A. B., McKinley, J. B., 2007. The relative contributions of aging, health, and lifestyle factors to serum testosterone decline in men. *J. Clin. Endocrinol. Metab.* 92, 549–555.

Trayhurn, P., Bing, C., Wood, I. S., 2006. Adipose tissue and adipokines—energy regulation from the human perspective. *J. Nutr.* 136, 1935s–1939s.

Tremblay, R. E., Schaal, B., Boulerice, B., Arseneault, L., Soussignan, R. G., Paquette, D., Laurent, D., 1998. Testosterone, physical aggression, dominance, and physical development in early adolescence. *Int. J. Behav. Dev.* 22, 753–777.

Tribollet, E., Audigier, S., Dubois-Dauphin, M., Dreifuss, J. J., 1990. Gonadal steroids regulate oxytocin receptors but not vasopressin receptors in the brain of male and female rats. An autoradiographical study. *Brain Res.* 511, 129–140.

Tribollet, E., Dubois-Dauphin, M., Dreifuss, J. J., Barberis, C., Jard, S., 1992. Oxytocin receptors in the central nervous system. Distribution, development, and species differences. *Ann. N. Y. Acad. Sci.* 652, 29–38.

Trivers, R. L., 1972. Parental investment and sexual selection. In: Campbell, B., (Ed.), *Sexual Selection and the Descent of Man*. Aldine de Gruyter, Hawthorne, NY, pp. 136–179.

———, 1974. Parent-offspring conflict. *Am. Zool.* 14, 249–264.

Tsai, L. W., Sapolsky, R. M., 1996. Rapid stimulatory effects of testosterone upon myotubule metabolism and sugar transport, as assessed by silicon microphysiometry. *Aggress. Behav.* 22, 357–364.

Tsutsui, K., Yamakazi, T., 1995. Avian neurosteroids. I. Pregnenolone synthesis in the quail brain. *Brain Res.* 678, 1–9.

Tu, M. T., Lupien, S. J., Walker, C. D., 2006a. Diurnal salivary cortisol levels in postpartum mothers as a function of infant feeding choice and parity. *Psychoneuroendocrinology* 31, 812–824.

———, 2006b. Multiparity reveals the blunting effect of breastfeeding on physiological reactivity to psychological stress. *J. Neuroendocrinol.* 18, 494–503.

Tupler, L. A., De Bellis, M. D., 2006. Segmented hippocampal volume in children and adolescents with posttraumatic stress disorder. *Biol. Psychiatry* 59, 523–529.

Turner, R., Altemus, M., Enos, T., Cooper, B., McGuinness, T., 1999. Preliminary research on plasma oxytocin in normal cycling women: Investigating emotion and interpersonal distress. *Psychiatry* 62, 97–113.

Tutin, C. E. G., 1979. Mating patterns and reproductive strategies in a community of wild chimpanzees *(Pan troglodytes schweinfurthii)*. *Behav. Ecol. Sociobiol.* 6, 29–38.

Tutin, C. E. G., McGinnis, P. R., 1981. Chimpanzee reproduction in the wild. In: Graham, C. E., (Ed.), *Reproductive Biology of the Great Apes: Comparative and Biomedical Perspectives*. Academic Press, New York, pp. 239–264.

Udry, J. R., Chantala, K., 2006. Masculinity-femininity predicts sexual orientation in men but not in women. *J. Biosoc. Sci.* 38, 797–809.

Udry, J. R., Talbert, L. M., 1988. Sex-hormone effects on personality at puberty. *J. Pers. Soc. Psychol.* 54, 291–295.

Ukena, K., Honda, Y., Inai, Y., Kohchi, C., Lea, R., Tsutsui, K., 1999. Expression and activity of 3b-hydroxysteroid dehydrogenase/Δ^5-Δ^4-isomerase in different regions of the avian brain. *Brain Res.* 818, 536–542.

Utami, S. S., Goossens, B., Bruford, M. W., de Ruiter, J. R., van Hooff, J. A. R. A. M., 2002. Male bimaturism and reproductive success in Sumatran orangutans. *Behav. Ecol.* 13, 643–652.

Uvnäs-Moberg, K., 1997. Physiological and endocrine effects of social contact. *Ann. N. Y. Acad. Sci.* 807, 146–163.

———, 1998. Oxytocin may mediate the benefits of positive social interaction and emotions. *Psychoneuroendocrinology* 23, 819–835.

———, 2003. *The Oxytocin Factor: Tapping the Hormone of Calm, Love, and Healing*. Da Capo Press, Cambridge, MA.

Uvnäs-Moberg, K., Alster, P., Lund, I., Lundeberg, T., Kurosawa, M., Ahlenius, S., 1996. Stroking of the abdomen causes decreased locomotor activity in conscious male rats. *Physiol. Behav.* 60, 1409–1411.

Valeggia, C., Ellison, P. T., 2004. Lactational amenorrhoea in well-nourished Toba women of Formosa, Argentina. *J. Biosoc. Sci.* 36, 573–595.

Vallée, M., Mayo, W., Dellu, F., Le Moal, M., Simon, H., Maccari, S., 1997. Prenatal stress induces high anxiety and postnatal handling induces low anxiety in adult offspring: Correlation with stress-induced corticosterone injection. *J. Neurosci.* 17, 2626–2636.

van Anders, S. M., Hamilton, L. D., Watson, N. V., 2007. Multiple partners are associated with higher testosterone in North American men and women. *Horm. Behav.* 41, 454–459.

van Anders, S. M., Hampson, E., 2005. Testing the prenatal androgen hypothesis: Measuring digit ratios, sexual orientation, and spatial abilities in adults. *Horm. Behav.* 47, 92–98.

van Anders, S. M., Hampson, E., Watson, N. V., 2006. Seasonality, waist-to-hip ratio, and salivary testosterone. *Psychoneuroendocrinology* 31, 895–899.

van Anders, S. M., Watson, N. V., 2006a. Relationship status and testosterone in North American heterosexual and non-heterosexual men and women: Cross-sectional and longitudinal data. *Psychoneuroendocrinology* 31, 715–723.

———, 2006b. Social neuroendocrinology: Effects of social contexts and behaviors on sex steroids in humans. *Hum. Nat.* 17, 212–237.

———, 2007. Testosterone levels in men and women who are single, in long-distance relationships, or same-city relationships. *Horm. Behav.* 51, 286–291.

van Anders, S. M., Wilbur, C. J., Vernon, P. A., 2006. Finger-length ratios show evidence of prenatal-hormone transfer between opposite-sex twins. *Horm. Behav.* 49, 315–319.

van Bokhoven, I., van Goozen, S. H., van Engeland, H., Schaal, B., Arseneault, L., Seguin, J. R., Assaad, J. M., Nagin, D. S., Vitaro, F., Tremblay, R. E., 2006. Salivary testosterone and aggression, delinquency, and social dominance in a population-based longitudinal study of adolescent males. *Horm. Behav.* 50, 118–125.

van de Beek, C., Thijssen, J. H. H., Cohen-Kettenis, P. T., van Goozen, S. H. M., Buitelaar, J. K., 2004. Relationships between sex hormones assessed in amniotic fluid and maternal and umbilical cord serum: What is the best source of information to investigate the effects of fetal hormonal exposure? *Horm. Behav.* 46, 663–669.

Van den Bergh, B. R. H., 1992. Maternal emotions during pregnancy and fetal and neonatal behavior. In: Nijhuis, J. G., (Ed.), *Fetal Behaviour: Developmental and Perinatal Aspects*. Oxford University Press, Oxford, pp. 157–178.

Van Den Berghe, P., 1979. *Human Family Systems: An Evolutionary View*. Elsevier, New York.

van der Werff ten Bosch, J. J., 1982. The physiology of reproduction of the orang-utan. In: De Boer, L. E. M., (Ed.), *The Orang-utan: Its Biology and Conservation*. W. Junk, The Hague, pp. 201–214.

van Honk, J., Peper, J. S., Schutter, D. J., 2005. Testosterone reduces unconscious fear but not consciously experienced anxiety: Implications for the disorders of fear and anxiety. *Biol. Psychiatry* 58, 218–225.

van Honk, J., Schutter, D., Hermans, E. J., Putman, P., 2004. Testosterone, cortisol, dominance, and submission: Biologically prepared motivation, no psychological mechanisms involved. *Behav. Brain Sci.* 27, 160–162.

van Honk, J., Tuiten, A., Verbaten, R., van den Hout, M., Koppeschaar, H., Thijssen, J., deHaan, E., 1999. Correlations among salivary testosterone, mood, and selective attention to threat in humans. *Horm. Behav.* 36, 17–24.

Van Horn, R. C., Engh, A. L., Scribner, K. T., Funk, S. M., Holekamp, K. E., 2004. Behavioral structuring of relatedness in the spotted hyena *(Crocuta crocuta)* suggests direct fitness benefits of clan-level cooperation. *Mol. Ecol.* 13, 449–458.

van Londen, L., Goekoop, J. G., van Kempen, G. M., Frankhuijzen-Sierevogel, A. C., Wiegant, V. M., van der Velde, E. A., de Wied, D., 1997. Plasma levels of arginine vasopressin elevated in patients with major depression. *Neuropsychopharmacology* 17, 284–292.

van Londen, L., Goekoop, J. G., Zwinderman, A. H., Lanser, J. B., Wiegant, V. M., De Wied, D., 1998. Neuropsychological performance and plasma cortisol, arginine vasopressin and oxytocin in patients with major depression. *Psychol. Med.* 28, 275–284.

van Noordwijk, M., van Schaik, C., 1985. Male migration and rank acquisition in wild long-tailed macaques *(Macaca fascicularis)*. *Anim. Behav.* 33, 849–861.

———, 2004. Sexual selection and the careers of primate males: Paternity concentration, dominance acquisition tactics and transfer decisions. In: Kappeler, P., van Schaik, C., (Eds.), *Sexual Selection in Primates*. Cambridge University Press, Cambridge, pp. 208–229.

Van Os, J., Selten, J. P., 1998. Pre-natal exposure to maternal stress and subsequent schizophrenia: The May 1940 invasion of the Netherlands. *Br. J. Psychiatry* 172, 324–326.

van Schaik, C. P., 1989. The ecology of social relationships amongst female primates. In: Standen, V., Foley, R. A., (Eds.), *Comparative Socioecology*. Blackwell, Oxford, pp. 195–218.

———, 2000. Social counterstrategies against infanticide by males in primates and other mammals. In: Kappeler, P., (Ed.), *Primate Males*. Cambridge University Press, Cambridge, pp. 34–54.

van Schaik, C. P., Janson, C., 2000. *Infanticide by Males*. Cambridge University Press, Cambridge.

van Schaik, C. P., van Noordwijk, M. A., Nunn, C. L., 1999. Sex and social evolution in primates. In: Lee, P. C., (Ed.), *Comparative Primate Socioecology*. Cambridge University Press, Cambridge, pp. 204–240.

Vandenbergh, J. G., Post, W., 1976. Endocrine coordination in rhesus monkeys: Female responses to the male. *Physiol. Behav.* 17, 979–984.

Vanson, A., Arnold, A. P., Schlinger, B. A., 1996. 3B-hydroxysteroid dehydrogenase/isomerase and aromatase activity in primary cultures of developing zebra finch telencephalon: Dehydroepiandrosterone as substrate for synthesis of androstendione and estrogens. *Gen. Comp. Endocrinol.* 102, 342–350.

Veith, J., Buck, M., Gertzlaf, S., Van Dolfsen, P., Slade, A., 1983. Exposure to men influences the occurrence of ovulation in women. *Physiol. Behav.* 31, 313–315.

Venners, S. A., Liu, X., Perry, M. J., Korrick, S. A., Li, Z., Yang, F., Yang, J., Lasley, B. L., Xu, X., Wang, X., 2006. Urinary estrogen and progesterone metabolite concentrations in menstrual cycles of fertile women with non-conception, early pregnancy loss or clinical pregnancy. *Hum. Reprod.* 21, 2272–2280.

Vermeulen, A., Goemaere, S., Kaufman, J. M., 1999. Testosterone, body composition and aging. *J. Endocrinol. Invest.* 22, 110–116.

Verona, E., Curtin, J. J., 2006. Gender differences in the negative affective priming of aggressive behavior. *Emotion* 6, 115–124.

Vigilant, L., Hofreiter, M., Siedel, H., Boesch, C., 2001. Paternity and relatedness in wild chimpanzee communities. *Proc. Natl. Acad. Sci. U. S. A.* 98, 12890–12895.

Vignozzi, L., Filippi, S., Luconi, M., Morelli, A., Mancina, R., Marini, M., Vannelli, G. B., Granchi, S., Orlando, C., Geimini, S., Ledda, F., Forti, G., Maggi, M., 2004. Oxytocin receptor is expressed in the penis and mediates an estrogen-dependent smooth muscle contractility. *Endocrinology.* 145, 1823–1834.

Vinovskis, M. A., 1990. Death and married life in the past. *Hum. Nat.* 1, 109–122.

Virgin, C. E., Sapolsky, R. M., 1997. Styles of male social behavior and their endocrine correlates among low-ranking baboons. *Am. J. Primatol.* 42, 25–39.

Vitzthum, V. J., Bentley, G. R., Spielvogel, H., Cacenes, E., Thornbung, J., Jones, L., Shore, S., Hodges, K. R., Chatterton, R. T., 2002. Salivary progesterone levels and rate of ovulation are significantly lower in poorer than in better-off urban-dwelling Bolivian women. *Hum. Reprod.* 17, 1906–1913.

Vitzthum, V. J., Spielvogel, H., Caceres, E., Gaines, J., 2000. Menstrual patterns and fecundity among non-lactating and lactating cycling women in rural highland Bolivia: Implications for contraceptive choice. *Contraception* 62, 181–187.

vom Saal, F. S., 1989. Sexual differentiation in litter-bearing mammals: Influence of sex of adjacent fetuses in utero. *J. Anim. Sci.* 67, 1824–1840.

von Eggelkraut-Gottanka, R., Beck-Sickinger, A. G., 2004. Biosynthesis of peptide hormones derived from precursor sequences. *Curr. Med. Chem.* 11, 2651–2665.

von Engelhardt, N., Kappeler, P. M., Heistermann, M., 2000. Androgen levels and female social dominance in *Lemur catta*. *Proc. R. Soc. Lond. B Biol. Sci.* 267, 1533–1539.

Voracek, M., Manning, J. T., Ponocny, I., 2005. Digit ratio (2D:4D) in homosexual and heterosexual men from Austria. *Arch. Sex. Behav.* 34, 335–340.

Waddington, C. H., 1959. Canalization of development and genetic assimilation of acquired characters. *Nature* 183, 1654–1655.

Wade, G. N., Jones, J. E., 2004. Neuroendocrinology of nutritional infertility. *Am. J. Physiol. Regul. Integr. Comp. Physiol.* 287, R1277–R1296.

Wadhwa, P. D., Porto, M., Garite, T. J., Chicz-DeMet, A., 1998. Maternal corticotropin-releasing hormone levels in the early third trimester predict length of gestation in human pregnancy. *Am. J. Obstet. Gynecol.* 179, 1079–1085.

Wagner, J. D., Flinn, M. V., England, B. G., 2002. Hormonal response to competition among male coalitions. *Evol. Hum. Behav.* 23, 437–442.

Walker, B. G., Boersma, P. D., Wingfield, J. C., 2005. Field endocrinology and conservation biology. *Integr. Comp. Biol.* 45, 12–18.

Wallen, K., 1990. Desire and ability: Hormones and the regulation of female sexual behavior. *Neurosci. Biobehav. Rev.* 14, 233–241.

———, 1996. Nature needs nurture: The interaction of hormonal and social influences on the development of behavioral sex differences in rhesus monkeys. *Horm. Behav.* 30, 364–378.

————, 2001. Sex and context: Hormones and primate sexual motivation. *Horm. Behav.* 40, 339–357.

————, 2005. Hormonal influences on sexually differentiated behavior in non-human primates. *Front. Neuroendocrinol.* 26, 7–26.

Wallen, K., Baum, M. J., 2002. Masculinization and defeminization in altricial and precocial mammals: Comparative aspects of steroid hormone action. In: Pfaff, D., Arnold, A., Etgen, A., Fahrbach, S., Rubin, R., (Eds.), *Hormones, Brain, and Behavior*, Vol. 4. Elsevier, Amsterdam, pp. 385–423.

Wallen, K., Schneider, J., 1999. *Reproduction in Context*. MIT Press, Cambridge, MA.

Wallis, J., 1997. A survey of reproductive parameters in the free-ranging chimpanzees of Gombe National Park. *J. Reprod. Fertil.* 109, 297–307.

Walther, G.-R., Post, E., Convey, P., Menzel, A., Permesan, C., Beebee, T. J. C., Fromentin, J.-M., Hoegh-Guldberg, O., Bairlein, F., 2002. Ecological responses to recent climate change. *Nature* 416, 389–395.

Wang, C., Alexander, G., Berman, N., Salehian, B., Davidson, T., McDonald, V., Steiner, B., Hull, L., Callegari, C., Swerdloff, R. S., 1996. Testosterone replacement therapy improves mood in hypogonadal men—a clinical research center study. *J. Clin. Endocrinol. Metab.* 81, 3578–3583.

Wang, C., Swerdloff, R. S., Iranmanesh, A., Dobs, A., Snyder, P., Cunningham, G., Matsumoto, A. M., Weber, T., Berman, N., Testosterone Gel Study Group, 2000. Transdermal testosterone gel improves sexual function, mood, muscle strength, and body composition parameters in hypogonadal men. *J. Clin. Endocrinol. Metab.* 85, 2839–2853.

Wang, Z. X., Ferris, C. F., De Vries, G. J., 1994. Role of septal vasopressin innervation in paternal behavior in prairie voles *(Microtus ochrogaster). Proc. Natl. Acad. Sci. U. S. A.* 91, 400–404.

Wang, Z. X., Insel, T. R., 1996. Parental behavior in voles. In: Rosenblatt, J. S., Snowdon, C. T., (Eds.), *Parental Care: Evolution, Mechanisms, and Adaptive Significance.* Academic Press, San Diego, pp. 361–384.

Wang, Z., Moody, K., Newman, J. D., Insel, T. R., 1997. Vasopressin and oxytocin immunoreactive neurons and fibers in the forebrain of male and female common marmosets *(Callithrix jacchus). Synapse* 27, 14–25.

Wang, Z. X., Young, L. J., De Vries, G. J., Insel, T. R., 1998. Voles and vasopressin: A review of molecular, cellular, and behavioral studies of pair bonding and paternal behaviors. *Prog. Brain Res.* 119, 483–499.

Washabaugh, K. F., Snowdon, C. T., Ziegler, T. E., 2002. Variations in care for cotton-top tamarin, *Saguinus oedipus,* infants as a function of parental experience and group size. *Anim. Behav.* 63, 1163–1174.

Wasser, S. K., Barash, D. P., 1983. Reproductive suppression among female mammals: Implications for biomedicine and sexual selection theory. *Q. Rev. Biol.* 58, 513–538.

Wasser, S. K., Place, N. J., 2001. Reproductive filtering and the social environment. In: Ellison, P. T., (Ed.), *Reproductive Ecology and Human Evolution.* Aldine de Gruyter, Hawthorne, NY, pp. 137–158.

Watson, J. B., Mednick, S. A., Huttunen, M., Wang, X., 1999. Prenatal teratogens and the development of adult mental illness. *Dev. Psychopathol.* 11, 457–466.

Watts, D. P., 1989. Infanticide in mountain gorillas: New cases and a reconsideration of the evidence. *Ethology* 81, 1–18.

———, 1990. Mountain gorilla life histories, reproductive competition, and sociosexual behavior and some implications for captive husbandry. *Zoo Biol.* 9, 185–200.

———, 1991. Mountain gorilla reproduction and sexual behavior. *Am. J. Primatol.* 24, 211–225.

———, 2000. Causes and consequences of variation in male mountain gorilla life histories and group membership. In: Kappeler, P., (Ed.), *Primate Males.* Cambridge University Press, Cambridge, pp. 169–179.

———, 2007. Effects of male group size, parity, and cycle stage on female chimpanzee copulation rates at Ngogo, Kibale National Park, Uganda. *Primates* 48, 222–231.

Watts, D. P., Mitani, J. C., 2001. Boundary patrols and intergroup encounters in wild chimpanzees. *Behaviour* 138, 299–327.

Weaver, I. C. G., Cervoni, N., Vhampagne, F. A., D'Alessio, A. C., Sharma, S., Seckl, J. R., Dymov, S., Szyf, M., Meaney, M. R., 2004. Epigenetic programming by maternal behavior. *Nat. Neurosci.* 7, 847–854.

Weaver, I. C. G., Diorio, J., Seckl, J. R., Szyf, M., Meaney, M. J., 2004. Early environmental regulation of hippocampal glucocorticoid receptor gene expression—characterization of intracellular mediators and potential genomic target sites. *Ann. N. Y. Acad. Sci.* 1024, 182–212.

Weaver, I. C. G., Meaney, M. J., Szyf, M., 2006. Maternal care effects on the hippocampal transcriptome and anxiety-mediated behaviors in the offspring that are reversible in adulthood. *Proc. Natl. Acad. Sci. U. S. A.* 103, 3480–3485.

Wedekind, C., Seebeck, T., Bettens, F., Paepke, A. J., 1995. MHC-dependent mate preferences in humans. *Proc. R. Soc. Lond. B Biol. Sci.* 260, 245–249.

Weekes-Shackelford, V. A., Easton, J. A., Stone, E. A., 2007. How having children affects mating psychology. In: Geher, G., Miller, G., (Eds.), *Mating Intelligence: Sex, Relationships, and the Mind's Reproductive System.* Erlbaum, Mahwah, NJ, pp. 159–170.

Weidong, M., Zhongshan, M., Novotny, M. V., 1998. Role of the adrenal gland and adrenal-mediated chemosignals in suppression of estrus in the house mouse: The Lee-Boot effect revisited. *Biol. Reprod.* 59, 1317–1320.

Weiner, H., 1992. *Perturbing the Organism.* University of Chicago Press, Chicago.

Weinrich, J. D., Grant, I., Jacobson, D. L., Robinson, S. R., McCutchan, J. A., Group, H., 1992. Effects of recalled childhood gender nonconformity on adult genitoerotic role and AIDS exposure. *Arch. Sex. Behav.* 21, 559–585.

Weinstock, M., 1997. Does pre-natal stress impair coping and regulation of the hypothalamic-pituitary-adrenal axis? *Neurosci. Biobehav. Rev.* 21, 1–10.

———, 2005. The potential influence of maternal stress hormones on development and mental health of the offspring. *Brain Behav. Immun.* 19, 296–308.

Weinstock, M., Matlina, E., Maor, G. I., Rosen, H., McEwen, B. S., 1992. Prenatal stress selectively alters the reactivity of the hypothalamic-pituitary adrenal system in the female rat. *Brain Res.* 595, 195–200.

Weise, K. L., Tuber, S., 2004. The self and object representations of narcissisti-
cally disturbed children: An empirical investigation. *Psychoanal. Psychol.*
21, 244–258.

Weisenfeld, A., Malatesta, C., Whitman, P., Grannose, C., Vile, R., 1985. Psychophysiological response of breast- and bottle-feeding mothers to their infants' signals. *Psychophysiology* 22, 79–86.

Welberg, L. A., Seckl, J. R., Holmes, M. C., 2000. Inhibition of 11beta-
hydroxysteroid dehydrogenase, the foeto-placental barrier to maternal glucocorticoids, permanently programs amygdala GR mRNA expression and
anxiety-like behaviour in the offspring. *Eur. J. Neurosci.* 12, 1047–1054.

Welling, L. L. M., Jones, B. C., DeBruine, L. M., Conway, C. A., Law Smith,
M. J., Little, A. C., Feinberg, D. R., Sharp, M. A., Al-Dujaili, E. A., 2007.
Raised salivary testosterone in women is associated with increased attrac-
tion to masculine faces. *Horm. Behav.* 52, 156–161.

West-Eberhard, M. J., 2003. *Developmental Plasticity and Evolution.* Oxford
University Press, Oxford.

Westergaard, G. C., Cleveland, A., Trenkle, M. K., Lussier, I. D., Higley, J. D.,
2003. CSF 5-HIAA concentration as an early screening tool for predicting
significant life history outcomes in female specific-pathogen-free (SPF) rhesus macaques (Macaca mulatta) maintained in captive breeding groups.
J. Med. Primatol. 32, 95–104.

Westergaard, G. C., Suomi, S. J., Chavanne, T. J., Houser, L., Hurley, A., Cleveland, A., Snoy, P. J., Higley, J. D., 2003. Physiological correlates of aggression
and impulsivity in free-ranging female primates. *Neuropsychopharmacology*
28, 1045–1055.

Westergaard, G. C., Suomi, S. J., Higley, J. D., Mehlman, P. T., 1999. CSF 5-
HIAA and aggression in female macaque monkeys: Species and interindividual differences. *Psychopharmacology* 146, 440–446.

Whitam, F. L., 1977. Childhood indicators of male homosexuality. *Arch. Sex.
Behav.* 6, 89–98.

Whitam, F. L., Mathy, R. M., 1986. *Male Homosexuality in Four Societies:
Brazil, Guatemala, the Philippines, and the United States.* Praeger, New
York.

Whiten, A., Byrne, R. W., 1997. *Machiavellian Intelligence II.* Cambridge University Press, Cambridge.

Whiting, B. B., Edwards, C., 1973. A cross-cultural analysis of sex differences in
the behavior of children aged 3 to 11. *J. Soc. Psychol.* 91, 171–188.

Whitten, P. L., 2000. Evolutionary endocrinology of the cercopithecoids. In:
Whitehead, P., Jolly, C. J., (Eds.), *Old World Monkeys.* Cambridge University Press, Cambridge, pp. 269–297.

Whitten, P. L., Brockman, D. K., Stavisky, R. C., 1998. Recent advances in noninvasive techniques to monitor hormone-behavior interactions. *Am. J. Phys.
Anthropol. Suppl.* 27, 1–23.

Whitten, P. L., Turner, T. R., 2004. Male residence and the patterning of serum
testosterone in vervet monkeys *(Cercopithecus aethiops). Behav. Ecol. Sociobiol.* 56, 565–578.

Wich, S. A., Geurts, M. L., Mitra Setia, T., Utami, S. S., 2006. Influence of fruit availability on Sumatran orangutan sociality and reproduction. In: Hohmann, G., Robbins, M. M., Boesch, C., (Eds.), *Feeding Ecology in Apes and Other Primates*. Cambridge University Press, Cambridge, pp. 337–358.

Widstrom, A., Wahlberg, V., Matthiesen, A. S., Eneroth, P., Uvnäs-Moberg, K., Werner, S., 1990. Short-term effects of early suckling and touch of the nipple on maternal behavior. *Early Hum. Dev.* 21, 153–163.

Wiedenmayer, C. P., 2004. Adaptations or pathologies? Long-term changes in brain and behavior after a single exposure to severe threat. *Neurosci. Biobehav. Rev.* 28, 1–12.

Wikelski, M., Hau, M., Wingfield, J. C., 1999. Social instability increases plasma testosterone in a year-round territorial neotropical bird. *Proc. Soc. Lond. B Biol. Sci.* 266, 551–556.

Wilczynski, W., Allison, J. D., Marler, C. A., 1993. Sensory pathways linking social and environmental cues to endocrine regions of amphibian forebrains. *Brain Behav. Evol.* 42, 252–264.

Williams, J. R., Catania, K. C., Carter, C. S., 1992. Development of partner preferences in female prairie voles *(Microtus ochrogaster)*: The role of social and sexual experience. *Horm. Behav.* 26, 339–349.

Williams, J. R., Insel, T. R., Harbaugh, C. R., Carter, C. S., 1994. Oxytocin centrally administered facilitates formation of a partner preference in female prairie voles *(Microtus ochrogaster)*. *J. Neuroendocrinol.* 6, 247–250.

Williams, L. E., Bernstein, I. S., 1983. Introduction and dominance manipulations involving old rhesus males. *Folia Primatol.* 40, 175–180.

Williams, N. I., Helmreich, D. L., Parfitt, D. B., Caston-Balderrama, A., Cameron, J. L., 2001. Evidence for a causal role of low energy availability in the induction of menstrual cycle disturbances during strenuous exercise training. *J. Clin. Endocrinol. Metab.* 86, 5184–5193.

Williams, T. J., Pepitone, M. E., Christensen, S. E., Cooke, B. M., Huberman, A. D., Breedlove, N. J., Breedlove, T. J., Jordan, C. L., Breedlove, S. M., 2000. Finger-length ratios and sexual orientation. *Nature* 404, 455–456.

Williams-Ashman, H. G., 1988. Perspectives on the male sexual physiology of eutherian mammals. In: Knobil, E., Neill, J. D., (Eds.), *The Physiology of Reproduction*. Raven Press, New York, pp. 727–751.

Willis, D., Franks, S., 1995. Insulin action in human granulosa cells from normal and polycystic ovaries is mediated by the insulin receptor and not the type-I insulin-like growth factor receptor. *J. Clin. Endocrinol. Metab.* 80, 3788–3790.

Willis, D., Mason, H., Gilling-Smith, C., Franks, S., 1996. Modulation by insulin of follicle-stimulating hormone and luteinizing hormone actions in human granulosa cells of normal and polycystic ovaries. *J. Clin. Endocrinol. Metab.* 81, 302–309.

Wilson, E. O., 1975. *Sociobiology: The New Synthesis*. Harvard. Cambridge, MA.

Wilson, M. L., Wrangham, R. W., 2003. Intergroup relations in chimpanzees. *Annu. Rev. Anthropol.* 32, 363–392.

Winberg, J., 2005. Mother and newborn baby: Mutual regulation of physiology and behavior—a selective review. *Dev. Psychobiol.* 47, 217–229.

Wingfield, J. C., 1985. Short-term changes in plasma levels of hormones during establishment and defense of a breeding territory in male song sparrows, *Melospiza melodia*. *Horm. Behav.* 19, 174–187.

———, 1994a. Control of territorial aggression in a changing environment. *Psychoneuroendocrinology* 19, 709–721.

———, 1994b. Hormone-behavior interactions and mating systems in male and female birds. In: Short, R. V., Balaban, E., (Eds.), *The Difference between the Sexes*. Cambridge University Press, Cambridge, pp. 303–330.

———, 2003. Control of behavioural strategies for capricious environments. *Anim. Behav.* 66, 807–816.

———, 2005. Flexibility in annual cycles of birds: Implications for endocrine control mechanisms. *J. Ornithol.* 146, 291–304.

———, 2006. Communicative behaviors, hormone-behavior interactions, and reproduction in vertebrates. In: Neill, J. D., (Ed.), *Physiology of Reproduction*. Academic Press, New York, pp. 1995–2040.

Wingfield, J. C., Breuner, C., Jacobs, J., Lynn, S., Maney, D., Ramenofsky, M., Richardson, R., 1998. Ecological bases of hormone-behavior interactions: The "Emergency Life History Stage." *Am. Zool.* 38, 191–206.

Wingfield, J. C., Hahn, T. P., 1993. Testosterone and territorial behavior in sedentary and migratory sparrows. *Anim. Behav.* 47, 77–89.

Wingfield, J. C., Hegner, R. E., Dufty, A. M., Ball, G. F., 1990. The "challenge hypothesis": Theoretical implications for patterns of testosterone secretion, mating systems, and breeding strategies. *Am. Nat.* 136, 829–846.

Wingfield, J. C., Jacobs, J., Hillgarth, N., 1997. Ecological constraints and the evolution of hormone-behavior interrelationships. *Ann. N. Y. Acad. Sci.* 807, 22–41.

Wingfield, J. C., Jacobs, J. D., Soma, K., Maney, D. L., Hunt, K., Wisti-Peterson, D., Meddle, S., Ramenofsky, M., Sullivan, K., 1999. Testosterone, aggression and communication: Ecological bases of endocrine phenomena. In: Hauser, M., Konishi, M., (Eds.), *The Design of Animal Communication*. MIT Press, Cambridge, MA, pp. 255–284.

Wingfield, J. C., Jacobs, J. D., Tramontin, A. D., Perfito, N., Meddle, S., Maney, D. L., Soma, K., 1999. Toward an ecological basis of hormone-behavior interactions in reproduction of birds. In: Wallen, K., Schneider, J., (Eds.), *Reproduction in Context*. MIT Press, Cambridge, MA, pp. 85–128.

Wingfield, J. C., Lynn, S. E., Soma, K. K., 2001. Avoiding the "costs" of testosterone: Ecological bases of hormone-behavior interactions. *Brain Behav. Evol.* 57, 239–251.

Wingfield, J. C., Moore, I. T., Goymann, W., Wacker, D. W., Sperry, T., 2005. Contexts and ethology of vertebrate aggression: Implications for the evolution of hormone-behavior interactions. In: Nelson, R. J., (Ed.), *Biology of Aggression*. Oxford University Press, New York, pp. 179–210.

Wingfield, J. C., Ramenofsky, M., 1985. Hormonal and environmental control of aggression in birds. In: Gilles, R., Balthazart, J., (Eds.), *Neurobiology*. Springer, Berlin, pp. 92–104,

————, 1999. Hormones and the behavioral ecology of stress. In: Balm, P. H. M., (Ed.), *Stress Physiology in Animals*. Sheffield Academic Press, Sheffield, UK, pp. 1–51.

Wingfield, J. C., Romero, L. M., 2001. Adrenocortical responses to stress and their modulation in free-living vertebrates. In: McEwen, B. S., (Ed.), *Handbook of Physiology, Section 7: The Endocrine System*, Vol. 4, *Coping with the Environment: Neural and Endocrine Mechanisms*. Oxford University Press, Oxford, pp. 211–236.

Wingfield, J. C., Sapolsky, R. M., 2003. Reproduction and resistance to stress: When and how. *J. Neuroendocrinol.* 15, 711–724.

Wingfield, J. C., Silverin, B., 2002. Ecophysiological studies of hormone-behavior relations in birds. In: Pfaff, D. W., Arnold, A. P., Etgen, A. M., Fahrbach, S. E., Rubin, R. T., (Eds.), *Hormones, Brain and Behavior*, Vol. 2. Elsevier, Amsterdam, pp. 587–647.

Wingfield, J. C., Soma, K. K., 2002. Spring and autumn territoriality: Same behavior, different mechanisms? *Integr. Comp. Biol.* 42, 11–20.

Wingfield, J. C., Whaling, C. S., Marler, P. R., 1994. Communication in vertebrate aggression and reproduction: The role of hormones. In: Knobil, E., Neill, J. D., (Eds.), *Physiology of Reproduction*, 2nd ed. Raven Press, New York, pp. 303–342.

Winking, J., 2006. Are men really that bad as fathers? The role of men's investments. *Soc. Biol.* 53, 100–115.

Winking, J., Kaplan, H., Gurven, M., Rucas, S., 2007. Why do men marry, and why do they stray? *Proc. R. Soc. Lond. B Biol. Sci.* 274, 1643–1649.

Winslow, J. T., Hastings, N., Carter, C. S., Harbaugh, C. R., Insel, T. R., 1993. A role for central vasopressin in pair bonding in monogamous prairie voles. *Nature* 365, 545–548.

Winslow, J. T., Insel, T. R., 2004. Neuroendocrine basis of social recognition. *Curr. Opin. Neurobiol.* 14, 248–253.

Wisenfeld, A., Klorman, R., 1978. The mother's psychophysiological reactions to contrasting affective expressions by her own and an unfamiliar infant. *Dev. Psychol.* 14, 294–304.

Wisniewski, A. B., Migeon, C. J., Meyer-Bahlburg, H. F., Gearhart, J. P., Berkovitz, G. D., Brown, T. R., Money, J., 2000. Complete androgen insensitivity syndrome: Long-term medical, surgical, and psychosexual outcome. *J. Clin. Endocrinol. Metab.* 85, 2664–2669.

Wisniewski, A. B., Nelson, R. J., 2000. Seasonal variation in human functional cerebral lateralization and free testosterone concentrations. *Brain Cogn.* 43, 429–438.

Withuhn, T. F., Kramer, K. M., Cushing, B. S., 2003. Early exposure to oxytocin affects the age of vaginal opening and first estrus in female rats. *Physiol. Behav.* 80, 135–138.

Witt, D. M., 1997. Mechanisms of oxytocin-mediated sociosexual behavior. *Ann. N. Y. Acad. Sci.* 807, 287–301.

Witt, D. M., Winslow, J. T., Insel, T. R., 1992. Enhanced social interactions in rats following chronic, centrally infused oxytocin. *Pharmacol. Biochem. Behav.* 43, 855–861.

Wolfe, L. D., Gray, J. P., 1982. A cross-cultural investigation into the sexual dimorphism of stature. In: Hall, R. L., (Ed.), *Sexual Dimorphism in Homo sapiens: A Question of Size.* Praeger, New York, pp. 197–230.

Wolfe, R., Ferrando, A., Sheffield-Moore, M., Urban, R., 2000. Testosterone and muscle protein metabolism. *Mayo Clin. Proc. Suppl.* 75 55–S59; discussion 59–S60.

Wong, R., Capoferro, C., Soldo, B. J., 1999. Financial assistance from middle-aged couples to parents and children. Racial and ethnic difference of inter-generational transfers by middle-aged adults: parents, children, or both? *J. Gerontol. Soc. Sci.* 54B, S145–S153.

Wood, J. W., Lai, D., Johnson, P. L., Campbell, K. L., Maslar, I. A., 1985. Lactation and birth spacing in highland New Guinea. *J. Biosoc. Sci. Suppl.* 9, 159–173.

Wood, R. I., 1996. Functions of the steroid-responsive neural network in the control of male hamster sexual behavior. *Trends Endocrinol. Metab.* 7, 338–344.

———, 1997. Thinking about networks in the control of male hamster sexual behavior. *Horm. Behav.* 32, 40–45.

Worthman, C. M., Jenkins, C. L., Stallings, J. F., Lai, D., 1993. Attenuation of nursing-related ovarian suppression and high fertility in well-nourished, intensively breast-feeding Amele women of lowland Papua New Guinea. *J. Biosoc. Sci.* 25, 425–443.

Worthman, C. M., Melby, M., 2002. Toward a comparative developmental ecology of human sleep. In: Carskadon, M. A., (Ed.), *Adolescent Sleep Patterns: Biological, Social, and Psychological Influences.* Cambridge University Press, New York, pp. 69–117.

Worthman, C. M., Stallings, J. F., 1994. Measurement of gonadotropins in dried blood spots. *Clin. Chem.* 40, 448–453.

Wrangham, R. W., 1980. An ecological model of female-bonded primate groups. *Behaviour* 75, 262–300.

———, 1999. Is military incompetence adaptive? *Evol. Hum. Behav.* 20, 3–17.

———, 2002. The cost of sexual attraction: Is there a trade-off in female Pan between sex appeal and received coercion? In: Boesch, C., Hohmann, G., Marchant, L. F., (Eds.), *Behavioural Diversity in Chimpanzees and Bonobos.* Cambridge University Press, Cambridge, pp. 204–215.

Wrangham, R. W., Wilson, M. L., Muller, M. N., 2006. Comparative rates of violence in chimpanzees and humans. *Primates* 47, 14–26.

Wright, P. C., 1984. Biparental care in *Aotus trivirgatus* and *Callicebus moloch.* In: Small, M. E., (Ed.), *Female Primates: Studies by Women Primatologists.* Alan R. Liss, New York, pp. 59–75.

Xiao, K., Kondo, Y., Sakuma, Y., 2004. Sex-specific effects of gonadal steroids on conspecific odor preference in the rat. *Horm. Behav.* 46, 356–361.

Yamamoto, M. E., 1993. From dependence to sexual maturity: The behavioral ontogeny of Callitrichidae. In: Rylands, A. B., (Ed.), *Marmosets and Tamarins: Systematics, Behaviour, and Ecology.* Oxford University Press, Oxford, pp. 235–254.

Yamamoto, Y., Carter, C. S., Cushing, B. S., 2006. Neonatal manipulation of oxytocin affects expression of estrogen receptor alpha. *Neuroscience* 137, 157–164.

Yamamoto, Y., Cushing, B. S., Kramer, K. M., Epperson, P. D., Hoffman, G. E., Carter, C. S., 2004. Neonatal manipulations of oxytocin alter expression of oxytocin and vasopressin immunoreactive cells in the paraventricular nucleus of the hypothalamus in a gender-specific manner. *Neuroscience* 125, 947–955.

Yang, C. J., Gray, P. B., Zhang, J. Pope, H. G., Jr., N. D., 2008. Second to fourth digit ratios, sex differences, and behavior in Chinese men and women. *Soc Neurosci.*

Yehuda, R., 2002. Post-traumatic stress disorder. *N. Engl. J. Med.* 346, 108–114.

Yehuda, R., Engel, S. M., Brand, S. R., Seckl, J., Marcus, S. M., Berkowitz, G. S., 2005. Transgenerational effects of posttraumatic stress disorder in babies of mothers exposed to the World Trade Center attacks during pregnancy. *J. Clin. Endocrinol. Metab.* 90, 4115–4118.

Yerkes, R. M., 1939. Sexual behavior in the chimpanzee. *Hum. Biol.* 211, 78–111.

Young, L. J., 1999. Oxytocin and vasopressin receptors and species-typical social behaviors. *Horm. Behav.* 36, 212–221.

Young, L. J., Wang, Z., 2004. The neurobiology of pair bonding. *Nat. Neurosci.* 7, 1048–1054.

Young, W. C., Dempsey, E. W., Myers, H. I., 1935. Cyclic reproductive behavior in the female guinea pig. *J. Comp. Psychol.* 19, 313–335.

Young, W. C., Goy, R. W., Phoenix, C. H., 1964. Hormones and sexual behavior. *Science* 143, 212–218.

Young, W. C., Myers, H. I., Dempsey, E. W., 1933. Some data from a correlated anatomical, physiological and behavioristic study of the reproductive cycle in the female guinea pig. *Am. J. Physiol.* 105, 393–398.

Yu, D. W., Shepard, G. H., 1998. Is beauty in the eye of the beholder? *Nature* 396, 321–322.

Zahed, S. R., Prudom, S. L., Snowdon, C. T., Ziegler, T. E., 2008. Male parenting and response to infant stimuli in the common marmoset *(Callithrix jacchus)*. *Am. J. Primatol.* 70, 84–92.

Zak, P. J., Kurzban, R., Matzner, W. T., 2004. The neurobiology of trust. *Ann. N. Y. Acad. Sci.* 1032, 224–227.

———, 2005. Oxytocin is associated with human trustworthiness. *Horm. Behav.* 48, 522–527.

Zala, S. M., Potts, W. K., Penn, D. J., 2004. Scent-marking displays provide honest signals of health and infection. *Behav. Ecol.* 15, 338–344.

Zhang, T.-Y., Parent, C., Weaver, I., Meaney, M. J., 2004. Maternal programming of individual differences in defensive responses in the rat. *Ann. N. Y. Acad. Sci.* 1032, 85–103.

Zhou, A., 2007. Hormonal profiles in captive male orangutans. B.A. thesis, Harvard University.

Ziegler, T. E., 2000. Hormones associated with non-maternal infant care: A review of mammalian and avian studies. *Folia Primatol.* 71, 6–21.

Ziegler, T. E., Bridson, W. E., Snowdon, C. T., Eman, S., 1987. Urinary gonadotropin and estrogen excretion during the postpartum estrous, conception and pregnancy in the cotton-top tamarin *(Saguinus Oedipus Oedipus)*. *Am. J. Primatol.* 12, 127–140.

Ziegler, T. E., Epple, G., Snowdon, C. T., Porter, T. A., Belcher, A., Kuederling, I., 1993. Detection of the chemical signals of ovulation in the cotton-top tamarin, *Saguinus oedipus. Anim. Behav.* 45, 313–322.

Ziegler, T. E., Jacoris, S., Snowdon, C. T., 2004. Sexual communication between breeding male and female cotton-top tamarins *(Saguinus oedipus)* and its relationship to infant care. *Am. J. Primatol.* 64, 57–69.

Ziegler, T. E., Mamanasiri, S., Prudom, S. L., Refetoff Zahed, S. K., Refetoff, S., 2006. Influence of genetic determinants on infant responsiveness in male marmosets. Presented as a poster at the Oxytocin, Vasopressin and Emotional Regulation: New Frontiers in Basic Neuroscience and Translational Opportunities conference, Atlanta, GA.

Ziegler, T. E., Prudom, S. L., Schultz-Darken, N. J., Kurian, A. V., Snowdon, C. T., 2006. Pregnancy weight gain: Marmoset and tamarin dads show it too. *Biol. Lett.* 2, 181–183.

Ziegler, T. E., Savage, A., Scheffler G., Snowdon, C. T., 1987. The endocrinology of puberty and reproductive functioning in female cotton-top tamarins *(Saguinus oedipus)* under varying social conditions. *Biol. Reprod.* 37, 618–627.

Ziegler, T. E., Scheffler, G., Snowdon, C. T., 1995. The relationship of cortisol levels to social environment and reproductive functioning in female cotton-top tamarins, *saguinus oedipus. Horm. Behav.* 29, 407–424.

Ziegler, T. E., Schultz-Darken, N. J., Scott, J. J., Snowdon, C. T., Ferris, C. F., 2005. Neuroendocrine response to female ovulatory odors depends upon social condition in male common marmosets, *Callithrix jacchus. Horm. Behav.* 47, 56–64.

Ziegler, T. E., Snowdon, C. T., 2000. Preparental hormone levels and parenting experience in male cotton-top tamarins, *Sagunis oedipus. Horm. Behav.* 38, 159–167.

Ziegler, T. E., Sousa, M. B., 2002. Parent-daughter relationships and social controls on fertility in female common marmosets, *Callithrix jacchus. Horm. Behav.* 42, 356–367.

Ziegler, T. E., Washabaugh, K. F., Snowdon, C. T., 2004. Responsiveness of expectant male cotton-top tamarins, *Saguinus oedipus,* to mate's pregnancy. *Horm. Behav.* 45, 84–92.

Ziegler, T. E., Wegner, F. H., Carlson, A. A., Lazaro-Perea, C., Snowdon, C. T., 2000. Prolactin levels during the periparturitional period in the biparental cotton-top tamarin *(Saguinus oedipus):* Interactions with gender, androgen levels, and parenting. *Horm. Behav.* 38, 111–122.

Ziegler, T. E., Wegner, F. H., Snowdon, C. T., 1996. Hormonal responses to parental and nonparental conditions in male cotton-top tamarins, *Saguinus oedipus,* a New World primate. *Horm. Behav.* 30, 287–297.

Ziegler, T. E., Widowski, T. M., Larson, M. L., Snowdon, C. T., 1990. Nursing does affect the duration of the post-partum to ovulation interval in cotton-top tamarins *(Sagunius oedipus). J. Reprod. Fertil.* 90, 563–570.

Ziegler, T. E., Wittwer, D. J., 2005. Fecal steroid research in the field and laboratory: Improved methods for storage, transport, processing and analysis. *Am. J. Primatol.* 67, 159–174.

Zinaman, M. J., Hughes, V., Queenan, J. T., Labbok, M. H., Albertson, B., 1992. Acute prolactin and oxytocin responses and milk yield to infant suckling and artificial methods of expression in lactating women. *Pediatrics* 89, 437–440.

Zingg, H. H., 1996. Vasopressin and oxytocin receptors. *Baillières Clin. Endocrinol. Metab.* 10, 75–96.

Zingg, H. H., Laporte, S. A., 2003. The oxytocin receptor. *Trends Endocrinol. Metab.* 14, 222–227.

Zingg, H. H., Rozen, F., Chu, K., Larcher, A., Arslan, A., Richard, S., Lefebvre, D., 1995. Gonadal steroid regulation of oxytocin and oxytocin receptor gene expression. *Recent Prog. Horm. Res.* 50, 255–273.

Zinner, D. P., Nunn, C. L., van Schaik, C. P., Kappeler, P., 2004. Sexual selection and exaggerated swellings of female primates. In: Kappeler, P., van Schaik, C., (Eds.), *Sexual Selection in Primates.* Cambridge University Press, Cambridge, pp. 71–89.

Zitzmann, M., Nieschlag, E., 2003. The CAG repeat polymorphism within the androgen receptor gene and maleness. *Int. J. Androl.* 26, 76–83.

Zucker, K. J., Bradley, S. J., Oliver, G., Blake, J., Fleming, S., Hood, J., 1996. Psychosexual development of women with congenital adrenal hyperplasia. *Horm. Behav.* 30, 300–318.

Contributors

KAREN L. BALES   Department of Psychology, University of California, Davis

ERICKA BOONE   Department of Psychiatry, University of Illinois, Chicago

BENJAMIN C. CAMPBELL   Department of Anthropology, University of Wisconsin, Milwaukee

C. SUE CARTER   Department of Psychiatry, University of Illinois, Chicago

JENNIE Y. CHEN   Department of Psychology, Texas A&M University

PETER T. ELLISON   Department of Anthropology, Harvard University

MELISSA EMERY THOMPSON   Department of Anthropology, University of New Mexico

LYNN A. FAIRBANKS   Department of Psychiatry and Biobehavioral Sciences, Semel Institute, University of California, Los Angeles

ALISON S. FLEMING   Department of Psychology, University of Toronto

MARK FLINN   Department of Anthropology, University of Missouri

ANDREA GONZALEZ   Department of Psychology, University of Toronto

PETER B. GRAY   Department of Anthropology and Ethnic Studies, University of Nevada, Las Vegas

ANGELA J. GRIPPO   Department of Psychiatry, University of Illinois, Chicago

JANICE HASSETT   Department of Psychology, Emory University

CAROLE K. HOOVEN   Department of Anthropology, Harvard University

HILLARD S. KAPLAN   Department of Anthropology, University of New Mexico

JANE B. LANCASTER   Department of Anthropology, University of New Mexico

PHYLLIS C. LEE   Behaviour and Evolution Research Group, Department of Psychology, University of Stirling

MATTHEW H. MCINTYRE   Department of Anthropology, University of Central Florida

PABLO NEPOMNASCHY   Faculty of Health Sciences, Simon Fraser University

JEFFREY C. PARKIN   Department of Anthropology and Ethnic Studies, University of Nevada, Las Vegas

JAMES R. RONEY   Department of Psychology, University of California, Santa Barbara

MICHAEL RUSCIO   Department of Psychology, College of Charleston

ROXANNE SANCHEZ   Department of Anthropology and Ethnic Studies, University of Nevada, Las Vegas

CHARLES T. SNOWDON   Department of Psychology, University of Wisconsin, Madison

SARI M. VAN ANDERS   Departments of Psychology and Women's Studies, University of Michigan, Ann Arbor

KIM WALLEN   Department of Psychology, Emory University

JOHN C. WINGFIELD   Section of Neurobiology, Physiology, and Behavior, University of California, Davis

TONI E. ZIEGLER   Wisconsin National Primate Research Center and the Department of Psychology, University of Wisconsin, Madison

Index